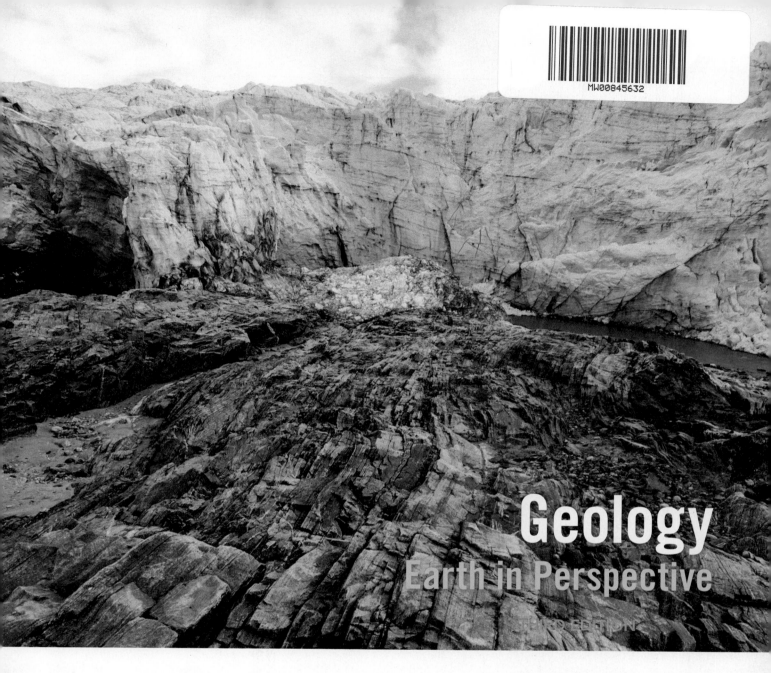

# Geology
## Earth in Perspective

THIRD EDITION

## Reed Wicander

Professor Emeritus
Central Michigan University

Adjunct Professor
The University of
Queensland, Australia

## James S. Monroe

Professor Emeritus
Central Michigan University

CENGAGE

Australia • Brazil • Mexico • Singapore • United Kingdom • United States

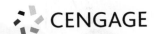

***Geology: Earth in Perspective,* Third Edition**
Reed Wicander, James S. Monroe

Product Director: Thais Alencar

Product Team Manager: Kelsey Churchman

Product Manager: Lauren Bakker

Production Manager: Julia White

Learning Designer: Linda Man

Content Manager: Nicole Evans

Product Assistant: Vanessa Desiato

Marketing Manager: Roxanne Wang

Digital Delivery Lead: Thomas Griffin

Art Director: Helen Bruno

Manufacturing Planner: Becky Cross

Intellectual Property

Analyst: Christine Myaskovsky

Project Manager: Erika Mugavin

Production Service: Lumina Datamatics, Inc.

Cover image(s): Jason Edwards/National Geographic Image Collection; Striations carved into metamorphic bedrock during retreat of the Russell Glacier, Kangerlussuaq, Greenland.

For product information and technology assistance, contact us at **Cengage Customer & Sales Support, 1-800-354-9706 or support.cengage.com.**

For permission to use material from this text or product, submit all requests online at **www.cengage.com/permissions.**

Library of Congress Control Number: 2018968240

Soft-cover Edition ISBN: 978-0-357-11733-0

Loose-leaf Edition ISBN: 978-0-357-12010-1

**Cengage**
20 Channel Center Street
Boston, MA 02210
USA

Cengage is a leading provider of customized learning solutions with employees residing in nearly 40 different countries and sales in more than 125 countries around the world. Find your local representative at **www.cengage.com**.

Cengage products are represented in Canada by Nelson Education, Ltd.

To learn more about Cengage platforms and services, visit **www.cengage.com**. To register or access your online learning solution or purchase materials for your course, visit **www.cengagebrain.com**.

Printed in the United States of America
Print Number: 01      Print Year: 2019

# BRIEF CONTENTS

**1** Understanding Earth  1

**2** Plate Tectonics  22

**3** Minerals  51

**4** Igneous Rocks and Intrusive Igneous Activity  69

**5** Volcanoes and Volcanism  89

**6** Weathering, Soil, and Sedimentary Rocks  108

**7** Metamorphism and Metamorphic Rocks  137

**8** Earthquakes and Earth's Interior  153

**9** Deformation, Mountain Building, and the Continents  179

**10** Mass Wasting  201

**11** Running Water  222

**12** Groundwater  243

**13** Glaciers and Glaciation  267

**14** The Work of Wind and Deserts  288

**15** Oceans, Shorelines, and Shoreline Processes  306

**16** Geologic Time  331

**17** Earth History  361

**18** Life History  398

# CONTENTS

## UNDERSTANDING EARTH: A DYNAMIC AND EVOLVING PLANET  1

Introduction  2
What Is Geology?  2
Geology and the Formulation of Theories  3
How Does Geology Relate to the Human Experience?  4
How Does Geology Affect Our Everyday Lives?  5
Global Geologic and Environmental Issues Facing Humankind  5

**GEO-FOCUS  Geology and Economic and Political Power  5**

Origin of the Universe and Solar System, and Earth's Place in Them  8
    Origin of the Universe: Did It Begin with a Big Bang?  8
    Our Solar System: Its Origin and Evolution  9
    Earth: Its Place in Our Solar System  11
Why Earth Is a Dynamic and Evolving Planet  11
    Plate Tectonic Theory  12
The Rock Cycle  14
    How Are the Rock Cycle and Plate Tectonics Related?  15
Organic Evolution and the History of Life  16
Geologic Time and Uniformitarianism  17
How Does the Study of Geology Benefit Us?  18

## PLATE TECTONICS: A UNIFYING THEORY  22

Introduction  23
Early Ideas about Continental Drift  23
    Alfred Wegener and the Continental Drift Hypothesis  24
What Is the Evidence for Continental Drift?  24
    Continental Fit  24
    Similarity of Rock Sequences and Mountain Ranges  25
    Glacial Evidence  25
    Fossil Evidence  26
Features of the Seafloor  27
    The Continental Shelf, Slope, and Rise  28
    Abyssal Plains, Oceanic Ridges, Submarine Hydrothermal Vents, and Oceanic Trenches  28
    Seamounts, Guyots, and Aseismic Ridges  29
    Continental Margins  31
Earth's Magnetic Field  31
Paleomagnetism and Polar Wandering  32
Magnetic Reversals and Seafloor Spreading  33
Plate Tectonics: A Unifying Theory  35
The Three Types of Plate Boundaries  36
    Divergent Boundaries  36
    Convergent Boundaries  37
    Transform Boundaries  41
Hot Spots and Mantle Plumes  42
Plate Movement and Motion  43
The Driving Mechanism of Plate Tectonics  43
Plate Tectonics and the Distribution of Natural Resources  45
    Petroleum  45
    Mineral Deposits  45

**GEO-FOCUS  Molybdenum Mining and Economic, Environmental, and Political Concerns  46**

Plate Tectonics and the Distribution of Life  47

## MINERALS: THE BUILDING BLOCKS OF ROCKS  51

Introduction  52
Matter: What Is It?  53
    Atoms and Elements  53
    Bonding and Compounds  54
Explore the World of Minerals  56
    Naturally Occurring Inorganic Substances  56
    Mineral Crystals  56

Chemical Composition of Minerals   57
Physical Properties of Minerals   57

**Mineral Groups Recognized by Geologists   58**
Silicate Minerals   59
Carbonate Minerals   61
Other Mineral Groups   61

**Physical Properties of Minerals   61**
Luster and Color   61
Crystal Form   62
Cleavage and Fracture   62
Hardness   63
Specific Gravity (Density)   63
Other Useful Mineral Properties   63

**Rock-Forming Minerals   64**

**How Do Minerals Form?   64**

**Natural Resources and Reserves   65**

**GEO-FOCUS  The Many Uses of Mica   66**

# IGNEOUS ROCKS AND INTRUSIVE IGNEOUS ACTIVITY   69

**Introduction   70**

**The Properties and Behavior of Magma and Lava   70**
Composition of Magma   70
How Hot Are Magma and Lava?   71
Viscosity: Resistance to Flow   71

**How Does Magma Originate and Change?   72**
Bowen's Reaction Series   72
The Origin of Magma at Spreading Ridges   74
Subduction Zones and the Origin of Magma   74
Hot Spots and the Origin of Magma   75
Compositional Changes in Magma   75

**Igneous Rocks: Their Characteristics and Classification   77**
Igneous Rock Textures   77
Composition of Igneous Rocks   78
Classifying Igneous Rocks   78

**GEO-FOCUS Granite—Common, Attractive, and Useful   82**

**Intrusive Igneous Bodies: Plutons   83**
Dikes, Sills, and Laccoliths   84
Volcanic Pipes and Necks   85
Batholiths and Stocks   85

**The Origin of Batholiths   85**

# VOLCANOES AND VOLCANISM   89

**Introduction   90**

**Volcanoes and Volcanism   91**
Volcanic Gases   91
Lava Flows   92
Pyroclastic Materials   94

**Types of Volcanoes   95**
Shield Volcanoes   95
Cinder Cones   95

**GEO-FOCUS  The Bronze Age Eruption of Santorini   96**
Composite Volcanoes (Stratovolcanoes)   98
Lava Domes   99

**Other Volcanic Landforms   100**
Fissure Eruptions and Basalt Plateaus   100
Pyroclastic Sheet Deposits   101

**Distribution of Volcanoes   101**

**Plate Tectonics, Volcanoes, and Plutons   102**
Igneous Activity at Divergent Plate Boundaries   103
Igneous Activity at Convergent Plate Boundaries   103
Intraplate Volcanism   103

**Volcanic Hazards, Volcano Monitoring, and Forecasting Eruptions   103**
How Large Is an Eruption, and How Long Do Eruptions Last?   103
Is It Possible to Forecast Eruptions?   105

# WEATHERING, SOIL, AND SEDIMENTARY ROCKS   108

**Introduction   109**

**How Are Earth Materials Altered?   109**
Mechanical Weathering   110
Chemical Weathering   112

**GEO-FOCUS  Industrialization and Acid Rain   112**

**Soil and its Origin   116**
The Soil Profile   116
Factors That Control Soil Formation   117
Soil Degradation   119

Weathering and Resources   120
Sediment and Sedimentary Rocks   120
    How Does Sediment Become Sedimentary
    Rock?   121
Types of Sedimentary Rocks   122
    Detrital Sedimentary Rocks   123
    Chemical and Biochemical Sedimentary
    Rocks   124
Sedimentary Facies   127
Reading the Story Preserved in
    Sedimentary Rocks   128
    Sedimentary Structures   128
    Fossils: Remains and Traces of Ancient Life   129
    Determining the Environment of
    Deposition   131
Important Resources in Sedimentary Rocks   132
    Coal   132
    Petroleum and Natural Gas   132
    Uranium   133
    Banded Iron Formation   134

# METAMORPHISM AND METAMORPHIC ROCKS   137

Introduction   138
The Agents of Metamorphism   138
    Heat   138
    Pressure   139
    Fluid Activity   139
The Three Types of Metamorphism   140
    Contact Metamorphism   140
    Dynamic Metamorphism   142
    Regional Metamorphism   142
    Index Minerals and Metamorphic
    Grade   142
How Are Metamorphic Rocks Classified?   143
    Foliated Metamorphic Rocks   143
    Nonfoliated Metamorphic Rocks   146
Metamorphic Zones and Facies   148
Plate Tectonics and Metamorphism   149
Metamorphism and Natural Resources   150
GEO-FOCUS  Asbestos: Good or Bad?   150

# EARTHQUAKES AND EARTH'S INTERIOR   153

Introduction   154
Elastic Rebound Theory   154
Seismology   156
    The Focus and Epicenter of an Earthquake   157
Where Do Earthquakes Occur, and How Often?   158
Seismic Waves   159
    Body Waves   159
    Surface Waves   160
Locating an Earthquake   160
Measuring the Strength of an Earthquake   162
    Intensity   162
    Magnitude   163
The Destructive Effects of Earthquakes   164
    Ground Shaking   164
    Fire   165
    Tsunami: Killer Waves   166
GEO-FOCUS  Designing and Building Earthquake-
             Resistant Structures   166
    Ground Failure   168
Earthquake Prediction   169
    Earthquake Prediction Programs   170
Earthquake Control   171
Earth's Interior   172
The Core   173
    Density and Composition of the Core   174
Earth's Mantle   175
    The Mantle's Structure, Density, and Composition   175
Earth's Internal Heat   176
Earth's Crust   176

# DEFORMATION, MOUNTAIN BUILDING, AND THE CONTINENTS   179

Introduction   180
GEO-FOCUS  Engineering and Geology   180
Rock Deformation: How Does it Occur?   181
    Stress and Strain   181
    Types of Strain   182

Strike and Dip: The Orientation of Deformed Rock Layers  183
Deformation and Geologic Structures  184
    Folded Rock Layers  184
    Joints  188
    Faults  188
    Dip-Slip Faults  189
    Strike-Slip Faults  189
    Oblique-Slip Faults  190
Deformation and the Origin of Mountains  191
    Mountain Building  191
    Plate Tectonics and Mountain Building  192
    Terranes and the Origin of Mountains  196
Earth's Crust  196
    Floating Continents?  196
    Principle of Isostasy  196
    Isostatic Rebound  197

# 10 MASS WASTING 201

Introduction  202
Factors That Influence Mass Wasting  202
    Slope Angle  202
    Weathering and Climate  204
    Water Content  204
    Vegetation  205
    Overloading  205
    Geology and Slope Stability  205
    Triggering Mechanisms  206
Types of Mass Wasting  206
    Falls  206
    Slides  207
GEO-FOCUS  Southern California Landslides  210
    Flows  211
    Complex Movements  217
Recognizing and Minimizing the Effects of Mass Wasting  217

# 11 RUNNING WATER 222

Introduction  223
Water on Earth  223
    The Hydrologic Cycle  223
    Fluid Flow  224

Running Water  224
    Sheet Flow and Channel Flow  224
    Gradient, Velocity, and Discharge  225
Running Water, Erosion, and Sediment Transport  226
Deposition by Running Water  227
    The Deposits of Braided and Meandering Streams  227
    Floodplain Deposits  229
    Deltas  230
    Alluvial Fans  231
GEO-FOCUS  The Mississippi River Delta—Past and Present  232
Can Floods Be Controlled and Predicted?  233
Drainage Systems  234
The Significance of Base Level  236
    What Is a Graded Stream?  238
The Evolution of Valleys  238
    Stream Terraces  239
    Incised Meanders  240
    Superposed Streams  240

# 12 GROUNDWATER 243

Introduction  244
Groundwater and the Hydrologic Cycle  244
Porosity and Permeability  244
The Water Table  245
Groundwater Movement  246
Springs, Water Wells, and Artesian Systems  247
    Springs  247
    Water Wells  248
    Artesian Systems  249
Groundwater Erosion and Deposition  250
    Sinkholes and Karst Topography  250
    Caves and Cave Deposits  252
Modifications of the Groundwater System and Its Effects  253
    Lowering the Water Table  254
    Saltwater Incursion  255
    Subsidence  255
GEO-FOCUS  Hydraulic Fracturing: Pros and Cons  256
    Groundwater Contamination  258
Hydrothermal Activity  261
    Hot Springs  261
    Geysers  262

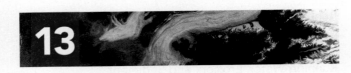

# 13

# GLACIERS AND GLACIATION    267

Introduction    268
The Kinds of Glaciers    268
    Valley Glaciers    268
    Continental Glaciers    269
Glaciers: Moving Bodies of Ice on Land    270
    Glaciers: Part of the Hydrologic Cycle    270
    How Do Glaciers Originate and Move?    270
    Distribution of Glaciers    271
The Glacial Budget    272
    How Fast Do Glaciers Move?    273
Erosion and Sediment Transport by Glaciers    274
    Erosion by Valley Glaciers    275
    Continental Glaciers and
    Erosional Landforms    278
Deposits of Glaciers    278
    Glacial Drift    278
    Landforms Composed of Till    278
    Landforms Composed of Stratified Drift    280
    Deposits in Glacial Lakes    280
What Causes Ice Ages?    282
    The Milankovitch Theory    283
    Short-Term Climatic Events    284

GEO-FOCUS  Glaciers and Global Warming    284

# 14

# THE WORK OF WIND AND DESERTS    288

Introduction    289
Sediment Transport by Wind    289
    Bed Load    289
    Suspended Load    290
Wind Erosion    290
    Abrasion    290
    Deflation    290
Wind Deposits    291
    The Formation and Migration of Dunes    292
    Dune Types    292
    Loess    295
Air-Pressure Belts and Global Wind Patterns    295
The Distribution of Deserts    297
Characteristics of Deserts    297
    Temperature, Precipitation, and Vegetation    298
    Weathering and Soils    298
    Mass Wasting, Streams, and Groundwater    298
    Wind    299
Desert Landforms    299

GEO-FOCUS  Windmills and Wind Power    300

# 15

# OCEANS, SHORELINES, AND SHORELINE PROCESSES    306

Introduction    307
Seawater, Oceanic Circulation, and Seafloor
    Sediments    308
    Seawater—Its Composition    308
    Oceanic Circulation    309
    Seafloor Sediments    309
Shorelines and Shoreline Processes    310
    Tides    310

GEO-FOCUS  Energy from the Oceans    312

    Waves    315
    Nearshore Currents    315
Shoreline Erosion and Deposition    318
    Erosion and Wave-Cut Platforms    318
    Sea Caves, Arches, and Stacks    319
    Deposition    319
    Beaches    319
    Seasonal Changes in Beaches    320
    Spits, Baymouth Bars, and Tombolos    321
    Barrier Islands    323
    The Nearshore Sediment Budget    323
Types of Coasts    324
    Depositional and Erosional Coasts    325
    Submergent and Emergent Coasts    325
The Perils of Living Along a Shoreline    326
    Storm Waves and Coastal Flooding    326
    Coastal Management as Sea Level Rises    327
Resources from the Oceans    328

# 16

# GEOLOGIC TIME: CONCEPTS AND PRINCIPLES    331

Introduction    332
How Is Geologic Time Measured?    332

Early Concepts of Geologic Time and Earth's Age  332

**GEO-FOCUS**  **The Anthropocene: A New Geologic Epoch?**  334

James Hutton and the Recognition of Geologic Time  335

Relative Dating Methods  335
  Fundamental Principles of Relative Dating  335
  Unconformities  339
  Applying the Principles of Relative Dating  340

Correlating Rock Units  344

Numerical Dating Methods  346
  Radioactive Decay and Half-Lives  348
  Sources of Uncertainty  350
  Long-Lived Radioactive Isotope Pairs  351
  Carbon-14 Dating Method  352

Development of the Geologic Time Scale  353

Stratigraphy and Stratigraphic Terminology  354

Geologic Time and Climate Change  355

# EARTH HISTORY  361

Introduction  362

Precambrian Earth History  362
  The Origin and Evolution of Continents  363
  Shields, Platforms, and Cratons  363
  Archean Earth History  364
  Proterozoic Earth History  366

The Paleozoic Geography of Earth  367

**GEO-FOCUS**  **The Devonian Old Red Sandstone**  369

The Paleozoic Evolution of North America  372
  The Sauk Sequence  372
  The Tippecanoe Sequence  372
  The Kaskaskia Sequence  374
  The Absaroka Sequence  375

The History of the Paleozoic Mobile Belts  378
  The Appalachian Mobile Belt  378
  The Cordilleran Mobile Belt  381
  The Ouachita Mobile Belt  381

The Role of Microplates  381

The Breakup of Pangaea  381

The Mesozoic History of North America  383
  Eastern Coastal Region  384
  Gulf Coastal Region  385
  Western Region  386

Cenozoic Earth History  389
  Cenozoic Plate Tectonics and Orogeny  389

The North American Cordillera  389
  The Continental Interior and the Gulf Coastal Plain  390
  Eastern North America  391
  Pleistocene Glaciation  393

# LIFE HISTORY  398

Introduction  399

Precambrian Life History  399

Paleozoic Life History  401
  Marine Invertebrates  401
  The Permian Mass Extinction  403
  Vertebrates  404
  Plants  407

Mesozoic Life History  408
  Marine Invertebrates  408
  The Diversification of Reptiles  408
  Birds  414

**GEO-FOCUS**  **Mary Anning's Contributions to Paleontology**  415
  Mammals  416
  Plants  416
  Cretaceous Mass Extinctions  417

Cenozoic Life History  418
  Marine Invertebrates and Phytoplankton  418
  Cenozoic Birds  418
  Diversification of Mammals  418
  Cenozoic Mammals  420
  Pleistocene Faunas  420
  Primate Evolution  422
  Hominids and Hominins  423

.....................................................

**APPENDICES**

**A:** English-Metric Conversion Chart  431
**B:** Mineral Identification Tables  432

**Answers**

Multiple-Choice Questions  434
Selected Short Answer Questions  435

**Glossary**  437

**Index**  449

# PREFACE

Earth is a dynamic planet that has changed continuously during its 4.6-billion-year history. The size, shape, and geographic distribution of the continents and ocean basins have changed through time, as have the atmosphere and biota. As scientists and concerned citizens, we have become increasingly aware of how fragile our planet is and, more importantly, how interdependent all of its various systems and subsystems are.

We also have learned that we cannot continually pollute our environment and that our natural resources are limited and, in most cases, nonrenewable. Furthermore, we are coming to realize just how central geology is to our everyday lives. For example, on January 12, 2010, a 7.0-magnitude earthquake struck the island nation of Haiti, killing more than 223,000 people. A major oil spill in the Gulf of Mexico in 2010 resulted in extensive ecological damage along the shorelines of the Gulf Coast of the United States. And in 2017, Hurricane Harvey inundated Houston, Texas, causing major flooding and tremendous damage to its infrastructure. For these and other reasons, geology is one of the most important college or university courses that a student can take.

*Geology: Earth in Perspective* is designed as an introductory course in geology that can serve both majors and nonmajors in geology and the Earth sciences. One of the problems with any introductory science course is that students are overwhelmed by the amount of material that must be learned. Furthermore, most of the material does not seem to be linked by any unifying theme and does not always appear to be relevant to their lives. This book, however, addresses that problem in its easy-to-read style and up-to-date examples and discoveries that elucidate the concepts and content in a way that makes it easy to relate geology to the world around us. In addition, numerous animations in *MindTap®* illustrate concepts, such as Bowen's Reaction Series, that are sometimes hard to understand only in text. By combining these two methods of teaching, we are able to show why geology is an exciting and ever-changing science that relates to all of us.

The overarching goal of this book is to provide students with a basic understanding of geology and its processes and, most importantly, with an understanding of how geology relates to the human experience—that is, how geology affects individuals, society, and nation-states. In both this textbook and the supplementary *MindTap®*, we also present the geologic and biologic history of Earth, not as a set of encyclopedic facts to memorize, but rather as a continuum of interrelated and interconnected events that reflect the underlying geologic and biologic principles and processes that have shaped our planet and life upon it.

Instead of emphasizing individual, and seemingly unrelated, facts and events, we provide students with the means to understand the underlying causes of why things happen the way they do and how all of Earth's systems and subsystems are interconnected. Using this systems approach, students gain a better understanding of how everything fits together and how it relates to the history of Earth, further emphasizing and reinforcing why the study of geology is important and relevant in today's world.

With these goals in mind, we introduce the major themes of this book in the first chapter, not only to provide an overview of the subject but also to enable students to see how the various systems, subsystems, and cycles of Earth are both interrelated and interconnected. We then cover the unifying theme of geology—plate tectonics—in the second chapter. Plate tectonic theory is central to geology because it links together many of its aspects and therefore is a theme that is woven throughout this book.

The economic and environmental aspects of geology are also stressed throughout both the text and in more focused examples in the Geo-Focus inserts. Rather than covering these two topics in separate chapters, we treat them as an integral part of geology, so that students can see, through topical and interesting examples, how geology affects our lives.

Climate change is an especially relevant and important topic that currently is in the news and being discussed and debated by scientists, politicians, and citizens alike. Because of its importance, we introduce the topic in the first chapter and integrate it throughout the book. Geology is unique in that it can provide the perspective of geologic time in this important debate, not only in regards to climate changes in the past but also the degree to which humans are contributing to climate change.

Environmental issues are another topic important to everyone, and these are covered throughout the book. Energy and industrial needs are closely related to environmental concerns and global economics. As economies shift from nonrenewable resources to renewable resources, geology is playing a central role. Current advances in electricity production from wind, waves, and tides are covered in the Geo-Focus sections in Chapters 14 and 15, respectively. We also address the issue of hydraulic fracturing, not only from an environmental aspect but also as an energy and economic issue. Because of the importance of this topic and its cross-disciplinary nature, we cover it in several chapters.

A feature not found in most introductory textbooks is the history of Earth and its biota. The last two chapters cover these topics showing how the principles and processes discussed in previous chapters are applied in a general overview to the history and evolution of our planet.

# FEATURES OF THIS BOOK

As Earth is dynamic, so too are the features in this book. We are committed to providing a textbook on a topic that is relevant to the student, but one that contains a high level of current information, presented in an engaging visual and text-friendly format. The following features highlight what we consider central to maximizing the goals of this textbook and the *MindTap®* supplement.

- Chapter content has been written to (1) clarify concepts, (2) emphasize underlying principles and processes, and (3) make the material more exploratory.
- Learning Objectives (LO) requiring an answer are provided after each first order heading to help ensure that the student is comprehending the important concepts and material in each section of the chapter.
- Many new, bold, and dramatic photos open each chapter, a number of which are recent geologic events, thus adding relevancy to the text and emphasizing the theme of how geology relates to humans.
- The numerous photos and illustrations in each chapter were chosen to highlight the visual nature of geology.
- Each figure has a blue dot in front of the first reference for easily locating the figure in the text.
- Each important term is shown in bold face type, and referenced, with definition, at the bottom of the page where it is first mentioned.
- *Critical Thinking Questions* are part of many of the figures. These questions are designed to encourage active student learning, guide observational skill development, and deepen understanding of geologic processes.
- There is one *Geo-Focus* feature per chapter, which emphasizes in greater depth environmental and economic topics such as climate change, hydraulic fracturing, alternate and renewable energy sources, industrial economics, building earthquake-resistant structures, and geologic engineering, to name a few.
- New end-of-chapter review sections include:
  - *Key Concepts Review*, which summarizes the important concepts covered in the chapter.
  - *Important Terms* from the chapter, and the page numbers on which they appear.
  - *Review Questions,* which include five multiple-choice and five short answer questions designed to test knowledge

of chapter content. Answers to all of the multiple-choice questions, and one of the short answer questions are provided in the *Answers* section at the end of the book.

- *Critical Thinking Visual Questions* that challenge students to describe the geologic process being depicted, engage in quantitative solutions, or address an issue using the information provided in the photo or graphic. These images and questions were chosen to encourage students to develop strong observational and critical thinking skills.
- A *MindTap®* icon directs the student to an animation that helps illustrate important and sometimes difficult concepts within the chapter.
- Lastly, a full *Glossary* of important terms appears at the end of the book and before the *Index*.

It is our strong belief that the features of *Geology: Earth in Perspective* and its ancillary materials make it an effective teaching tool that engages students in the learning process, thereby fostering a better understanding of the material and how it relates to Earth in the 21st century.

# ACKNOWLEDGMENTS

As the authors, we are, of course, responsible for the organization, style, and accuracy of the text, and any mistakes, omissions, or errors are our responsibility. We benefited over the years from the numerous comments and advice from many geologists who reviewed all or parts of our previous geology text books. And we thank them for their many insightful suggestions.

Special thanks must go to product manager Lauren Bakker at Cengage Learning, who initiated this textbook, and to our learning designer Lauren Oliveira and content manager Nicole Evans, who not only kept us on task but also superbly edited and managed the content for this edition, as well as provided a fresh perspective on this edition. We also thank senior designer Helen Bruno for the fresh design. Additional thanks goes to Dr. Cynthia Liutkus-Pierce for all of her insightful comments, suggestions, and help in reviewing portions of the text and animations. Lastly, we would like to recognize marketing manager Andrew Stock and market development manager Roxanne Wang for their contributions.

As always, our families were very patient and encouraging when much of our spare time and energy were devoted to this book. We again thank them for their continued support and understanding.

Reed Wicander would especially like to thank his wife Linda for all of her encouragement and support over the years, especially during the revision of this third edition of *Geology: Earth in Perspective*.

*Reed Wicander*

*James S. Monroe*

# AUTHOR BIOGRAPHY

**Reed Wicander** is a professor emeritus of geology at Central Michigan University, where he taught physical geology, historical geology, prehistoric life, and invertebrate paleontology. He is currently an adjunct professor in the School of Earth and Environmental Sciences at The University of Queensland, Brisbane, Australia. Reed earned his B.S. degree in geology from San Diego State University and his Ph.D. from UCLA. His main research focuses on various aspects of Paleozoic palynology, specifically the study of acritarchs, a group of organic-walled microphytoplankton, on which he has published many papers. In addition, he has coauthored numerous geology textbooks with James S. Monroe. He is a past president of the American Association of Stratigraphic Palynologists – The Palynological Society, and Commission Internationale de Microflore du Paléozoïque, as well as a former Councillor of the International Federation of Palynological Societies.

**James S. Monroe** is Professor Emeritus of Geology at Central Michigan University, where he taught Physical Geology, Historical Geology, Prehistoric Life, and Stratigraphy and Sedimentology beginning in 1975 and served as chair of the Geology department. He received his Ph.D. from the University of Montana. He has coauthored several textbooks with Reed Wicander and has interests in Cenozoic geology and geologic education. Monroe now lives in Chico, California, where he remains active in geology by teaching courses to a large group of retirees.

# 1

# UNDERSTANDING EARTH

## A Dynamic and Evolving Planet

Color satellite image of Earth as it stands out against the black void of outer space. In this book, we examine Earth as a system of interconnected components that interact with each other. This complex interaction between Earth's four major subsystems—the atmosphere, biosphere, hydrosphere, and lithosphere, as well as its interior—results in a dynamically changing planet.

# INTRODUCTION

A major benefit of the space age has been the ability to look back from space and view our planet in its entirety. Every astronaut has remarked in one way or another on how Earth stands out as an inviting oasis in the otherwise black void of space (see Chapter Opening photo). We are able to see not only the beauty of our planet but also its fragility and how humans are affecting the environment. And although we do not witness it firsthand, we can still read the story of Earth's long and turbulent 4.6-billion-year history by deciphering the clues preserved in the geologic record.

A major theme of this book is that Earth is a complex, dynamic planet that has changed continuously since its origin some 4.6 billion years ago. These changes and the present-day features we observe result from the interactions between Earth's internal and external systems, subsystems, and cycles. By viewing Earth as a whole—that is, thinking of it as a *system*—we not only see how its various components are interconnected but can also better appreciate its complex and dynamic nature.

The system concept makes it easier for us to study a complex subject such as Earth because it divides the whole into smaller components that we can easily understand, without losing sight of how the components fit together as a whole. In the same way, you can think of this book as a large, panoramic landscape painting. Each chapter fills in the details of the landscape, thereby enhancing the overall enjoyment and understanding of the entire painting.

A **system** is a combination of related parts that interact in an organized manner. An automobile is a good example of a system. Its various components, or subsystems, such as the engine, transmission, steering, and brakes, are all interconnected in such a way that a change in any one of them affects the others.

We can examine Earth in the same way we view an automobile—that is, as a system of interconnected components that interact and affect each other in many ways. The principal subsystems of Earth are the *atmosphere, biosphere, hydrosphere, lithosphere, mantle,* and *core* (●**FIGURE 1.1**). The complex interactions between these subsystems result in a dynamically changing planet in which matter and energy are continuously recycled into different forms. Examined in this manner, the continuous evolution of Earth and its life is not a series of isolated and unrelated events, but rather it is a dynamic interplay between its various subsystems.

We also must not forget that humans are part of the Earth system, and our activities can produce changes with potentially wide-ranging consequences. When people discuss and debate such environmental issues as pollution and global warming, it is important to remember that these are not isolated issues but are part of the larger Earth system. Furthermore, remember that Earth goes through longer time cycles than humans are used to. Although global warming may have deleterious short-term effects on Earth's biota, climate change is part of longer-term cycles that have resulted in large-scale periods of soaring global temperature and numerous episodes of glaciations.

As you study the various topics covered in this book, keep in mind the themes discussed in this chapter and how, like the parts of a system, they are interrelated. By relating each chapter's topic to its place in the entire Earth system, you will gain a greater appreciation of why geology is so integral to our lives.

# WHAT IS GEOLOGY?

L01   Define geology

**Geology**, from the Greek *geo* and *logos*, is defined as the study of Earth, but now it must also include the study of the planets and moons in our solar system. The discipline of geology is generally divided into two broad areas—physical geology and historical geology. *Physical geology* is the study of Earth materials, such as **minerals** and **rocks**, as well as the processes operating within Earth and on its surface. *Historical geology* examines the origin and evolution of Earth and its continents, oceans, atmosphere, and biota.

Although the discipline of geology is broad and subdivided into numerous fields, or specialties, nearly every aspect of geology has some economic or environmental relevance. For example, many geologists are involved in exploration for mineral and energy resources, using their specialized knowledge to locate the natural resources on which our industrialized society is based.

Other geologists use their expertise to address various environmental and societal problems. Not only is finding adequate sources of groundwater for the ever-burgeoning needs of communities and industries important, but so too is the monitoring and prevention of surface and groundwater pollution, and when necessary, its cleanup. Geologic engineers help find safe locations for dams, waste-disposal sites, and power plants, as well as designing earthquake-resistant structures.

Geologists are increasingly asked to make short- and long-range predictions about earthquakes and volcanic eruptions and the potential destruction that may result. In fact, geologists are involved in working with various governmental agencies and civil defense planners to ensure that contingency plans are in place and timely warnings are given for areas potentially affected by natural disasters such as earthquakes, volcanic eruptions, and tsunami.

---

**system**  A combination of related parts that interact in an organized fashion; Earth systems include the atmosphere, hydrosphere, biosphere, and solid Earth.

**geology**  The study of Earth, as well as the planets and moons in our solar system. It is generally divided into two broad areas—*physical geology,* which is the study of Earth's materials, such as minerals and rocks, as well as the processes operating within Earth and on its surface, and *historical geology,* which examines the origin and evolution of Earth, its continents, oceans, atmosphere, and biota.

**mineral**  A naturally occurring, inorganic, crystalline solid that has characteristic physical properties and a narrowly defined chemical composition.

**rock**  A solid aggregate of one or more minerals, as in limestone and granite, or a consolidated aggregate of rock fragments, as in conglomerate, or masses of rocklike materials, such as coal and obsidian.

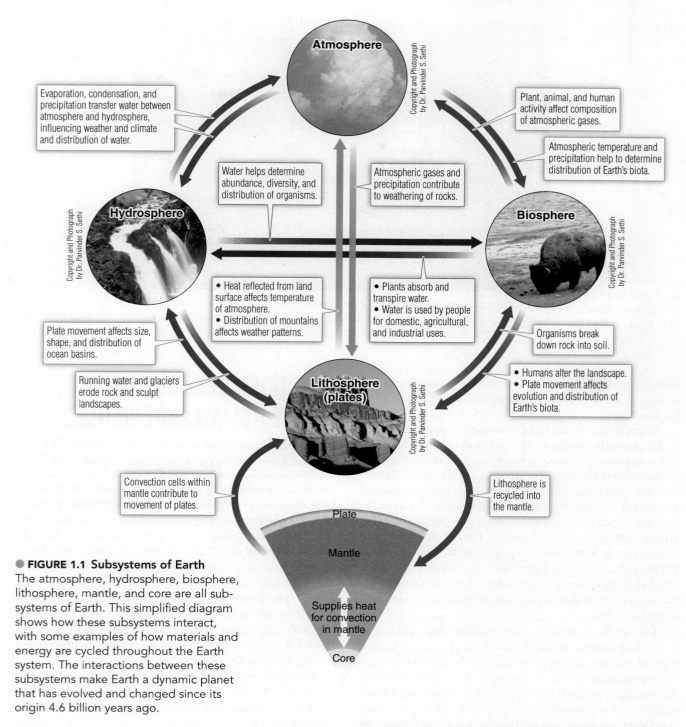

**● FIGURE 1.1 Subsystems of Earth**
The atmosphere, hydrosphere, biosphere, lithosphere, mantle, and core are all sub-systems of Earth. This simplified diagram shows how these subsystems interact, with some examples of how materials and energy are cycled throughout the Earth system. The interactions between these subsystems make Earth a dynamic planet that has evolved and changed since its origin 4.6 billion years ago.

# GEOLOGY AND THE FORMULATION OF THEORIES

**LO2    Describe how the scientific method and the formulation of theories has impacted the study of geology**

The term **theory** has various meanings and is frequently misunderstood and consequently misused. In collo-quial usage, it means a speculative or conjectural view of something—hence, the widespread belief that scientific theories are little more than unsubstantiated wild guesses. In scientific usage, however, a theory is a coherent explanation for one or several related natural phenomena supported by a large body of objective evidence. From a theory, sci-entists derive predictive statements that they test by obser-vations and/or experiments so that their validity can be assessed. The law of universal gravitation is an example of

**theory** An explanation for some natural phenomenon that has a large body of supporting evidence. To be scientific, a theory must be testable—for example, plate tectonic theory.

● **FIGURE 1.2 Geology and Art** Ferde Grofé's *Grand Canyon Suite* was inspired by the beauty of Arizona's Grand Canyon, where sedimentary rock layers grandly document some of Earth's past history.

a theory that describes the attraction between masses (an apple and Earth in the popularized account of Newton and his discovery).

Theories are formulated through the process known as the **scientific method**. This method is an orderly, logical approach that involves gathering and analyzing facts or data about the problem under consideration. Tentative explanations, or **hypotheses**, are then formulated to explain the observed phenomena. Next, the hypotheses are tested to see whether what was predicted actually occurs in a given situation. Finally, if one of the hypotheses is found, after repeated tests, to explain the phenomena, then that hypothesis is proposed as a theory. Remember, however, that in science, even a theory is still subject to further testing and refinement as new data become available.

The fact that a scientific theory can be tested and is subject to such testing separates it from other forms of human inquiry. Because scientific theories can be tested, they have the potential for being supported or even proven wrong. Accordingly, science must proceed without any appeal to beliefs or supernatural explanations, not because such beliefs or explanations are necessarily untrue but because we have no way to investigate or test them. For this reason, science makes no claim about the existence or nonexistence of a supernatural or spiritual realm.

Each scientific discipline has certain theories that are particularly important. In geology, the formulation of *plate tectonic theory* has changed the way geologists view Earth. For example, geologists now consider Earth from a global perspective in which all of its subsystems and cycles are interconnected, and Earth history is seen to be a continuum of interrelated events that are part of a global pattern of change.

# HOW DOES GEOLOGY RELATE TO THE HUMAN EXPERIENCE?

**L03** **Explain, with some examples, the relationship between geology and the human experience**

You would probably be surprised at the numerous references to geology in the arts, music, and literature. Many sketches and paintings by famous artists depict rocks and landscapes realistically. Examples include Leonardo da Vinci's *Virgin of the Rocks*, Giovanni Bellini's *Saint Francis in Ecstasy*, and Asher Brown Durand's *Kindred Spirits*.

In the field of music, Ferde Grofé's *Grand Canyon Suite* was no doubt inspired by the grandeur and timelessness of Arizona's Grand Canyon and its vast rock exposures (●**FIGURE 1.2**). The rocks on the Island of Staffa in the Inner Hebrides provided the inspiration for Felix Mendelssohn's famous *Hebrides Overture*.

References to geology abound in *The German Legends of the Brothers Grimm*. Jules Verne's novel *Journey to the Center of the Earth* describes an expedition into Earth's interior. There is even a series of mystery books by Sarah Andrews that features the fictional geologist Em Hansen, who uses her knowledge of geology to solve crimes.

Geology has also played an important role in the history and culture of humankind. Empires throughout history

**scientific method** A logical, orderly approach that involves gathering data, formulating and testing hypotheses, and proposing theories.

**hypothesis** A provisional explanation for observations that is subject to continual testing. If well supported by evidence, a hypothesis may be called a theory.

have risen and fallen because mineral and energy resources are unequally distributed, thus resulting in wars to secure, for example, such natural resources as oil and gas (see GEO-FOCUS). Natural barriers, such as mountain ranges and rivers, which have formed by geologic agents, have frequently served as political boundaries.

## HOW DOES GEOLOGY AFFECT OUR EVERYDAY LIVES?

**L04**  Discuss the impact geology has, not only on our standard of living, but also in terms of economic and political power

The most obvious connection between geology and our everyday lives is when natural disasters such as volcanic eruptions, earthquakes, landslides, tsunami, and floods strike. Less apparent, but equally significant, are the connections between geology and economic, social, and political issues.

Consider, for example, just how dependent we are on geology in our daily routines (●FIGURE 1.3). Much of the electricity for our appliances comes from the burning of coal, oil, natural gas, or uranium consumed in nuclear-generating plants. Geologists locate the coal, petroleum (oil and natural gas), and uranium. The copper or other metal wires through which electricity travels are manufactured from materials found as the result of mineral exploration. The concrete foundation (concrete is a mixture of clay, sand, or gravel, and limestone), drywall (made largely from the mineral gypsum), and windows (the mineral quartz is the principal ingredient in the manufacture of glass) of the buildings we live and work in owe their very existence to geologic resources.

The majority of cars or public transportation we use to go to work are powered and lubricated by some type of petroleum

by-product and are constructed of metal alloys and plastics. In addition, the roads or rails we ride over come from geologic materials such as gravel, asphalt, concrete, or steel. All of these items are the result of processing geologic resources.

As individuals and societies, we enjoy a standard of living that is directly dependent on the consumption of geologic materials. We therefore need to be aware of how our use and misuse of geologic resources may affect the environment. We need to support the development of policies that not only encourage management of our natural resources but also allow for continuing economic development among all the world's nations.

## GLOBAL GEOLOGIC AND ENVIRONMENTAL ISSUES FACING HUMANKIND

**L05**  Describe some of the major environmental issues facing humankind

Most scientists would argue that overpopulation is the greatest environmental problem facing the world today (●FIGURE 1.4). The world's population in 2017 was 7.6 billion, and projections indicate that this number will reach 9.8 billion people by 2050. Although this may not seem to be a geologic problem, remember that these people must be fed, housed, and clothed, and all with a minimal impact on the environment. Much of this population growth will be in areas that are already at risk from such hazards as earthquakes, tsunami, volcanic eruptions, and floods. Adequate water supplies must be found and protected from pollution. Additional energy resources will be needed to help fuel the economies of nations with ever-increasing populations

# GEO-FOCUS

## Geology and Economic and Political Power

Geology can be closely connected to economic and political power. The configuration of Earth's surface, or its topography, which is shaped by geologic agents, has often played a critical role in military tactics. For example, Napoleon included two geologists in his expeditionary forces when he invaded Egypt in 1798, and the Russians used geologists as advisors in selecting fortification sites during the Russo-Japanese war of 1904–1905. Natural barriers such as mountain ranges and rivers have frequently

served as political boundaries, and the shifting of river channels has sparked numerous border disputes.

Mineral and energy resources are not equally distributed, and no country is self-sufficient in all of them. Throughout history, people have fought wars to secure these resources. The United States was involved in the 1990–1991 Gulf War largely because it needed to protect its oil interests in that region. Many foreign policies and treaties develop from the need to acquire and maintain adequate

supplies of mineral and energy resources.

In addition to such energy resources as coal, petroleum, and natural gas, today's technological society is becoming increasingly dependent on such minerals as lithium and cobalt, which are used in electric cars and cell-phone batteries. Without new exploration and production of these minerals, a potential shortage will occur in the years ahead as the demand for these materials increases.

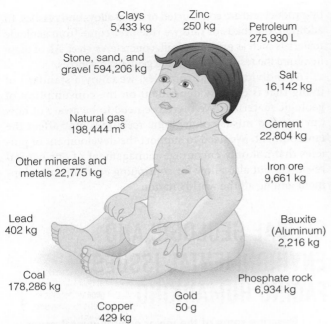

Clays
5,433 kg

Zinc
250 kg

Petroleum
275,930 L

Stone, sand, and
gravel 594,206 kg

Salt
16,142 kg

Natural gas
198,444 m³

Cement
22,804 kg

Other minerals and
metals 22,775 kg

Iron ore
9,661 kg

Lead
402 kg

Bauxite
(Aluminum)
2,216 kg

Coal
178,286 kg

Phosphate rock
6,934 kg

Copper
429 kg

Gold
50 g

● **FIGURE 1.3 Lifetime Mineral Usage** According to the Mineral Information Institute in Golden, Colorado, the average American born in 2017 has a life expectancy of 79.1 years and will need 859,528 kg of minerals, metals, and fuels to sustain his or her standard of living over a lifetime. This is an average of 10,866 kg of mineral and energy resources per year for every man, woman, and child in the United States.

**Critical Thinking Question** Every year the life expectancy of the average American increases as well as our usage of minerals, metals, and fuels needed during a lifetime to maintain our standard of living. Is this increase sustainable, and is there anything that can be done to balance the depletion of natural resources but still maintain a high standard of living? How does our increasing consumption of these natural resources impact the rest of the world's population? Will the recycling of goods and materials significantly affect the amount of mineral and energy resources used per year?

(●**FIGURE 1.5**). New techniques must be developed to reduce the use of our dwindling nonrenewable resource base and to increase our recycling efforts so that we can decrease our dependence on new sources of these materials.

The problems of overpopulation and how it affects the global ecosystem vary from country to country. For many poor and nonindustrialized countries, the problem is too many people and not enough food. For the more developed and industrialized countries, it is too many people rapidly depleting both the nonrenewable and the renewable natural resource base. And in the most industrially developed countries, it is people producing more pollutants than the environment can safely recycle on a human time scale. The common thread tying these varied situations together is an environmental imbalance created by a human population exceeding Earth's short-term carrying capacity.

● **FIGURE 1.4 Overpopulation** Overpopulation is perhaps the greatest environmental problem facing the world today. Until the world's increasing population is brought under control, people will continue to strain Earth's limited resources. Shown here is the crush of people at the Sadar Bazaar in New Delhi, India. With a population of 1.35 billion people (2018), India represents 17.7% of the world's population.

● **FIGURE 1.5 Offshore Oil Drilling** With increasing demand for energy, offshore drilling for oil and natural gas has increased in recent years.

Other global environmental issues affecting all of us include the greenhouse effect, global warming, and climate change. The relationship between the greenhouse effect and global warming is an excellent example of how Earth's various subsystems are interrelated. As a by-product of respiration and the burning of organic material, carbon dioxide is a component of the global ecosystem and is constantly being recycled as part of the *carbon cycle*. The concern in recent years over the increase in atmospheric carbon dioxide levels is related to its role in the greenhouse effect.

The recycling of carbon dioxide between Earth's crust and atmosphere is an important climate regulator because carbon dioxide and other gases, such as methane, nitrous oxide, chlorofluorocarbons, and water vapor, allow sunlight to pass through them, but they trap the heat reflected back from Earth's surface. This retention of heat is called

a) Short-wavelength radiation from the Sun that is not reflected back into space penetrates the atmosphere and warms Earth's surface.

b) Earth's surface radiates heat in the form of long-wavelength radiation back into the atmosphere, where some of it escapes into space.

The rest is absorbed by greenhouse gases and water vapor and reradiated back toward Earth.

c) Increased concentrations of greenhouse gases trap more heat near Earth's surface, causing a general increase in surface and atmospheric temperatures, which leads to global warming.

● **FIGURE 1.6 The Greenhouse Effect and Global Warming**

the **greenhouse effect**. It results in an increase in the temperature of Earth's surface and, more importantly, its atmosphere, thus producing global warming (●**FIGURE 1.6**). The issue is not whether we have a greenhouse effect, because we do, but rather the degree to which human activity, such as the burning of fossil fuels, is increasing the greenhouse effect, and thus contributing to global warming.

Because of the increase in human-produced greenhouse gases during the past 200 years, many scientists are concerned that a global warming trend has already begun and will result in severe global climatic shifts. Presently, most climate researchers use a range of scenarios for greenhouse gas emissions when predicting future warming rates. Climate model simulations published in the *2013 Fifth Intergovernmental Panel on Climate Change* show a predicted increase in global average temperature that is likely to exceed 1.5°C by 2100 as compared to the period between 1850 and 1900. This predicted increase in temperature is based on various scenarios that explore different global development pathways.

Regardless of which scenario is followed, the global temperature change will be uneven, with the greatest warming occurring in the higher latitudes of the Northern Hemisphere. As a consequence, rainfall patterns will shift dramatically and have a major effect on the largest grain-producing areas of the world, such as the American Midwest. Drier and hotter conditions will intensify the severity and frequency of droughts, leading to more crop failures and higher food prices. With such shifts in climate, Earth's deserts will expand, with a resulting decrease in valuable crop and grazing land. Just as many regions will experience longer and hotter summers, other areas will suffer from intense and increased rainfall, resulting in severe flooding and landslides.

Moreover, continued global warming will result in a rise in mean sea level, as icecaps and glaciers melt and contribute their water to the world's oceans. Based on coastal tide gauges and satellite measurements of sea level rise from 1870 to 2018, it is projected that sea level will continue to rise due to icecap and glacial melting, thus increasing the number of people at risk from flooding in coastal areas.

We would be remiss, however, if we did not point out that many other scientists are not convinced that the global warming trend is the direct result of increased human activity related to industrialization. In fact, there has been much heated debate concerning the data and statistics used in the various models that are employed to make climate change predictions. These scientists indicate that although the level of greenhouse gases has increased, we are still uncertain about their rate of generation and rate of removal and about whether the rise in global temperatures during the past century resulted from normal climatic variations through time or from human activity. Furthermore, they conclude that even if a general global warming trend occurs during the next 100 years, it is not certain that the dire predictions made by proponents of global warming will come true.

Earth, as we know, is a remarkably complex system, with many feedback mechanisms and interconnections throughout its various subsystems and cycles. It is very difficult to predict with any certainty all of the possible consequences

**greenhouse effect** The retention of heat in the atmosphere that results when carbon dioxide and other gases, such as methane, nitrous oxide, chlorofluorocarbons, and water vapor, allow sunlight to pass through them but traps the heat that is reflected back from Earth's surface.

that global warming would have for atmospheric and oceanic circulation patterns and its ultimate effect on Earth's biota. It is, however, important to remember that although everyone is vulnerable to weather-related disasters, large-scale changes brought about by climate change will have a far greater impact on people in poor countries than those in the more industrialized nations. Whether these climate changes are part of a natural global cycle taking place over thousands or hundreds of thousands of years—that is, on a geologic time scale—or are driven, in part, by human activities is immaterial. The bottom line is that we already are, or eventually will be, affected by them in some way, be it economic or social.

# ORIGIN OF THE UNIVERSE AND SOLAR SYSTEM, AND EARTH'S PLACE IN THEM

**LO6**   Discuss the origin and evolution of the universe, solar system, and Earth's place in them

How did the universe begin? What has been its history? What is its eventual fate, or is it infinite? These are just some of the basic questions people have asked and wondered about since they first looked into the nighttime sky and saw the vastness of the universe beyond Earth.

## Origin of the Universe: Did It Begin with a Big Bang?

Most scientists think that the universe originated about 14 billion years ago in what is popularly called the **Big Bang**. The Big Bang is a model for the evolution of the universe in which a dense, hot state was followed by expansion, cooling, and a less-dense state.

According to modern *cosmology* (the study of the origin, evolution, and nature of the universe), the universe has no edge and therefore no center. Thus, when the universe began, all matter and energy were compressed into an infinitely small high-temperature and high-density state in which both time and space were set at zero. Therefore, there is no "before the Big Bang," only what occurred after it. As demonstrated by Einstein's Theory of Relativity, space and time are unalterably linked to form a *space–time continuum*; that is, without space, there can be no time.

How do we know that the Big Bang took place approximately 14 billion years ago? Why couldn't the universe have always existed as we know it today? Two fundamental phenomena indicate that the Big Bang occurred:

1. the universe is expanding and
2. it is permeated by background radiation.

When astronomers look beyond our own solar system, they observe that everywhere in the universe galaxies are moving away from each other at tremendous speeds. Edwin Hubble first recognized this phenomenon in 1929.

By measuring the optical spectra of distant galaxies, Hubble noted that the velocity at which a galaxy moves away from Earth increases proportionally to its distance from Earth. He observed that the spectral lines (wavelengths of light) of the galaxies are shifted toward the red end of the spectrum; that is, the lines are shifted toward longer wavelengths. Galaxies receding from each other at tremendous speeds would produce such a redshift. This is an example of the *Doppler effect*, which is a change in the frequency of a sound, light, or other wave caused by movement of the wave's source relative to the observer, such as the analogy to the sound of a passing train's whistle. As the train approaches, the sound waves are slightly compressed, so an individual hears a shorter-wavelength, higher-pitched sound. As the train passes, and recedes from the individual, the sound waves are slightly expanded, and a longer-wavelength, lower-pitched sound is heard.

An easy way to envision how velocity increases with increasing distance is by reference to the analogy of a rising loaf of raisin bread, in which the raisins are evenly distributed throughout the loaf (●**FIGURE 1.7**). As the dough rises, the raisins are uniformly pushed away from each other at velocities directly proportional to the distance between any two raisins. The farther away a given raisin is to begin with, the farther it must move to maintain the regular spacing during the expansion, and hence the greater its velocity must be.

In the same way that raisins move apart in a rising loaf of bread, galaxies are receding from each other at a rate proportional to the distance between them, which is exactly what astronomers see when they observe the universe. By measuring this expansion rate, astronomers can calculate how long ago the galaxies were all together at a single point, which turns out to be about 14 billion years, the currently accepted age of the universe.

In 1965, Arno Penzias and Robert Wilson of Bell Telephone Laboratories made the second important observation that provided evidence of the Big Bang. They discovered that there is a pervasive background radiation of 2.7 Kelvin (K) above absolute zero (absolute zero equals $-273°C$; 2.7 K $= -270.3°C$) everywhere in the universe. This background radiation is thought to be the fading afterglow of the Big Bang.

Currently, cosmologists cannot say what it was like at time zero of the Big Bang because they do not understand the physics of matter and energy under such extreme conditions. However, it is thought that during the first second following the Big Bang, the four basic forces—

1. *gravity* (the attraction of one body toward another),
2. *electromagnetic force* (combines electricity and magnetism into one force and binds atoms into molecules),
3. *strong nuclear force* (binds protons and neutrons together), and
4. *weak nuclear force* (responsible for the breakdown of an atom's nucleus, producing radioactive decay)—separated, and the universe experienced enormous expansion.

**Big Bang** A model for the evolution of the universe in which a dense, hot state was followed by expansion, cooling, and a less-dense state.

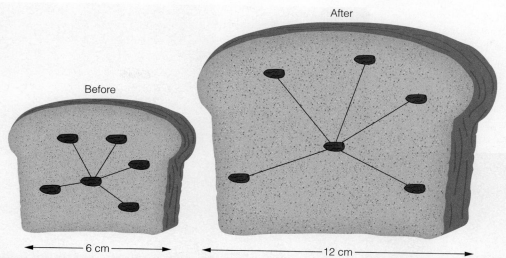

● **FIGURE 1.7 The Expanding Universe** The motion of raisins in a rising loaf of raisin bread illustrates the relationship that exists between distance and speed and is analogous to an expanding universe. In this diagram, adjacent raisins are located 2 cm apart before the loaf rises. After one hour, any raisin is now 4 cm away from its nearest neighbor and 8 cm away from the next raisin over, and so on. Therefore, from the perspective of any raisin, its nearest neighbor has moved away from it at a speed of 2 cm per hour, and the next raisin over has moved away from it at a speed of 4 cm per hour. In the same way that raisins move apart in a rising loaf of bread, galaxies are receding from each other at a rate proportional to the distance between them.

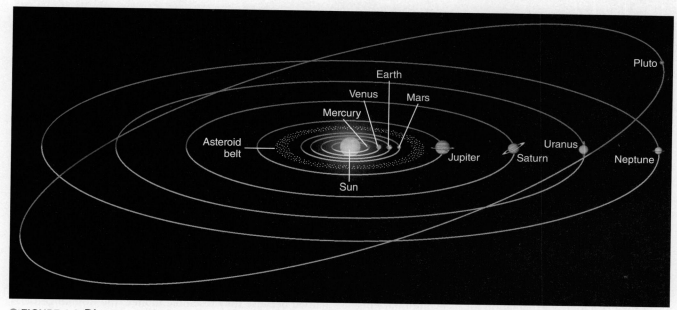

● **FIGURE 1.8 Diagrammatic Representation of the Solar System** This representation of the solar system shows the planets and the dwarf planet Pluto, and their orbits around the Sun. Pluto orbits among the icy debris of the Kuiper Belt, a disc-shaped region beyond Neptune, in which it and four other dwarf planets are known.

As the universe continued expanding and cooling, stars and galaxies began to form, and the chemical makeup of the universe changed. Initially, the universe was 100% hydrogen and helium, whereas today it is 98% hydrogen and helium and 2% all other elements by weight.

## Our Solar System: Its Origin and Evolution

Our solar system, which is part of the Milky Way galaxy, consists of the Sun, eight planets, and their moons; five dwarf planets (Ceres, Eris, Haumea, Makemake, and Pluto); a tremendous number of asteroids, most of which orbit in a zone between Mars and Jupiter; the Kuiper Belt—a disc-shaped region beyond Neptune; and the even more distant Oort Cloud, which is thought to be the source of many of the millions of comets that orbit our Sun (●**FIGURE 1.8**). Any theory formulated to explain the origin and evolution of our solar system must, therefore, take into account all of its various features and characteristics.

Many scientific theories for the origin of the solar system have been proposed, modified, and discarded

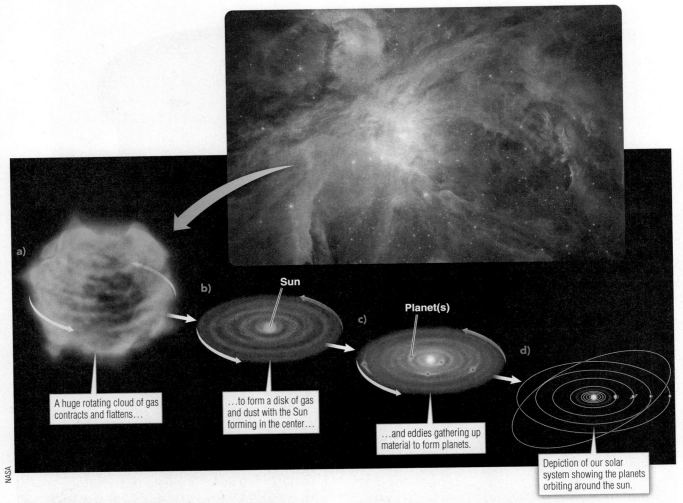

a)

A huge rotating cloud of gas contracts and flattens…

b)    **Sun**

…to form a disk of gas and dust with the Sun forming in the center…

c)    **Planet(s)**

…and eddies gathering up material to form planets.

d)

Depiction of our solar system showing the planets orbiting around the sun.

NASA

● **FIGURE 1.9 Solar Nebula Theory** According to the currently accepted theory for the origin of our solar system, the planets and the Sun formed from a rotating cloud of gas.

since the French scientist and philosopher René Descartes first proposed, in 1644, that the solar system formed from a gigantic whirlpool within a universal fluid. Today, the **solar nebula theory** for the formation of the solar system not only best explains the features of the solar system but also provides a logical explanation for its evolutionary history (●**FIGURE 1.9**).

According to the solar nebula theory, the condensation and subsequent collapse of interstellar material in a spiral arm of the Milky Way galaxy resulted in a counterclockwise-rotating disk of gases and small grains. About 90% of this material was concentrated in the central part of the disk, thus forming an embryonic Sun, around which swirled a rotating cloud of material called a *solar nebula*. Within this solar nebula were localized eddies in which gases and solid particles condensed. During the condensation process, gaseous, liquid, and solid particles began to accrete into ever-larger masses called *planetesimals*, which collided and grew in size and mass until they eventually became planets.

The composition and evolutionary history of the planets are a consequence, in part, of their distance from the Sun. The

**terrestrial planets**—Mercury, Venus, Earth, and Mars—so named because they are similar to *terra*, Latin for "earth," are all small and composed of rock and metallic elements that condensed at the high temperatures of the inner nebula. The **Jovian planets**—Jupiter, Saturn, Uranus, and Neptune—so named because they resemble Jupiter (the Roman god was also called *Jove*), all have small, rocky cores compared to their overall size and are composed mostly of hydrogen, helium, ammonia, and methane, which condense at low temperatures.

**solar nebula theory** A theory for the evolution of the solar system from a rotating cloud of gas.

**terrestrial planet** Any of the four innermost planets (Mercury, Venus, Earth, and Mars). They are all small and have high mean densities, indicating that they are composed of rock and metallic elements.

**Jovian planet** Any of the four planets (Jupiter, Saturn, Uranus, and Neptune) that resemble Jupiter. All are large and have low mean densities, indicating that they are composed mostly of lightweight gases, such as hydrogen and helium, and frozen compounds, such as ammonia and methane.

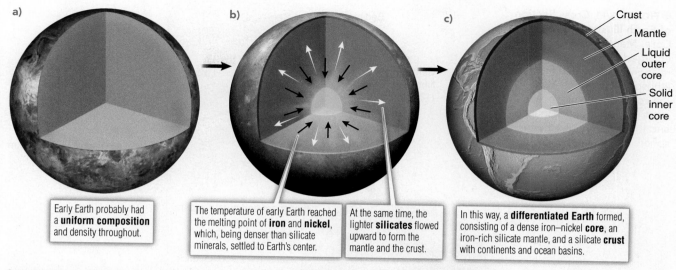

a)

Early Earth probably had a **uniform composition** and density throughout.

b)

The temperature of early Earth reached the melting point of **iron** and **nickel**, which, being denser than silicate minerals, settled to Earth's center.

At the same time, the lighter **silicates** flowed upward to form the mantle and the crust.

c)

Crust

Mantle

Liquid outer core

Solid inner core

In this way, a **differentiated Earth** formed, consisting of a dense iron–nickel **core**, an iron-rich silicate mantle, and a silicate **crust** with continents and ocean basins.

● **FIGURE 1.10 Homogenous Accretion Theory for the Formation of a Differentiated Earth**

While the planets were accreting, material that had been pulled into the center of the nebula also condensed, collapsed, and was heated to several million degrees by gravitational compression. The result was the birth of a star, our Sun.

During the early accretionary phase of the solar system's history, collisions between various bodies were common, as indicated by the numerous craters on many planets and moons. Asteroids probably formed as planetesimals in a localized eddy between what eventually became Mars and Jupiter in much the same way that other planetesimals formed the terrestrial planets. The tremendous gravitational field of Jupiter, however, prevented this material from ever accreting into a planet. Comets, which are interplanetary bodies composed of loosely bound rocky and icy materials, are thought to have condensed beyond the orbit of Neptune.

## Earth: Its Place in Our Solar System

Some 4.6 billion years ago, planetesimals in our solar system gathered enough material together to form Earth and the other planets. Scientists think that this early Earth was probably cool, of generally uniform composition and density throughout, and composed mostly of silicates (compounds consisting of silicon and oxygen), iron and magnesium oxides, and smaller amounts of all the other chemical elements (●**FIGURE 1.10A**). Subsequently, when the combination of meteorite impacts, gravitational compression, and heat from radioactive decay increased the temperature of Earth enough to melt iron and nickel, this homogeneous composition disappeared (●**FIGURE 1.10B**) and was replaced by a series of concentric layers of differing composition and density, resulting in a differentiated planet (●**FIGURE 1.10C**).

This differentiation into a layered planet is probably the most significant event in Earth's history. Not only did it lead to the formation of a crust and eventually continents, but it also was probably responsible for the emission of gases from the interior that eventually led to the origin of the oceans and atmosphere.

# WHY EARTH IS A DYNAMIC AND EVOLVING PLANET

**LO7    Explain why Earth is a dynamic and evolving planet**

Earth is a dynamic planet that has continuously changed during its 4.6-billion-year existence. The size, shape, and geographic distribution of continents and ocean basins have changed throughout time; the composition of the atmosphere has evolved; and life-forms existing today differ from those that lived during the past. Mountains and hills have been worn away by erosion, and the forces of wind, water, and ice have sculpted a diversity of landscapes. Volcanic eruptions and earthquakes reveal an active interior, and folded and fractured rocks are testimony to the tremendous power of Earth's internal forces.

Earth consists of three concentric layers: the core, the mantle, and the crust (●**FIGURE 1.11**). This orderly division results from density differences between the layers as a function of variations in composition, temperature, and pressure.

The **core** has a calculated density of 10–13 grams per cubic centimeter ($g/cm^3$) and occupies about 16% of Earth's total volume. Seismic (earthquake) data indicate that the core consists of a small, solid inner region and a larger, apparently liquid, outer portion. Both are thought to consist mostly of iron and a small amount of nickel.

The **mantle** surrounds the core and comprises about 83% of Earth's volume. It is less dense than the core

---

**core** The interior part of Earth beginning at a depth of 2,900 km that probably consists mostly of iron and nickel.

**mantle** The thick layer between Earth's crust and core.

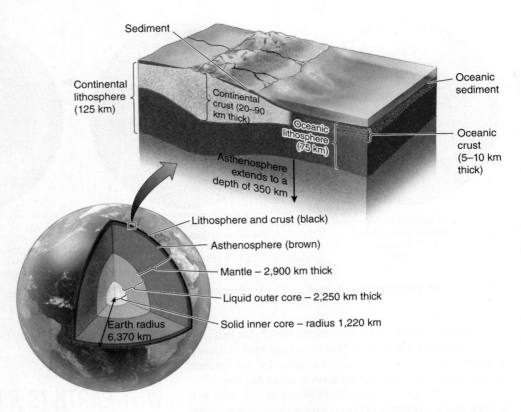

● **FIGURE 1.11 Cross Section of Earth Illustrating the Core, Mantle, and Crust** The enlarged portion shows the relationship between the lithosphere (composed of the continental crust, oceanic crust, and solid upper mantle) and the underlying asthenosphere and lower mantle.

$(3.3-5.7 \text{ g/cm}^3)$ and is thought to be composed mostly of *peridotite*, a dark, dense, igneous rock containing abundant iron and magnesium. The mantle is divided into three distinct zones based on physical characteristics. The *lower mantle* is solid and forms most of the volume of Earth's interior. The **asthenosphere** surrounds the lower mantle. It has the same composition as the lower mantle but behaves plastically and flows slowly. Partial melting within the asthenosphere generates **magma** (molten material), some of which rises to the surface, because it is less dense than the rock from which it was derived. The *upper mantle* surrounds the asthenosphere. The solid upper mantle and the overlying crust constitute the **lithosphere**, which is broken into numerous individual pieces called **plates** that move over the asthenosphere, partially as a result of underlying *convection cells* (●**FIGURE 1.12**). Interactions of these plates are responsible for such phenomena as earthquakes, volcanic eruptions, and the formation of mountain ranges and ocean basins.

The **crust**, Earth's outermost layer, consists of two types. *Continental crust* is thick (20–90 km), has an average density of 2.7 $\text{g/cm}^3$, and contains considerable silicon and aluminum. *Oceanic crust* is thin (5–10 km), denser than continental crust (3.0 $\text{g/cm}^3$), and composed of the dark igneous rocks *basalt* and *gabbro*.

## Plate Tectonic Theory

The recognition that the lithosphere is divided into rigid plates that move over the asthenosphere (●**FIGURE 1.13**) forms the foundation of **plate tectonic theory**, a unifying theory of geology holding that large segments of Earth's outer part (lithospheric plates) move relative to one another

(discussed in greater detail in Chapter 2). Zones of volcanic activity, earthquakes, or both, mark most plate boundaries. Along these boundaries, plates separate (diverge), collide (converge), or slide sideways past each other.

The acceptance of plate tectonic theory is recognized as a major milestone in the geologic sciences, comparable to the revolution that Darwin's theory of evolution caused in biology. Plate tectonics has provided a framework for interpreting the composition, structure, and internal processes of Earth on a global scale. It has led to the realization that the continents and ocean basins are part of a lithosphere–atmosphere–hydrosphere system that evolved together with Earth's interior.

A revolutionary concept when it was proposed in the 1960s, plate tectonic theory has had far-reaching consequences in all fields of geology because it provides the basis for relating many seemingly unrelated phenomena, such as the formation and occurrence of Earth's natural resources,

---

**asthenosphere** The part of the mantle that lies below the lithosphere; it behaves plastically and flows slowly.

**magma** Molten rock material generated within Earth.

**lithosphere** Earth's outer, rigid part, consisting of the upper mantle, oceanic crust, and continental crust.

**plate** An individual segment of the lithosphere that moves over the asthenosphere.

**crust** Earth's outermost layer; the upper part of the lithosphere that is separated from the mantle by the Moho; divided into continental and oceanic crust.

**plate tectonic theory** The theory holding that large segments of Earth's outer part (lithospheric plates) move relative to one another.

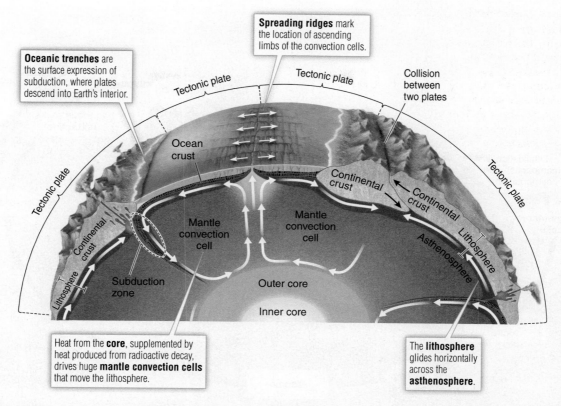

**Oceanic trenches** are the surface expression of subduction, where plates descend into Earth's interior.

**Spreading ridges** mark the location of ascending limbs of the convection cells.

Collision between two plates

Tectonic plate

Tectonic plate

Tectonic plate

Tectonic plate

Ocean crust

Continental crust

Continental crust

Lithosphere

Mantle convection cell

Mantle convection cell

Asthenosphere

Continental crust

Subduction zone

Outer core

Inner core

T Lithosphere

Heat from the **core**, supplemented by heat produced from radioactive decay, drives huge **mantle convection cells** that move the lithosphere.

The **lithosphere** glides horizontally across the **asthenosphere**.

● **FIGURE 1.12 Movement of Earth's Plates** Earth's plates are thought to move partially as a result of underlying mantle convection cells in which warm material from deep within Earth rises toward the surface, cools, and then, upon losing heat, descends back into the interior as shown in this diagrammatic cross section.

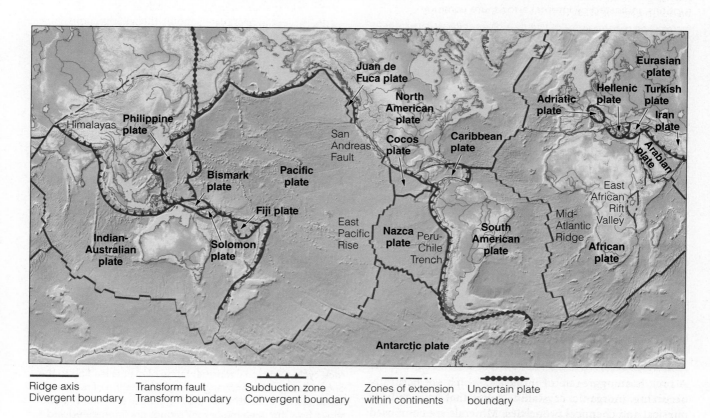

Eurasian plate

Hellenic plate

Turkish plate

Adriatic plate

Iran plate

Arabian plate

Juan de Fuca plate

North American plate

Philippine plate

Himalayas

San Andreas Fault

Cocos plate

Caribbean plate

Bismark plate

Pacific plate

East African Rift Valley

Mid-Atlantic Ridge

Fiji plate

East Pacific Rise

Nazca plate

Peru-Chile Trench

South American plate

African plate

Indian-Australian plate

Solomon plate

Antarctic plate

Ridge axis
Divergent boundary

Transform fault
Transform boundary

Subduction zone
Convergent boundary

Zones of extension
within continents

Uncertain plate
boundary

● **FIGURE 1.13 Earth's Plates** Earth's lithosphere is divided into rigid plates of various sizes that move over the asthenosphere.

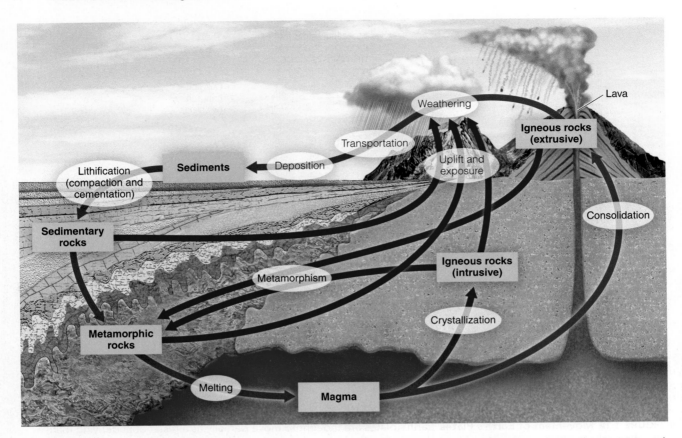

● **FIGURE 1.14 The Rock Cycle** This cycle shows the interrelationship between Earth's internal and external processes and how the three major rock groups are related. An ideal cycle includes the events on the outer margin of the cycle, but interruptions, indicated by internal arrows, are common.

**Critical Thinking Question** What types of geologic events would cause interruptions to the idealized cycle shown on the margin of the rock cycle? What do you think is more common and why—completion of an idealized cycle or one in which there are interruptions?

as well as the distribution and evolution of the world's biota. Furthermore, the impact of plate tectonic theory has been particularly notable in the interpretation of Earth's history. For example, the Appalachian Mountains in eastern North America and the mountain ranges of Greenland, Scotland, Norway, and Sweden are not the result of unrelated mountain-building episodes but, rather, are part of a larger mountain-building event that involved the closing of an ancient Atlantic Ocean and the formation of the supercontinent Pangaea approximately 252 million years ago.

# THE ROCK CYCLE

**L08**  Describe the rock cycle and the interrelationship between Earth's internal and external processes

A rock is an aggregate of minerals, which are naturally occurring, inorganic, crystalline solids that have definite physical and chemical properties. Minerals are composed of elements such as oxygen, silicon, and aluminum, and

elements are made up of atoms, the smallest particles of matter that retain the characteristics of an element. More than 5,000 minerals have been identified and described, but only about a dozen make up the bulk of the rocks in Earth's crust.

Geologists recognize three major groups of rocks— *igneous, sedimentary,* and *metamorphic*—each of which is characterized by its mode of formation. The **rock cycle** is a pictorial representation of events leading to the origin, destruction or changes, and reformation of rocks as a consequence of Earth's internal and surface processes (●**FIGURE 1.14**). Furthermore, it shows that the three major rock groups are interrelated; that is, any rock type can be derived from any of the others.

**rock cycle**  A pictorial representation of events leading to the origin, destruction, or changes, and reformation of rocks as a consequence of Earth's internal and surface processes. It also shows how the three major rock groups are interrelated and how any rock type can be derived from any other rock type.

Copyright and Photograph by Dr. Parvinder S. Sethi

● **FIGURE 1.15 Hand Specimens of Common Igneous, Sedimentary, and Metamorphic Rocks** (a) **Granite,** an intrusive igneous rock. (b) **Basalt,** an extrusive igneous rock. (c) **Conglomerate,** a sedimentary rock formed by the consolidation of rounded rock fragments. (d) **Limestone,** a sedimentary rock formed by the extraction of mineral matter from seawater by organisms or by the inorganic precipitation of the mineral calcite from seawater. (e) **Gneiss,** a foliated metamorphic rock. (f) **Quartzite,** a nonfoliated metamorphic rock.

**Igneous rocks** result when magma or lava crystallizes, or when volcanic ejecta (pyroclastic materials), such as ash, accumulate and consolidate. As magma cools, minerals crystallize, and the resulting rock is characterized by interlocking mineral grains. Magma that cools slowly beneath the surface produces *intrusive igneous rocks* (●**FIGURE 1.15A**); magma that cools at the surface produces *extrusive igneous rocks* (●**FIGURE 1.15B**).

Rocks exposed at Earth's surface are broken into particles and dissolved by various weathering processes. The particles and dissolved materials may be transported by wind, water, or ice and eventually deposited as *sediment*. This sediment may then be compacted, or cemented (lithified), into sedimentary rock.

**Sedimentary rocks** form in one of the following three ways (●**FIGURE 1.15C, D**):

1. consolidation of mineral or rock fragments,
2. precipitation of mineral matter from solution, or
3. compaction of plant or animal remains.

Because sedimentary rocks form at or near Earth's surface, geologists can infer some information about the environment in which they were deposited, the transporting agent, and perhaps even something about the source from which the sediments were derived (see Chapter 6). Accordingly, sedimentary rocks are especially useful for interpreting Earth history.

**Metamorphic rocks** result from the alteration of other rocks, usually beneath the surface, by heat, pressure, and the chemical activity of fluids. For example, marble—a rock preferred by many sculptors and builders—is a metamorphic rock formed when the agents of metamorphism are applied to the sedimentary rocks limestone or dolostone. Metamorphic rocks are either *foliated* (●**FIGURE 1.15E**) or *nonfoliated* (●**FIGURE 1.15F**). Foliation, the parallel alignment of minerals caused by the application of pressure, gives the rock a layered, or banded, appearance.

## How Are the Rock Cycle and Plate Tectonics Related?

Interactions between plates determine, to some extent, which of the three rock groups will form (●**FIGURE 1.16**). For example, when plates converge, heat and pressure generated along the plate boundary may lead to igneous activity and metamorphism within the descending oceanic plate, thus producing various igneous and metamorphic rocks.

Some of the sediments and sedimentary rocks on the descending plate are melted, whereas other sediments

---

**igneous rock** Any rock formed by cooling and crystallization of magma or lava or the consolidation of pyroclastic materials.

**sedimentary rock** Any rock composed of sediment that forms at or near Earth's surface.

**metamorphic rock** Any rock that has been changed from its original condition by heat, pressure, and the chemical activity of fluids, as in marble and slate.

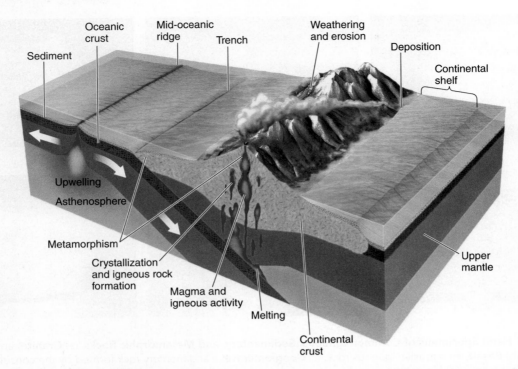

● **FIGURE 1.16 Plate Tectonics and the Rock Cycle** Plate movement provides the driving mechanism that recycles Earth materials. This block diagram shows how the three major rock groups—igneous, sedimentary, and metamorphic—are recycled through both the continental and oceanic regions. Subducting plates are partially melted to produce magma, which rises and either crystallizes beneath Earth's surface as intrusive igneous rock or spills out on the surface, solidifying as extrusive igneous rock. Rocks exposed at the surface are weathered and eroded to produce sediments that are transported and eventually lithified into sedimentary rocks. Metamorphic rocks result from pressure generated along converging plates or adjacent to rising magma.

and sedimentary rocks along the boundary of the nondescending plate are metamorphosed by the heat and pressure produced along the converging plate boundary. Later, the mountain range or chain of volcanic islands formed along the convergent plate boundary will be weathered and eroded, and the new sediments will be transported to the ocean, where they will be deposited, to begin yet another cycle.

# ORGANIC EVOLUTION AND THE HISTORY OF LIFE

**LO9    Define organic evolution and discuss its role in the history of life**

Plate tectonic theory provides us with a model for understanding the internal workings of Earth and their effect on Earth's surface. The theory of **organic evolution** (whose central thesis is that all present-day organisms

are related and that they have descended with modifications from organisms that lived in the past) provides the conceptual framework for understanding the history of life. Together, the theories of plate tectonics and organic evolution have changed the way we view our planet, and we should not be surprised at the intimate association between them. Although the relationship between plate tectonic processes and the evolution of life is incredibly complex, paleontologic data provide indisputable evidence of the influence of plate movement on the distribution of organisms.

The publication in 1859 of Darwin's *On the Origin of Species by Means of Natural Selection* revolutionized biology and marked the beginning of modern evolutionary biology. Upon its publication, most naturalists recognized that evolution provided a unifying theory that explained an otherwise encyclopedic collection of biologic facts.

**organic evolution** The theory holding that all living organisms are related and that they descended with modification from other organisms that lived during the past.

When Darwin proposed his theory of organic evolution, he cited a wealth of supporting evidence, including the way organisms are classified, embryology, comparative anatomy, the geographic distribution of organisms, and, to a limited extent, the fossil record. Furthermore, Darwin proposed that *natural selection*, which results in the survival to reproductive age of those organisms best adapted to their environment, is the mechanism that best accounts for evolution.

Perhaps the most compelling evidence in favor of evolution can be found in the fossil record. Just as the geologic record allows geologists to interpret physical events and conditions in the geologic past, **fossils**, which are the remains or traces of once-living organisms, not only provide evidence that evolution has occurred but also demonstrate that Earth has a history extending beyond that recorded by humans. The succession of fossils in the rock record provides geologists with a means for dating rocks and allowed for a relative geologic time scale to be constructed in the 1800s.

# GEOLOGIC TIME AND UNIFORMITARIANISM

**LO10** Define geologic time, and explain the difference between the human perspective of time and the immensity of geologic time

**LO11** Discuss the importance of the principle of uniformitarianism as one of the cornerstones of geology

An appreciation of the immensity of geologic time is central to understanding the evolution of Earth and its biota. Indeed, time is one of the main aspects that sets geology apart from the other sciences, except astronomy. Most people have difficulty comprehending geologic time because they tend to think in terms of the human perspective—seconds, hours, days, and years. Ancient history is what occurred hundreds or even thousands of years ago. When geologists talk of ancient geologic history, however, they are referring to events that happened hundreds of millions or even billions of years ago.

It is therefore important to remember that Earth goes through cycles of much longer duration than the human perspective of time. Although some of these cycles, such as global warming and cooling, may have disastrous effects on the human species, they are, nonetheless, part of the larger cycle of global change that has, for example, resulted in numerous glacial advances and retreats during the past 2.6 million years.

The **geologic time scale** (●**FIGURE 1.17**) subdivides geologic time into a hierarchy of increasingly shorter time

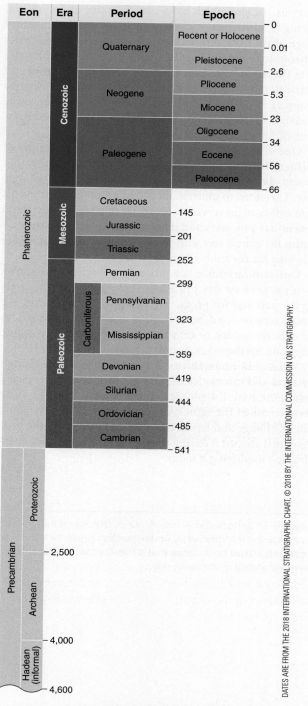

● **FIGURE 1.17 The Geologic Time Scale** The numbers to the right of the columns are ages in millions of years before the present.

**fossil** Remains or traces of prehistoric organisms preserved in rocks.

**geologic time scale** A chart that subdivides geologic time into a hierarchy of increasingly shorter time intervals, each of which has a specific name and duration.

intervals, each of which has a specific name and duration. During the 19th century, many geologists pieced together information from numerous rock exposures to construct a chronology, or time scale, based on changes in Earth's biota through time. Subsequently, with the discovery of radioactivity in 1895, and the development of various radiometric-dating techniques, geologists have been able to assign numerical age dates, in years, to the subdivisions of the geologic time scale.

One of the cornerstones of geology is the **principle of uniformitarianism**, which is based on the premise that present-day processes have operated throughout geologic time. Therefore, to understand and interpret geologic events from evidence preserved in rocks, we must first understand present-day processes and their results. In fact, uniformitarianism fits completely with the system approach that we are following for the study of Earth.

Uniformitarianism is a powerful principle that allows us to use present-day processes as the basis for interpreting the past and for predicting potential future events. We should keep in mind, however, that uniformitarianism does not exclude sudden or catastrophic events such as volcanic eruptions, earthquakes, tsunami, landslides, or floods.

What uniformitarianism does mean is that even though the rates and intensities of geologic processes have varied during the past, the physical and chemical laws of nature have remained the same. Although Earth is in a dynamic state of change and has been ever since it formed, the processes that shaped it during the past are the same ones operating and modifying it today.

---

**principle of uniformitarianism** A principle holding that past events can be interpreted by understanding present-day processes; based on the idea that natural processes have always operated in the same way.

# HOW DOES THE STUDY OF GEOLOGY BENEFIT US?

**LO12**    Summarize how the study of geology is beneficial and an integral part of our lives

The most meaningful lesson to learn from the study of geology is that Earth is an extremely complex planet in which interactions are taking place between its various subsystems and have been for the past 4.6 billion years. If we want to ensure the survival of the human species, we must first understand how the various subsystems work and interact with each other and, more importantly, how our actions affect the delicate balance between these systems. We can do this, in part, by studying what has happened in the past, particularly on the global scale, and use that information to try and predict how our actions might affect the balance between Earth's various subsystems now and in the future.

The study of geology goes beyond learning numerous facts about Earth. In reality, we do not just study geology—we *live* it. Geology is an integral part of our lives. Our standard of living depends directly on our consumption of natural resources that formed millions and billions of years ago. However, the way that we consume those natural resources and interact with the environment—as individuals and as a society—determines our ability to pass on this standard of living to the next generation.

As you study the various subjects covered in this book, keep in mind the themes and topics discussed in this chapter and how, like the parts of a system, they are interrelated—and how they are responsible for the events that occurred during Earth's 4.6-billion-year history. By relating each chapter's topic to its place in the Earth system, you will gain a greater appreciation of why geology is so integral to our lives.

# Key Concepts Review

- Earth can be viewed as a system of interconnected components that interact with and affect each other. The principal subsystems of Earth are the atmosphere, hydrosphere, biosphere, lithosphere, mantle, and core.
- Earth is a continually changing and dynamic planet because of the interactions between its various subsystems and cycles.
- Geology, the study of Earth, is divided into (1) physical geology, which is the study of Earth materials and the processes that operate both within Earth and on its surface, and (2) historical geology, which examines the origin and evolution of Earth and its continents, oceans, atmosphere, and life.
- The scientific method is an orderly, logical approach that involves gathering and analyzing facts about a particular phenomenon, formulating hypotheses to explain the phenomenon, testing the hypotheses, and finally proposing a theory. A theory is a testable explanation for some natural phenomenon that has a large body of supporting evidence.
- Geology is not only part of the human experience, examples of which can be found in art, music, and literature, but it also affects our daily lives as individuals, societies, and nation-states. A basic understanding of geology, and science in general, is critical for dealing with and finding solutions to the many environmental problems and issues facing humankind.
- The universe began, in what is popularly called the Big Bang, approximately 14 billion years ago. Astronomers have deduced this age by observing that celestial objects are moving away from each other in an ever-expanding universe. Furthermore, the universe has a pervasive background radiation of 2.7 K above absolute zero (2.7 K = −270.3°C), which is thought to be the faint afterglow of the Big Bang.
- About 4.6 billion years ago, our solar system formed from a rotating cloud of interstellar matter. As this cloud condensed, it eventually collapsed under the influence of gravity and flattened into a counterclockwise-rotating disk. Within this rotating disk, the Sun, planets, and moons formed from the turbulent eddies of nebular gases and solids.
- Earth formed from a swirling eddy of nebular material 4.6 billion years ago, accreting as a solid body and soon thereafter differentiating into a layered planet.

- Earth's outermost layer is the crust, which is divided into continental and oceanic portions. The crust and underlying solid upper mantle, together known as the lithosphere, overlie the asthenosphere, a zone that behaves plastically and flows slowly. The asthenosphere is underlain by the solid lower mantle. Earth's core consists of an outer liquid portion and an inner solid portion.
- The lithosphere is divided into a series of plates that diverge, converge, and slide sideways past one another.
- Plate tectonic theory provides a unifying explanation for many geologic features and events. The interaction between plates is responsible for volcanic eruptions, earthquakes, the formation of mountain ranges and ocean basins, and the recycling of rock materials.
- The three major rock groups are igneous, sedimentary, and metamorphic. Igneous rocks result from the crystallization of magma or the consolidation of volcanic ejecta. Sedimentary rocks are typically formed by the consolidation of rock fragments, precipitation of mineral matter from solution, or compaction of plant or animal remains. Metamorphic rocks result from the alteration of other rocks, usually beneath Earth's surface, by heat, pressure, and chemically active fluids.
- The rock cycle illustrates the interactions between Earth's internal and external processes and how the three rock groups are interrelated.
- The central thesis of the theory of organic evolution is that all living organisms evolved (descended with modifications) from organisms that existed in the past.
- Time sets geology apart from the other sciences except astronomy, and an appreciation of the immensity of geologic time is central to understanding Earth's evolution. The geologic time scale is the calendar geologists use to date past events.
- The principle of uniformitarianism is basic to the interpretation of Earth history. This principle holds that the laws of nature have been constant through time and that the same processes operating today have operated in the past, although not necessarily at the same rates.
- Geology is an integral part of our lives. Our standard of living depends directly on our consumption of natural resources, most of which formed millions and billions of years ago.

# Important Terms

asthenosphere   12

Big Bang   8

core   11

crust   12

fossil   17

geologic time scale   17

geology   2

greenhouse effect   7

hypothesis   4

igneous rock   15

Jovian planet   10

lithosphere   12

magma   12

mantle   11

metamorphic rock   15

mineral   2

organic evolution   16

plate   12

plate tectonic theory   12

principle of uniformitarianism   18

rock   2

rock cycle   14

scientific method   4

sedimentary rock   15

solar nebula theory   10

system   2

terrestrial planet   10

theory   3

# Review Questions

1. The movement of plates is thought to result from
   a. _____ density differences between the inner and outer core.
   b. _____ rotation of the mantle around the core.
   c. _____ gravitational forces.
   d. _____ the Coriolis effect.
   e. _____ convection cells.

2. Which of the following statements about a scientific theory is not true?
   a. _____ It is an explanation for some natural phenomenon.
   b. _____ Predictive statements can be derived from it.
   c. _____ It is a conjecture or guess.
   d. _____ It has a large body of supporting evidence.
   e. _____ It is testable.

3. The study of the origin and evolution of Earth and its continents, oceans, atmosphere, and biota is
   a. _____ meteorology.
   b. _____ cosmology.
   c. _____ historical geology.
   d. _____ geomorphology.
   e. _____ physical geology.

4. A combination of related parts interacting in an organized fashion is
   a. _____ a cycle.
   b. _____ a theory.
   c. _____ uniformitarianism.
   d. _____ a hypothesis.
   e. _____ a system.

5. That present-day processes have operated throughout geologic time is the premise for the principle of
   a. _____ organic evolution.
   b. _____ plate tectonics.
   c. _____ uniformitarianism.
   d. _____ geologic time.
   e. _____ scientific deduction.

6. Why is plate tectonics called the unifying theory of geology? How does it fit into a systems approach to the study of Earth?

7. How does the solar nebula theory account for the formation of our solar system, its features, and evolutionary history?

8. Why do most scientists think overpopulation is the greatest environmental problem facing the world today? Do you agree? If not, what do you think is the greatest threat to our existence as a species?

9. Using plate movement as the driving mechanism of the rock cycle, explain how the three rock groups are related and how each rock group can be converted into a different rock group.

# Creative Thinking Visual Question

One of the major environmental issues facing humankind today is global warming and the effect, if any, human activity has had on raising global average temperatures. Proponents on both sides of the debate point to increasing mean global temperature changes from 1880 to 2008, which is a historical time interval (●FIGURE 1). On the other hand, mean global temperature and precipitation have changed, in some cases significantly, when viewed from a geologic time perspective (●FIGURE 2).

Discuss why an understanding of geologic time is crucial to the global warming debate and its consequences, and whether it is possible to extrapolate changes based on a historical time frame to one of a geologic time frame.

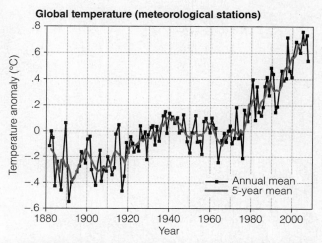

**● FIGURE 1 Mean Global Temperature Changes from 1880 to 2008** The zero line represents the average from 1951 to 1980. The plus and minus values represent deviations (in °C) from the average for the annual mean (in black) and 5-year mean (in red). Note the dramatic increase in temperature from approximately 1970 to 2008.

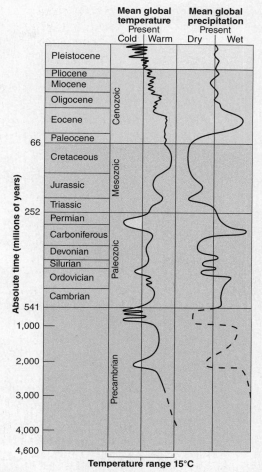

**● FIGURE 2 Mean Global Temperature and Precipitation Changes Throughout Earth History** There have been several episodes of global cooling throughout Earth's history, notably at the end of the Precambrian (Proterozoic Eon), the end of the Ordovician Period, the Late Carboniferous to Permian periods, and a series of glacial and interglacial episodes during the Pleistocene Epoch.

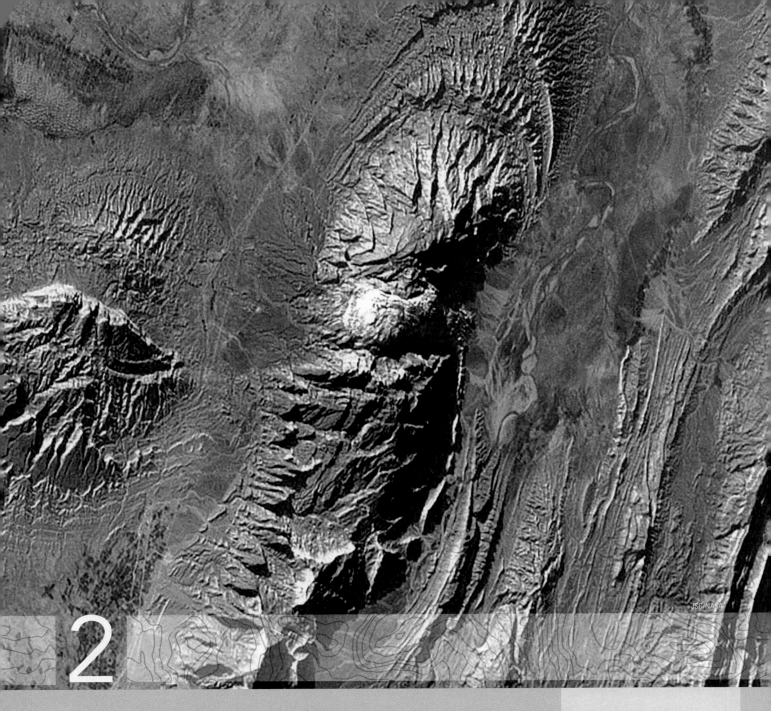

JSC/NASA

# 2

# PLATE TECTONICS

## A Unifying Theory

The Zagros Mountains in western Iran consist of a series of anticlines and synclines, formed by compressive forces resulting from the collision between the Arabian and Eurasian plates.

# INTRODUCTION

Imagine it is the day after Christmas, December 26, 2004, and you are vacationing on a beautiful beach in Thailand. You look up from the book you're reading to see the sea suddenly retreat from the shoreline. Within minutes of this unusual event, a powerful tsunami will sweep over your resort and everything in its path for several kilometers inland. In the next few hours, the coasts of Indonesia, Sri Lanka, India, Thailand, Somalia, Myanmar, Malaysia, and the Maldives will be inundated by the deadliest tsunami in history and more than 280,000 people will die.

Seven years later, on March 11, 2011, a catastrophic 9.0-magnitude earthquake struck Japan, killing more than 13,000 people and leaving thousands injured and homeless. This destructive earthquake also generated a tsunami, one that sent walls of water crashing into the northeastern shores of the island, causing further damage and causalities.

Now go forward to June 2018, when the Fuego volcano, located approximately 44 km southwest of Guatemala City, Guatemala, erupted numerous times with devastating results. These eruptions sent dense ash plumes as high as 6 km into the atmosphere, blanketing the surrounding area with thick layers of volcanic ash. Heavy rains turned this ash into hot (as high as 150°C), raging, volcanic mudflows, some as much as 40 m wide and 5 m deep, flowing down rivers and valleys, and burying everything in their paths. As a result of these mudflows, at least 110 people were killed and 12,578 people evacuated from areas around the volcano.

What do these three tragic events have in common? They are part of the dynamic interactions involving Earth's plates. When two plates come together, one plate is pushed or pulled under the other plate, triggering large earthquakes such as the 7.0-magnitude earthquake that killed an estimated 223,000 people in Haiti in 2010 and the 7.1-magnitude earthquake in Mexico in 2017, resulting in 370 deaths, mostly from collapsing buildings. In fact, from 2000 to 2016, an estimated 802,000 people have died from 5.0- to >8.0-magnitude earthquakes.

As the descending plate moves downward and is assimilated into Earth's interior, magma is generated. Being less dense than the surrounding material, the magma rises toward the surface, where it may erupt as a volcano. It therefore should not be surprising that the distribution of volcanoes and earthquakes closely follows plate boundaries.

If you are like most people, you probably have only a vague notion of what plate tectonic theory is. Yet, plate tectonics affects us all. Volcanic eruptions, earthquakes, and tsunami are the result of interactions between plates. Global weather patterns and oceanic currents are caused, in part, by the configuration of the continents and ocean basins. The formation and distribution of many natural resources are related to plate movement, and thus have an impact on the economic well-being and political decisions of nations. It is therefore important to understand this unifying theory, not only because it affects us as individuals, and as citizens of nation-states, but also because it ties together many aspects of geology.

# EARLY IDEAS ABOUT CONTINENTAL DRIFT

The idea that Earth's past geography was different from today is not new. The earliest maps showing the east coast of South America and the west coast of Africa probably provided people with the first evidence that continents may have once been joined, then broken apart and moved to their present positions. As far back as 1620, Sir Francis Bacon commented on the similarity of the shorelines of western Africa and eastern South America. However, he did not make the connection that the Old and New Worlds might once have been joined together.

During the late 19th century, the Austrian geologist Edward Suess noted the similarities between the Late Paleozoic plant fossils of India, Australia, South Africa, and South America, as well as evidence of glaciation in the rock sequences of these continents. The plant fossils comprise a unique flora that occurs in the coal layers just above the glacial deposits of these southern continents. This flora is very different from the contemporaneous coal swamp flora of the northern continents and is collectively known as the ***Glossopteris* flora** after its most conspicuous genus (●FIGURE 2.1).

Suess also proposed the name Gondwanaland (or **Gondwana** as we will use here) for a supercontinent composed of the aforementioned southern continents. Abundant

*Copyright and Photograph by Dr. Parvinder S. Sethi*

● **FIGURE 2.1 Fossil *Glossopteris* Leaves** Plant fossils, such as these *Glossopteris* leaves from the Upper Permian Dunedoo Formation in Australia, are found on all five of the Gondwana continents. Their presence on continents with widely varying climates today is evidence that the continents were at one time connected. The distribution of the plants at that time was in the same climatic latitudinal belt.

---

***Glossopteris* flora** A Late Paleozoic association of plants found only on the Southern Hemisphere continents and India; named for its best-known genus, *Glossopteris*.

**Gondwana** A major Paleozoic continent composed of South America, Africa, Australia, India, and Antarctica.

fossils of the *Glossopteris* flora are found in coal beds in Gondwana, a province in India. Suess thought these southern continents were at one time connected by land bridges over which plants and animals migrated. Thus, in his view, the similarities of fossils on these continents were due to the appearance and disappearance of the connecting land bridges.

## Alfred Wegener and the Continental Drift Hypothesis

Alfred Wegener, a German meteorologist, is generally credited with developing the hypothesis of **continental drift**. In his monumental book, *The Origin of Continents and Oceans* (first published in 1915), Wegener proposed that all landmasses were originally united in a single supercontinent that he named **Pangaea**, from the Greek, meaning "all land." Wegener portrayed his grand concept of continental movement in a series of maps showing the breakup of Pangaea and the movement of the various continents to their present-day locations. Wegener amassed a tremendous amount of geologic, paleontologic, and climatologic evidence in support of continental drift; however, the initial reaction of scientists to his then-heretical ideas can best be described as mixed.

Nevertheless, Alexander du Toit, a South African geologist, and one of Wegener's more ardent supporters, further developed Wegener's arguments and gathered more geologic and paleontologic evidence in support of his continental drift hypothesis. In 1937, du Toit published *Our Wandering Continents*, in which he contrasted the glacial deposits of Gondwana with coal deposits of the same age found in the continents of the Northern Hemisphere. To resolve this apparent climatologic paradox, du Toit moved the Gondwana continents to the South Pole and brought the northern continents together such that the coal deposits were located at the equator. He named this northern landmass **Laurasia**. It consisted of present-day North America, Greenland, Europe, and Asia (except for India).

# WHAT IS THE EVIDENCE FOR CONTINENTAL DRIFT?

**LO1    List and explain the evidence for continental drift used by its early proponents**

What then was the evidence Wegener, du Toit, and others used to support the hypothesis of continental drift? It included the fit of the shorelines of continents, the appearance of the same rock sequences and mountain ranges of the same age on continents now widely separated, the matching of glacial deposits and paleoclimatic zones, and the similarities of many extinct plant and animal groups whose fossil remains are found today on widely separated continents. Wegener and his supporters argued that this vast amount of evidence from a variety of sources surely indicates that the continents must have been close together in the past.

## Continental Fit

Wegener, like some before him, was impressed by the close resemblance between the coastlines of continents on opposite sides of the Atlantic Ocean, particularly South America and Africa. He cited these similarities as partial evidence that the continents were at one time joined together as a supercontinent that subsequently split apart. As his critics pointed out, though, the configuration of coastlines results from erosional and depositional processes and therefore is continuously being modified. So, even if the continents had separated during the Mesozoic Era, as Wegener proposed, it is not likely that the coastlines would fit exactly.

A more realistic approach is to fit the continents together along the continental slope where erosion would be minimal. In 1965, Sir Edward Bullard, an English geophysicist, and two associates showed that the best fit between the continents occurs at a depth of about 2,000 m (●**FIGURE 2.2**).

iStock.com/Jan Rysavy

●**FIGURE 2.2 Continental Fit**   When continents are placed together based on their outlines, the best fit is not along their present-day coastlines, but rather along the continental slope at a depth of about 2,000 m, where erosion would be minimal.

**Critical Thinking Question**   Why is the best fit between continents along the continental slope and not along the current coastline?

**continental drift**   The theory that continents were joined into a single landmass that broke apart, with the various fragments (continents) moving with respect to one another.

**Pangaea**   The name Alfred Wegener proposed for a supercontinent consisting of all of Earth's landmasses at the end of the Paleozoic Era.

**Laurasia**   A Late Paleozoic Northern Hemisphere continent made up of North America, Greenland, Europe, and Asia.

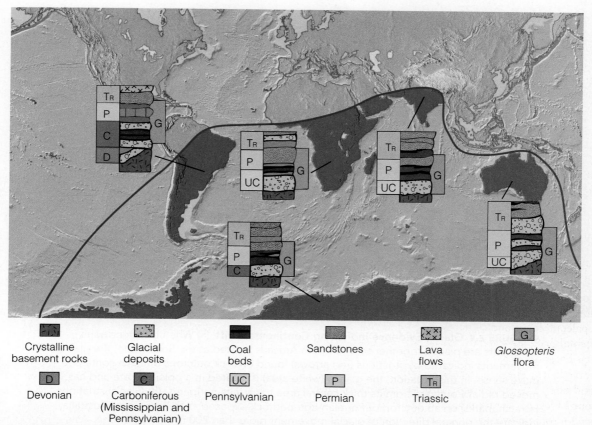

Crystalline basement rocks

Glacial deposits

Coal beds

Sandstones

Lava flows

*Glossopteris* flora

D — Devonian

C — Carboniferous (Mississippian and Pennsylvanian)

UC — Pennsylvanian

P — Permian

TR — Triassic

● **FIGURE 2.3 Similarity of Rock Sequences on the Gondwana Continents** Sequences of marine, nonmarine, and glacial rocks of Pennsylvanian (UC) to Triassic (TR) age are nearly the same on all five Gondwana continents (South America, Africa, India, Australia, and Antarctica). These continents are widely separated today and have different environments and climates ranging from tropical to polar. Thus, the rocks forming on each continent are very different. When the continents were all joined together in the past, however, the environments of adjacent continents were similar, and the rocks forming in those areas were similar. The range indicated by G in each column is the age range of the *Glossopteris* flora.

Since then, other reconstructions, using the latest ocean basin data, have confirmed the close fit between continents when they are reassembled to form Pangaea.

## Similarity of Rock Sequences and Mountain Ranges

If the continents were at one time joined, then the rocks and mountain ranges of the same age in adjoining locations on the opposite continents should closely match. Such is the case for the Gondwana continents (●FIGURE 2.3). Marine, nonmarine, and glacial rock sequences of the Pennsylvanian to Jurassic periods are almost identical on all five Gondwana continents, strongly indicating that they were united at one time.

Furthermore, the trends of several major mountain ranges also support the hypothesis of continental drift. These mountain ranges seemingly end at the coastline of one continent only to apparently continue on another continent across the ocean. The folded Appalachian Mountains of North America, for example, trend northeastward through the eastern United States and Canada and terminate abruptly at the Newfoundland coastline. Mountain ranges of the same age and deformational style are found in eastern Greenland, Ireland, Great Britain, and Norway. So even though the Appalachian Mountains and their equivalent-age mountain ranges in Great Britain are currently separated by the Atlantic Ocean, they form an essentially continuous mountain range when the continents are positioned next to each other as they were during the Paleozoic Era.

## Glacial Evidence

During the Late Paleozoic Era, massive glaciers covered large continental areas of the Southern Hemisphere. Evidence for this glaciation includes layers of *till* (sediments deposited by glaciers) and *striations* (scratch marks) in the bedrock beneath the till (●FIGURE 2.4). Fossils and sedimentary rocks of the same age from the Northern Hemisphere, however, give no indication of glaciation. Fossil plants found in coals indicate that the Northern Hemisphere had a tropical climate during the time that the Southern Hemisphere was glaciated.

All of the Gondwana continents except Antarctica are currently located near the equator in subtropical to

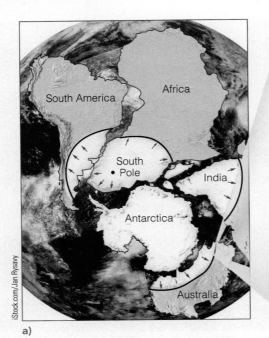

a)                                                                                    b)

● **FIGURE 2.4  Glacial Evidence Indicating Continental Drift**  (a) When the Gondwana continents are placed together so that South Africa is located at the South Pole, the glacial movements indicated by striations (red arrows) found on rock outcrops on each continent make sense. In this situation, the glacier (white area) is located in a polar climate and has moved radially outward from its thick central area toward its periphery. (b) Glacial striations (scratch marks) on an outcrop of Permian-age bedrock exposed at Hallet's Cove, Australia, indicate the general direction of glacial movement more than 200 million years ago. As a glacier moves over a continent's surface, it grinds and scratches the underlying rock. The glacial striations that are preserved on a rock's surface thus provide evidence of the direction (red arrows) the glacier moved at that time.

tropical climates. Mapping of glacial striations in bedrock in Australia, India, and South America indicates that the glaciers moved from the areas of the present-day oceans onto land. Yet, this would be highly unlikely because large continental glaciers (such as occurred on the Gondwana continents during the Late Paleozoic Era) flow outward from their central area of accumulation toward the sea.

Therefore, if the continents did not move during the past, one would have to explain how glaciers moved from the oceans onto land and how large-scale continental glaciers formed near the equator. But if the continents are reassembled as a single landmass with South Africa located at the South Pole, the direction of movement of Late Paleozoic continental glaciers makes sense (Figure 2.4). Furthermore, this geographic arrangement places the northern continents nearer the tropics, which is consistent with the fossil and climatologic evidence from Laurasia.

## Fossil Evidence

Some of the most compelling evidence for continental drift comes from the fossil record (●**FIGURE 2.5**). Fossils of the *Glossopteris* flora, which include the seed fern *Glossopteris* (Figure 2.1), as well as many other distinctive and easily identified plants, are found in equivalent Pennsylvanian- and Permian-age coal deposits on all five Gondwana continents. Whereas plant pollen and spores can be dispersed

over great distances by wind, the *Glossopteris*-type plants produced seeds that are too large to have been carried by winds. Even if the seeds had floated across the ocean, they probably would not have remained viable for any length of time in saltwater.

The present-day climates of South America, Africa, India, Australia, and Antarctica range from tropical to polar and thus are too diverse to support the type of plants in the *Glossopteris* flora. Wegener therefore reasoned that these continents must once have been joined such that these now widely separated localities were at one time all in the same latitudinal climatic belt (Figure 2.5).

The fossil remains of animals also provide strong evidence for continental drift. One of the best examples is *Mesosaurus*, a freshwater reptile whose fossils are found in Permian-age rocks in certain regions of Brazil and South Africa and nowhere else in the world (Figure 2.5). Because the physiologies of freshwater and marine animals are completely different, it is hard to imagine how a freshwater reptile could have swum across the Atlantic Ocean and found a freshwater environment nearly identical to its former habitat. Moreover, if *Mesosaurus* could have swum across the ocean, its fossil remains should be widely dispersed. It is more logical to assume that *Mesosaurus* lived in lakes in what were once adjacent areas of South America and Africa when they were united into a single continent. Discoveries of fossils from additional land-dwelling animals on these and

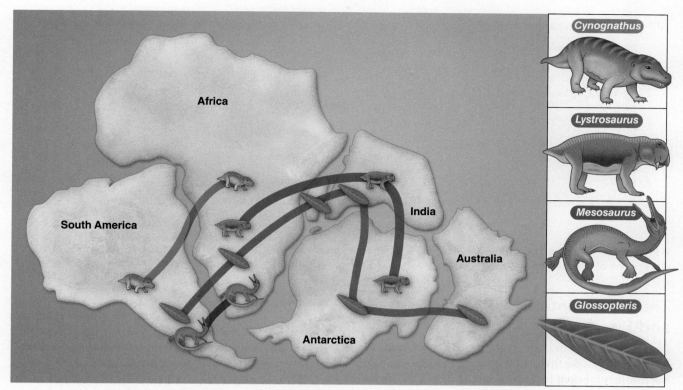

● **FIGURE 2.5 Fossil Evidence Supporting Continental Drift** Some of the plants and animals whose fossils are found today on the widely separated continents of South America, Africa, India, Australia, and Antarctica. During the Late Paleozoic Era, these continents were joined together to form Gondwana, the southern landmass of Pangaea. Plants of the *Glossopteris* flora are found on all five continents, which today have widely different climates; however, during the Pennsylvanian and Permian periods, they were all located in the same general climatic belt. *Mesosaurus* is a freshwater reptile whose fossils are found only in similar nonmarine Permian-age rocks in Brazil and South Africa. *Cynognathus* and *Lystrosaurus* are land reptiles that lived during the Early Triassic Period. Fossils of *Cynognathus* are found in South America and Africa, whereas fossils of *Lystrosaurus* have been recovered from Africa, India, and Antarctica. It is hard to imagine how a freshwater reptile and land-dwelling reptiles could have swum across the wide oceans that presently separate these continents. It is more logical to assume that the continents were once connected.

other Gondwana continents further solidify the argument that these landmasses were at one time in close proximity to each other.

Notwithstanding all of the empirical evidence presented by Wegener and later by du Toit and others, most geologists simply refused to entertain the idea that continents might have moved in the past. The geologists were not necessarily being obstinate about accepting new ideas; rather, they found the evidence for continental drift inadequate and unconvincing. In part, this was because no one could provide a suitable mechanism to explain how continents could move over Earth's surface.

# FEATURES OF THE SEAFLOOR

**LO2**   Describe the various features of the seafloor

At this point, it is useful to discuss some of the various features of Earth's seafloor. Many of the topographic features found on the seafloor and along the continental

margins are the manifestations of Earth's internal processes and activities taking place along plate margins. Thus, it is important to know how these features relate to plate tectonic theory.

Most people think of continents as land areas outlined by the oceans, but the true geologic margin of a continent—where granitic continental crust changes to basalt and gabbro oceanic crust—is below sea level. A **continental margin** is made up of a slightly sloping continental shelf, a more steeply inclined continental slope, and, in some cases, a deeper, gently sloping continental rise (●**FIGURE 2.6**). Thus, the continental margins extend to increasingly greater depths until they merge with the deep seafloor. Continental crust changes to oceanic crust somewhere beneath the continental rise, so that part of the continental slope and the continental rise actually rest on oceanic crust.

---

**continental margin** The area separating the part of a continent above sea level from the deep seafloor.

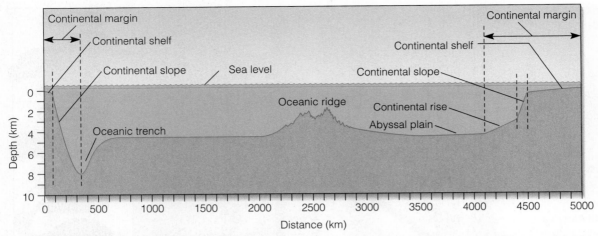

● **FIGURE 2.6 Features of Continental Margins** A generalized profile showing features of the continental margins. The vertical dimensions of the features in this profile are greatly exaggerated because the vertical and horizontal scales differ.

## The Continental Shelf, Slope, and Rise

As one proceeds seaward from the shoreline across the continental margin, the first area encountered is the gently sloping **continental shelf** lying between the shore and the more steeply dipping continental slope (Figure 2.6). The width of the continental shelf varies considerably, ranging from a few tens of meters to more than 1,000 km; the shelf terminates where the inclination of the seafloor increases abruptly from 1 degree or less to several degrees.

The seaward margin of the continental shelf is marked by the *shelf–slope break* (at an average depth of 135 m), where the more steeply inclined **continental slope** begins (Figure 2.6). In most areas around the margins of the Atlantic, the continental slope merges with a more gently sloping **continental rise**. This rise is absent around the margins of the Pacific, where continental slopes descend directly into an oceanic trench (Figure 2.6).

## Abyssal Plains, Oceanic Ridges, Submarine Hydrothermal Vents, and Oceanic Trenches

Beyond the continental rises are **abyssal plains**—flat surfaces covering vast areas of the seafloor. In some areas, they are interrupted by peaks rising more than 1 km, but abyssal plains are nevertheless the flattest, most featureless areas on Earth (Figure 2.6). Their flatness is a result of sediment deposition covering the rugged topography of the seafloor.

A renewed interest in oceanographic research led to extensive mapping of the ocean basins during the 1960s. Such mapping revealed an **oceanic ridge** system more than 65,000 km long, constituting the most extensive mountain range in the world (Figure 2.6). This system runs from the Arctic Ocean through the middle of the Atlantic and curves around South Africa, where the Indian Ridge continues into the Indian Ocean. The Atlantic–Pacific Ridge extends

eastward, and a branch of it, the East Pacific Rise, trends northeast until it reaches the Gulf of California. Perhaps the best-known part of the ridge system is the Mid-Atlantic Ridge, which divides the Atlantic Ocean basin into two nearly equal parts (see Figure 1.13).

Oceanic ridges are composed almost entirely of the igneous rocks basalt and gabbro and possess features produced by tensional forces. Thus, they are the sites where new oceanic crust is generated and plates move away from each other along divergent plate boundaries.

First seen on the seafloor in 1979, **submarine hydrothermal vents** are found at or near spreading ridges. Here, cold seawater seeps through oceanic crust, is heated by the hot rocks at depth, and then rises and discharges into the seawater as plumes of hot water with temperatures as high as 400°C. Many of the plumes are black because dissolved minerals give them the appearance of black smoke—hence the name *black smoker* (●**FIGURE 2.7**).

Submarine hydrothermal vents are interesting from the biologic, geologic, and economic points of view. Near the vents live communities of organisms—such as bacteria,

---

**continental shelf** The very gently sloping part of the continental margin between the shoreline and the continental slope.

**continental slope** The relatively steeply inclined part of the continental margin between the continental shelf and the continental rise or between the continental shelf and an oceanic trench.

**continental rise** The gently sloping part of the continental margin between the continental slope and the abyssal plain.

**abyssal plain** The flat surface covering vast areas of the seafloor. Abyssal plains are the flattest, most featureless areas on Earth, and their flatness is a result of sediment deposition covering the usually rugged topography of the seafloor.

**oceanic ridge** A mostly submarine mountain system composed of basalt found in all ocean basins.

**submarine hydrothermal vent** A crack, or fissure, in the seafloor through which superheated water issues.

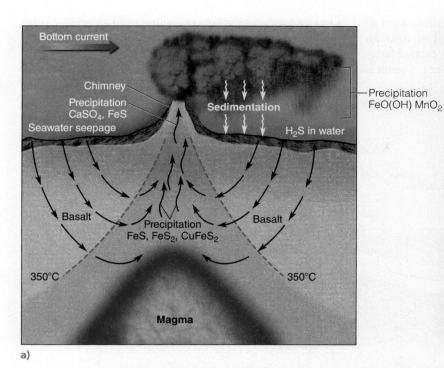

a)                                                                                   b)

● **FIGURE 2.7  Submarine Hydrothermal Vents** (a) Cross section showing the origin of a submarine hydrothermal vent called a black smoker. (b) This black smoker on the East Pacific Rise is at a depth of 2,800 m. The plume of "black smoke" is heated seawater with dissolved minerals.

crabs, mussels, starfish, and tube worms—many of which have never been seen before. No sunlight is available, so these organisms depend on bacteria that oxidize sulfur compounds for their ultimate source of nutrients. The vents are also interesting because of their economic potential. The heated seawater reacts with oceanic crust, transforming it into a metal-rich solution that discharges into seawater and cools, precipitating iron, copper, and zinc sulfides as well as other minerals (●**FIGURE 2.7A**). A chimney-like vent forms that eventually collapses and forms a mound of sediments rich in the elements just mentioned (●**FIGURE 2.7B**).

**Oceanic trenches** are long, steep-sided depressions on the seafloor near convergent plate boundaries and constitute no more than 2% of the seafloor (Figure 2.6). It is here, however, that oceanic lithosphere is consumed by subduction; that is, oceanic lithosphere plunges into Earth's interior along convergent plate boundaries (see Figure 1.13). It should also be noted that the greatest oceanic depths are found in trenches; the Challenger Deep of the Mariana Trench in the Pacific is more than 11,000 m deep!

## Seamounts, Guyots, and Aseismic Ridges

Except for the abyssal plains, the seafloor is not a flat, featureless expanse. In fact, a large number of volcanic hills, seamounts, and guyots rise above the seafloor in all ocean basins and are particularly abundant in the Pacific. All are of volcanic origin and differ mostly in size. **Seamounts** rise more than 1 km above the seafloor, and if flat-topped, they are called **guyots** (●**FIGURE 2.8**). Guyots are volcanoes that originally extended above sea level. However, as the plate upon which they were located continued to move, they were carried away from a spreading ridge, and as the oceanic crust cooled, it descended to greater depths. Thus, what was once an island slowly sank beneath the sea, and as it did, wave erosion produced the typical flat-topped appearance (Figure 2.8).

Other common features in the ocean basins are long, narrow ridges and broad, plateau-like features rising as much as 2 to 3 km above the surrounding seafloor. These **aseismic ridges** are so called because they lack seismic (earthquake) activity. A few of these ridges are probably small fragments separated from continents during rifting and are referred to as *microcontinents*. Avalonia is a good example of a Paleozoic microcontinent (see Figure 17.9b).

**oceanic trench** A long, narrow feature restricted to active continental margins and along which subduction occurs.

**seamount** A submarine volcanic mountain rising at least 1 km above the seafloor.

**guyot** A flat-topped seamount of volcanic origin rising more than 1 km above the seafloor.

**aseismic ridge** A ridge, or broad area, rising above the seafloor that lacks seismic activity.

Most aseismic ridges form as a linear succession of hot-spot volcanoes. These may develop at or near an oceanic ridge, but each volcano so formed is carried laterally with the plate upon which it originated. The net result is a line of seamounts/guyots extending from an oceanic ridge (Figure 2.8). Aseismic ridges also form over hot spots unrelated to ridges—the Hawaiian–Emperor chain in the Pacific, for example (●FIGURE 2.9).

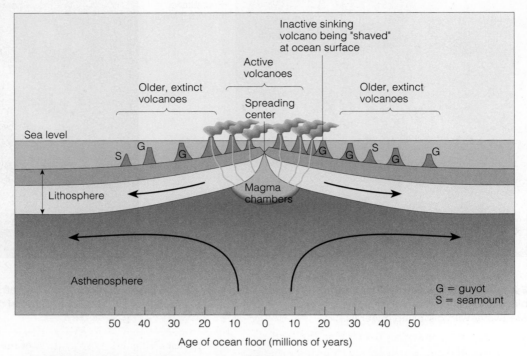

● **FIGURE 2.8 The Origin of Seamounts and Guyots** As the plate on which a volcano sits moves into greater water depths, the submerged volcanic island is called a seamount. Those that are flat-topped are called guyots.

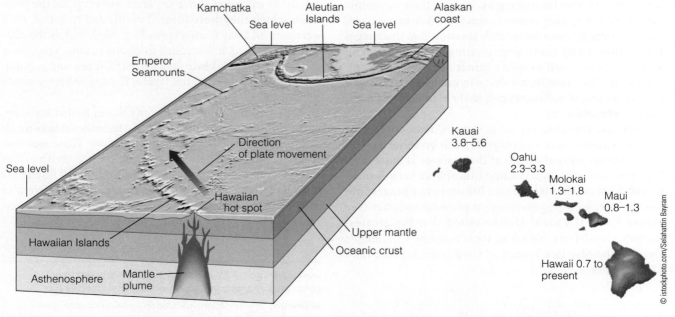

● **FIGURE 2.9 Hot Spots** A hot spot is the location where a stationary mantle plume has risen to the surface and formed a volcano. The Emperor Seamount–Hawaiian Island chain formed as a result of the Pacific plate moving over a mantle plume, and the line of volcanic islands in this chain traces the direction of plate movement. The island of Hawaii and the Loihi Seamount are the only current hot spots of this island chain. The numbers indicate the ages of the Hawaiian Islands in millions of years.

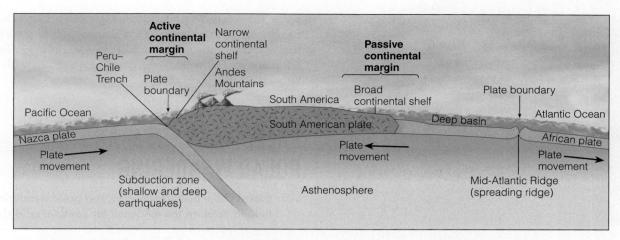

● **FIGURE 2.10 Passive and Active Continental Margins** Active and passive continental margins along the west and east coasts of South America. Notice that the passive margins are much wider than active margins. Seafloor sediment is not shown.

## Continental Margins

Although we will be discussing convergent and divergent plate boundaries later in this chapter, here is a good place to cover the two types of continental margins associated with the respective aforementioned plate boundaries.

**Active continental margins** develop at the leading edge of a continental plate where oceanic lithosphere is subducted. The western margin of South America is a good example of where an oceanic plate is subducted beneath the continent, resulting in seismic activity, a geologically young mountain range (Andes Mountains), and active volcanism (● **FIGURE 2.10**). In addition, the continental shelf is narrow, and the continental slope descends directly into an oceanic trench, so sediment is dumped into the trench and no continental rise develops (Figure 2.6, left side).

The western margin of North America is also considered an active continental margin, although much of it is now bounded by transform faults (see the section on transform boundaries later in this chapter) rather than a subduction zone. However, plate convergence and subduction continue in the Pacific Northwest along the continental margins of northern California, Oregon, and Washington.

The continental margins of eastern North America and South America differ considerably from their western counterparts. For one thing, they possess broad continental shelves, as well as a continental slope and rise, with abyssal plains adjacent to the rises (Figure 2.6, right side). Furthermore, these **passive continental margins** are within a plate rather than along a plate boundary, and they lack the volcanic and seismic activity found at active continental margins (Figure 2.10). Nevertheless, earthquakes do take place there occasionally, such as the magnitude 5.8 earthquake that struck the east coast of the Unites States on August 23, 2011, causing slight damage to the Washington National Cathedral.

## EARTH'S MAGNETIC FIELD

**LO3** Define magnetism and magnetic field and use these definitions to explain how these phenomena relate to Earth's magnetic field

Having just discussed those features of the seafloor that are related to plate movement, we now turn our attention to the phenomenon of magnetism and its role in the formulation of plate tectonic theory.

**Magnetism** is a physical phenomenon resulting from the spin of electrons in some solids—particularly those of iron—and moving electricity. A **magnetic field** is an area in which magnetic substances such as iron are affected by lines of magnetic force emanating from Earth (●**FIGURE 2.11**). The magnetic field shown in Figure 2.11 is *dipolar*, meaning that it possesses two unlike magnetic poles referred to as the north and south poles.

A useful analogy is to think of Earth as a giant dipole magnet in which the magnetic poles are in proximity to the geographic poles (●**FIGURE 2.12**). This arrangement means that the strength of the magnetic field is

MINDTAP
From Cengage

Animation:
Earth's Magnetic Field
and Paleomagnetism

---

**active continental margin** A continental margin with volcanism and seismicity at the leading edge of a continental plate where oceanic lithosphere is subducted.

**passive continental margin** A continental margin within a tectonic plate, as along the East Coast of North America, where little seismic activity and no volcanism occur; characterized by a broad continental shelf and a continental slope and rise.

**magnetism** A physical phenomenon resulting from moving electricity and the spin of electrons in some solids in which magnetic substances are attracted toward one another.

**magnetic field** The area in which magnetic substances are affected by lines of magnetic force emanating from Earth.

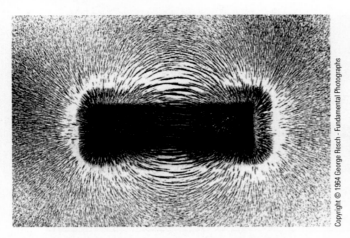

● **FIGURE 2.11  Magnetic Field** Iron filings align along the lines of magnetic force radiating from a bar magnet.

not constant, but varies. Notice in Figure 2.12 that the lines of magnetic force around Earth parallel its surface only near the equator, just as the iron filings do around a bar magnet (Figure 2.11). As the lines of force approach the poles, they are oriented at increasingly larger angles with respect to the surface, and the strength of the magnetic field increases; it is strongest at the poles and weakest at the equator.

Another important aspect of the magnetic field is that the magnetic poles, where the lines of force leave and enter Earth, do not coincide with the geographic (rotational)

poles. Currently, an 11.5-degree angle exists between the two (Figure 2.12). Studies of Earth's magnetic field show that the locations of the magnetic poles vary slightly over time, but that they still correspond closely, on average, with the locations of the geographic poles.

# PALEOMAGNETISM AND POLAR WANDERING

**L04  Explain how paleomagnetism and polar wandering helped validate the evidence for continental drift**

Interest in continental drift revived during the 1950s as a result of evidence from paleomagnetic studies, a relatively new discipline at the time. **Paleomagnetism** is the remnant magnetism in ancient rocks recording the direction and intensity of Earth's magnetic poles at the time of the rock's formation.

When magma cools, the magnetic iron-bearing minerals align themselves with Earth's magnetic field, recording both its direction and strength. The temperature at which iron-bearing

**Paleomagnetism**  Residual magnetism in rocks, studied to determine the intensity and direction of Earth's past magnetic field.

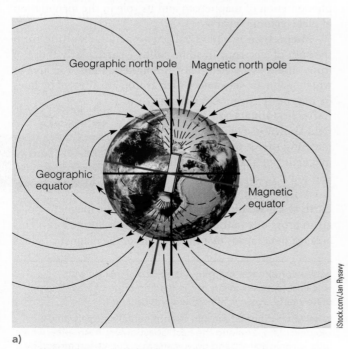

a)

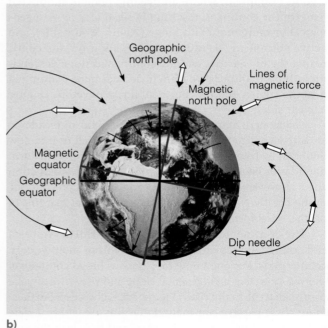

b)

● **FIGURE 2.12  Earth's Magnetic Field** (a) Earth's magnetic field has lines of force like those of a bar magnet. (b) The strength of the magnetic field changes from the magnetic equator to the magnetic poles. This change in strength causes a dip needle (a magnetic needle that is balanced on the tip of a support so that it can freely move vertically) to be parallel to Earth's surface only at the magnetic equator, where the strength of the magnetic north and south pole is equally balanced. Its inclination, or dip, with respect to Earth's surface increases as it moves toward the magnetic poles until it is at 90 degrees, or perpendicular, to Earth's surface at the magnetic poles.

minerals gain their magnetization is called the **Curie point**. As long as the rock is not subsequently heated above the Curie point, it will preserve that magnetism. Thus, an ancient lava flow provides a record of the orientation and strength of Earth's magnetic field at the time the lava flow cooled.

As paleomagnetic research progressed during the 1950s, some unexpected results emerged. When geologists measured the paleomagnetism of geologically recent rocks from different continents, they found that it was generally consistent with Earth's current magnetic field.

The paleomagnetism of ancient rocks, though, showed different orientations. For example, paleomagnetic studies of Silurian lava flows in North America indicated that the north magnetic pole was located in the western Pacific Ocean at that time, whereas the paleomagnetic evidence from Permian lava flows pointed to yet another location in Asia. When plotted on a map, the paleomagnetic readings of numerous lava flows from all ages in North America trace the apparent movement of the magnetic pole (called *polar wandering*) through time (●**FIGURE 2.13**).

Furthermore, paleomagnetic readings from European Silurian and Permian lava flows pointed to a different magnetic pole location than those of the same age from North America (Figure 2.13). Analysis of lava flows from all continents indicated that each continent seemingly had its own series of magnetic poles! How could this be? Does it really mean there were different north magnetic poles for each continent?

The best explanation for such data is that the magnetic poles have remained near their present locations at the geographic north and south poles, and that the continents have moved. When the continental margins are fit together so that the paleomagnetic data point to only one magnetic pole, we find, just as Wegener did, that the rock sequences and glacial deposits match, and that the fossil evidence is consistent with the reconstructed paleogeography.

# MAGNETIC REVERSALS AND SEAFLOOR SPREADING

**LO5**  Discuss how the discovery of magnetic reversals and its application to seafloor spreading accounts for continental movement

**LO6**  Explain how the pattern of striped magnetic anomalies in oceanic crust confirms that ocean basins are geologically young features whose openings and closings are partially responsible for continental movement

Geologists refer to Earth's present magnetic field as being normal—that is, with the north and south magnetic poles located approximately at the north and south geographic poles. At various times in the geologic past, however, Earth's magnetic field has completely reversed. The magnetic north and south poles have switched positions so that the magnetic north pole became the magnetic south pole, and the magnetic south pole became the magnetic north pole. During such a reversal, the magnetic field weakens until it temporarily disappears. When the magnetic field returns, the magnetic poles have reversed their position. The existence of such **magnetic reversals** was discovered by dating and determining the orientation of the remnant magnetism in lava flows on land (●**FIGURE 2.14**). Although the cause of magnetic reversals is still uncertain, their occurrence in the geologic record is well documented.

As a result of oceanographic research conducted during the 1950s, Harry Hess of Princeton University proposed, in a 1962 landmark paper, the theory of **seafloor spreading**

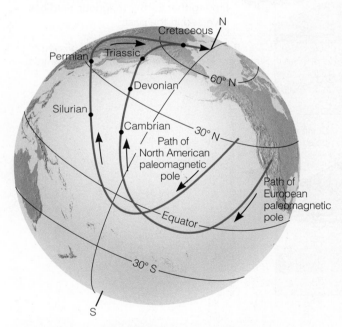

●**FIGURE 2.13 Polar Wandering** The apparent paths of polar wandering for North America and Europe. The apparent location of the north magnetic pole is shown for different periods on each continent's polar wandering path. Because Earth has only one magnetic north pole, the paleomagnetic readings taken on different continents for the same time in the past should all point to the same location if the continents have not moved. However, the north magnetic pole has different locations for the same time in the past when measured on different continents, indicating multiple north magnetic poles. The logical explanation for this dilemma is that the magnetic north pole has remained at the same approximate geographic location during the past, and the continents have moved.

---

**Curie point**  The temperature at which iron-bearing minerals in cooling magma or lava attain their magnetism.

**magnetic reversal**  The phenomenon involving the complete reversal of the north and south magnetic poles.

**seafloor spreading**  The theory that the seafloor moves away from spreading ridges and is eventually consumed at subduction zones.

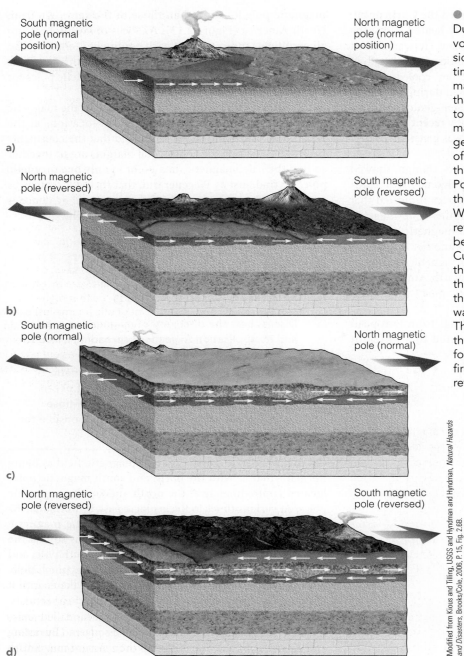

a)

South magnetic
pole (normal
position)

North magnetic
pole (normal
position)

b)

North magnetic
pole (reversed)

South magnetic
pole (reversed)

c)

South magnetic
pole (normal)

North magnetic
pole (normal)

d)

North magnetic
pole (reversed)

South magnetic
pole (reversed)

Modified from Kious and Tilling, USGS and Hyndman and Hyndman, *Natural Hazards and Disasters*, Brooks/Cole, 2006, P. 15, Fig. 2.6B.

● **FIGURE 2.14  Magnetic Reversals**
During the time period shown (a–d), volcanic eruptions produced a succession of overlapping lava flows. At the time of these volcanic eruptions, Earth's magnetic field completely reversed— that is, the magnetic north pole moved to the geographic south pole, and the magnetic south pole moved to the geographic north pole. Thus, the end of the needle of a magnetic compass that today would point to the North Pole would point to the South Pole if the magnetic field should again reverse. We know that Earth's magnetic field has reversed numerous times in the past because when lava flows cool below the Curie point, magnetic minerals within the flow orient themselves parallel to the magnetic field at the time. They thus record whether the magnetic field was normal or reversed at that time. The white arrows in this diagram show the direction of the north magnetic pole for each individual lava flow, thus confirming that Earth's magnetic field has reversed in the past.

to account for continental movement. He suggested that continents do not move through oceanic crust as do ships plowing through sea ice, but rather that the continents and oceanic crust move together as a single unit. Thus, the theory of seafloor spreading answered a major objection of the opponents of continental drift—namely, how could continents move through oceanic crust? The answer is that they do not. In fact, the continents moved with the oceanic crust as part of a lithospheric system.

As a mechanism to drive this system, Hess revived the idea of a heat transfer mechanism—or **thermal convection cells**— within the mantle to move the plates. According to Hess,

hot magma rises from the mantle, intrudes along fractures defining oceanic ridges, and thus forms new crust. Cold crust is subducted back into the mantle at oceanic trenches, where it is heated and recycled, thus completing a thermal convection cell (see Figure 1.12).

How could Hess's hypothesis be confirmed? Magnetic surveys of the oceanic crust revealed a pattern of striped

**thermal convection cell** A type of circulation of material in the asthenosphere during which hot material rises, moves laterally, cools and sinks, and is reheated and continues the cycle.

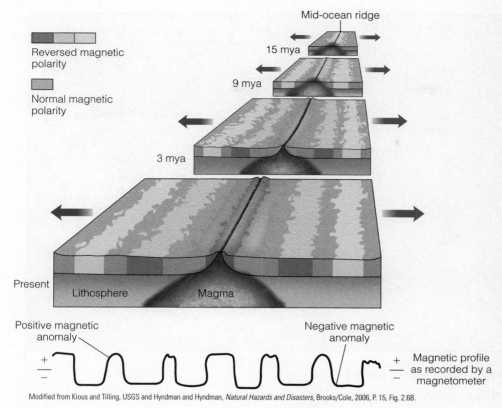

Mid-ocean ridge

15 mya

9 mya

3 mya

Present    Lithosphere    Magma

Reversed magnetic polarity

Normal magnetic polarity

Positive magnetic anomaly

Negative magnetic anomaly

+
−

+    Magnetic profile
−    as recorded by a
     magnetometer

Modified from Kious and Tilling, USGS and Hyndman and Hyndman, *Natural Hazards and Disasters*, Brooks/Cole, 2006, P. 15, Fig. 2.6B.

● **FIGURE 2.15 Magnetic Anomalies and Seafloor Spreading** The sequence of magnetic anomalies preserved within the oceanic crust is both parallel to and symmetric around oceanic ridges. Basaltic lava intruding into an oceanic ridge today and spreading laterally away from the ridge records Earth's current magnetic field, or polarity (considered by convention to be normal). Basaltic intrusions 3, 9, and 15 million years ago record Earth's reversed magnetic field at those times. This schematic diagram shows how the solidified basalt moves away from the oceanic ridge (or spreading ridge), carrying with it the magnetic anomalies that are preserved in the oceanic crust. Magnetic anomalies are magnetic readings that are either higher (positive magnetic anomalies) or lower (negative magnetic anomalies) than Earth's current magnetic field strength. The magnetic anomalies are recorded by a magnetometer, which measures the strength of the magnetic field.

**magnetic anomalies** (deviations from the average strength of Earth's present-day magnetic field) in the rocks that are both parallel to and symmetric around the oceanic ridges (●**FIGURE 2.15**). A positive magnetic anomaly results when Earth's magnetic field at the time of oceanic crust formation along an oceanic ridge summit was the same as today, thus yielding a stronger than normal (positive) magnetic signal. A negative magnetic anomaly results when Earth's magnetic field at the time of oceanic crust formation was reversed, therefore yielding a weaker than normal (negative) magnetic signal.

Thus, as new oceanic crust forms at oceanic ridge summits and records Earth's magnetic field at the time, the previously formed crust moves laterally away from the ridge. These magnetic stripes therefore represent times of normal and reversed polarity at oceanic ridges (where upwelling magma forms new oceanic crust), and conclusively confirm Hess's theory of seafloor spreading.

One of the consequences of the seafloor spreading theory is its confirmation that ocean basins are geologically young features whose openings and closings are partially responsible for continental movement. Radiometric dating reveals that the oldest oceanic crust is somewhat younger than 180 million years old, whereas the oldest continental crust is about 4 billion years old. Although geologists do not universally accept the idea of thermal convection cells as the sole driving mechanism for plate movement, most accept that plates are created at oceanic ridges and destroyed at deep-sea trenches (Figure 2.10), regardless of the driving mechanism involved.

# PLATE TECTONICS: A UNIFYING THEORY

**L07** Discuss and explain why plate tectonic theory is a unifying theory for explaining how seemingly unrelated geologic features and events are interrelated on a global scale

**Plate tectonic theory** is based on a simple model of Earth. The rigid lithosphere, composed of both oceanic and continental crust, as well as the underlying upper mantle, consists of numerous variable-sized pieces called *plates* (●**FIGURE 2.16**). There are seven major plates (Eurasian, Indian-Australian, Antarctic, North American, South American, Pacific, and African), and numerous smaller ones ranging from only a few tens to several hundreds of kilometers in width. Plates also vary in thickness; those composed of upper mantle and continental crust are as much as 250 km thick, whereas those of upper mantle and oceanic crust are up to 100 km thick.

The lithosphere overlies the hotter and weaker semiplastic asthenosphere. It is thought that movement resulting

**magnetic anomaly** Any deviation, such as a change in average strength, in Earth's magnetic field.

**plate tectonic theory** The theory holding that large segments of Earth's outer part (lithospheric plates) move relative to one another.

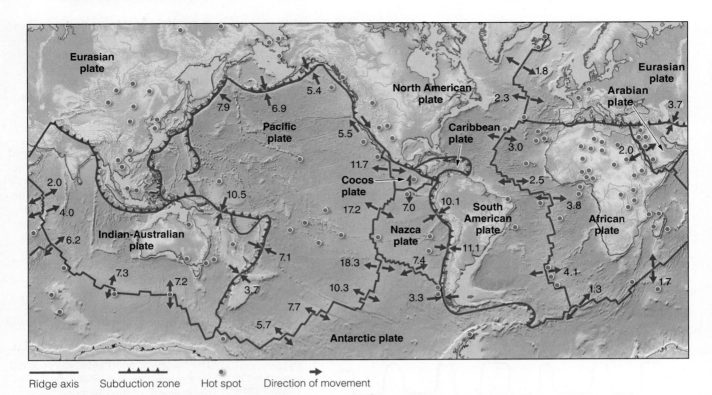

Ridge axis     Subduction zone     Hot spot     Direction of movement

● **FIGURE 2.16 Earth's Plates** A world map showing Earth's plates, their boundaries, their relative motion, average rates of movement in centimeters per year, and hot spots.

from some type of heat-transfer system within the asthenosphere causes the overlying plates to move.

As plates move over the asthenosphere, they separate, mostly at oceanic ridges; in other areas, such as at oceanic trenches, they collide and are subducted back into the mantle.

An easy way to visualize plate movement is to think of a conveyor belt moving luggage from an airplane's cargo hold to a baggage cart. The conveyer belt represents convection currents within the mantle, and the luggage represents Earth's lithospheric plates. The luggage is moved along by the conveyer belt until it is dumped into the baggage cart—in the same way that plates are moved by convection cells until they are subducted into Earth's interior.

Although this analogy allows you to visualize how the mechanism of plate movement takes place, remember that the analogy is limited. The major limitation is that, unlike the luggage, plates consist of continental and oceanic lithosphere, which have different densities, and only oceanic lithosphere is subducted into Earth's interior.

Most geologists accept plate tectonic theory because the evidence for it is overwhelming and because it ties together many seemingly unrelated geologic features and events and shows how they are interrelated. Consequently, geologists now view many geologic processes from the global perspective of plate tectonic theory, in which plate interaction along plate margins is responsible for such phenomena as mountain building, earthquakes, and volcanism.

# THE THREE TYPES OF PLATE BOUNDARIES

**LO8**   Define and describe the three types of plate boundaries with examples of each type, and how each plate boundary can be recognized in the rock record

Because it appears that plate tectonics has operated since at least the Proterozoic Eon, it is important that we understand how plates move and interact with each other and how ancient plate boundaries are recognized. After all, the movement of plates has profoundly affected the geologic and biologic history of our planet.

Geologists recognize three major types of plate boundaries: *divergent*, *convergent*, and *transform*. Along these boundaries, new plates are formed, consumed, or slide laterally past one another. Interaction of plates at their boundaries accounts for most of Earth's volcanic eruptions and earthquakes, as well as the formation and evolution of its mountain systems.

## Divergent Boundaries

**Divergent plate boundaries,** or *spreading ridges*, occur where plates are separating and new oceanic lithosphere is

**divergent plate boundary** The boundary between two plates that are moving apart.

forming. Divergent boundaries are places where the crust is extended, thinned, and fractured as magma, derived from the partial melting of the mantle, rises to the surface. The magma is almost entirely basaltic and intrudes into vertical fractures to form dikes and pillow lava flows (see Figure 5.6). As successive injections of magma cool and solidify, they form new oceanic crust and record the intensity and orientation of Earth's magnetic field (Figure 2.15).

Divergent boundaries most commonly occur along the crests of oceanic ridges—for example, the Mid-Atlantic Ridge. Oceanic ridges are thus characterized by rugged topography with high relief resulting from displacement of rocks along large fractures, shallow-depth earthquakes, high heat flow, and basaltic flows or pillow lavas.

Divergent boundaries are also present under continents during the early stages of continental breakup. When magma wells up beneath a continent, the crust is initially elevated, stretched, and thinned, producing fractures, faults, rift valleys, and volcanic activity (●FIGURE 2.17A). As magma intrudes into faults and fractures, it solidifies or flows out onto the surface as lava flows; the latter often covering the rift valley floor (●FIGURE 2.17B). The East African Rift Valley is an excellent example of continental breakup at this stage (●FIGURE 2.18).

As spreading proceeds, some rift valleys continue to lengthen and deepen until the continental crust eventually breaks and a narrow linear sea is formed, separating two continental blocks (●FIGURE 2.17C). The Red Sea, separating the Arabian Peninsula from Africa (Figure 2.18), and the Gulf of California, which separates Baja California from mainland Mexico, are good examples of this more advanced stage of rifting.

As a newly created narrow sea continues to enlarge, it may eventually become an expansive ocean basin such as the Atlantic Ocean basin is today, separating North and South America from Europe and Africa by thousands of kilometers (●FIGURE 2.17D). The Mid-Atlantic Ridge is the boundary between these diverging plates; the American plates are moving westward, and the Eurasian and African plates are moving eastward.

Extending outward from the eastern coasts of North and South America, as well as the western coasts of Europe and Africa, are broad continental shelves, continental slopes, and continental rises. These features, discussed earlier, are referred to as *passive continental margins* (Figure 2.10). And, although they are found within a plate, rather than at a plate boundary, they result from the breakup of a continent and the subsequent movement away from a divergent plate boundary (Figure 2.17).

**AN EXAMPLE OF ANCIENT RIFTING** What features in the geologic record can geologists use to recognize ancient rifting? Associated with regions of continental rifting are faults, dikes (vertical intrusive igneous bodies), sills (horizontal intrusive igneous bodies), lava flows, and thick sedimentary sequences within rift valleys, all features that are preserved in the geologic record. The Triassic fault basins of the eastern

United States are a good example of ancient continental rifting (see Figure 17.23). These fault basins mark the zone of rifting that occurred when North America split apart from Africa. The basins contain thousands of meters of continental sediment and are riddled with dikes and sills (see Chapter 17).

Pillow lavas, in association with deep-sea sediment, are also evidence of ancient rifting. The presence of pillow lavas marks the formation of a spreading ridge in a narrow linear sea (Figure 2.17c). Magma, intruding into the sea along this newly formed spreading ridge, solidifies as pillow lavas, which are preserved in the geologic record along with the sediment being deposited on them.

## Convergent Boundaries

Whereas new crust forms at divergent plate boundaries, older crust must be destroyed and recycled in order for the entire surface area of Earth to remain the same. Otherwise, we would have an expanding Earth. Such plate destruction occurs along **convergent plate boundaries** (●FIGURE 2.19), where two plates collide and the leading edge of one plate is subducted beneath the margin of the other plate and eventually incorporated into the asthenosphere. A dipping plane of earthquake foci, called a *Benioff zone*, defines subduction zones (see Figure 8.5). Most of these planes dip from oceanic trenches beneath adjacent island arcs or continents, marking the surface of slippage between the converging plates.

Deformation, volcanism, mountain building, metamorphism, earthquake activity, and deposits of valuable minerals characterize convergent boundaries. Three types of convergent plate boundaries are recognized: *oceanic–oceanic*, *oceanic–continental*, and *continental–continental*.

**OCEANIC–OCEANIC BOUNDARIES** When two oceanic plates converge, one is subducted beneath the other along an **oceanic–oceanic plate boundary** (●FIGURE 2.19A). The subducting plate bends downward to form the outer wall of an oceanic trench. A subduction complex, composed of wedge-shaped slices of highly folded and faulted marine sediments and oceanic lithosphere scraped off the descending plate, forms along the inner wall of the oceanic trench. As the subducting plate descends into the mantle, it is heated and partially melted, generating magma that is commonly of andesitic composition (see Chapter 4). This magma is less dense than the surrounding mantle rocks and rises to the surface of the nonsubducted plate to form a curved chain of volcanic islands called a *volcanic island arc* (any plane intersecting a sphere makes an arc). This arc is nearly parallel to the oceanic trench and is separated from it by a distance of up to several hundred kilometers—the distance depending on the angle of dip of the subducting plate (Figure 2.19a).

**convergent plate boundary** The boundary between two plates that move toward each other.

**oceanic–oceanic plate boundary** A convergent plate boundary along which two oceanic plates collide and one is subducted beneath the other.

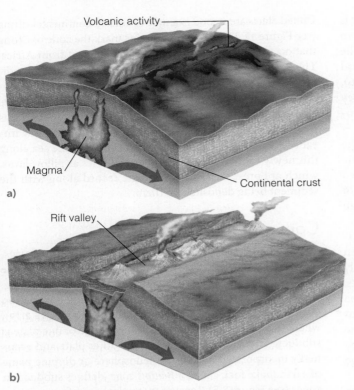

Volcanic activity

Magma

Continental crust

a)

Rift valley

b)

Coastal mountain range

Narrow fault-bounded sea

c)

● **FIGURE 2.17 History of a Divergent Plate Boundary** (a) Rising magma beneath a continent pushes the crust up, producing numerous fractures, faults, and volcanic activity. (b) As the crust is stretched and thinned, rift valleys develop, and lava flows onto the valley floors, such as seen today in the East African Rift Valley. (c) Continued spreading further separates the continent until it splits apart and a narrow seaway develops. The Red Sea, which separates the Arabian Peninsula from Africa, is a good example of this stage of development. (d) As spreading continues, an oceanic ridge system forms, and an ocean basin develops and grows. The Mid-Atlantic Ridge illustrates this stage in a divergent plate boundary's history.

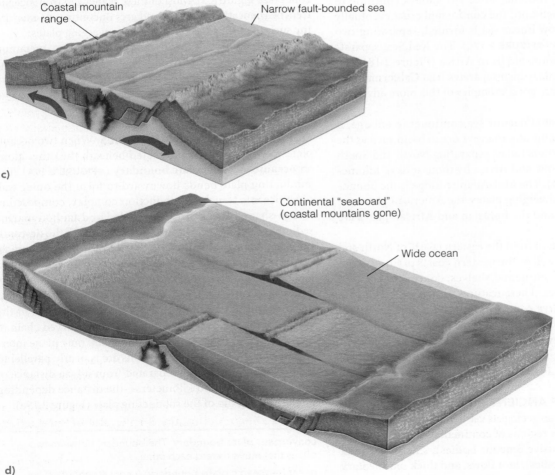

Continental "seaboard" (coastal mountains gone)

Wide ocean

d)

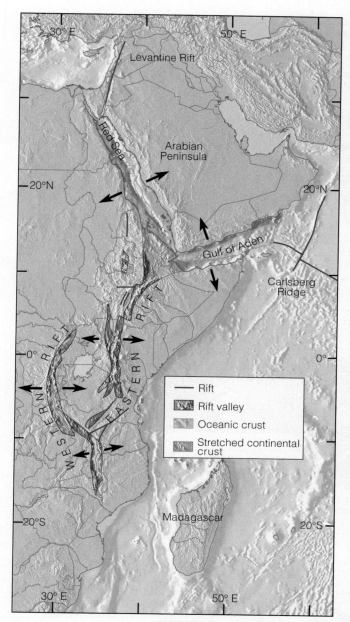

● **FIGURE 2.18 East African Rift Valley and the Red Sea—Present-Day Examples of Divergent Plate Boundaries** The East African Rift Valley and the Red Sea represent different stages in the history of a divergent plate boundary. The East African Rift Valley is being formed by the separation of eastern Africa from the rest of the continent along a divergent plate boundary. The Red Sea represents a more advanced stage of rifting, in which two continental blocks (Africa and the Arabian Peninsula) are separated by a narrow sea.

In those areas where the rate of subduction is faster than the forward movement of the overriding plate, the lithosphere on the landward side of the volcanic island arc may be subjected to tensional stress and be stretched and thinned, resulting in the formation of a *back-arc basin*. This back-arc basin may grow by spreading if magma breaks through

the thin crust and forms new oceanic crust (Figure 2.19a). A good example of a back-arc basin associated with an oceanic–oceanic plate boundary is the Sea of Japan between the Asian continent and the islands of Japan.

Most present-day active volcanic island arcs are in the Pacific Ocean basin and include the Aleutian Islands, the Kermadec–Tonga arc, and the Japanese (Figure 2.19a) and Philippine Islands. The Scotia and Antillean (Caribbean) island arcs are in the Atlantic Ocean basin.

**OCEANIC–CONTINENTAL BOUNDARIES** When an oceanic and a continental plate converge, the denser oceanic plate is subducted under the continental plate along an **oceanic–continental plate boundary** (●**FIGURE 2.19B**). Just as at oceanic–oceanic plate boundaries, the descending oceanic plate forms the outer wall of an oceanic trench.

The magma generated by subduction rises beneath the continent and either crystallizes as large intrusive bodies, called *plutons*, before reaching the surface or erupts at the surface to produce a chain of andesitic volcanoes, also called a *volcanic arc*. An excellent example of an oceanic–continental plate boundary is the Pacific Coast of South America, where the oceanic Nazca plate is currently being subducted under South America (Figure 2.19b; see also Chapter 9). The Peru–Chile Trench marks the site of subduction, and the Andes Mountains are the resulting volcanic mountain chain on the nonsubducting plate.

Just as there are passive continental margins, there are also active continental margins (Figure 2.10). The aforementioned oceanic–continental plate boundary between the west coast of the South American plate and the eastern side of the oceanic Nazca plate is an excellent example of an active continental margin. Here, the continental shelf is narrow, and the continental slope descends directly into the Peru–Chile Trench, where sediment is rapidly dumped into the trench (Figure 2.10).

**CONTINENTAL–CONTINENTAL BOUNDARIES** Two continents approaching each other are initially separated by an ocean floor that is being subducted under one continent. The edge of that continent displays the features characteristic of oceanic–continental convergence. As the ocean floor continues to be subducted, the two continents come closer together until they eventually collide. Because continental lithosphere, which consists of continental crust and the upper mantle, is less dense than oceanic lithosphere (oceanic crust and upper mantle), it cannot sink into the asthenosphere. Although one continent may partially slide under the other, it cannot be pulled or pushed down into a subduction zone (●**FIGURE 2.19C**).

When two continents collide, they are welded together along a zone marking the former site of subduction.

---

**oceanic–continental plate boundary** A convergent plate boundary along which oceanic lithosphere is subducted beneath continental lithosphere.

**Oceanic–oceanic plate boundary**

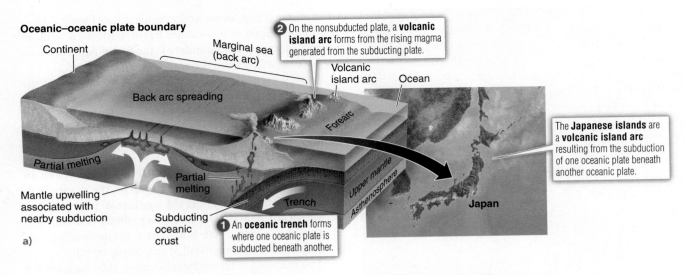

Continent

Marginal sea (back arc)

**2** On the nonsubducted plate, a **volcanic island arc** forms from the rising magma generated from the subducting plate.

Volcanic island arc

Ocean

Back arc spreading

Forearc

The **Japanese islands** are a **volcanic island arc** resulting from the subduction of one oceanic plate beneath another oceanic plate.

Partial melting

Partial melting

Upper mantle

Asthenosphere

Mantle upwelling associated with nearby subduction

Trench

Subducting oceanic crust

**1** An **oceanic trench** forms where one oceanic plate is subducted beneath another.

**Japan**

a)

**Oceanic–continental plate boundary**

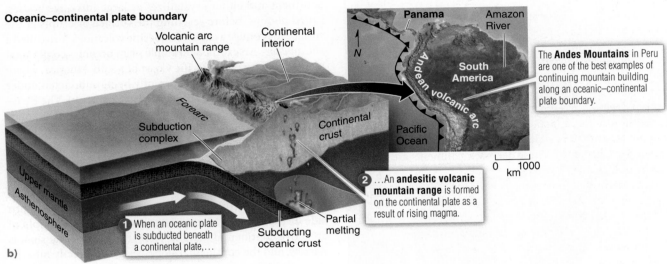

Volcanic arc mountain range

Continental interior

Panama

Amazon River

N

Forearc

Subduction complex

Continental crust

Andean volcanic arc

South America

The **Andes Mountains** in Peru are one of the best examples of continuing mountain building along an oceanic–continental plate boundary.

Upper mantle

Asthenosphere

Pacific Ocean

0   1000
km

**2** …An **andesitic volcanic mountain range** is formed on the continental plate as a result of rising magma.

**1** When an oceanic plate is subducted beneath a continental plate,…

Subducting oceanic crust

Partial melting

b)

**Continental–continental plate boundary**

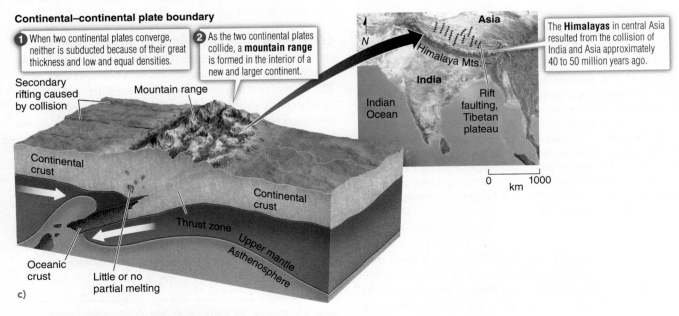

**1** When two continental plates converge, neither is subducted because of their great thickness and low and equal densities.

**2** As the two continental plates collide, a **mountain range** is formed in the interior of a new and larger continent.

Asia

N

Himalaya Mts.

The **Himalayas** in central Asia resulted from the collision of India and Asia approximately 40 to 50 million years ago.

Secondary rifting caused by collision

Mountain range

**India**

Continental crust

Indian Ocean

Rift faulting, Tibetan plateau

Continental crust

Thrust zone

Upper mantle

0   1000
km

Oceanic crust

Asthenosphere

Little or no partial melting

c)

● **FIGURE 2.19 Three Types of Convergent Plate Boundaries**

At this **continental–continental plate boundary**, an interior mountain belt is formed consisting of deformed sediments and sedimentary rocks, igneous intrusions, metamorphic rocks, and fragments of oceanic crust. In addition, the entire region is subjected to numerous earthquakes. The Himalayas in central Asia, the world's youngest and highest mountain system, resulted from the collision between India and Asia that began 40 to 50 million years ago and is still continuing (Figure 2.19c; see Chapter 9).

**RECOGNIZING ANCIENT CONVERGENT PLATE BOUNDARIES** How can former subduction zones be recognized in the geologic record? Igneous rocks provide one such clue. The magma erupted at the surface, forming island arc volcanoes and continental volcanoes, is of andesitic composition. Another clue is the zone of intensely deformed rocks between the deep-sea trench where subduction is taking place and the area of igneous activity. Here, sediments and submarine rocks are folded, faulted, and metamorphosed into a chaotic mixture of rocks termed a *mélange*.

During subduction, pieces of oceanic lithosphere are sometimes incorporated into the mélange and accreted onto the edge of the continent. Slices of oceanic crust and upper mantle are called *ophiolites* (●FIGURE 2.20). They consist of a layer of deep-sea sediments that include graywackes (poorly sorted sandstones containing abundant feldspars and rock fragments, usually in a clay-rich matrix), black shales, and cherts (see Chapter 6). These deep-sea sediments are underlain by pillow lavas, a sheeted dike complex, massive gabbro (a dark intrusive igneous rock), and layered gabbro, all of which form the oceanic crust. Beneath the gabbro is peridotite (a dark, intrusive igneous rock composed of the mineral olivine), which probably represents the upper mantle. The presence of ophiolites in an outcrop or drilling core is a key indicator of plate convergence along a subduction zone.

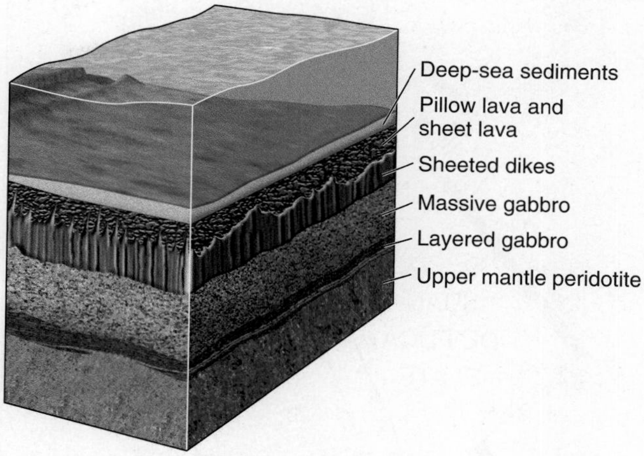

## Transform Boundaries

The third type of plate boundary is a **transform plate boundary,** which mostly occur along fractures in the seafloor, known as *transform faults*, where plates slide laterally past one another roughly parallel to the direction of plate

● **FIGURE 2.20 Ophiolites** Ophiolites are sequences of rock on land consisting of deep-sea sediments, oceanic crust, and upper mantle. Ophiolites are one feature used to recognize ancient convergent plate boundaries.

movement. Although lithosphere is neither created nor destroyed along a transform fault boundary, the movement between plates results in a zone of intensely shattered rock and numerous shallow-depth earthquakes.

**Transform faults** "transform," or change, one type of motion between plates into another type of motion. Most commonly, transform faults connect two oceanic ridge segments; however, they can also connect ridges to trenches and trenches to trenches (●FIGURE 2.21). Although the majority of transform faults are in oceanic crust and are marked by distinct fracture zones, they may also extend into continents.

One of the best-known transform faults is the San Andreas Fault in California. It separates the Pacific plate from

**continental–continental plate boundary** A convergent plate boundary along which two continental lithospheric plates collide.

**transform plate boundary** A plate boundary along which plates slide past one another and crust is neither produced nor destroyed.

**transform fault** A fault along which one type of motion is transformed into another; commonly displaces oceanic ridges; on land, recognized as a strike-slip fault, such as the San Andreas Fault.

● **FIGURE 2.21 Transform Plate Boundaries** Horizontal movement between plates occurs along transform faults. Extensions of transform faults on the seafloor form fracture zones. Most transform faults connect two oceanic ridge segments.

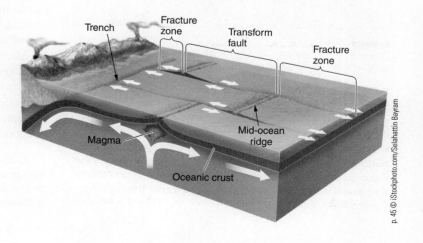

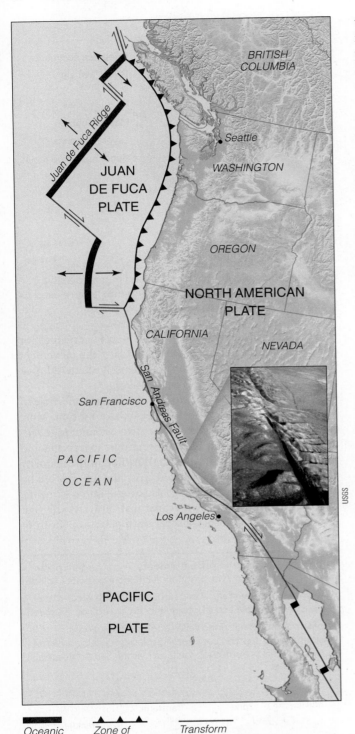

Oceanic ridge ▪ Zone of subduction ▲▲▲ ▪ Transform faults

● **FIGURE 2.22 The San Andreas Fault—A Transform Plate Boundary** The San Andreas Fault is a transform fault separating the Pacific plate from the North American plate. It connects the spreading ridges in the Gulf of California with the Juan de Fuca and Pacific plates off the coast of northern California. Movement along the San Andreas Fault has caused numerous earthquakes. The insert photo shows a segment of the San Andreas Fault as it cuts through the Carrizo Plain, California.

the North American plate and connects spreading ridges in the Gulf of California with the Juan de Fuca and Pacific plates off the coast of northern California (●FIGURE 2.22). Many of the earthquakes that affect California are the result of movement along this fault (see Chapter 8).

Unfortunately, transform faults generally do not leave any characteristic or diagnostic features except for the obvious displacement of the rocks with which they are associated. This displacement is usually large, on the order of tens to hundreds of kilometers. Such large displacements in ancient rocks can sometimes be related to transform fault systems.

# HOT SPOTS AND MANTLE PLUMES

LO9   Define hot spots and mantle plumes and how they can be used to determine both the direction and rate of plate movement

Before leaving the topic of plate boundaries, we should mention an intraplate feature found beneath both oceanic and continental plates. A **hot spot** (Figures 2.9 and 2.16) is the location on Earth's surface where a stationary column of magma, originating deep within the mantle (*mantle plume*), has slowly risen to the surface and formed a volcano. Because the mantle plumes seemingly remain stationary (although some evidence suggests that they might not) within the mantle while the plates move over them, the resulting hot spots leave a trail of extinct and progressively older volcanoes (*aseismic ridges*) that record the movement of the plate.

One of the best examples of aseismic ridges and hot spots is the Emperor Seamount–Hawaiian Island chain (Figure 2.9). This chain of islands and seamounts extends from the island of Hawaii to the Aleutian Trench off Alaska, a distance of some 6,000 km, and consists of more than 80 volcanic structures.

Currently, the only active volcanoes in this island chain are on the island of Hawaii and the Loihi Seamount. The rest of the islands are extinct volcanic structures that become progressively older toward the north and northwest. This means that the Emperor Seamount–Hawaiian Island chain records the direction that the Pacific plate traveled as it moved over an apparently stationary mantle plume.

Mantle plumes and hot spots help geologists explain some of the geologic activity occurring within plates as opposed to activity occurring at or near plate boundaries. In addition, if mantle plumes are essentially fixed with respect to Earth's rotational axis, they can be used to determine not only the direction of plate movement but also the rate of movement.

**hot spot** A localized zone of melting below the lithosphere that probably overlies a mantle plume; detected by volcanism at the surface.

# PLATE MOVEMENT AND MOTION

**LO10** Explain how geologists use magnetic anomalies to calculate not only the average rate of plate movement but also plate positions in the geologic past

How fast and in what direction are Earth's plates moving? Do they all move at the same rate? Rates of plate movement can be calculated in several ways. The least accurate method is to determine the age of the sediments immediately above any portion of the oceanic crust and then divide the distance from the spreading ridge by that age. Such calculations give an average rate of movement.

A more accurate method of determining both the average rate of movement and relative motion is by dating the magnetic anomalies in the crust of the seafloor. The distance from an oceanic ridge axis to any magnetic anomaly indicates the width of new seafloor that formed during that time interval. For example, if the distance between the present-day Mid-Atlantic Ridge and anomaly 31 is 2,010 km, and anomaly 31 formed 67 million years ago (●**FIGURE 2.23**), then the average rate of movement during the past 67 million years has been 3 cm per year (2,010 km, which equals 201 million cm divided by 67 million years; 201,000,000 cm/67,000,000 years = 3 cm/year). Thus, for a given interval of time, the wider the strip of seafloor, the faster the plate has moved. In this way, not only can the present average rate of movement and relative motion be determined (Figure 2.16) but the average rate of movement in the past can also be calculated by dividing the distance between anomalies by the amount of time elapsed between anomalies.

Geologists use magnetic anomalies not only to calculate the average rate of plate movement but also to determine plate positions at various times in the past. Because magnetic anomalies are parallel and symmetric with respect to spreading ridges, all one must do to determine the position of continents when particular anomalies formed is to move the anomalies back to the spreading ridge, which will also move the continents with them (Figure 2.23). Because subduction destroys oceanic crust and the magnetic record that it carries, we have an excellent record of plate movements since the breakup of Pangaea, but not as good an understanding of plate movement before that time.

The average rate of movement, as well as the relative motion between any two plates, can also be determined by satellite-laser ranging techniques. Laser beams from a station on one plate are bounced off a satellite (in geosynchronous orbit) and returned to a station on a different plate. As the plates move away from each other, the laser beam takes more time to go from the sending station to the stationary satellite and back to the receiving station. This difference in elapsed time is used to calculate the rate of movement and the relative motion between plates.

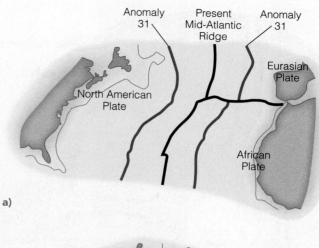

a)

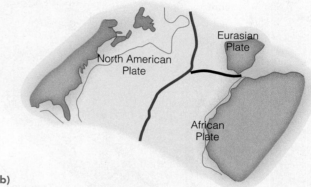

b)

●**FIGURE 2.23 Reconstructing Plate Positions Using Magnetic Anomalies** (a) The present North Atlantic, showing the Mid-Atlantic Ridge and magnetic anomaly 31, which formed 67 million years ago. (b) The Atlantic Ocean 67 million years ago. Anomaly 31 marks the plate boundary 67 million years ago. By moving the anomalies back together, along with the plates they are on, we can reconstruct the former positions of the continents.

# THE DRIVING MECHANISM OF PLATE TECTONICS

**LO11** Explain what geologists think is the driving mechanism of plate movement

A major obstacle to the acceptance of the continental drift hypothesis was the lack of a driving mechanism to explain continental movement. When it was shown that continents and ocean floors moved together, not separately, and that new crust is formed at spreading ridges by rising magma, most geologists accepted some type of convective heat system (convection cells) as the basic process responsible for plate motion.

Most of the heat from Earth's interior results from the decay of radioactive elements, such as uranium (see Chapter 16), in the core and lower mantle. The most efficient way for this heat to escape Earth's interior is through some type of slow convection system. Heat from the core, supplemented by heat generated from radioactive decay, thus drives large mantle convection cells (●**FIGURE 2.24**).

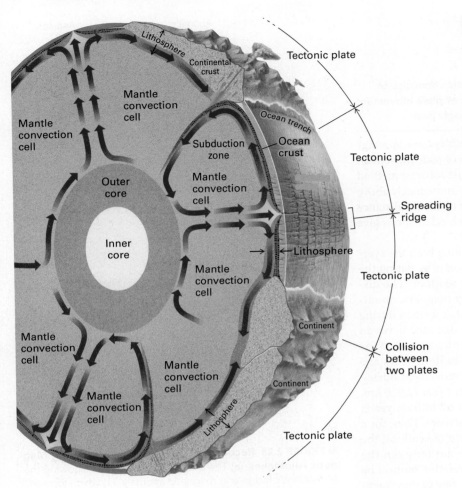

**● FIGURE 2.24 Thermal Convection Cells as the Driving Force of Plate Movement** A cutaway view of Earth shows that the lithosphere glides horizontally across the asthenosphere. Heat from the core, supplemented by heat produced from radioactive decay, drives huge mantle convection cells that move the lithosphere. Spreading ridges mark the location of ascending limbs of the convection cells, and oceanic trenches are the surface expression of subduction, where plates descend into Earth's interior.

**● FIGURE 2.25 Convection in a Pot of Stew** Heat from the stove is applied to the base of the stew pot, causing the stew to heat up. As heat rises through the stew, pieces of the stew are carried to the surface, where the heat is dissipated, the pieces of stew cool, and then sink back to the bottom of the pot. The bubbling seen at the surface of the stew is the result of convection cells churning the stew. In the same manner, heat from the decay of radioactive elements produces convection cells with Earth's interior.

In this manner, hot rock from the interior rises toward the surface, loses heat to the overlying lithosphere, becomes denser as it cools, and then sinks back into the interior, where it is heated and the process repeats itself. This type of convective heat system is analogous to a pot of stew cooking on a stove (●**FIGURE 2.25**).

In this mantle convection cell model, spreading ridges mark the ascending limbs of adjacent convection cells, and trenches are present where convection cells descend back into Earth's interior. The convection cells therefore determine the location of spreading ridges and trenches, with the lithosphere lying above the thermal convection cells. Thus, each plate corresponds to a single convection cell that moves as a result of the convective movement of the cell itself (Figure 2.24).

Although mantle convection is not completely understood, most geologists agree that some type of heat convective mechanism is involved in plate movement. The fact that plates have moved in the past, and are still moving today, has been proven beyond a doubt. Furthermore, continuing research into Earth's interior is providing new insights into explaining Earth's convective heat system.

# PLATE TECTONICS AND THE DISTRIBUTION OF NATURAL RESOURCES

LO12 Discuss and provide examples of how plate movement affects the formation and distribution of many natural resources

In addition to being responsible for the major features of Earth's crust and influencing the distribution and evolution of the world's biota, plate movement also affects the formation and distribution of some natural resources. Consequently, geologists are using plate tectonic theory in their search for petroleum and mineral deposits and in explaining the occurrence of these natural resources. It is becoming increasingly clear that if we are to keep up with the continuing demands of a global industrialized society, the application of plate tectonic theory to the origin and distribution of natural resources is essential.

## Petroleum

Although significant concentrations of petroleum occur in many areas of the world, 29% of all proven reserves are in the Persian Gulf region. It should therefore not be surprising that many of the conflicts and disputes in the Middle East have shared as their underlying cause the desire to control these vast amounts of petroleum. Most people, however, are not aware of *why* there is so much oil in this region of the world. The answer lies in the paleogeography and plate movement of this region during the Mesozoic and Cenozoic eras.

During the Mesozoic Era, and particularly the Cretaceous Period, when most of the petroleum formed, the Persian Gulf area was a broad marine shelf extending eastward from Africa. This passive continental margin lay near the equator, where countless microorganisms lived in the surface waters. The remains of these organisms accumulated with the bottom sediments and as they were covered and subsequently buried by continuing sedimentation, they began the complex process of petroleum generation and the formation of source beds in which petroleum forms.

During the Cenozoic Era, convergent plate movement in the Persian Gulf area resulted in the sediments of the passive continental margin being subducted, which generated heat that broke down the organic molecules, leading to the formation of petroleum. The continued subduction and collisions in the area folded the rocks, thus creating traps for petroleum to accumulate, so much so that this vast area is now a major oil-producing region. Elsewhere in the world, plate tectonics is also responsible for concentrations of petroleum.

## Mineral Deposits

Many metallic mineral deposits such as copper, gold, lead, silver, tin, and zinc are related to igneous and associated hydrothermal (hot water) activity (see GEO-FOCUS). So it is not surprising that a close relationship exists between plate boundaries and the occurrence of these valuable deposits.

The magma generated by partial melting of a subducting plate rises toward the surface, and as it cools, it precipitates and concentrates various metallic ores. Many of the world's major metallic ore deposits are associated with convergent plate boundaries, including those in the Andes of South America, the Coast Ranges and Rockies of North America, Japan, the Philippines, Russia, and a zone extending from the eastern Mediterranean region to Pakistan. In addition, the majority of the world's gold is associated with deposits located at ancient convergent plate boundaries in such areas as Canada, Alaska, California, Venezuela, Brazil, Russia, southern India, and western Australia.

The copper deposits of western North and South America are an excellent example of the relationship between convergent plate boundaries and the distribution, concentration, and exploitation of valuable metallic ores (●FIGURE 2.26A). The world's largest copper deposits are found along this belt. The majority of the copper deposits in the Andes and the southwestern United States were formed less than 60 million years ago when oceanic plates were subducted under the North and South American plates. The rising magma and associated hydrothermal fluids carried minute amounts of copper, which were originally widely disseminated but eventually became concentrated in the cracks and fractures of the surrounding andesites. These low-grade copper deposits contain from 0.2% to 2% copper and are extracted from large open-pit mines (●FIGURE 2.26B).

# GEO-FOCUS

## Molybdenum Mining and Economic, Environmental, and Political Concerns

Molybdenum is an element that is important in many industrial metallurgical applications. It has an extremely high melting point, making it useful as an alloy for high-strength and high-temperature steels, as well as in other products requiring resistance to very high temperatures. Like many metals, it is mined from porphyry-style deposits, which typically form in high-temperature igneous environments, such as those associated with convergent plate boundaries. China is the leading producer of

molybdenum, followed by the United States. Currently, the world's largest molybdenum development project is at Mount Hope, approximately 50 km northwest of Eureka, Nevada.

In 2012, General Moly, Inc., a Colorado-based mining company, received the necessary permits from the Bureau of Land Management and the Nevada Division of Environmental Protection to proceed with its Mount Hope Project. This joint venture (General Moly 80% interest), along with South Korea's POSXO (20% interest), the world's largest

steel producer, is the largest pure-play molybdenum development in the world, with proven and probable ore reserves of 1,050 million tons (2017). A conventional open pit method of mining will be utilized to extract the molybdenum ore.

Numerous economic, environmental, and political factors come into play when dealing with metallic ores critical to an industrial society. Discuss some of these factors that you think might have been involved in the decision to go ahead and exploit this important ore deposit.

The location and view of Mount Hope, near Eureka, Nevada, is the site of the world's largest molybdenum ore deposit.

Mount Lewis Field Offie/U.S. Department of the Interior
Bureau of Land Management

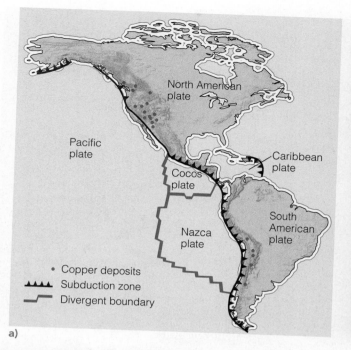

● **FIGURE 2.26 Copper Deposits and Convergent Plate Boundaries** (a) Valuable copper deposits are located along the west coasts of North and South America in association with convergent plate boundaries. Through time, the rising magma and associated hydrothermal activity resulting from subduction carried small amounts of copper that then became trapped and concentrated in the surrounding rocks. (b) Bingham Copper Mine, near Salt Lake City, Utah, is a huge open-pit copper mine with reserves estimated at 1.7 billion tons. More than 400,000 tons of rock are removed for processing each day. Note the small specks toward the middle of the photograph that are the four-meter-high dump trucks!

# PLATE TECTONICS AND THE DISTRIBUTION OF LIFE

**LO13 Illustrate and explain with examples, the role of plate tectonics in the distribution of Earth's biota**

Plate tectonic theory is as revolutionary and far-reaching in its implications for geology as the theory of evolution was for biology when it was proposed. Interestingly, it was the fossil evidence that convinced Wegener, Suess, and du Toit, as well as many other geologists, of the correctness of the theory of continental drift. Together, the theories of plate tectonics and evolution have changed the way we view our planet, and we should not be surprised at the intimate association between them. Although the relationship between plate tectonic processes and the evolution of life is incredibly complex, paleontologic data provide convincing evidence of the influence of plate movement on the distribution of organisms.

The present distribution of plants and animals is not random, but is controlled mostly by climate and geographic barriers. The world's biota occupy *biotic provinces*, which are regions characterized by a distinctive assemblage of plants and animals. Organisms within a province have similar ecological requirements, and the boundaries separating provinces are therefore natural ecological breaks. Climatic or

geographic barriers are the most common province boundaries, and these are mostly controlled by plate movement.

The complex interaction between wind and ocean currents has a strong influence on the world's climates. Wind and ocean currents, in turn, are strongly influenced by the number, distribution, topography, and orientation of continents. The distribution of continents and ocean basins not only influences wind and ocean currents but also affects provinciality by creating physical barriers to, or pathways for, the migration of organisms. Intraplate volcanoes, island arcs, mid-oceanic ridges, mountain ranges, and subduction zones all result from the interaction of plates, and their orientation and distribution strongly influence the number of provinces and hence total global diversity. Thus, provinciality and diversity will be highest where numerous small continents are spread across many zones of latitude.

When a geographic barrier separates a once-uniform fauna, species must adapt to the new conditions, migrate, or become extinct. Adaptation to the new environment by various species may involve enough change that new species eventually evolve. The formation of the Isthmus of Panama provides an excellent example of how plate tectonics influences evolution. Prior to the rise of this connection between North and South America, a homogeneous population of bottom-dwelling invertebrates inhabited the shallow seas of the area. After the formation of the Isthmus of Panama by subduction of the Pacific plate approximately 5 million years

● **FIGURE 2.27 Plate Tectonics and the Distribution of Organisms** The Isthmus of Panama forms a barrier that divides a once-uniform fauna of molluscs that inhabited the shallow seas of both the Pacific Ocean and Caribbean Sea. Its creation also formed a land corridor in which migration between the two continents took place. Prior to the formation of this isthmus, South America was isolated from all other landmasses during much of the Cenozoic, and its mammalian fauna consisted of marsupials (pouched mammals) and placentals that lived nowhere else. When the Isthmus of Panama formed during the Late Pliocene, many placental mammals migrated south, resulting in numerous South American mammals becoming extinct. A few South American mammals migrated north and successfully occupied North America.

**Critical Thinking Question** Why is the mammalian fauna of Australia so different from elsewhere?

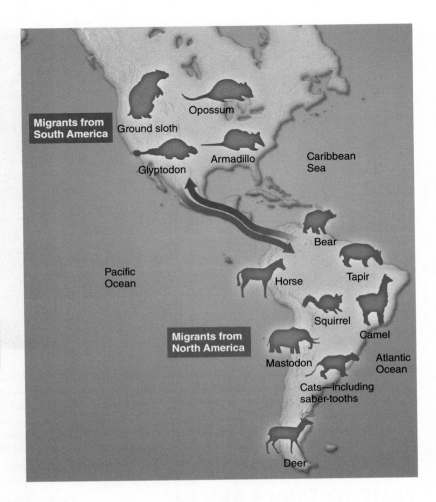

ago, the original population was divided. In response to the changing environment, new species evolved on opposite sides of the Isthmus.

The formation of the Isthmus of Panama also influenced the evolution of the North and South American mammalian faunas (●**FIGURE 2.27**). During most of the Cenozoic Era,

South America was an island continent, and its mammalian fauna evolved in isolation. When North and South America were connected by the Isthmus, most of the indigenous South American mammals were replaced by migrants from North America. Yet surprisingly, only a few South American mammal groups migrated northward.

# Key Concepts Review

- The concept of continental movement is not new. The earliest maps showing the similarity between the east coast of South America and the west coast of Africa provided the first evidence that continents may once have been united and subsequently separated from each other.

- Alfred Wegener is generally credited with developing the hypothesis of continental drift. He provided abundant geologic and paleontologic evidence to show that the continents were once united in one supercontinent,

which he named Pangaea. Unfortunately, Wegener could not explain how the continents moved, and most geologists ignored his ideas.

- Various features of the continental margins and the sea floor are a reflection of plate movement. Continental margins are active or passive, depending on their relationship to plate boundaries. Oceanic trenches are long, steep-sided depressions on the sea floor near convergent plate boundaries where oceanic lithosphere is consumed by subduction. Submarine hydrothermal vents are found at or near spreading ridges and are associated with divergent plate boundaries.

- The hypothesis of continental drift was revived during the 1950s when paleomagnetic studies of rocks indicated the presence of multiple magnetic north poles instead of just one as there is today. This paradox was resolved by moving the continents so that the paleomagnetic data became consistent with a single magnetic north pole. When this was done, the rock sequences, glacial deposits and striations, and fossil distributions aligned with the reconstructed paleogeography.
- Seafloor spreading was confirmed by the discovery of magnetic anomalies in the oceanic crust that were both parallel to and symmetric around oceanic ridges, indicating that new oceanic crust must have formed as the sea floor was spreading. The pattern of oceanic magnetic anomalies matched the pattern of magnetic reversals already known from continental lava flows and showed that Earth's magnetic field has reversed itself numerous times during the past.
- Radiometric dating reveals that the oldest oceanic crust is less than 180 million years old, whereas the oldest continental crust is approximately 4 billion years old. Fossil evidence and the thickness of sediments overlying the oceanic crust further support and confirm that ocean basins are recent geologic features.
- Plate tectonic theory became widely accepted by the 1970s because the evidence overwhelmingly supports it and because it provides geologists with a powerful theory for explaining such phenomena as volcanism, earthquake activity, mountain building, global climatic changes, and the distribution of many mineral resources as well as the world's.
- Geologists recognize three types of plate boundaries: divergent boundaries, where plates move away from each other; convergent boundaries, where two plates collide; and transform boundaries, where two plates slide past each other.
- Ancient plate boundaries can be recognized by their associated rock assemblages and geologic structures. For divergent boundaries, these may include rift valleys with thick sedimentary sequences and numerous dikes and sills. For convergent boundaries, ophiolites and andesitic rocks are two characteristic features. Transform plate boundaries generally do not leave any characteristic or diagnostic features in the geologic record.
- The average rate of movement and relative motion of the plates can be calculated in several ways. The results of these different methods all agree and indicate that the plates move at different average velocities.
- The absolute motion of plates can be determined by the movement of plates over mantle plumes. A mantle plume is a seemingly stationary column of magma that rises to the surface from deep within the mantle and either forms a subsurface mushroom-shaped plume head or erupts at the surface as a volcano.
- Although a comprehensive theory of plate movement has yet to be developed, geologists think that some type of convective system is involved in plate movement.
- A close relationship exists between the formation of petroleum, as well as some mineral deposits, and plate boundaries. Furthermore, the formation and distribution of many natural resources are related to plate movements.
- The relationship between plate tectonic processes and the evolution of life is complex. The distribution of plants and animals is not random, but rather is controlled mostly by climate and geographic barriers, which, in turn, are influenced, to a great extent, by the movement of plates.

# Important Terms

abyssal plain   28

active continental margin   31

aseismic ridge   29

continental–continental plate boundary   41

continental drift   24

continental margin   27

continental rise   28

continental shelf   28

continental slope   28

convergent plate boundary   37

Curie point   33

divergent plate boundary   36

*Glossopteris* flora   23

Gondwana   23

guyots   29

hot spot   42

Laurasia   24

magnetic anomaly   35

magnetic field   31

magnetic reversal   33

magnetism   31

oceanic–continental plate boundary   39

oceanic–oceanic plate boundary   37

oceanic ridge   28

oceanic trench   29

paleomagnetism   32

Pangaea   24

passive continental margin   31

plate tectonic theory   35

seafloor spreading   33

seamount   29

submarine hydrothermal vent   28

thermal convection cell   34

transform fault   41

transform plate boundary   41

# Review Questions

1. Magnetic surveys of the ocean basins indicate that
   a. ___ the oceanic crust is youngest adjacent to mid-oceanic ridges.
   b. ___ the oceanic crust is oldest adjacent to mid-oceanic ridges.
   c. ___ the oceanic crust is youngest adjacent to the continents.
   d. ___ the oceanic crust is the same age everywhere.
   e. ___ answers b and c.

2. The most common biotic province boundaries are
   a. ___ geographic barriers.
   b. ___ biologic barriers.
   c. ___ climatic barriers.
   d. ___ answers a and b.
   e. ___ answers a and c.

3. The man credited with developing the continental drift hypothesis is
   a. ___ Wilson.
   b. ___ Wegener.
   c. ___ Hess.
   d. ___ du Toit.
   e. ___ Vine.

4. The driving mechanism of plate movement is thought to be largely the result of
   a. ___ isostasy.
   b. ___ Earth's rotation.
   c. ___ thermal convection cells.
   d. ___ magnetism.
   e. ___ polar wandering.

5. Along what type of plate boundary does subduction occur?
   a. ___ divergent.      b. ___ transform.
   c. ___ convergent.     d. ___ answers a and b.
   e. ___ answers a and c.

6. What evidence convinced Wegener and others that continents must have moved in the past and at one time formed a supercontinent?

7. Why is some type of thermal convection mechanism thought to be the major force driving plate movement?

8. In addition to the volcanic eruptions and earthquakes associated with convergent and divergent plate boundaries, why are these boundaries also associated with the formation and accumulation of various metallic ore deposits?

9. Plate tectonic theory builds on the continental drift hypothesis and the theory of sea floor spreading. As such, it is a unifying theory of geology. Explain why it is a unifying theory.

# Creative Thinking Visual Question

Using the ages (the numbers represent ages in millions of years) for each of the Hawaiian Islands, as well as the scale given in the figure, calculate the average rate of movement per year for the Pacific plate since each island formed (●**FIGURE 1**). Is the average rate of movement the same for each island? Would you expect it to be? Explain why it may not be and why there are different ages for some of the islands.

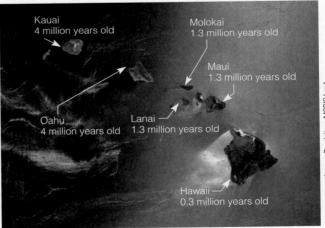

Image courtesy Jacques Descloitres, MODIS Land Rapid Response Team at NASA GSFC

●**FIGURE 1  Hawaiian Islands** Image of the Hawaiian islands with the age of each island in millions of years.

Amethyst crystals from
Banska Stiavnica, Slovakia.

3

# MINERALS
## The Building Blocks of Rocks

# INTRODUCTION

A **mineral** is a naturally occurring, inorganic, crystalline solid, with a narrowly defined chemical composition, and characteristic physical properties (●FIGURE 3.1A). Although lengthy, the only part of this definition that may be unfamiliar is "crystalline solid," which means that a mineral's constituent atoms are arranged in a specific three-dimensional framework. For example, ice is a mineral because it is inorganic, that is, it is not composed of the complex carbon-based molecules of organisms, and it is certainly naturally occurring. Chemically, it consists of hydrogen and oxygen atoms ($H_2O$), and it has distinguishing features such as color, hardness, and density. And because its hydrogen and oxygen atoms are arranged in an orderly fashion, as opposed to the atoms in glass, which have no such orderly arrangement, ice is, by definition, a mineral.

We cannot overstate the essential role that minerals play in all societies, especially in industrial ones. Our economic well-being requires finding and extracting ores such as iron, copper, aluminum, and tin, as well as other minerals and rocks needed in an industrialized economy, such as, for example, glass, fertilizer, computers, and cell phones. Additionally, we depend on energy resources such as petroleum, natural gas, coal, and uranium. The fact that the distribution of natural resources is uneven, and no country is self-sufficient in all resources, accounts for global economic ties between nations and has important implication in foreign relations.

One of the many reasons to study minerals is their economic importance, but for students, it is important to know that minerals are essential constituents of *rocks* (see Figures 1.14 and 1.15). Granite, for example, is made up of specific percentages of the minerals quartz, potassium feldspars, and plagioclase feldspars, plus a few others in minor amounts (●FIGURE 3.1B). Some minerals are attractive and eagerly sought by collectors and for museum displays, whereas others, known as *gemstones*—that is, precious and semiprecious minerals and rocks—are used for decorative purposes, especially jewelry.

From our discussion so far, we have a formal definition of the term *mineral*, and we know that minerals are the basic constituents of rocks. Now let's delve deeper into what minerals are made of by considering matter, atoms, elements, and bonding.

---

**mineral** A naturally occurring, inorganic, crystalline solid that has characteristic physical properties and a narrowly defined chemical composition.

a)

b)

● **FIGURE 3.1 Minerals and Rocks** (a) A spectacular specimen of tourmaline (the elongate minerals) and quartz (colorless) from the Himalaya Mine in San Diego County, California. Notice the change in color along the lengths of the tourmaline crystals. (b) The igneous rock granite (see Chapter 4) is made up of mostly three minerals—quartz, potassium feldspar, and plagioclase feldspar—but it may also contain small amounts of biotite, muscovite, and hornblende.

**Critical Thinking Question** You must explain to an interested audience the distinction between minerals and rocks. How would you do so, and can you think of any analogies that might clarify the points that you make?

# MATTER: WHAT IS IT?

**LO1**  Define matter, chemical elements, atoms, isotopes, and compounds

**LO2**  Explain what the atomic number and atomic mass number of an atom means

**LO3**  Define the two major types of chemical bonding and give an example of each type of bonding

Anything that has mass and occupies space is **matter** and accordingly includes water, plants, all organisms, the atmosphere, as well as minerals and rocks. Physicists recognize four states, or phases, of matter: *liquids, gases, solids,* and *plasma* (composed of ionized gases, as in the Sun and stars). Liquids and gases are important in our understanding of several surface processes such as running water and wind, but here our main concern is solids because, by definition, minerals are solids.

## Atoms and Elements

Matter is made up of chemical **elements**, which, in turn, are composed of **atoms**, the smallest units of matter that retain the characteristics of a particular element (●**FIGURE 3.2**). That is, elements cannot be changed into different substances except through radioactive decay (discussed in Chapter 16). Thus, an element is made up of atoms, all of which have the same properties. Scientists have discovered 92 naturally occurring elements, some of which are listed in Figure 3.2. All naturally occurring elements have a name and a symbol—for example, oxygen (O), aluminum (Al), and potassium (K).

At the center of an atom is a **nucleus** composed of one or more **protons**, which have a positive electrical charge, and **neutrons**, which are electrically neutral. The nucleus is only about 1/100,000th of the diameter of an atom, yet it contains virtually all of the atom's mass. **Electrons**, particles with a negative electrical charge, orbit rapidly around the nucleus at specific distances in one or more **electron shells** (Figure 3.2). The electrons control how an atom interacts with other atoms, but the nucleus determines how many electrons an atom has, because the positively charged protons attract and hold the negatively charged electrons in their orbits.

The number of protons in an atom's nucleus determines an atom's identity and its **atomic number**. Hydrogen (H), for instance, has one proton in its nucleus and thus has an atomic number of 1. The nuclei of helium (He) atoms possess 2 protons, whereas those of carbon (C) have 6, and uranium (U) 92, so their atomic numbers are 2, 6, and 92, respectively. Atoms also have an **atomic mass number** which is the sum of protons and neutrons in the nucleus (electrons

---

**matter**  Anything that has mass and occupies space.

**element**  A substance composed of atoms that all have the same properties; atoms of one element can change to atoms of another element by radioactive decay, but otherwise they cannot be changed by ordinary chemical means.

**atom**  The smallest unit of matter that retains the characteristics of an element.

**nucleus**  The central part of an atom, consisting of protons and neutrons.

**proton**  A positively charged particle found in the nucleus of an atom.

**neutron**  An electrically neutral particle found in the nucleus of an atom.

**electron**  A negatively charged particle of very little mass that encircles the nucleus of an atom.

**electron shell**  Electrons orbit an atom's nucleus at specific distances in electron shells.

**atomic number**  The number of protons in the nucleus of an atom.

**atomic mass number**  The number of protons plus neutrons in the nucleus of an atom.

---

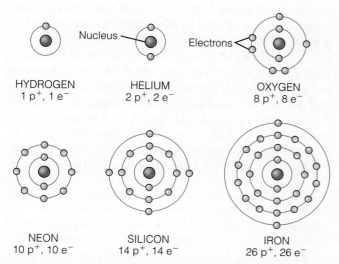

HYDROGEN
1 p⁺, 1 e⁻

HELIUM
2 p⁺, 2 e⁻

OXYGEN
8 p⁺, 8 e⁻

NEON
10 p⁺, 10 e⁻

SILICON
14 p⁺, 14 e⁻

IRON
26 p⁺, 26 e⁻

| Element | Symbol | Atomic Number | Distribution of Electrons | | | |
|---|---|---|---|---|---|---|
| | | | First Shell | Second Shell | Third Shell | Fourth Shell |
| Hydrogen | H | 1 | 1 | — | — | — |
| Helium | He | 2 | 2 | — | — | — |
| Carbon | C | 6 | 2 | 4 | — | — |
| Oxygen | O | 8 | 2 | 6 | — | — |
| Neon | Ne | 10 | 2 | 8 | — | — |
| Sodium | Na | 11 | 2 | 8 | 1 | — |
| Magnesium | Mg | 12 | 2 | 8 | 2 | — |
| Aluminum | Al | 13 | 2 | 8 | 3 | — |
| Silicon | Si | 14 | 2 | 8 | 4 | — |
| Phosphorus | P | 15 | 2 | 8 | 5 | — |
| Sulfur | S | 16 | 2 | 8 | 6 | — |
| Chlorine | Cl | 17 | 2 | 8 | 7 | — |
| Potassium | K | 19 | 2 | 8 | 8 | 1 |
| Calcium | Ca | 20 | 2 | 8 | 8 | 2 |
| Iron | Fe | 26 | 2 | 8 | 14 | 2 |

●**FIGURE 3.2 Shell Models for Common Atoms** The shell model for several atoms and their electron configurations. A blue circle represents the nucleus of each atom, but remember that atomic nuclei are composed of protons and neutrons, as shown in Figure 3.3.

● **FIGURE 3.3 Carbon Isotopes** Schematic representation of the isotopes of carbon. Carbon has an atomic number of 6 and an atomic mass number of 12, 13, or 14, depending on the number of neutrons in its nucleus.

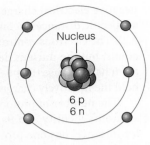

$^{12}$C (Carbon 12)

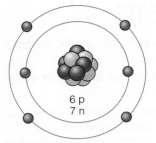

$^{13}$C (Carbon 13)

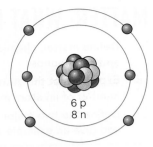

$^{14}$C (Carbon 14)

contribute negligible mass to atoms). However, atoms of the same chemical element might have different atomic mass numbers, because the number of neutrons can vary. All carbon (C) atoms have six protons—otherwise they would not be carbon—but the number of neutrons is 6, 7, or 8. Thus we recognize three types of carbon, or what are known as *isotopes* (●FIGURE 3.3), each with a different atomic mass number.

The isotopes of carbon, or those of any other element, behave the same chemically; carbon 12 and carbon 14 are both present in carbon dioxide ($CO_2$), for example. However, some isotopes are radioactive, meaning that they spontaneously decay or change to other elements. Carbon 14 is radioactive, whereas both carbon 12 and carbon 13 are not. Radioactive isotopes are important for determining the numeric ages of rocks (see Chapter 16).

## Bonding and Compounds

Interactions between electrons around atoms can result in two or more atoms joining together, a process known as **bonding**. If atoms of two or more elements bond, the resulting substance is a **compound**. Gaseous oxygen consists of only oxygen atoms, and is thus an element, whereas the mineral quartz, consisting of silicon and oxygen atoms, is a compound. Most minerals are compounds, although gold, platinum, and several others are important exceptions.

To understand bonding, it is necessary to delve deeper into the structure of atoms. Recall that negatively charged electrons orbit the nuclei of atoms in electron shells. With the exception of hydrogen, which has only one proton and one electron, the innermost electron shell of an atom contains only two electrons. The other shells contain various numbers of electrons, but the outermost shell never has more than eight (Figure 3.2). The electrons in the outermost shell are usually involved in chemical bonding.

*Ionic* and *covalent bonding* are important in minerals, and many minerals contain both types of bonds. Two other types of bonds, *metallic* and *van der Waals*, are much less common, but are important in determining the properties of some useful minerals.

**IONIC BONDING** Most atoms have fewer than eight electrons in their outermost electron shell, but some, including

neon and argon, have complete outer shells with eight electrons (Figure 3.2). Because of this electron configuration, these elements, known as the *noble gases*, do not react readily with other elements to form compounds. Interactions between atoms tend to produce electron configurations similar to those of the noble gases. That is, atoms interact so that their outermost electron shell is filled with eight electrons, unless the first shell (with two electrons) is also the outermost electron shell, as in helium.

One way that the noble gas configuration is attained is by the transfer of one or more electrons from one atom to another. Common salt is composed of the elements sodium (Na) and chlorine (Cl); each element is poisonous, but when combined chemically, they form the mineral halite, a compound of sodium chloride (NaCl). Sodium has 11 protons and 11 electrons; thus the positive electrical charges of the protons are exactly balanced by the negative charges of the electrons, and the atom is electrically neutral (●FIGURE 3.4A). Likewise, chlorine, with 17 protons and 17 electrons, is electrically neutral. However, neither sodium nor chlorine has eight electrons in its outermost electron shell; sodium has only one, whereas chlorine has seven. To attain a stable configuration, sodium loses the electron in its outermost electron shell, leaving its next shell with eight electrons as the outermost one (Figure 3.4a). Sodium now has one less electron (negative charge) than it has protons (positive charge), so it is an electrically charged **ion** and is symbolized $Na^+$.

The electron lost by sodium is transferred to the outermost electron shell of chlorine, which had seven electrons to begin with. The addition of one more electron gives chlorine an outermost electron shell of eight electrons, the configuration of a noble gas. But its total number of electrons is now 18, which exceeds the number of protons by 1. Accordingly, chlorine also becomes an ion, but it is negatively charged ($Cl^-$). An **ionic bond** forms between sodium

---

**bonding**  The process whereby atoms join to other atoms.

**compound**  Any substance resulting from the bonding of two or more different elements (e.g., water, $H_2O$, and quartz, $SiO_2$).

**ion**  An electrically charged atom produced by adding or removing electrons from the outermost electron shell.

**ionic bond**  A chemical bond resulting from the attraction between positively and negatively charged ions.

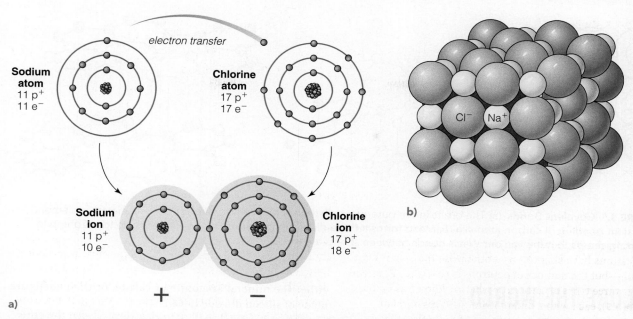

electron transfer

Sodium
atom
11 p⁺
11 e⁻

Chlorine
atom
17 p⁺
17 e⁻

Sodium
ion
11 p⁺
10 e⁻

Chlorine
ion
17 p⁺
18 e⁻

Cl⁻  Na⁺

+  −

a)

b)

● **FIGURE 3.4 Ionic Bond to Form the Mineral Halite (NaCl)** (a) Transfer of the electron in the outermost shell of sodium to the outermost shell of chlorine. After electron transfer, the sodium and chlorine atoms are positively and negatively charged ions, respectively. (b) This diagram shows the relative sizes of the sodium and chlorine atoms and their locations in a crystal of halite.

and chlorine because of the attractive force between the positively charged sodium ion and the negatively charged chlorine ion (Figure 3.4a).

In ionic compounds, such as sodium chloride (the mineral halite, also known as salt), the ions are arranged in a three-dimensional framework that results in overall electrical neutrality. In halite, sodium ions are bonded to chlorine ions on all sides, and chlorine ions are surrounded by sodium ions (●**FIGURE 3.4B**).

**COVALENT BONDING  Covalent bonds** form between atoms when their electron shells overlap and they share electrons. For example, atoms of the same element, such as carbon, cannot bond by transferring electrons from one atom to another. Carbon (C), which forms the minerals graphite and diamond, has four electrons in its outermost electron shell (●**FIGURE 3.5A**). If these four electrons were transferred to another carbon atom, the atom receiving the electrons would have the noble gas configuration of eight electrons in its outermost electron shell, but the atom contributing the electrons would not.

In such situations, adjacent atoms share electrons by overlapping their electron shells. A carbon atom in diamond shares all four of its outermost electrons with a neighbor to produce a stable noble gas configuration (Figure 3.5a). Diamond and graphite, both crystalline forms of carbon, have covalent bonds, but in diamond the atoms form a three-dimensional framework, whereas in graphite, the carbon atoms form sheets that are weakly bonded together (●**FIGURE 3.5B** and **C**).

It should be noted that covalent bonds are not restricted to substances composed of atoms of a single kind. Among the most common minerals, the silicates (discussed later in this chapter), the element silicon forms partly covalent and partly ionic bonds with oxygen.

**METALLIC AND VAN DER WAALS BONDS** *Metallic bonding* involves an extreme type of electron sharing. The electrons of the outermost electron shell of metals such as gold, silver, and copper readily move from one atom to another. This electron mobility accounts for the fact that metals have a metallic luster (their appearance in reflected light), provide good electrical and thermal conductivity, and can easily be reshaped. Only a few minerals possess metallic bonds, but those that do are very useful; copper, for example, is used for electrical wiring because of its high electrical conductivity.

Some electrically neutral atoms and molecules* have no electrons available for ionic, covalent, or metallic bonding. They nevertheless have a weak attractive force between them, called a *van der Waals* or *residual bond*, when in proximity. The carbon atoms in the mineral graphite are covalently bonded to form sheets, but the sheets are weakly held together by van der Waals bonds (Figure 3.5c). This type of bonding makes graphite useful for pencil lead; when a pencil point is moved across a piece of paper, small pieces of graphite flake off along the planes held together by van der Waals bonds and adhere to the paper.

---

**covalent bond** A chemical bond formed by the sharing of electrons between atoms.

*A molecule is the smallest unit of a substance that has the properties of that substance. A water molecule ($H_2O$), for example, possesses two hydrogen atoms and one oxygen atom.

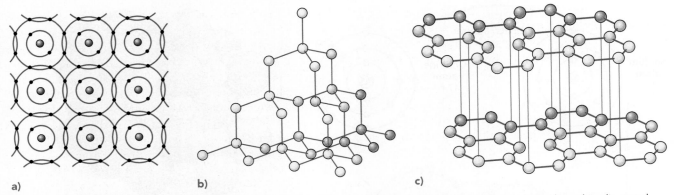

● **FIGURE 3.5  Covalent Bonds** (a) The orbits in the outermost electron shell overlap, so electrons are shared in diamond. (b) Covalent bonding of carbon atoms in diamond forms a three-dimensional framework. (c) Covalent bonds in graphite form strong sheets, but the van der Waals bonds between sheets are weak.

# EXPLORE THE WORLD OF MINERALS

**LO4**   **Define a mineral**

**LO5**   **Name and discuss each part of the definition of a mineral**

We defined a *mineral* as a naturally occurring, inorganic crystalline solid with a narrowly defined chemical composition and characteristic physical properties. In the following sections, we examine each part of the definition of the term mineral.

## Naturally Occurring Inorganic Substances

The criterion *naturally occurring* excludes from minerals all substances manufactured by humans, such as synthetic diamonds and rubies. Some geologists think the term *inorganic* in the mineral definition is unnecessary, but it does remind us that animal matter and vegetable matter are not minerals. Nevertheless, some organisms, including corals, clams, and several other animals and some plants, construct their shells of calcium carbonate ($CaCO_3$), which is either the mineral aragonite or calcite, or their shells are made of silicon dioxide ($SiO_2$).

## Mineral Crystals

By definition, minerals are **crystalline solids**, in which the constituent atoms are arranged in a regular, three-dimensional framework (Figure 3.5b). Under ideal conditions, crystalline solids grow and form perfect **crystals** that possess planar surfaces (crystal faces), sharp corners, and straight edges (●**FIGURE 3.6**). In other words, the regular geometric shape of a mineral crystal is the exterior manifestation of an ordered internal atomic arrangement. Not all rigid substances are crystalline solids; natural and manufactured glass lack the ordered arrangement of atoms and are said to be *amorphous*, meaning "without form." Minerals are crystalline solids, but crystalline solids are not always well-formed crystals. The reason is that when crystals form, they may grow in proximity and form an interlocking mosaic in which individual crystals are not apparent or easily discerned.

**crystalline solid**  A solid in which the constituent atoms are arranged in a regular, three-dimensional framework.

**crystal**  A naturally occurring solid of an element or compound with a specific internal structure that is manifested externally by planar faces, sharp corners, and straight edges.

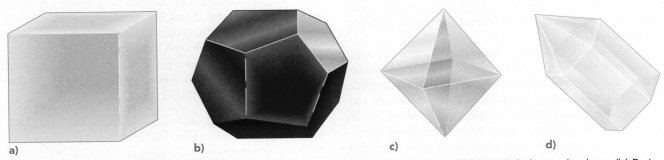

● **FIGURE 3.6  A Variety of Mineral Crystal Shapes** (a) Cubic crystals are typical of the minerals halite and galena. (b) Pyritohedron crystals such as those of pyrite have 12 pentagonal faces. (c) Diamond has octahedral, or eight-sided, crystals. (d) A prism terminated by a pyramid is found in quartz.

So how do we know that minerals with no obvious crystals are actually crystalline? X-ray beams and light transmitted through mineral crystals behave in a predictable manner, which provides compelling evidence for an orderly internal structure. Another way to determine that minerals are actually crystalline is by their *cleavage*, the property of breaking or splitting repeatedly along smooth, closely spaced planes. Not all minerals have cleavage planes, but many do, and such regularity certainly indicates that cleavage is controlled by internal structure.

As early as 1669, the Danish scientist Nicholas Steno determined that the angles of intersection of equivalent crystal faces on different specimens of quartz are identical. Since then, this *constancy of interfacial angles* has been demonstrated for many other minerals, regardless of their size, shape, age, or geographic occurrence. Steno postulated that mineral crystals are made up of very small, identical building blocks, and that the arrangement of these building blocks determines the external form of mineral crystals, a proposal that has since been verified.

## Chemical Composition of Minerals

A mineral's chemical composition is shown by a formula, which is a shorthand way of indicating the numbers of atoms of different elements in a mineral. Quartz, for example, consists of one silicon (Si) atom for every two oxygen (O) atoms and thus has the formula $SiO_2$; the subscript number indicates the number of atoms. Orthoclase, for example, is composed of one potassium, one aluminum, three silicon, and eight oxygen atoms, so its formula is $KAlSi_3O_8$. Some minerals known as *native elements* consist of a single element and include silver (Ag), platinum (Pt), gold (Au), and graphite and diamond, both of which are composed of carbon (C).

For many minerals, the chemical composition does not vary. Quartz is composed of only silicon and oxygen ($SiO_2$), and halite contains only sodium and chlorine (NaCl). However, some minerals have a range of compositions because one element can substitute for another if the atoms of two or more elements are nearly the same size and the same charge. Thus, the definition of the term mineral has the phrase *narrowly defined chemical composition*. Iron and magnesium meet these criteria and can substitute for one another (●**FIGURE 3.7**), so that the chemical formula for olivine is $(Mg, Fe)_2SiO_4$, meaning that in addition to silicon and oxygen, it may contain magnesium, iron, or both. A number of other minerals also have ranges of composition, so they are actually mineral groups with several members.

## Physical Properties of Minerals

The last criterion in our definition of a mineral, *characteristic physical properties*, refers to such properties as hardness, color, and crystal form. These properties are controlled by composition and structure. The physical properties of minerals are further discussed later in this chapter.

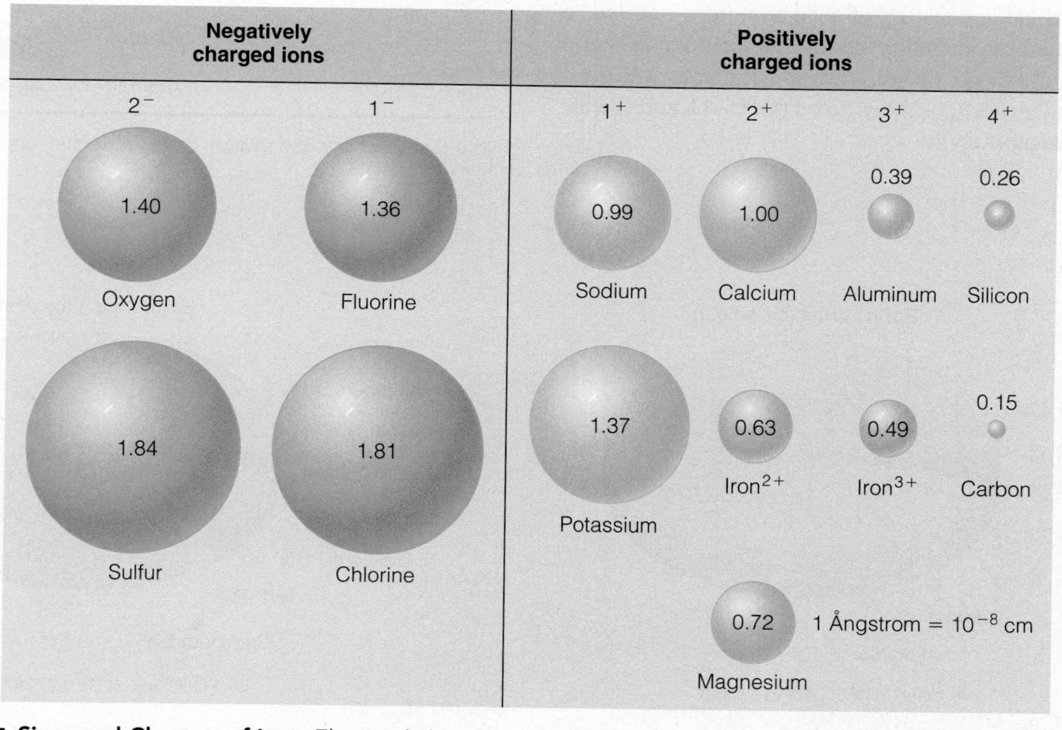

● **FIGURE 3.7 Sizes and Charges of Ions** Electrical charges and relative sizes of ions common in minerals. The numbers within the ions are the radii shown in Ångstrom units.

# MINERAL GROUPS RECOGNIZED BY GEOLOGISTS

**LO6**  Define the basic building block of silicate minerals

**LO7**  Define the four structures that silicate tetrahedra can form

**LO8**  Explain how the two subgroups of silicates differ from each other

**LO9**  Define the carbonate mineral group

**LO10**  Provide examples of other mineral groups

Geologists have identified and described more than 5,000 minerals, but only a few—perhaps two dozen—are common. One might think that an extremely large number of minerals could form from 92 naturally occurring elements; however, the bulk of Earth's crust is made up of only eight chemical elements, and even among these, silicon and oxygen are by far the most abundant. In fact, the most common minerals in Earth's crust consist of silicon, oxygen, and one or more of the elements in ●**FIGURE 3.8**.

Geologists recognize mineral classes or groups, each with members that share the same negatively charged ion or ion group (●**TABLE 3.1**). We have mentioned that ions are atoms that have either a positive or negative electrical charge resulting from the loss or gain of electrons in their outermost shell. In addition to ions, some minerals contain tightly bonded, complex groups of different atoms known as *radicals* that act as single units. A good example is the carbonate radical, consisting of a carbon atom bonded to three oxygen atoms and thus having the formula $CO_3$ and a −2 electrical charge. Other common radicals and their charges are sulfate ($SO_4$, −2), hydroxyl (OH, −1), and silicate ($SiO_4$, −4) (●**FIGURE 3.9**).

**TABLE 3.1** Mineral Groups Recognized by Geologists

| Mineral Group | Negatively Charged Ion or Radical | Examples | Composition |
|---|---|---|---|
| Carbonate | $(CO_3)^{-2}$ | Calcite | $CaCO_3$ |
|  |  | Dolomite | $CaMg(CO_3)_2$ |
| Halide | $Cl^{-1}$, $F^{-1}$ | Halite | $NaCl$ |
|  |  | Fluorite | $CaF_2$ |
| Hydroxide | $(OH)^{-1}$ | Brucite | $Mg(OH)_2$ |
| Native element | — | Gold | $Au$ |
|  |  | Silver | $Ag^*$ |
|  |  | Diamond | $C$ |
| Phosphate | $(PO_4)^{-3}$ | Apatite | $Ca_5(PO_4)_3(F, Cl)$ |
| Oxide | $O^{-2}$ | Hematite | $Fe_2O_3$ |
|  |  | Magnetite | $Fe_3O_4$ |
| Silicate | $(SiO_4)^{-4}$ | Quartz | $SiO_2$ |
|  |  | Potassium feldspar | $KAlSi_3O_8$ |
|  |  | Olivine | $(Mg, Fe)_2SiO_4$ |
| Sulfate | $(SO_4)^{-2}$ | Anhydrite | $CaSO_4$ |
|  |  | Gypsum | $CaSO_4 \cdot 2H_2O$ |
| Sulfide | $S^{-2}$ | Galena | $PbS$ |
|  |  | Pyrite | $FeS_2$ |
|  |  | Argentite | $Ag_2S^*$ |

*Note that silver is found as both a native element and a sulfide mineral.

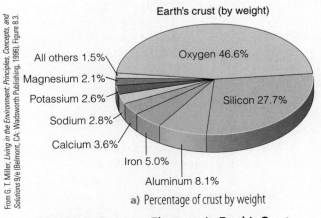

From G. T. Miller, *Living in the Environment: Principles, Concepts, and Solutions* 9/e (Belmont, CA: Wadsworth Publishing, 1996), Figure 8.3.

Earth's crust (by weight)

All others 1.5%
Magnesium 2.1%
Potassium 2.6%
Sodium 2.8%
Calcium 3.6%
Iron 5.0%
Aluminum 8.1%
Oxygen 46.6%
Silicon 27.7%

**a)** Percentage of crust by weight

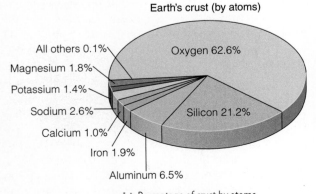

Earth's crust (by atoms)

All others 0.1%
Magnesium 1.8%
Potassium 1.4%
Sodium 2.6%
Calcium 1.0%
Iron 1.9%
Aluminum 6.5%
Oxygen 62.6%
Silicon 21.2%

**b)** Percentage of crust by atoms

●**FIGURE 3.8 Common Elements in Earth's Crust**

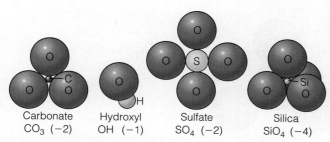

Carbonate
$CO_3$ (−2)

Hydroxyl
OH (−1)

Sulfate
$SO_4$ (−2)

Silica
$SiO_4$ (−4)

● **FIGURE 3.9 Radicals** Many minerals contain radicals, which are complex groups of atoms tightly bonded together. The silica and carbonate radicals are particularly common in many minerals, such as quartz ($SiO_2$) and calcite ($CaCO_3$).

## Silicate Minerals

Because silicon and oxygen are the two most abundant elements in Earth's crust, it is not surprising that many minerals contain these elements. A combination of silicon and oxygen is known as **silica**, and minerals that contain silica are **silicates**. Quartz ($SiO_2$) is pure silica, because it is composed entirely of silicon and oxygen. But most silicates have one or more additional elements, as in orthoclase ($KAlSi_3O_8$) and olivine [$(Mg,Fe)_2SiO_4$]. Silicate minerals include about one-third of all known minerals, but their abundance is even more impressive when one considers that they make up perhaps 95% of Earth's crust.

The basic building block of all silicate minerals is the **silica tetrahedron**, consisting of one silicon atom and four oxygen atoms (●**FIGURE 3.10A**). These atoms are arranged so that the four oxygen atoms surround a silicon atom, which occupies the space between the oxygen atoms, thus forming a four-faced pyramidal structure (●**FIGURE 3.10B**). The silicon atom has a positive charge of 4, and each of the four oxygen atoms has a negative charge of 2, resulting in a radical with a total negative charge of 4 $(SiO_4)^{-4}$.

Because the silica tetrahedron has a negative charge, it does not exist in nature as an isolated ion group; rather, it combines with positively charged ions or shares its oxygen atoms with other silica tetrahedra. In the simplest silicate minerals, the silica tetrahedra exist as single units bonded to positively charged ions. In minerals that contain isolated tetrahedra, the silicon-to-oxygen ratio is 1:4, and the negative charge of the silica ion is balanced by positive ions (●**FIGURE 3.10C**). Olivine [$(Mg,Fe)_2SiO_4$], for example, has either two magnesium ($Mg^{+2}$) ions, two iron ($Fe^{+2}$) ions, or one of each to offset the −4 charge of the silica ion.

Silica tetrahedra may also join together to form chains of indefinite length (●**FIGURE 3.10D**). Single chains, as in the pyroxene minerals, form when each tetrahedron shares two of its oxygen atoms with an adjacent tetrahedron, resulting in a silicon-to-oxygen ratio of 1:3. Enstatite, a pyroxene-group mineral, reflects this ratio in its chemical formula $MgSiO_3$. Individual chains, however, possess a net −2 electrical charge, so they are balanced by positive ions, such as $Mg^{+2}$, that link parallel chains together (Figure 3.10d).

The amphibole group of minerals is characterized by a double-chain structure in which alternate tetrahedra in two parallel rows are cross-linked (Figure 3.10d). The formation of double chains results in a silicon-to-oxygen ratio of 4:11, so each double chain possesses a −6 electrical charge. $Mg^{+2}$, $Fe^{+2}$, and $Al^{+3}$ are usually involved in linking the double chains together.

In sheet-structure silicates, three oxygen atoms of each tetrahedron are shared by adjacent tetrahedra (●**FIGURE 3.10E**). Such structures result in continuous sheets of silica tetrahedra with silicon-to-oxygen ratios of 2:5. Continuous sheets also possess a negative electrical charge satisfied by positive ions located between the sheets. This particular structure accounts for the characteristic sheet structure of the *micas*, such as biotite and muscovite, and the *clay minerals*.

Three-dimensional networks of silica tetrahedra form when all four oxygen atoms of the silica tetrahedra are shared by adjacent tetrahedra (●**FIGURE 3.10F**). Such sharing of oxygen atoms results in a silicon-to-oxygen ratio of 1:2, which is electrically neutral. Quartz is a common framework silicate.

Geologists define two subgroups of silicates: ferromagnesian and nonferromagnesian silicates. The **ferromagnesian silicates** are those that contain iron (Fe), magnesium (Mg), or both. These minerals are commonly dark and more dense than nonferromagnesian silicates. Some of the common ferromagnesian silicate minerals are olivine, the pyroxenes, the amphiboles, and biotite (●**FIGURE 3.11A**).

The **nonferromagnesian silicates** lack iron and magnesium, are generally light colored, and are less dense than ferromagnesian silicates (●**FIGURE 3.11B**). The most common minerals in Earth's crust are nonferromagnesian silicates known as *feldspars*. Feldspar is a general name, however, that includes two distinct groups, each of which includes several species. The *potassium feldspars* are represented by microcline and orthoclase ($KAlSi_3O_8$). The second group of feldspars, the *plagioclase feldspars*, range from calcium-rich ($CaAl_2Si_2O_8$) to sodium-rich ($NaAlSi_3O_8$) varieties.

Quartz ($SiO_2$) is another common nonferromagnesian silicate. It is a framework silicate that can usually be recognized by its glassy appearance and hardness. Another fairly common nonferromagnesian silicate is muscovite, which is a mica (Figure 3.11b).

---

**silica** A compound of silicon and oxygen.

**silicate** A mineral that contains silica, such as quartz ($SiO_2$).

**silica tetrahedron** The basic building block of all silicate minerals; consists of one silicon atom and four oxygen atoms.

**ferromagnesian silicate** Any silicate mineral that contains iron, magnesium, or both.

**nonferromagnesian silicate** A silicate mineral that has no iron or magnesium.

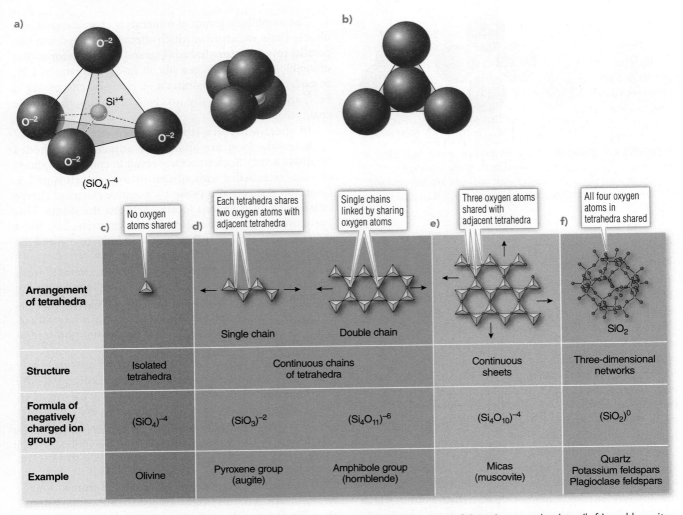

| Arrangement of tetrahedra | No oxygen atoms shared | Each tetrahedra shares two oxygen atoms with adjacent tetrahedra | Single chains linked by sharing oxygen atoms | Three oxygen atoms shared with adjacent tetrahedra | All four oxygen atoms in tetrahedra shared |
|---|---|---|---|---|---|
| | c) | d) Single chain | Double chain | e) | f) $SiO_2$ |
| **Structure** | Isolated tetrahedra | Continuous chains of tetrahedra | | Continuous sheets | Three-dimensional networks |
| **Formula of negatively charged ion group** | $(SiO_4)^{-4}$ | $(SiO_3)^{-2}$ | $(Si_4O_{11})^{-6}$ | $(Si_4O_{10})^{-4}$ | $(SiO_2)^0$ |
| **Example** | Olivine | Pyroxene group (augite) | Amphibole group (hornblende) | Micas (muscovite) | Quartz Potassium feldspars Plagioclase feldspars |

● **FIGURE 3.10  The Silica Tetrahedron and Silicate Materials**  (a) Expanded view of the silica tetrahedron (left) and how it actually exists with its oxygen atoms touching (right). (b) View of the silica tetrahedron from above. Only the oxygen atoms are visible. (c–f) Structures of the common silicate minerals shown by various arrangements of the silica tetrahedra.

● **FIGURE 3.11  Common Rock-Forming Silicate Minerals**  (a) The ferromagnesian silicates. (b) The nonferromagnesian silicates.

## Carbonate Minerals

**Carbonate minerals** are those that contain the negatively charged carbonate radical $(CO_3)^{-2}$ and include calcium carbonate $(CaCO_3)$ as the minerals *aragonite* or *calcite* (●**FIGURE 3.12A**). Aragonite is unstable and commonly changes to calcite, the main constituent of the sedimentary rock *limestone*. Several other carbonate minerals are known, but only one of these need concern us: *dolomite* $[CaMg(CO_3)_2]$, formed by the chemical alteration of calcite by the addition of magnesium. Sedimentary rock composed of the mineral dolomite is *dolostone*.

## Other Mineral Groups

Even though minerals from the other groups in Table 3.1 are less common, many are found in rocks in small quantities and others are important resources. In the oxides, an element combines with oxygen as in hematite $(Fe_2O_3)$ and magnetite $(Fe_3O_4)$. Rocks with high concentrations of these minerals are sources of iron ores for the manufacture of steel. The related hydroxides form mostly by the chemical alteration of other minerals.

We have noted that the *native elements* are minerals composed of a single element, such as diamond and graphite (C) and the precious metals gold (Au), silver (Ag), and platinum (Pt). Some elements, such as silver and copper, are found both as native elements and as compounds and are thus also included in other mineral groups; argentite $(Ag_2S)$, a silver sulfide, is an example (Table 3.1).

Several minerals and rocks that contain the phosphate radical $(PO_4)^{-3}$ are important sources of phosphorus for fertilizers. The sulfides, such as galena (PbS), the ore of lead, have a positively charged ion combined with sulfur $(S^{-2})$ (●**FIGURE 3.12B**), whereas the sulfates have an element combined with the complex radical $(SO_4)^{-2}$, as in gypsum $(CaSO_4 \cdot 2H_2O)$ (●**FIGURE 3.12C**). The halides contain the halogen elements, fluorine $(F^{-1})$ and chlorine $(Cl^{-1})$; examples are halite (NaCl) (●**FIGURE 3.12D**) and fluorite $(CaF_2)$.

# PHYSICAL PROPERTIES OF MINERALS

**LO11** Define the seven main physical properties used to identify minerals

**LO12** Explain how each property is useful in identifying minerals

Many physical properties of minerals are remarkably constant for a given mineral species, but some, especially color, may vary. Although professional geologists use sophisticated techniques to study and identify various minerals, you can identify most common minerals by using the physical properties described next.

## Luster and Color

**Luster** (not to be confused with *color*) is the quality and intensity of light reflected from a mineral's surface. Geologists define luster as *metallic*, having the appearance of a metal, and *nonmetallic*. Of the four minerals shown in Figure 3.12, only galena has a metallic luster. Among the several types of nonmetallic luster are glassy or vitreous (as in quartz), dull or earthy, waxy, greasy, and brilliant (as in diamond).

Beginning geology students are distressed by the fact that the color of some minerals varies, making the most obvious physical property of little use for identifying a number of minerals. In any case, we can make several helpful generalizations about color. Ferromagnesian silicates are typically black, brown, or dark green, although olivine is olive green (Figure 3.11a). Nonferromagnesian silicates vary in color but are rarely very dark. White, cream, colorless, and shades of pink and pale green are typical (Figure 3.11b).

●**FIGURE 3.12 Representative Specimens from Four Mineral Groups** (a) Calcite $(CaCO_3)$ is the most common carbonate mineral. (b) The sulfide mineral galena (PbS) is the ore of lead. (c) Gypsum $(CaSO_4 \cdot 2H_2O)$ is a common sulfate mineral. (d) Halite (NaCl) is a good example of a halide mineral.

**Critical Thinking Question** Which of these minerals has a metallic luster?

---

**carbonate mineral** A mineral with the carbonate radical $(CO_3)^{-2}$, as in calcite $(CaCO_3)$ and dolomite $[CaMg(CO_3)_2]$.

**luster** The appearance of a mineral in reflected light. Luster is metallic or nonmetallic, although the latter has several subcategories.

Another helpful generalization is that the color of minerals with a metallic luster is more consistent than is the color of nonmetallic minerals. For example, galena is always lead-gray (Figure 3.12b), and pyrite is invariably brassy yellow. In contrast, quartz, a nonmetallic mineral, may be colorless, smoky brown to almost black, rose, yellow-brown, milky white, blue, or violet to purple.

## Crystal Form

As we noted, many mineral specimens do not show the perfect crystal form typical of that mineral species. Nevertheless, some minerals do typically occur as crystals (Figure 3.6). For example, 12-sided crystals of garnet are common, as are 6- and 12-sided crystals of pyrite. Minerals that grow in cavities or are precipitated from circulating hot water (hydrothermal solutions) in cracks and crevices in rocks also commonly occur as crystals.

Crystal form is useful for mineral identification, but some minerals have the same crystal form. Pyrite ($FeS_2$), fluorite ($CaF_2$), and halite ($NaCl$) all occur as cubic crystals, but they can be easily identified by other properties, such as color, luster, hardness, and density (●FIGURE 3.13).

## Cleavage and Fracture

Not all minerals possess **cleavage**, but those that do break or split along smooth planes of weakness determined by the strength of their chemical bonds. Cleavage is characterized

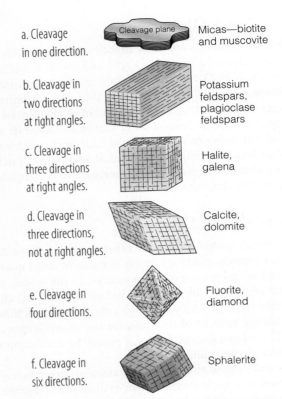

a. Cleavage in one direction.    Micas—biotite and muscovite

b. Cleavage in two directions at right angles.    Potassium feldspars, plagioclase feldspars

c. Cleavage in three directions at right angles.    Halite, galena

d. Cleavage in three directions, not at right angles.    Calcite, dolomite

e. Cleavage in four directions.    Fluorite, diamond

f. Cleavage in six directions.    Sphalerite

● **FIGURE 3.14 Several Types of Mineral Cleavage**

in terms of quality (perfect, good, poor), direction, and angles of intersection of cleavage planes. Biotite, a common ferromagnesian silicate, has perfect cleavage in one direction (●FIGURE 3.14A). Biotite is a sheet silicate, with the sheets of silica tetrahedra weakly bonded to one another by iron and magnesium ions.

Feldspars possess two directions of cleavage that intersect at right angles (●FIGURE 3.14B), and the mineral halite has three directions of cleavage, all of which intersect at right angles (●FIGURE 3.14C). Calcite also possesses three directions of cleavage, but none of the intersection angles is a right angle, so cleavage fragments of calcite are rhombohedrons (●FIGURE 3.14D). Minerals with four directions of cleavage include fluorite and diamond (●FIGURE 3.14E). Ironically, diamond, the hardest mineral, can be easily cleaved. A few minerals, such as sphalerite, an ore of zinc, have six directions of cleavage (●FIGURE 3.14F).

Cleavage is an important diagnostic property of minerals, and recognizing it is essential in distinguishing between some minerals. The pyroxene mineral augite and the amphibole mineral hornblende, for example, look much alike. Both are dark green to black, have the same hardness, and possess two directions of cleavage. However, the cleavage planes of

a) Pyrite ($FeS_2$).

b) Fluorite ($CaF_2$).

c) Halite ($NaCl$).

● **FIGURE 3.13 Minerals with the Same Kind of Crystals** Mineral crystals are found in a variety of shapes (Figure 3.6), but different minerals may have the same kinds of crystals. Differentiating one from the other is easy. Pyrite (a) is brassy yellow (the silvery color results from reflected light), and it is much denser than the other two. You can identify fluorite (b) and halite (c) by their cleavage and taste—halite tastes salty whereas fluorite has no taste.

**cleavage** The breaking or splitting of mineral crystals along planes of internal weakness.

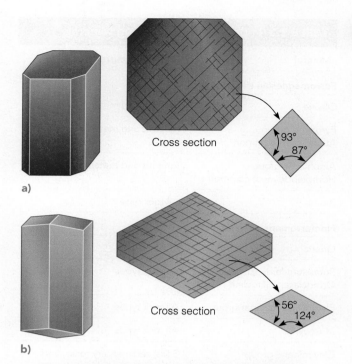

**FIGURE 3.15 Cleavage in Augite and Hornblende**
(a) Augite crystal and cross section of crystal showing cleavage. (b) Hornblende crystal and cross section of crystal showing cleavage.

| Hardness | Mineral | Hardness of Some Common Objects |
|---|---|---|
| 10 | Diamond | |
| 9 | Corundum | |
| 8 | Topaz | |
| 7 | Quartz | |
| | | Steel file (6½) |
| 6 | Orthoclase | |
| | | Glass (5½–6) |
| 5 | Apatite | |
| 4 | Fluorite | |
| 3 | Calcite | Copper penny (3) |
| | | Fingernail (2½) |
| 2 | Gypsum | |
| 1 | Talc | |

**TABLE 3.2** Mohs Hardness Scale

augite intersect at about 90 degrees, whereas the cleavage planes of hornblende intersect at angles of 56 degrees and 124 degrees (**FIGURE 3.15**).

In contrast to cleavage, *fracture* is mineral breakage along irregular surfaces. Any mineral will fracture if enough force is applied, but the fracture surfaces are uneven or conchoidal (curved) rather than smooth.

## Hardness

An Austrian geologist, Friedrich Mohs (1773–1839), devised a relative hardness scale for 10 minerals. He arbitrarily assigned a hardness value of 10 to diamond, the hardest mineral known, and lesser values to the other minerals. Relative hardness is easily determined by the use of Mohs hardness scale (**TABLE 3.2**). Quartz will scratch fluorite but cannot be scratched by fluorite, gypsum can be scratched by a fingernail, and so on. Thus, **hardness** is defined as a mineral's resistance to abrasion and is controlled mostly by internal structure. Both graphite and diamond are composed of carbon, but the former has a hardness of 1 to 2, whereas the latter has a hardness of 10.

## Specific Gravity (Density)

A mineral's **specific gravity** is the ratio of its weight to the weight of an equal volume of pure water at 4°C. Therefore, a mineral with a specific gravity of 3.0 is three times as heavy as water. **Density**, in contrast, is a mineral's mass (weight) per unit of volume expressed in grams per cubic centimeter. So the specific gravity of galena (Figure 3.12b) is 7.58, and its density is 7.58 g/cm$^3$. In most instances, we will refer to a mineral's density, and in some of the following chapters, we will mention the density of rocks.

Because ferromagnesian silicates contain iron, magnesium, or both, they tend to be denser than nonferromagnesian silicates. In general, the metallic minerals, such as galena and hematite, are denser than nonmetals. Pure gold, with a density of 19.3 g/cm$^3$, is about 2.5 times as dense as lead. Diamond and graphite, both of which are composed of carbon (C), illustrate how structure controls specific gravity or density. The specific gravity of diamond is 3.5, whereas that of graphite varies from 2.09 to 2.33.

## Other Useful Mineral Properties

Talc has a distinctive soapy feel, graphite writes on paper, halite tastes salty, and magnetite is magnetic. Calcite possesses the property of *double refraction*, meaning that an object when viewed through a transparent piece of calcite

**hardness** A term used to express the resistance of a mineral to abrasion.

**specific gravity** The ratio of a substance's weight, especially a mineral, to an equal volume of water at 4°C.

**density** The mass of an object per unit volume; usually expressed in grams per cubic centimeter (g/cm$^3$).

will have a double image. Some sheet silicates are plastic and, when bent into a new shape, will retain that shape; others are flexible and, if bent, will return to their original position when the forces that bent them are removed.

A simple chemical test to identify the minerals calcite and dolomite involves applying a drop of dilute hydrochloric acid to the mineral specimen. If the mineral is calcite, it will react vigorously with the acid and release carbon dioxide, which causes the acid to bubble or effervesce. Dolomite, in contrast, will not react with hydrochloric acid unless it is powdered.

# ROCK-FORMING MINERALS

**LO13**   **Define a rock**

**LO14**   **Explain why only a few minerals are sufficiently common for rock identification and classification**

A **rock** is a solid aggregate of one or more minerals, but the term also refers to masses of mineral-like matter, as in the natural glass obsidian, and masses of solid organic matter, as in coal. And even though some rocks may contain many minerals, only a few, designated **rock-forming minerals**, are sufficiently common for rock identification and classification (●**TABLE 3.3** and Figure 3.1b). Others, known as *accessory minerals*, are commonly present, but in such small quantities that they can be disregarded.

Given that silicate minerals are by far the most common in Earth's crust, it follows that most rocks are composed of these minerals. Indeed, feldspar minerals (plagioclase feldspars and potassium feldspars) and quartz make up more than 60% of Earth's crust. So, even though there are about 2,500 known silicate minerals, only a few are particularly common in rocks.

The most common nonsilicate rock-forming minerals are the carbonates calcite ($CaCO_3$) and dolomite [$CaMg(CO_3)_2$], the main constituents of the sedimentary rocks limestone and dolostone, respectively. Among the sulfates and halides, gypsum ($CaSO_4 \cdot 2H_2O$) in rock gypsum and halite (NaCl) in rock salt are common enough to qualify as rock-forming minerals. Even though these minerals and their corresponding rocks might be common in some areas, their overall abundance is limited compared to the silicate and carbonate rock-forming minerals.

# HOW DO MINERALS FORM?

**LO15**   **Explain the different ways that minerals can form**

Thus far, we have discussed the composition, structure, and physical properties of minerals, but we have not addressed how they originate. One phenomenon that accounts for minerals is the cooling of molten rock

**TABLE 3.3** Important Rock-Forming Minerals

| Mineral | Primary Occurrence |
| --- | --- |
| **Ferromagnesian silicates** | |
| Olivine | Igneous and metamorphic rocks |
| Pyroxene group Augite most common | Igneous and metamorphic rocks |
| Amphibole group Hornblende most common | Igneous and metamorphic rocks |
| Biotite | All rock types |
| **Nonferromagnesian silicates** | |
| Quartz | All rock types |
| Potassium feldspar group Orthoclase, microcline | All rock types |
| Plagioclase feldspar group | All rock types |
| Muscovite | All rock types |
| Clay mineral group | Soils, sedimentary rocks, and some metamorphic rocks |
| **Carbonates** | |
| Calcite | Sedimentary rocks |
| Dolomite | Sedimentary rocks |
| **Sulfates** | |
| Anhydrite | Sedimentary rocks |
| Gypsum | Sedimentary rocks |
| **Halides** | |
| Halite | Sedimentary rocks |

material known as *magma* (magma that reaches the surface is called *lava*).

As magma or lava cools, minerals crystallize and grow, thereby determining the mineral composition of igneous rocks such as basalt (dominated by ferromagnesian silicates) and granite (dominated by nonferromagnesian silicates). Hot-water solutions derived from magma commonly invade cracks and crevasses in adjacent rocks, and from these solutions, several minerals crystallize, some of which have economic importance. Minerals also originate when water in

**rock** A solid aggregate of one or more minerals, as in limestone and granite, or a consolidated aggregate of rock fragments, as in conglomerate, or masses of rocklike materials, such as coal and obsidian.

**rock-forming mineral** Any mineral common in rocks that is important in their identification and classification.

hot springs cools, and when hot, mineral-rich water discharges onto the seafloor at *hydrothermal vents* (see Figure 2.7).

Dissolved materials in seawater, more rarely in lake water, combine to form minerals such as halite (NaCl), gypsum ($CaSO_4 \cdot 2H_2O$), and several others when the water evaporates. Aragonite and/or calcite, both varieties of calcium carbonate ($CaCO_3$), might also form from evaporating water, but most originate when organisms such as clams, oysters, corals, and floating microorganisms use this compound to construct their shells. A few plants and animals (such as radiolarians) use silicon dioxide ($SiO_2$) for their skeletons, which accumulate as mineral matter on the seafloor when the organisms die.

Some clay minerals form when chemical processes compositionally and structurally alter other minerals, and others originate when rocks are changed during metamorphism. In fact, the agents that cause metamorphism—heat, pressure, and chemically active fluids—are responsible for the origin of many minerals.

# NATURAL RESOURCES AND RESERVES

**LO16**  Define a resource and reserve

**LO17**  Explain how resources differ from reserves

**LO18**  Discuss the importance of natural resources in ensuring a high standard of living

**LO19**  Explain how the distribution of natural resources and reserves impacts on global politics and international trade

Geologists at the U.S. Geological Survey define a **resource** as a naturally occurring concentration of solid, liquid, or gaseous material in, or on, Earth's crust in such form and amount that economic extraction of a commodity is currently or potentially feasible. Many natural resources are concentrations of minerals, rocks, or both, but liquid petroleum and natural gas are also included. In fact, we refer to *metallic resources* (copper, tin, iron ore, etc.), *nonmetallic resources* (sand and gravel, crushed stone, sulfur, salt, etc.), and *energy resources* (petroleum, natural gas, coal, and uranium). All are resources, but it is important to make a distinction between a *resource*, the total amount of a commodity whether discovered or not, and a **reserve**, which is only that part of the resource base that is known and can be economically recovered.

In principle, the distinction between a resource and a reserve is simple, but in practice it depends on several factors, not all of which remain constant. For instance, a resource in a remote region might not be mined because transport expenses are too high, and what may be deemed a resource rather than a reserve in a developed nation might be mined in a developing country where labor costs are low. The value of gold and diamonds makes it profitable to mine them just about anywhere, whereas deposits of sand and gravel for construction must be close to their market areas. Changes in technology are important, too. For example, iron ore mined in the Great Lakes region is not as rich as that mined in decades past, but a technique for separating the iron from its host rock and shaping it into pellets has made mining of low-grade deposits profitable.

In addition to resources such as petroleum, gold, and ores of iron, copper, and aluminum, some common minerals and rocks are important. Pure quartz sand is used to make glass, sandpaper, and optical instruments; clay minerals are needed to make ceramics and paper; feldspars are used for porcelain, ceramics, enamel, and glass; and phosphate-bearing rocks are used for fertilizers. Muscovite is an ingredient in wallboard joint compound, lipstick, glitter, and eye shadow, as well as the lustrous paints on appliances and automobiles (see GEO-FOCUS). In 2017, the extraction of nonfuel mineral resources in the United States totaled $75.2 billion.

Access to natural resources is essential to all societies, but especially to industrialized countries with their high standard of living. However, most resources are *nonrenewable*, meaning that there is a limited supply and they cannot be replenished by natural processes as fast as they are depleted. Accordingly, as the amount of a resource diminishes, suitable substitutes, if available, must be found, or we continue to exploit the resource and simply pay more and more as further depletion takes place.

The United States has become increasingly reliant on imports from numerous countries, some of which are in politically unstable regions of the world. For example, in 2017, the United States imported 100% of 21 mineral commodities. These include the rare earths, manganese (an alloy that converts iron into steel), gallium (microelectronic components), cesium (atomic clocks, DNA separation techniques), and rubidium (electronic and medical applications). Furthermore, the United States was more than 50% import reliant on an additional 24 economically important minerals, such as cobalt (used in hybrid electric vehicles, smart phones, missile guidance systems), platinum-group elements (catalytic converters), lithium (batteries), aluminum (transportation, packaging, construction, machinery), and chromium (stainless steel, refractories), to name a few.

---

**resource** A concentration of naturally occurring solid, liquid, or gaseous material in, or on, Earth's crust in such form and amount that economic extraction of a commodity from the concentration is currently or potentially feasible.

**reserve** The part of the resource base that can be extracted economically.

Given that the United States uses such large quantities of resources and must import all or some of them, what can be done? Of course one partial solution is to use resources more efficiently and to recycle. Indeed, much of all aluminum is recycled. Another approach is the use of newly developed technology to find and develop both new and previously uneconomical mineral resources.

To ensure the continued supply of mineral and energy resources, scientists, government agencies, and executives in business and industry continually assess the status of resources in view of changing economic and political conditions and changes in science and technology. The United States Geological Survey keeps detailed statistical records of mine production, imports, and exports, and regularly publishes reports on the status of numerous resources in its *Mineral Commodity Summaries*. Similar reports also appear in the *Canadian Minerals Yearbook*.

# GEO-FOCUS

## The Many Uses of Mica

Industrial societies depend on finding, extracting, processing, and using many minerals and rocks, not the least of which are ores of metals, such as iron, lead, tin, and zinc, as well as minerals and/or rocks used as energy sources, such as coal or uranium. Most people have heard of these mineral commodities. However, how many people have given any thought as to why the paint on cars and appliances is so lustrous, or why many cosmetics glitter and shine? The answer is that they all contain the mineral muscovite (●FIGURE 1A), one of 37 sheet silicate minerals that have one direction of cleavage, such that when cleaved, they split into thin, flexible sheets (see Figures 3.10e and 3.14a). Of these 37 micas, only a few are common in rocks (biotite and muscovite), and only muscovite has several commercial uses.

When ground up dry, muscovite loses much of its luster, but retains its platy nature and is ideal for wallboard joint compound and as an additive to paint. It is important in wallboard joint compound because it makes the compound smoother and easier to work with and it prevents cracking. In addition, dry ground muscovite is used in plastics, roofing, rubber, and welding rods.

If muscovite is ground up wet, it retains its sparkling shine and is used in many cosmetics, including body powder, eye shadow, lipstick, blush, and nail polish (●FIGURE 1B). Fortunately, muscovite is chemically inert and poses no risk when applied to the skin. The brilliant, resinous sheen of some paints on automobiles and their changing colors are the result of muscovite additives in its paint (●FIGURE 1C).

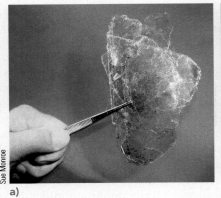
a)

c)

b)

●**FIGURE 1 Muscovite** (a) Muscovite mica is colorless, white, pale red, or pale green. It is a sheet silicate that has many industrial uses. (b) Muscovite is used in several kinds of cosmetics, including nail polish. (c) Muscovite is used as a paint additive, providing the lustrous sheen seen in automobile paints.

# Key Concepts Review

- Matter is anything that has mass and occupies space. It is composed of chemical elements, each of which consists of atoms. Protons and neutrons are present in an atom's nucleus, and electrons orbit around the nucleus in electron shells.
- The number of protons in an atom's nucleus determines its atomic number. The atomic mass number is the number of protons plus neutrons in the nucleus.
- Bonding results when atoms join with other atoms; different elements bond to form compounds. With few exceptions, minerals are compounds.
- Ionic and covalent bonds are the most common types of bonds in minerals, but metallic and van der Waals bonds are also found in some.
- Minerals are crystalline solids, which means that they possess an ordered internal arrangement of atoms.
- Mineral composition is indicated by a chemical formula, such as $SiO_2$ for quartz.
- Some minerals have a range of compositions because some elements can substitute for one another if their atoms are about the same size and have the same electrical charge.
- More than 5,000 minerals are known, and most of them are silicates. The two types of silicates are ferromagnesian and nonferromagnesian.
- In addition to silicates, geologists recognize carbonates, native elements, hydroxides, oxides, phosphates, halides, sulfates, and sulfides.
- Structure and composition control the physical properties of minerals, such as luster, crystal form, hardness, color, cleavage, fracture, and specific gravity.
- A few minerals, designated rock-forming minerals, are common enough in rocks to be essential in their identification and classification. Most rock-forming minerals are silicates, but some carbonates are also common.
- Several processes account for the origin of minerals, including cooling magma and lava, weathering, evaporation of seawater, metamorphism, and organisms using dissolved substances in seawater to build their shells.
- Many resources are concentrations of minerals or rocks of economic importance. They are further characterized as metallic resources, nonmetallic resources (industrial minerals), and energy resources.
- Reserves are that part of the resource base that can be extracted profitably. Distinguishing a resource from a reserve depends on market price, labor costs, geographic location, and developments in science and technology.
- The United States has become increasingly reliant on imports to maintain its industrial capacity.
- To ensure the continued supply of mineral and energy resources, countries must use resources more efficiently by recycling where possible, developing new techniques to find and develop previously uneconomical deposits, and continue to assess current and future usage.

# Important Terms

atom   53

atomic mass number   53

atomic number   53

bonding   54

carbonate mineral   61

cleavage   62

compound   54

covalent bond   55

crystal   56

crystalline solid   56

density   63

electron   53

electron shell   53

element   53

ferromagnesian silicate   59

hardness   63

ion   54

ionic bond   54

luster   61

matter   53

mineral   52

neutron   53

nonferromagnesian silicate   59

nucleus   53

proton   53

reserve   65

resource   65

rock   64

rock-forming mineral   64

silica   59

silica tetrahedron   59

silicate   59

specific gravity   63

# Review Questions

1. By far the most common minerals in Earth's crust are
   a. silicates.
   b. carbonates.
   c. oxides.
   d. sulfides.
   e. sulfates.

2. In ionic bonding, electrons
   a. are shared by adjacent atoms.
   b. freely migrate from atom to atom.
   c. change from a negative to a positive electrical charge.
   d. are transferred from one atom to another.
   e. double in number and size.

3. A naturally occurring solid with its atoms arranged in a specific three-dimensional framework is said to be
   a. amorphous.
   b. crystalline.
   c. covalent.
   d. ionic.
   e. electronic.

4. Resources are characterized as metallic resources, nonmetallic resources, and _____ resources.
   a. oxide.
   b. carbonate.
   c. commercial.
   d. lithologic.
   e. energy.

5. The atomic mass number of an atom having 10 protons, 16 neutrons, and 10 electrons is
   a. 22.
   b. 26.
   c. 32.
   d. 38.
   e. 48.

6. What accounts for the fact that some minerals, such as plagioclase feldspars, have a range of chemical compositions rather than one specific composition?

7. One part of the definition of the term *mineral* is that minerals are crystalline solids. What are crystalline solids and how do they differ from noncrystalline solids?

8. What factors determine whether a mineral or rock commodity is a resources or a reserve?

9. Although about 5,000 minerals have been named and described, only a few are very common. Why is this, and which ones are they?

# Creative Thinking Visual Question

If diamond is perfectly cleaved, it would yield geometric figures like those in ● **FIGURE 1** but this mineral is fluorite. From the image alone you should be able to tell that these specimens are not diamond. How? What other mineral properties do you think can be used to differentiate fluorite from diamond?

● **FIGURE 1 Fluorite** This mineral has four directions of cleavage as in Figure 3.14e.

Sue Monroe

# 4

The Harney Peak Granite in the Black Hills of South Dakota with the images of presidents George Washington, Thomas Jefferson, Theodore Roosevelt, and Abraham Lincoln. This large body of granite, which formed about 1.7 billion years ago, forms the central part of the Black Hills. The 18-m-high images of the presidents were carved between 1927 and 1941 and are now the primary attraction at Mount Rushmore National Memorial.

# IGNEOUS ROCKS AND INTRUSIVE IGNEOUS ACTIVITY

# INTRODUCTION

From our discussions so far, you know that the term *rock* applies to aggregates of one or more minerals (see Chapter 3), mineral-like matter as in obsidian (natural glass), and coal (altered organic matter). You also know from Chapter 1 that the three groups of rocks are igneous, sedimentary, and metamorphic (see Figure 1.14). Here we are concerned only with **igneous rocks**, which are rocks made up of minerals that crystallized from **magma** (molten rock matter below the surface), and **lava** (molten rock material at the surface), and from particulate matter known as pyroclastic materials that is ejected from volcanoes during explosive eruptions. Magma and lava are the same material, although the gas content of lava is less because of decreased pressure.

Most people have given little or no thought to magma but are familiar with volcanoes, which are simply hills or mountains built up where magma reaches the surface. In any case, volcanism is an observable phenomenon, but most magma never reaches the surface. In fact it cools and crystallizes underground, a phenomenon called *intrusive igneous activity*, thus forming several types of igneous rock bodies known as *plutons*, named for Pluto, the Roman god of the underworld. For example, granite, composed of potassium feldspars, plagioclase feldspars, and quartz, is common in many of the larger plutons.

Even though we discuss intrusive igneous activity and the origin of plutons in this chapter and volcanism in Chapter 5, both processes are related. Moreover, the same kinds of magmas are involved, but magma varies in its mobility, so only some of it reaches the surface. Furthermore, plutons lie beneath areas of volcanism and are the source of the lava flows and particulate matter erupted at volcanoes or along fissures, a topic we cover more fully in Chapter 5.

An important reason to study igneous rocks and plutons is that these rocks make up parts of the continents and all of the oceanic crust, which forms continuously at divergent plate boundaries. As a matter of fact, most plutons and volcanoes are found at or near divergent and convergent plate boundaries, so the presence of igneous rocks in the geologic record is one criterion for recognizing ancient plate boundaries. Another reason to study these topics is their economic importance. When these large masses of magma begin to cool and crystallize to form minerals, fluids emanating from them follow cracks and crevasses in adjacent rocks where important mineral resources such as copper may form.

# THE PROPERTIES AND BEHAVIOR OF MAGMA AND LAVA

LO1   **Define lava flows and pyroclastic materials**

LO2   **Discuss how volcanic rocks differ from plutonic rocks**

LO3   **Explain how ultramafic, mafic, intermediate, and felsic magmas differ from each other**

LO4   **Explain why silica content and temperature are important controls on the viscosity of magma and lava**

Magma is less dense than the rock from which it formed, so it tends to rise toward the surface, but much of it solidifies deep underground, thereby accounting for the origin of plutons. The magma that does reach the surface issues forth as **lava flows**, and some of it is forcefully ejected into the atmosphere as particles known as **pyroclastic materials** (from the Greek *pyro* for "fire" and *klastos* for "broken").

All igneous rocks derive ultimately from magma; however, two separate processes account for them. They form when (1) magma or lava cools and crystallizes to form aggregates of minerals, or (2) pyroclastic materials are consolidated. Those igneous rocks derived from lava flows and pyroclastic materials, both of which are extruded onto the surface, are known as **volcanic rocks** or **extrusive igneous rocks**. When magma cools below the surface it forms **plutonic rocks** or **intrusive igneous rocks**.

## Composition of Magma

By far the most abundant minerals in Earth's crust are silicates such as quartz, feldspars, and several ferromagnesian silicates, all made up of silicon and oxygen, and other elements shown in Figure 3.8. As a result, melting of the crust yields mostly silica-rich magmas that also contain considerable aluminum, calcium, sodium, iron, magnesium, and potassium, and several other elements in lesser quantities. Another source of magma is Earth's upper mantle, which is composed of rocks that contain mostly ferromagnesian silicates. Thus, magma from this source contains comparatively less silicon and oxygen (silica) and more iron and magnesium.

With few exceptions, the primary constituent of magma is silica, which varies enough to distinguish magmas called **ultramafic** (less than 45% silica), **mafic** (45–52% silica),

---

**igneous rock**  Any rock formed by cooling and crystallization of magma or lava or the consolidation of pyroclastic materials.

**magma**  Molten rock material generated within Earth.

**lava**  Magma that reaches Earth's surface.

**lava flow**  A stream of magma flowing over Earth's surface.

**pyroclastic material**  Fragmental substances, such as ash, explosively ejected from a volcano.

**volcanic (extrusive igneous) rock**  An igneous rock formed when magma is extruded onto Earth's surface where it cools and crystallizes, or when pyroclastic materials become consolidated.

**plutonic (intrusive igneous) rock**  Igneous rock that formed from magma intruded into or formed in place within the crust.

**ultramafic magma**  Magma with less than 45% silica.

**mafic magma**  Magma with between 45% and 52% silica and proportionately more calcium, iron, and magnesium than intermediate and felsic magma.

**TABLE 4.1** The Most Common Types of Magmas and Their Characteristics

| Type of Magma | Silica Content (%) | Sodium, Potassium, and Aluminum | Calcium, Iron, and Magnesium |
|---|---|---|---|
| Ultramafic | <45 | | Increase ↑ |
| Mafic | 45–52 | | |
| Intermediate | 53–65 | ↓ | |
| Felsic | >65 | Increase | |

**intermediate** (53–65% silica), and **felsic** (more than 65% silica) (●TABLE 4.1). According to Table 4.1, mafic magma is silica-poor but contains more calcium, iron, and magnesium than silica-rich felsic magma, which has more sodium, potassium, and aluminum. And, as you would expect, intermediate magma has a composition between mafic and felsic.

## How Hot Are Magma and Lava?

Erupting lava has a temperature in the range of 700°C to 1,200°C, although a temperature of 1,350°C was recorded above a lava lake in Hawaii where volcanic gases reacted with the atmosphere. Magma must be even hotter than lava, but no direct measurements of magma temperatures have ever been made.

Most lava temperatures are taken at volcanoes that show little or no explosive activity, so our best information comes from mafic lava flows such as those in Hawaii (●FIGURE 4.1). In contrast, eruptions of felsic lava are not as common, and the volcanoes from which these flows issue tend to be explosive and thus cannot be approached safely.

● **FIGURE 4.1** A geologist using a thermal measuring device determines the temperature of an active lava flow in Hawaii.

Nevertheless, the temperatures of some bulbous masses of felsic lava in lava domes have been measured at a distance with an optical pyrometer. The surfaces of these lava domes are as hot as 900°C, but their interiors must surely be even hotter.

The reason that lava and magma retain heat so well is that rock poorly conducts heat. Accordingly, the interiors of thick lava flows and pyroclastic flows may remain hot for months or years. In 1959, lava filled a crater to 85 m deep in Hawaii, and when it was drilled in 1988, it still had not completely solidified near its base. Plutons, depending on their size and depth, may not cool completely for thousands to millions of years.

## Viscosity: Resistance to Flow

**Viscosity**, or resistance to flow, is a property of all liquids. Water's viscosity is very low, so it is highly fluid and flows readily. For other liquids, such as cold motor oil and syrup, viscosity is so high that they flow much more slowly. But when these liquids are heated, their viscosity is much lower and they flow more easily. Accordingly, you might suspect that temperature controls the viscosity of magma and lava, and this inference is partly correct. We can generalize and say that the hotter the magma or lava, the more readily it moves, but we must qualify this statement by noting that temperature is not the only control of viscosity.

Silica content also strongly controls magma and lava viscosity. With increasing silica content, numerous networks of silica tetrahedra form and retard flow, because for flow to take place, the strong bonds of the networks must be ruptured. Mafic magma and lava with 45% to 52% silica have fewer silica tetrahedra networks and, as a result, are more mobile than felsic magma and lava flows (●FIGURE 4.2). One mafic flow in 1783 in Iceland flowed nearly 80 km, and geologists in Washington State have traced ancient lava flows for more than 500 km. Felsic magma, because of its higher viscosity, does not reach the surface as often as mafic magma. And when felsic lava flows do occur, they tend to be slow moving and thick, and move only short distances.

Temperature and silica content are important controls on the viscosity of magma and lava, but other factors include gases, mostly water vapor and $CO_2$, as well as the presence of mineral crystals and friction from the surface over which lava flows. Lava with a high content of dissolved gases flows more readily than one with a lesser amount of gases, whereas lava that has many crystals or that flows over a rough surface tends to be more viscous.

---

**intermediate magma**  Magma with a silica content between 53% and 65% and an overall composition intermediate between mafic and felsic magma.

**felsic magma**  Magma with more than 65% silica and considerable sodium, potassium, and aluminum, but little calcium, iron, and magnesium.

**viscosity**  A fluid's resistance to flow.

a)                                    b)

● **FIGURE 4.2  Viscosity of Magma and Lava** Temperature is an important control of viscosity, but so is composition. Mafic lava tends to be fluid, whereas felsic lava is much more viscous. (a) A mafic lava flow in 1984 on Mauna Loa Volcano in Hawaii. These flows move rapidly and form thin layers. (b) The Novarupta lava dome in Katmai National Park in Alaska, photographed in 1987. The lava is felsic and viscous, so it was extruded as a bulbous mass.

**Critical Thinking Question** Which one of these volcanoes would you expect to erupt explosively? Explain.

# HOW DOES MAGMA ORIGINATE AND CHANGE?

LO5    Discuss how magmas originate

LO6    Explain how Bowen's reaction series accounts for the derivation of felsic and intermediate magmas from mafic magmas

LO7    Explain how magmas form at spreading ridges, subduction zones, and at hot spots

LO8    Explain how crystal settling, assimilation, and magma mixing can change the composition of magma

Even if you have not witnessed a volcanic eruption, you are probably familiar with lava flows and pyroclastic eruptions seen in news accounts. In any case, we know about some aspects of igneous activity, but most people are unaware of how and where magma originates, how it rises, and how it might change. Indeed, there is a misconception that lava comes from a continuous layer of molten rock beneath the crust or that it comes from Earth's molten outer core.

First, let us address how magma originates. We know that the atoms in a solid are in constant motion and that, when a solid becomes hot enough, the energy of motion exceeds the binding forces and the solid melts. We are all familiar with

this phenomenon, and we are also aware that not all solids melt at the same temperature. Likewise, if heated, the minerals in rocks begin to melt, but not all at the same time. Once magma forms, it tends to rise, because as rocks become hotter, they expand and their density decreases.

Magma may come from 100 to 300 km deep, but most forms at much shallower depths in the upper mantle or lower crust and accumulates in reservoirs known as **magma chambers**. Beneath spreading ridges, where the crust is thin, these chambers lie only a few kilometers deep, but along convergent plate boundaries, they are commonly a few tens of kilometers deep. The volume of a magma chamber ranges from a few to many hundreds of cubic kilometers of molten rock within the otherwise solid lithosphere. Some simply cools and crystallizes within Earth's crust, thus accounting for the origin of plutons, whereas some rises to the surface and is erupted as lava flows or pyroclastic materials.

## Bowen's Reaction Series

During the early 1900s, N. L. Bowen proposed that minerals crystallize in a predictable sequence from cooling magma. Based on his observations and laboratory experiments,

---

**magma chamber** A reservoir of magma within Earth's upper mantle or lower crust.

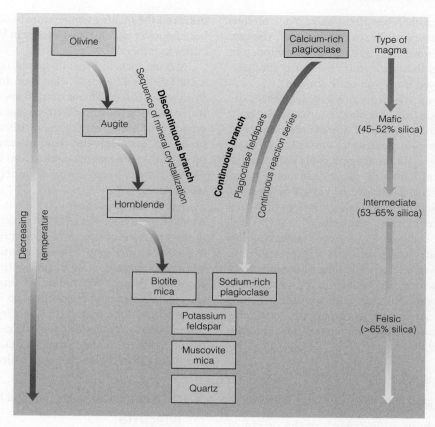

● **FIGURE 4.3 Bowen's Reaction Series** Bowen's reaction series consists of a *discontinuous branch*, along which a succession of ferromagnesian silicates crystallize as the magma's temperature decreases, and a *continuous branch*, along which plagioclase feldspars with increasing amounts of sodium crystallize. Notice also that the composition of the initial mafic magma changes as crystallization takes place along the two branches.

MINDTAP
From Cengage

Animation:
Crystallization Series
of Silicate Minerals
from a Magma

Bowen put forth a mechanism, now called **Bowen's reaction series**, to account for the derivation of intermediate and felsic magmas from mafic magma. Bowen's reaction series consists of two branches: a *discontinuous branch* and a *continuous branch* (●FIGURE 4.3). As the temperature of magma decreases, minerals crystallize along both branches simultaneously, but for convenience, we will discuss them separately.

In the discontinuous branch, which contains only ferromagnesian silicates—those silicates containing iron (Fe), magnesium (Mg), or both—minerals will crystallize out of a magma along specific temperature ranges (Figure 4.3). As the magma cools, a temperature range is reached at which a mineral begins to crystallize, incorporating certain elements from the magma into its crystal lattice, and the composition of the remaining magma changes with continued crystallization. As the magma continues to cool, it reaches a temperature range where a second mineral can form from the elements that remain in the magma. This succession of mineral formation and the changing amounts of elements available in the residual melt to form other minerals continues through the discontinuous branch of Bowen's reaction series.

At the highest temperature in the discontinuous branch, olivine $[(Mg, Fe)_2 SiO_4]$ is the first ferromagnesian silicate to crystallize from the magma, removing a significant amount of iron and magnesium from the melt as it forms. As the magma continues to cool, it reaches a temperature range at which augite $[Ca(Mg, Fe, Al)(Al, Si)_2O_6]$ is stable, and if the necessary elements are present in the remaining melt, augite will form.

A helpful analogy in explaining Bowen's reaction series is to think of the elements in a magma chamber like dollars in your bank account. At the beginning of the month (at the highest temperature), the first thing you need to do is pay your rent (make olivine). So you withdraw the money needed to pay your rent, thus reducing the money left in your bank account (you used Fe, Mg, and $SiO_4$ to make olivine and it is now gone from your magma chamber). Now that the rent is paid, you need to pay for food (make augite). You don't have as much money as you had at the beginning of the month (less Fe, Mg, and $SiO_4$), so the only money left to buy food is what's left after you paid your rent (the elements

**Bowen's reaction series** A series of minerals that form in a specific sequence in cooling magma or lava; originally proposed to explain the origin of intermediate and felsic magma from mafic magma.

left over in the magma chamber after you've made olivine are the ones that are now available to make augite). You can then continue to pay your bills until all of your money is gone, but you can't buy food with your rent money (because it's already been removed from your bank account).

Plagioclase feldspars, which are nonferromagnesian silicates, that is, they lack iron and magnesium, are the only minerals in the continuous branch of Bowen's reaction series (Figure 4.3). Calcium-rich plagioclase crystallizes first at the highest temperature. As the magma cools and calcium is depleted from the residual melt, plagioclase incorporates proportionately more sodium into its crystal lattice. As minerals crystallize simultaneously along the two branches of Bowen's reaction series, iron, magnesium, calcium, and sodium are depleted, because they are used in making the ferromagnesian silicates and plagioclase feldspars. At this point, any leftover magma is enriched in potassium, aluminum, and silicon, which can combine to form orthoclase ($KAlSi_3O_8$), a potassium feldspar, and, if water pressure is high, the sheet silicate muscovite forms. Any remaining magma is enriched in silicon and oxygen (silica) and forms the mineral quartz ($SiO_2$).

## The Origin of Magma at Spreading Ridges

One fundamental observation regarding the origin of magma is that Earth's temperature, or *geothermal gradient,* increases with depth (see Chapter 8). Accordingly, rocks at depth are hot but remain solid, because their melting temperature rises with increasing pressure. However, beneath spreading ridges, the temperature locally exceeds the melting temperature, at least in part because pressure decreases. That is, plate separation at ridges probably causes a decrease in pressure on the already hot rocks at depth, thus initiating melting (●**FIGURE 4.4**). In addition, the presence of water decreases the melting temperature beneath spreading ridges, because water aids thermal energy in breaking the chemical bonds in minerals.

Magma formed beneath spreading ridges is invariably mafic (45% to 52% silica). However, the upper-mantle rocks from which this magma is derived are ultramafic (<45% silica), consisting mostly of ferromagnesian silicates and lesser amounts of nonferromagnesian silicates. To explain how mafic magma originates from ultramafic rock, geologists propose that the magma comes from source rock that only partially melts because not all minerals melt at the same temperature.

Recall the sequence of minerals in Bowen's reaction series (Figure 4.3). The order in which these minerals melt is the opposite of their order of crystallization. Accordingly, rocks made up of quartz, potassium feldspar, and sodium-rich plagioclase begin melting at lower temperatures than those composed of ferromagnesian silicates and the calcic varieties of plagioclase. So when ultramafic rock starts to melt, the minerals richest in silica melt first, followed by those containing less silica. Therefore, if melting is not complete, mafic magma containing proportionately more silica than the source rock results.

## Subduction Zones and the Origin of Magma

Another fundamental observation regarding magma is that where an oceanic plate is subducted beneath either a continental plate or another oceanic plate, a belt of volcanoes and plutons is found near the leading edge of the overriding plate (Figure 4.4). It would seem, then, that subduction and the origin of magma must be related in some way, and indeed they are. Furthermore, magma at these convergent plate boundaries is mostly intermediate (53% to 65% silica) or felsic (>65% silica).

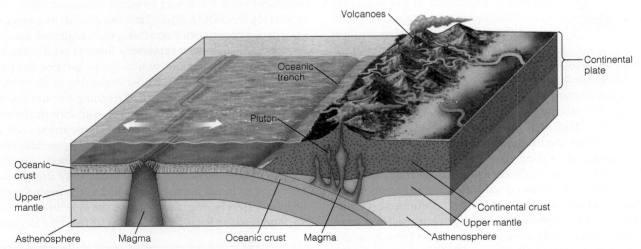

● **FIGURE 4.4 The Origin of Magma** Magma forms beneath spreading ridges, because as plates separate, pressure is reduced on the hot rocks and partial melting of the upper mantle begins. Invariably, the magma formed is mafic. Magma also forms at subduction zones, where water from the subducted plate causes partial melting of the upper mantle. This magma is also mafic, but as it rises, melting of the lower crust makes it more felsic.

Once again, geologists invoke the phenomenon of partial melting to explain the origin and composition of magma at subduction zones. As a subducted plate descends toward the asthenosphere, it eventually reaches the depth where the temperature is high enough to initiate partial melting. In addition, the oceanic crust descends to a depth at which dewatering of hydrous minerals takes place, and as the water rises into the overlying mantle, it enhances melting and magma forms.

Recall that partial melting of ultramafic rock at spreading ridges yields mafic magma. Similarly, partial melting of mafic rocks of the oceanic crust yields intermediate (53% to 65% silica) and felsic (>65% silica) magmas, both of which are richer in silica than the source rock. Moreover, some of the silica-rich sediments and sedimentary rocks of continental margins are probably carried downward with the subducted plate and contribute their silica to the magma. Also, mafic magma rising through the lower continental crust is contaminated with silica-rich materials, which changes its composition.

## Hot Spots and the Origin of Magma

Most volcanism occurs at divergent and convergent plate boundaries, but there are some chains of volcanic outpourings in the ocean basins and on continents that are not near either of these boundaries. The Emperor Seamount–Hawaiian Islands, for instance, form a chain of volcanic islands 6,000 km long, and the volcanic rocks become progressively older toward the northwest (see Figure 2.9). In 1963, Canadian geologist J. Tuzo Wilson proposed that the Hawaiian Islands and other areas showing similar trends lay above a **hot spot** over which a plate moves, thereby yielding a succession of volcanoes (●**FIGURE 4.5A**).

Many geologists now think that hot-spot volcanism results from a rising **mantle plume**, a cylindrical plume of hot mantle rock that rises from perhaps near the core-mantle boundary. As it rises toward the surface, the pressure decreases on the hot rock and melting begins, thus yielding magma. Hot-spot volcanism may also account for vast flat-lying areas of overlapping lava flows on the continents or what geologists call *flood basalts* (●**FIGURE 4.5B**). Figure 2.16 shows the locations of many hot spots.

## Compositional Changes in Magma

Obviously the composition of any magma depends on what was melted in the first place. Should mafic rock melt completely, the resulting magma would also be mafic. However, as we discussed previously, partial melting yields magma that differs from its parent rock. In any case, once magma forms, it may change by **crystal settling**, which involves gravitational settling of minerals as they crystallize (●**FIGURE 4.6**). A good example is olivine, the first ferromagnesian silicate in the discontinuous branch of Bowen's reaction series, which has a density greater than the

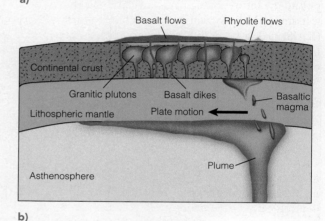

● **FIGURE 4.5 Mantle Plume and Hot Spot** (a) A mantle plume beneath oceanic crust with a hot spot. Rising magma forms a series of volcanoes that become younger in the direction of plate movement. (b) A mantle plume with an overlying hot spot yields flood basalts, and some of the continental crust melts to form felsic magma.

remaining melt and tends to sink. As a result, the remaining melt becomes richer in silica, sodium, and potassium, because much of the iron and magnesium were removed as olivine and perhaps pyroxene minerals crystallized. In other words, the remaining melt becomes more felsic.

In some thick, sheet-like plutons called *sills*, the first-formed ferromagnesian silicates are in fact concentrated in their lower parts, thus making their upper parts less mafic.

---

**hot spot** A localized zone of melting below the lithosphere that probably overlies a mantle plume; detected by volcanism at the surface.

**mantle plume** A cylindrical mass of magma rising from the mantle toward the surface; recognized at the surface by a hot spot, an area such as the Hawaiian Islands where volcanism takes place.

**crystal settling** The physical separation and concentration of minerals in the lower part of a magma chamber or pluton by crystallization and gravitational settling.

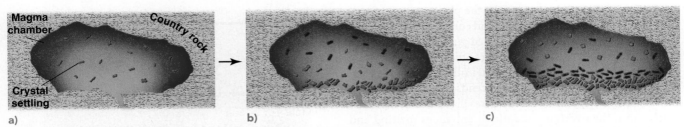

a)    b)    c)

● **FIGURE 4.6  Crystal Settling Is a Process That Changes the Composition of Magma**  (a) Early formed ferromagnesian silicates such as olivine crystallize, and because of their density, they settle to the bottom of the magma chamber. (b) Ferromagnesian silicates continue to form and settle. (c) The remaining melt becomes richer in silicon, sodium, and potassium because much of the iron and magnesium originally present is now in the ferromagnesian minerals that settled.

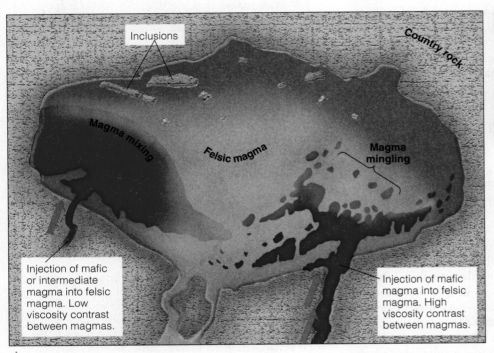

a)

● **FIGURE 4.7  Assimilation and Magma Mixing**  (a) Fragments of rock dislodged by rising magma may melt and become incorporated into the magma, a process called assimilation, or they may remain as inclusions. Magma mixing is shown on the left side where mafic magma is injected into felsic magma. (b) Dark inclusions in granitic rocks in the Sierra Nevada in California.

But even so, very little felsic magma forms by this process. Calculations show that to yield a given volume of granite (a felsic igneous rock), about 10 times as much mafic magma would have to be present initially for crystal settling to yield the volume of granite in question. If this were so, mafic igneous rocks should be much more common than felsic ones, but just the opposite is true, so it seems that something other than crystal settling must account for the large volume of felsic magma.

Once again, geologists refer to partial melting to solve this apparent dilemma. Remember that partial melting of oceanic crust and silica-rich sediments of continental margins yields magma richer in silica (more felsic) than the source rock. In addition, magma rising through continental crust changes in composition by **assimilation** as it reacts with preexisting rock, called **country rock**, with which it comes in contact (●**FIGURE 4.7A**). Country rock may be heated to 1,300°C and partially or completely melt,

provided that its melting temperature is lower than that of the magma. Because assimilated rocks rarely have the same composition as the magma, the composition of the magma changes.

The fact that assimilation takes place is clearly indicated by *inclusions*, which are incompletely melted pieces of country rock that are fairly common in many igneous rocks (●**FIGURE 4.7B**). As magma rises, it forces its way into cracks and crevasses and wedges loose pieces of country rock. Assimilation certainly takes place, but its effect on the bulk composition of magma must be slight. The reason is that the heat for melting must come from the magma, which has the

---

**assimilation**  A process whereby magma changes composition as it reacts with country rock.

**country rock**  Any preexisting rock that has been intruded by a pluton or altered by metamorphism.

effect of cooling the magma, so that only a limited amount of country rock can be assimilated.

Neither crystal settling nor assimilation can produce a significant amount of felsic magma from a mafic one. But both processes, if operating concurrently, can bring about greater changes than either process acting alone. Some geologists think that this is one way that intermediate magma (53% to 65% silica) forms where oceanic lithosphere is subducted beneath continental lithosphere.

A single volcano can erupt lavas of different composition, indicating that magmas of differing composition are present. It seems likely that some of these magmas would come into contact and mix with one another. If this is the case, we would expect that the composition of the magma resulting from **magma mixing** would be a modified version of the parent magmas (Figure 4.7a).

# IGNEOUS ROCKS: THEIR CHARACTERISTICS AND CLASSIFICATION

LO9    Define the four types of igneous rock textures

LO10   Explain how texture and composition are used to classify igneous rocks

LO11   Define ultramafic rocks

LO12   Discuss how the following igneous rock pairs are related, but also differ from each other: basalt-gabbro, andesite-diorite, and rhyolite-granite

LO13   Define pegmatite, tuff, obsidian, pumice, and scoria

We have already defined *plutonic*, or *intrusive igneous rocks*, and *volcanic*, or *extrusive igneous rocks*. Here, we have considerably more to say about the texture, composition, and classification of these rocks, which constitute one of the three major rock groups depicted in the rock cycle (see Figure 1.14).

## Igneous Rock Textures

*Texture* refers to the size, shape, and arrangement of the minerals that make up igneous rocks. Size is the most important characteristic, because mineral crystal size is related to the cooling history of magma or lava and usually indicates whether an igneous rock is plutonic or volcanic. The atoms in magma and lava are in constant motion, but when cooling begins, some atoms bond to form small nuclei. As other atoms in the liquid chemically bond to these nuclei, they do so in an orderly geometric arrangement, and the nuclei grow into crystalline *mineral grains*, the individual particles that make up igneous rocks.

If cooling takes place rapidly, as in lava flows, the rate at which mineral nuclei form exceeds the rate of growth and an aggregate of many small mineral grains forms. The result is a fine-grained or **aphanitic texture**, in which individual minerals are too small to be seen without magnification (●**FIGURE 4.8A**). If cooling is slow, the rate of growth exceeds the rate of nuclei formation, and large mineral grains form, thus yielding a coarse-grained, or **phaneritic texture**, in which minerals are clearly visible (●**FIGURE 4.8B**). Aphanitic textures usually indicate an extrusive origin, whereas rocks with phaneritic textures are commonly intrusive. However, shallow plutons might have an aphanitic texture, and the rocks that form in the interiors of thick lava flows might be phaneritic.

Another common texture in igneous rocks is one termed **porphyritic**, in which minerals of markedly different size are present in the same rock. The larger minerals are *phenocrysts* and the smaller ones collectively make up the *groundmass*, which is simply the grains between phenocrysts (●**FIGURE 4.8C**).

The only requirement for a porphyritic texture is that the phenocrysts be considerably larger than the minerals in the groundmass, which may be either aphanitic or phaneritic. Igneous rocks with porphyritic textures are designated *porphyry*, as in basalt porphyry. These rocks have more complex cooling histories than those with aphanitic or phaneritic textures and might involve, for example, magma partly cooling beneath the surface, followed by its eruption and rapid cooling at the surface.

Lava may cool so rapidly that its constituent atoms do not have time to become arranged in the ordered, three-dimensional frameworks of minerals. As a consequence, a *natural glass*, such as *obsidian* forms (●**FIGURE 4.8D**).

Some magmas contain large amounts of water vapor and other gases. These gases may be trapped in cooling lava, where they form numerous small holes or cavities known as **vesicles**; rocks with many vesicles are termed *vesicular*, as in vesicular basalt (●**FIGURE 4.8E**).

A **pyroclastic**, or **fragmental**, **texture** characterizes igneous rocks formed by explosive volcanic activity (●**FIGURE 4.8F**).

---

**magma mixing** The process whereby magmas of different composition mix together to yield a modified version of the parent magmas.

**aphanitic texture** A texture in igneous rocks in which individual mineral grains are too small to be seen without magnification; results from rapid cooling of magma and generally indicates an extrusive origin.

**phaneritic texture** Igneous rock texture in which minerals are easily visible without magnification.

**porphyritic texture** An igneous texture with minerals of markedly different sizes; results from slow cooling of magma and generally indicates an intrusive origin.

**vesicle** A small hole or cavity formed by gas trapped in cooling lava.

**pyroclastic (fragmental) texture** A fragmental texture characteristic of igneous rocks composed of pyroclastic materials.

● **FIGURE 4.8  Textures of Igneous Rocks** (a) Rapid cooling, as in lava flows, results in many small minerals and an aphanitic (fine-grained) texture. (b) Slower cooling in plutons yields a phaneritic (coarse-grained) texture. (c) These porphyritic textures indicate a complex cooling history. (d) Obsidian has a glassy texture because magma cooled too quickly for mineral crystals to form. (e) Gases expand in lava to yield a vesicular texture. (f) Microscopic view of a rock with a fragmental texture. The colorless, angular particles of volcanic glass measure up to 2 mm.

For example, ash discharged high into the atmosphere eventually settles to the surface, where it accumulates; if consolidated, it forms pyroclastic igneous rock.

## Composition of Igneous Rocks

Most igneous rocks, like the magma from which they originate, are mafic (45% to 52% silica), intermediate (53% to 65% silica), or felsic (<65% silica). A few are *ultramafic* (<45% silica), but these are probably derived from mafic magma by a process we discuss later. The parent magma plays an important role in determining the mineral composition of igneous rocks, yet it is possible for the same magma to yield a variety of igneous rocks, because its composition can change as a result of the sequence in which minerals crystallize, or by crystal settling, assimilation, and magma mixing (Figures 4.3, 4.6, and 4.7).

## Classifying Igneous Rocks

Geologists use texture and composition to classify most igneous rocks, although a few are classified mostly by texture. Notice in ●**FIGURE 4.9** that all rocks except peridotite are in pairs; the members of a pair have the same composition but different textures. Basalt and gabbro, andesite and diorite, and rhyolite and granite are compositional (mineralogical) pairs, but basalt, andesite, and rhyolite are aphanitic and most commonly volcanic, whereas gabbro, diorite, and granite are phaneritic and mostly plutonic. The volcanic and plutonic members of each pair can usually be distinguished by texture, but remember that rocks in some shallow plutons may be aphanitic, and rocks that formed in thick lava flows may be phaneritic. In other words, all of these rocks exist in a textural continuum.

The igneous rocks in Figure 4.9 are also differentiated by their mineral content. Reading across the chart from rhyolite

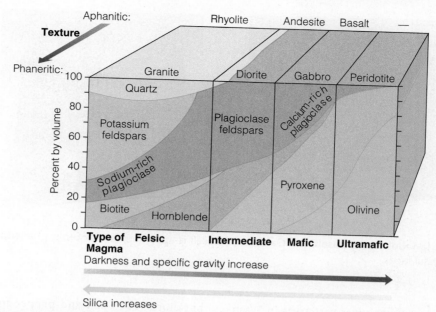

● **FIGURE 4.9 Classification of Igneous Rocks** This diagram shows the percentages of minerals, as well as the textures of common igneous rocks. For example, an aphanitic (fine-grained) rock of mostly calcium-rich plagioclase and pyroxene is basalt, whereas a phaneritic (coarse-grained) rock of the same composition is gabbro.

**Critical Thinking Question** Of the magmas that crystallized to form rhyolite, andesite, and basalt, which one would have been the most viscous? How do you know?

to andesite to basalt, for example, we see that the proportions of nonferromagnesian and ferromagnesian silicates change. The differences in composition, however, are gradual along a compositional continuum. In other words, there are rocks with compositions that correspond to the lines between granite and diorite, basalt and andesite, and so on.

**ULTRAMAFIC ROCKS** *Ultramafic rocks* (<45% silica) are composed mostly of ferromagnesian silicates. *Peridotite* contains olivine, lesser amounts of pyroxene, and usually a little plagioclase feldspar (●**FIGURE 4.10**), and *pyroxenite* is composed predominantly of pyroxene. Because these minerals are dark, the rocks are black or green. Ultramafic rocks in Earth's crust probably originate by concentration of the early-formed ferromagnesian minerals that separated from mafic magmas.

Ultramafic lava flows, called *komatiites*, are known in rocks older than 2.5 billion years but are rare or absent in younger ones. The reason is that in order to erupt, ultramafic lava must have a near-surface temperature of about 1,600°C; the surface temperatures of present-day mafic lava flows are rarely more than 1,200°C. During early Earth history, though, more radioactive decay heated the mantle to as much as 300°C hotter than now, and ultramafic lavas could erupt onto the surface. Because the amount of radiogenic heat has decreased over time, Earth has cooled, and eruptions of ultramafic lava flows are rare.

● **FIGURE 4.10 Peridotite** This specimen of the ultramafic rock peridotite is made up mostly of olivine. Notice in Figure 4.9 that peridotite is the only phaneritic rock that does not have an aphanitic counterpart. Peridotite is rare at Earth's surface, but is very likely the rock that makes up the mantle.

**BASALT-GABBRO** *Basalt* and *gabbro* are the aphanitic and phaneritic rocks that crystallize from mafic magma (45% to 52% silica) (●**FIGURE 4.11**). Both have the same composition—mostly calcium-rich plagioclase and pyroxene, with smaller amounts of olivine and amphibole (Figure 4.9). Because they contain a large proportion of ferromagnesian silicates, basalt and gabbro are dark; those that are porphyritic contain calcium plagioclase or olivine phenocrysts.

a)                                                    b)

● **FIGURE 4.11  Mafic Igneous Rocks Basalt and Gabbro**  (a) Basalt is aphanitic. (b) Gabbro is phaneritic. Notice the light reflected from the crystal faces.

Extensive basalt lava flows cover vast areas in Washington, Oregon, Idaho, and northern California (see Chapter 5). Oceanic islands such as Iceland, the Galápagos, the Azores, and the Hawaiian Islands are composed mostly of basalt, which also makes up the upper part of the oceanic crust.

Gabbro is much less common than basalt, at least in the continental crust, where it can be easily observed. Small intrusive bodies of gabbro are present in the continental crust, but intermediate to felsic intrusive rocks are much more common. However, the lower part of the oceanic crust is composed of gabbro.

**ANDESITE-DIORITE**  Intermediate-composition magma (53% to 65% silica) crystallizes to form *andesite* and *diorite*, which are compositionally equivalent fine- and coarse-grained igneous rocks (●**FIGURE 4.12**). Andesite and diorite are composed predominantly of plagioclase feldspar, with the typical ferromagnesian component being amphibole or biotite (Figure 4.9). Andesite is medium to dark gray,

but diorite has a salt-and-pepper appearance because of its white to light-gray plagioclase and dark ferromagnesian silicates.

Andesite is a common extrusive igneous rock that cooled from lava erupted in volcanic chains at convergent plate boundaries. The volcanoes of the Andes Mountains of South America and the Cascade Range in western North America are composed, in part, of andesite. Intrusive bodies of diorite are fairly common in the continental crust.

**RHYOLITE-GRANITE**  *Rhyolite* and *granite* crystallize from felsic magma (>65% silica) and are therefore silica-rich rocks (●**FIGURE 4.13**). They consist mostly of potassium feldspar, sodium-rich plagioclase, and quartz, with perhaps some biotite and rarely hornblende (Figure 4.9). Because nonferromagnesian silicates predominate, rhyolite and granite are typically light colored. Rhyolite is fine grained, although most often it contains phenocrysts of potassium feldspar or quartz, and granite is coarse grained. Granite porphyry is also fairly common.

a)                                                    b)

● **FIGURE 4.12  Intermediate Igneous Rocks Andesite and Diorite**  (a) This specimen of andesite has hornblende phenocrysts that are so numerous that the rock is classified as an andesite-hornblende porphyry. (b) Diorite has a salt-and-pepper appearance because it contains light-colored nonferromagnesian silicates and dark-colored ferromagnesian silicates.

a)    Sue Monroe

b)    Sue Monroe

● **FIGURE 4.13 Felsic Igneous Rocks Rhyolite and Granite** These rocks are typically light-colored because they contain mostly nonferromagnesian silicate minerals. The dark spots in the granite specimen are biotite mica. The white and pinkish minerals are feldspar, whereas the glassy-appearing minerals are quartz. (a) Rhyolite. (b) Granite.

Rhyolite lava flows are much less common than andesite and basalt flows. Recall that one control of magma viscosity is silica content. Thus, if felsic magma rises to the surface, it begins to cool, the pressure on it decreases, and gases are released explosively, usually yielding rhyolitic pyroclastic materials. The rhyolitic lava flows that do occur are thick and highly viscous and move only short distances.

Granite is a coarsely crystalline igneous rock with a composition corresponding to that of the field shown in Figure 4.9. Strictly speaking, not all rocks in this field are granites. For example, a rock with a composition close to the line separating granite and diorite is called *granodiorite*. To avoid the confusion that might result from introducing more rock names, we will follow the practice of referring to rocks to the left of the granite-diorite line in Figure 4.9 as *granitic*.

Granitic rocks are by far the most common plutonic igneous rocks, and they are restricted to the continents (see GEO-FOCUS ). Most granitic rocks were intruded at or near convergent plate margins during mountain-building episodes. When these mountainous regions are uplifted and eroded, the vast bodies of granitic rocks forming their cores are exposed.

**PEGMATITE** The term *pegmatite* refers to a particular texture rather than a specific composition, but most pegmatites are composed primarily of quartz, potassium feldspar, and sodium-rich plagioclase, thus corresponding closely to granite (●**FIGURE 4.14**). The most remarkable feature of pegmatites is the size of their minerals, which measure at least 1 cm across, and in some pegmatites they measure tens of centimeters or meters (see Figure 3.1a). Many pegmatites are adjacent to large granite plutons and are composed of minerals that formed from the water-rich magma that remained after most of the granite crystallized.

Courtesy of Steve Stahl

● **FIGURE 4.14 Pegmatite** This pegmatite, the light-colored rock, is exposed in the Black Hills of South Dakota.

When felsic magma cools and forms granite, the remaining water-rich magma has a lower density and viscosity and invades cracks in the nearby rocks where minerals crystallize. This water-rich magma also contains elements that rarely enter into the common minerals that form granite. Pegmatites that are essentially very coarsely crystalline granite are simple pegmatites, whereas those with minerals containing elements such as lithium, beryllium, cesium, boron, and several others are complex pegmatites.

The formation and growth of mineral-crystal nuclei in pegmatites are similar to those processes in other magmas, but with one critical difference: The water-rich magma from which pegmatites crystallize inhibits the formation of nuclei. However, some nuclei do form, and because the appropriate ions in the liquid can move easily and attach themselves to a growing crystal, individual minerals may grow very large.

# GEO-FOCUS

## Granite—Common, Attractive, and Useful

Granite and closely related rocks are common in the large batholiths of the world. Indeed, if you live in the Northeast you are probably aware of New Hampshire's Old Man of the Mountain, a rugged granite outcrop that collapsed in 2003. Mount Rushmore in South Dakota is another mass of granite with the carved images of Presidents Washington, Jefferson, Lincoln, and Theodore Roosevelt (see Chapter Opening photo). The Sierra Nevada batholith, mostly in California, is a huge composite body of granitic rocks with the highest peak in the continental United States.

Today, no one doubts that granite and related rocks formed when magma cooled beneath the surface, but this has not always been the case. In 1787, a German professor of mineralogy, Abraham Gottlob Werner (1749–1817), proposed that all rocks, including basalt and granite, had precipitated in an orderly sequence from a primeval, worldwide ocean. Werner further claimed that metamorphic rocks and granite were the oldest rocks on Earth. His concept, called Neptunism, was popular for many years, but it eventually became apparent that basalt cooled from lava, whereas granite cooled from magma as we have discussed. Furthermore, rock type is not an indication of its age as Werner proposed.

Granitic rocks are quarried, cut and polished, and otherwise shaped to meet specific purposes. Any stone used in this fashion is called *dimension stone*, and in the case of granitic rocks, they may be used for facing stones on buildings, walkways (●**FIGURE 1**), tombstones, mantelpieces, countertops, and monuments. A 23-m-high, red granite obelisk weighing 250 metric tons was carved from a single piece of granite at Aswan, Egypt, about 3,300 years ago and transported nearly 200 km to its present location at Luxor, Egypt.

Plymouth Rock at Plymouth, Massachusetts, has great symbolic value, but otherwise it is a rather ordinary rock (●**FIGURE 2**). Actually, it is a rather small stone from the 600-million-year-old Dedham Granodiorite, which was transported by a glacier to its present location. As its name indicates, granodiorite has a composition between granite and diorite. Legend has it that Plymouth Rock is the landing place where the Pilgrims first set foot in the New World in 1620. In fact, the Pilgrims first landed at Provincetown on Cape Cod and later went to the Plymouth area, but even then they probably landed north of Plymouth.

● **FIGURE 1 Plymouth Rock** Plymouth Rock is the first piece of land on which the Pilgrims supposedly set foot when they arrived in Massachusetts in 1620. The rock first became a patriotic icon during the Revolutionary War.

● **FIGURE 2 Granite Dimension Stone** Any rock that is shaped for a specific purpose is called dimension stone. The granite used here for the walkway, steps, and facing for the plant enclosure is at the Mena House Oberia, a luxury hotel near Cairo, Egypt.

The term *granite* has a specific meaning in geology, but in commercial use the term is applied to several rocks, such as basalt and gabbro, as in black granite. True granite is, in fact, used for a variety of purposes, although many granite-like substances are manufactured. This *cultured granite*, as it is called, is made by chemically bonding material and molding them into various shapes, which cannot be done with natural stones.

**OTHER IGNEOUS ROCKS** Geologists classify the igneous rocks in Figure 4.9 by texture and composition, but a few others are identified primarily by their textures. Much of the fragmental material erupted by volcanoes is *ash*, a designation for pyroclastic materials measuring less than 2.0 mm, most of which consists of pieces of minerals or shards of volcanic glass (Figure 4.8f). The consolidation of ash forms the pyroclastic rock *tuff* (●**FIGURE 4.15A**). Most tuff is silica-rich and light colored and is appropriately called *rhyolite tuff*. Some ash flows are so hot that as they come to rest, the ash particles fuse together and form a *welded tuff*. Consolidated deposits of larger pyroclastic materials, such as cinders, blocks, and bombs, are *volcanic breccia*.

Both *obsidian* and *pumice* are varieties of volcanic glass (●**FIGURE 4.15B, C**). Obsidian may be black, dark gray, red, or brown, depending on the presence of iron. Obsidian breaks with a conchoidal (smoothly curved) fracture typical of glass. Analyses of many samples indicate that most obsidian has a high silica content and is compositionally similar to rhyolite.

*Pumice* is a variety of volcanic glass containing numerous vesicles that develop when gas escapes through lava and forms a froth (Figure 4.15c). If pumice falls into water, it can be carried great distances, because it is so porous and light

that it floats. Another vesicular rock is *scoria*. It is more crystalline and denser than pumice, but it has more vesicles than solid rock (●**FIGURE 4.15D**).

# INTRUSIVE IGNEOUS BODIES: PLUTONS

**LO14**  Define pluton
**LO15**  Explain the differences between dikes, sills, and laccoliths
**LO16**  Explain how batholiths differ from stocks

Unlike volcanism and the origin of volcanic rocks, we can study intrusive igneous bodies, collectively called **plutons**, only indirectly, because intrusive rocks form when magma cools and crystallizes within Earth's crust (●**FIGURE 4.16A**). We can observe these rock bodies only when uplift and deep erosion have taken place, thereby exposing them at

**pluton**  An intrusive igneous body that forms when magma cools and crystallizes within the crust, such as a batholith or sill.

●**FIGURE 4.15 Examples of Igneous Rocks Classified Primarily by Their Texture** (a) Tuff is composed of pyroclastic materials such as those in Figure 4.8f. (b) The natural glass obsidian. (c) Pumice is glassy and extremely vesicular. (d) Scoria is also vesicular, but it is darker, heavier, and more crystalline than pumice.

**Critical Thinking Question** Why do pumice and scoria have so many vesicles, but obsidian has none?

the surface. Furthermore, geologists cannot duplicate the conditions under which intrusive rocks form except in small laboratory experiments.

Geologists recognize several types of plutons based on their geometry (three-dimensional shape) and relationships to the country rocks. In terms of their geometry, plutons are tabular, cylindrical, or irregular (massive). Furthermore, they may be **concordant**, meaning they have boundaries that parallel the layering in the country rock, or **discordant**, with boundaries that cut across the country rock's layering (Figure 4.16a).

## Dikes, Sills, and Laccoliths

**Dikes** and **sills** are tabular, or sheetlike, igneous bodies that differ only in that dikes are discordant and sills are concordant (Figure 4.16a). Dikes are quite common and range from a few centimeters to more than 100 m thick (**FIGURE 4.16B**). Invariably, they are intruded into preexisting fractures or where fluid pressure is great enough for them to form their own fractures as they move upward into country rock.

Sills are tabular, just as dikes are, but they are concordant. Many sills are a meter or less thick, although some are much thicker. Most sills were intruded into

---

**concordant** Igneous body whose boundaries parallel the layering in the country rock.

**discordant** Igneous body with boundaries that cut across the layering in the country rock.

**dike** A tabular, or sheetlike, discordant pluton.

**sill** A tabular, or sheetlike, concordant pluton.

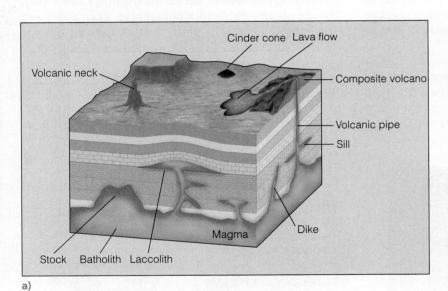

a)

c)

b)

● **FIGURE 4.16 Plutons** (a) Block diagram showing various plutons. Some plutons cut across the layering in country rock and are discordant, whereas others parallel the layering and are concordant. (b) The dark material in this image is a dike that cuts through Cenozoic-age sedimentary rocks near Dulce, New Mexico. (c) A volcanic neck in Monument Valley Navajo Tribal Park, Arizona. This landform is 457 m high. Most of the original volcano was eroded, leaving only this remnant.

sedimentary rocks, but eroded volcanoes also reveal that sills are injected into piles of volcanic rocks. In fact, some of the inflation of a volcano that precedes an eruption may be caused by the injection of sills (see Chapter 5). In contrast to dikes, which follow zones of weakness, sills are intruded between layers in country rock when the fluid pressure is great enough for the magma to actually lift the overlying rocks.

Under some circumstances, a sill inflates and causes the overlying rocks to bow upward, forming an igneous body called a **laccolith** (Figure 4.16a). A laccolith has a flat floor and is domed up in its central part, giving it a mushroom-like geometry. Like sills, laccoliths are rather shallow intrusions that lift the overlying rocks.

## Volcanic Pipes and Necks

Volcanoes have a cylindrical conduit known as a **volcanic pipe** that connects to an underlying magma chamber. Magma rises through this structure; however, when a volcano ceases to erupt, its slopes are attacked by weathering and erosion, but the magma in the pipe is commonly more resistant to erosion and is left as a remnant called a **volcanic neck** (●**FIGURE 4.16C**). Several volcanic necks are found in the southwestern United States, especially in Arizona and New Mexico, and others are recognized elsewhere.

## Batholiths and Stocks

A **batholith**, the largest of all plutons, must have at least 100 km² of surface area, and most are far larger (Figure 4.16a). A **stock**, in contrast, is similar but smaller. Some stocks are simply parts of large plutons that, once exposed by erosion, are parts of batholiths. Both batholiths and stocks are mostly discordant, although locally they may be concordant, and batholiths, especially, consist of multiple intrusions. In other words, a batholith is a large composite body produced by repeated, voluminous intrusions of magma in the same region.

The igneous rocks that make up batholiths are typically granitic, although diorite may also be present. Batholiths and stocks are emplaced mostly near convergent plate boundaries during episodes of mountain building. One example is the Sierra Nevada batholith of California, which formed during millions of years. Other large batholiths in North America include the Idaho batholith, the Boulder batholith in Montana, and the Coast Range batholith in British Columbia, Canada.

Mineral resources are found in rocks of batholiths and stocks, and in the adjacent country rocks. Near Salt Lake City, Utah, copper is mined from the mineralized rocks of the Bingham stock, a composite pluton composed of granite and granite porphyry. Granitic rocks also are the primary source of gold, which forms from mineral-rich solutions moving through cracks and fractures of the igneous body.

# THE ORIGIN OF BATHOLITHS

**LO17  Discuss the origin of batholiths and their emplacement into the crust**

Geologists realized long ago that the origin of batholiths posed a space problem. What happened to the rock that was once in the space now occupied by a batholith? One solution to the space problem is that these large igneous bodies melted their way into the crust. In other words, they simply assimilated the country rock as they moved upward (Figure 4.7). The presence of inclusions, especially near the tops of some plutons, indicates that assimilation does occur. Nevertheless, as we noted, assimilation is a limited process, because magma cools as country rock is assimilated. Calculations indicate that far too little heat is available in magma to assimilate the huge quantities of country rock necessary to make room for a batholith.

Geologists are now in general agreement that batholiths were emplaced by *forceful injection* as magma moved upward (●**FIGURE 4.17A**). Recall that granite is derived from viscous felsic magma and therefore rises slowly. It appears that the magma deforms and shoulders aside the country rock, and as it rises farther, some of the country rock fills the space beneath the magma.

Some batholiths do show evidence of having been emplaced by forceful injection, but this mechanism probably occurs in the deeper parts of the crust where temperature and pressure are high and the country rocks are easily deformed in the manner described. At shallower depths, the crust is more rigid and tends to deform by fracturing. In this environment, batholiths may move upward by **stoping**, a process in which rising magma detaches and engulfs pieces of country rock (●**FIGURE 4.17B**). According to this concept, magma moves up along fractures and the planes separating layers of country rock. Eventually, pieces of country rock detach and settle into the magma. No new room is created during stoping; the magma simply fills the space formerly occupied by country rock.

---

**laccolith**  A concordant pluton with a mushroomlike geometry.

**volcanic pipe**  The conduit connecting the crater of a volcano with an underlying magma chamber.

**volcanic neck**  An erosional remnant of the material that solidified in a volcanic pipe.

**batholith**  An irregularly shaped, discordant pluton with at least 100 km² of exposed surface area.

**stock**  An irregularly shaped, discordant pluton with a surface area smaller than 100 km².

**stoping**  A process in which rising magma detaches and engulfs pieces of the country rock.

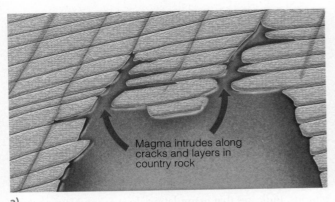

Magma intrudes along cracks and layers in country rock

a)

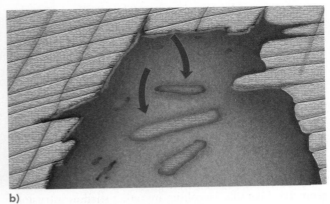

b)

● **FIGURE 4.17 Emplacement of a Batholith by Stoping** (a) Stoping takes place when magma rises into the crust by detaching and engulfing pieces of country rock. (b) Some of the detached blocks may be assimilated, and some may remain as inclusions (Figure 4.7b).

# Key Concepts Review

- Magma is the term for molten rock below Earth's surface, whereas the same material at the surface is lava.
- Geologists recognize two broad categories of igneous rocks: volcanic or extrusive and plutonic or intrusive.
- Silica content is what differentiates ultramafic (<45% silica), mafic (45%–52% silica), intermediate (53%–65% silica), and felsic (>65% silica) magmas.
- Magma and lava viscosity depends mostly on temperature and composition: the higher the temperature, the lower the viscosity; the more silica, the greater the viscosity.
- Texture and composition are the criteria used to classify igneous rocks, although a few are defined mostly by texture.
- Minerals crystallize from magma and lava when small crystal nuclei form and grow.
- Rapid cooling accounts for the aphanitic textures of volcanic rocks, whereas comparatively slow cooling yields the phaneritic textures of plutonic rocks. Igneous rocks with markedly different-sized minerals are porphyritic.
- Igneous rock composition is determined mostly by the composition of the parent magma, but magma composition can change so that the same magma may yield more than one kind of igneous rock.
- According to Bowen's reaction series, cooling mafic magma yields a sequence of minerals, each of which is stable within specific temperature ranges. Only ferromagnesian silicates are found in the discontinuous branch of Bowen's reaction series. The continuous branch yields only plagioclase feldspars that become increasingly enriched with sodium as cooling occurs.
- A chemical change in magma may take place as early ferromagnesian silicates form and, because of their density, settle in the magma.
- Compositional changes can also take place in magma when it assimilates country rock or one magma mixes with another.
- Intrusive igneous bodies known as plutons vary in their geometry and their relationship to country rock: some are concordant, whereas others are discordant.
- The largest plutons, known as batholiths, consist of multiple intrusions of magma during long periods of time.
- Most plutons, including batholiths, are emplaced mostly near convergent plate boundaries.

# Important Terms

aphanitic texture    77

assimilation    76

batholith    85

Bowen's reaction series    73

concordant    84

country rock    76

crystal settling    75

dike    84

discordant    84

felsic magma    71

hot spot    75

igneous rock    70

intermediate magma    71

laccolith    85

lava    70

lava flow    70

mafic magma    70

magma    70

magma chamber    72

magma mixing    77

mantle plume    75

phaneritic texture    77

pluton    83

plutonic (intrusive igneous) rock    70

porphyritic texture    77

pyroclastic (fragmental) texture    77

pyroclastic material    70

sill    84

stock    85

stoping    85

ultramafic magma    70

vesicle    77

viscosity    71

volcanic (extrusive igneous) rock    70

volcanic neck    85

volcanic pipe    85

# Review Questions

1. A laccolith is what type of pluton?
   a. discordant/massive
   b. concordant/porphyritic
   c. felsic/gabbroic
   d. concordant/mushroom-shaped
   e. ultramafic/viscous

2. Which pair of the following igneous rocks has the same mineral composition?
   a. andesite-diorite
   b. granite-basalt
   c. tuff-gabbro
   d. obsidian-pegmatite
   e. peridotite-rhyolite

3. Which one of the following statements about batholiths is correct?
   a. They are composed of a single, huge mass of magma.
   b. They consist primarily of basalt and gabbro.
   c. They form at divergent plate boundaries.
   d. They are mostly concordant.
   e. They are composed mostly of granitic rocks.

4. The phenomenon by which pieces of country rock are detached and engulfed by rising magma is
   a. crystal settling.
   b. magma mixing.
   c. assimilation.
   d. stoping.
   e. forceful injection.

5. The order in which minerals crystallized from mafic magma is called
   a. crystal settling.
   b. Bowen's reaction series.
   c. Richard's assimilation index.
   d. plutonic mixing.
   e. crustal differentiation.

6. How do assimilation and stoping contribute to the emplacement of batholiths in the continental crust?

7. What are the controls on the viscosity of a lava flow?

8. Two phaneritic rocks have the following compositions: Specimen 1: 5% biotite, 20% sodium-rich plagioclase, 65% potassium feldspar, and 10% quartz; Specimen 2: 10% olivine, 55% pyroxene, 5% hornblende, and 30% calcium-rich plagioclase. Use Figure 4.9 and classify these rocks. Which one would you expect to be the darkest and densest? Explain.

9. How do aphanitic and phaneritic textures form, and what do they tell us about rocks with these textures?

# Creative Thinking Visual Question

●**FIGURE 1** shows an imaginary landscape made up of sandstone (SS) and the following igneous rocks: basalt (B), diorite (D), granite (GR), and pegmatite (PG). What type of igneous body is composed of gabbro? Is the diorite older or younger than the sandstone, and is the granite older or younger than the diorite? How do you know?

● **FIGURE 1** Imaginary landscape showing the relationships between several bodies of igneous rocks and sedimentary rocks.

# 5

# VOLCANOES AND VOLCANISM

Anak Krakatau, which means "daughter of Krakatau," is a small volcano in the Sundra Strait between Java and Sumatra that is the remnant of a much larger volcano known as Krakatau. In 1883, Krakatau erupted with such explosive force that most of the island disappeared and the volcano collapsed in on itself forming a huge depression, known as a caldera. Renewed volcanism built up Anak Krakatau, which rose from the ocean in 1927. It is shown here erupting on November 3, 2010.

# INTRODUCTION

Volcanic eruptions are certainly one of the most awe-inspiring geologic phenomena, especially when glowing streams of lava pour from volcanic vents, or when volcanoes erupt explosively and eject pyroclastic materials high into the atmosphere. Lava flows do, in fact, destroy buildings and cover what may have been productive land, but overall, they are the least dangerous manifestation of volcanic eruptions.

It is difficult to get a precise number of *active volcanoes*, that is, volcanoes that have erupted during historic time, but a common estimate is about 550, 10 to 20 of which are erupting at any one time. We also know of several active submarine volcanoes, but there are certainly many more. Most volcanic eruptions are minor and go unreported in the popular press unless they occur in populated areas or have tragic consequences. For example, the April 2010 eruptions of Eyjafjallajökull in Iceland were not very large, but they took place beneath 200 m of glacial ice, resulting in steam explosions that ejected huge amounts of volcanic ash into the atmosphere, disrupting air traffic over the North Atlantic for several days (●**FIGURE 5.1**).

In addition to active volcanoes, Earth has many *dormant volcanoes*, meaning they have not erupted in historic time but may do so again. The distinction between active and dormant is not easy to make, because some volcanoes that have not erupted for centuries become active again. For instance, in 1991, Mount Pinatubo in the Philippines erupted for the first time in 600 years, whereas Mount Vesuvius in Italy had not erupted in human memory until its A.D. 79 eruptions that destroyed the towns of Herculaneum, Pompeii, and Stabiae (●**FIGURE 5.2**). Thousands of volcanoes have not erupted during historic time and show no signs of doing so in the future, so they are *extinct* or *inactive* volcanoes. The Sutter Buttes in northern California last erupted about 1.5 million years ago and now show no activity, and will probably not erupt again.

Erupting volcanoes do cause fatalities and considerable damage, but when considered in the context of Earth history, volcanism is actually a constructive process. Volcanic emissions during Earth's early history played an important part in the origin and evolution of the atmosphere and hydrosphere. The continents are composed, in part, of volcanic and plutonic rocks, and oceanic crust is produced continuously by extrusive and intrusive igneous activity at divergent plate boundaries. Volcanic islands such as the Hawaiian Islands, Iceland, the Azores, and many others owe their existence, and in some cases, their continuing

● **FIGURE 5.1 Eyjafjallajökull Erupting, Iceland** Eyjafjallajökull in Iceland erupting a huge cloud of volcanic ash and steam on April 17, 2010. The eruption, although not very large, took place beneath a glacier that added to the cloud over the volcano. Air traffic over the North Atlantic was disrupted for several days.

bl3/ZUMA Press/Newscom

Sue Monroe

● **FIGURE 5.2 The A.D. 79 Eruption of Mount Vesuvius** Body casts of some of the volcano's victims in Pompeii. Approximately 2,000 bodies in this city of 20,000 have been found, but certainly far more were killed.

evolution to volcanic eruptions. In addition, weathering of lava, pyroclastic materials, and volcanic mudflows in areas such as Indonesia converts them to productive soils.

An excellent reason to study volcanism is that it illustrates the complex interactions between Earth's systems. The emission of gases and pyroclastic materials has an immediate and profound impact on the atmosphere, hydrosphere, and biosphere, at least in the vicinity of an eruption. And in some cases, the effects are worldwide, as they were following the eruptions of Tambora in 1815, Krakatau in 1883, and Pinatubo in 1991.

# VOLCANOES AND VOLCANISM

**LO1** Define a volcano

**LO2** List the composition of present-day volcanic gases

**LO3** List and discuss the dangers from volcanic gases

**LO4** Define the two types of basalt lava flows

**LO5** Explain how pillow lavas and columnar joint form

**LO6** Define each of the different types of pyroclastic materials

A **volcano** is a hill or mountain that forms around a vent where lava, pyroclastic materials, and gases erupt, so it is a *landform*—a feature on Earth's surface. The term **volcanism** refers to all processes related to the rise and discharge of magma and gases at the surface or into the atmosphere. Thus, volcanism accounts for the origin of all volcanic (extrusive igneous) rocks, such as basalt, tuff, and obsidian, as well as volcanoes, but not all volcanism results in the origin of volcanoes. Some eruptions take place along fissures and build up basalt plateaus (discussed later in this chapter).

All of the other terrestrial planets (Mercury, Venus, and Mars) and Earth's Moon were volcanically active during their early histories, but now the only one with possible active volcanoes is Venus. In addition, volcanoes are known on three moons in the outer solar system. Triton, a moon of Neptune, and Enceladus, a moon of Saturn, have ice volcanoes, and volcanoes on Io, a moon of Jupiter, erupt sulfur compounds.

## Volcanic Gases

Volcanic gases from present-day volcanoes are 50% to 80% water vapor, with lesser amounts of carbon dioxide, nitrogen, sulfur gases—especially sulfur dioxide and hydrogen sulfide—and very small amounts of carbon monoxide, hydrogen, and chlorine. In areas of recent volcanism, emission of gases from fumaroles (volcanic vents) continues, and one cannot help noticing the rotten-egg odor of hydrogen sulfide gas (●**FIGURE 5.3A**).

As magma rises toward the surface, pressure is reduced and the contained gases begin to expand. In highly viscous, felsic magma, expansion is inhibited and gas pressure increases. Eventually, the pressure may become great enough to cause an explosion and produce pyroclastic materials. In contrast, low-viscosity mafic magma allows gases to expand and escape easily, so mafic magma usually erupts rather quietly.

Most volcanic gases quickly dissipate in the atmosphere and pose little danger to humans, but on several occasions, they have had far-reaching climatic effects or have caused fatalities. In 1783, sulfur gases erupting from Laki fissure in Iceland were responsible for the Blue Haze Famine, crop failures, and the deaths of 24% of Iceland's population. In 1816, a persistent "dry fog" caused unusually cold spring and summer weather in Europe and eastern North America. In North America, the year of 1816 was called "The Year Without a Summer" or "Eighteen Hundred and Froze-to-Death." Killing frost during the summer resulted in crop failures and food shortages. This particularly cold summer was attributed to the huge eruption of Tambora in 1815 in Indonesia, although the eruption of Mayon volcano in the Philippines during the previous year also probably contributed.

In 1986, in the African nation of Cameroon, 1,746 people died when a cloud of carbon dioxide engulfed them. The gas accumulated in the waters of Lake Nyos, which occupies a volcanic caldera. Scientists disagree about what caused the gas to suddenly burst forth from the lake, but once it did, it flowed downhill along the surface because it was denser than air. In fact, the density and velocity of the gas cloud were great enough to flatten vegetation, including trees, a few kilometers from the lake. Unfortunately, thousands of animals and many people, some as far as 23 km from the lake, were asphyxiated.

---

**volcano** A hill or mountain formed around a vent as a result of the eruption of lava and pyroclastic materials.

**volcanism** The processes whereby magma and its associated gases rise through the crust and are extruded onto the surface or into the atmosphere.

a)

b)

● **FIGURE 5.3 Volcano Gases** (a) This fumarole called the Black Growler is in Norris Geyser Basin in Yellowstone National Park, Wyoming. (b) Volcanic gases are mostly water vapor, but carbon dioxide, sulfur dioxide, and several others are emitted as well. Sulfur dioxide produces a haze and the unpleasant odor of sulfur, or what people in Hawaii call *vog*. Seen here are gases billowing from Halema 'uma'u Crater on Kilauea Volcano in Hawaii.

**Critical Thinking Question**   Are volcanic gases dangerous to humans, and if so, why?

Residents of the island of Hawaii have coined the term *vog* for volcanic smog. Kilauea Volcano has been erupting continuously since 1983, releasing lava and copious quantities of carbon dioxide and sulfur dioxide (●**FIGURE 5.3B**). Carbon dioxide is no problem because it dissipates quickly in the atmosphere, but sulfur dioxide produces a haze and the unpleasant odor of sulfur. Vog probably poses little or no health risk for tourists, but a long-term threat exists for residents of the west side of the island where vog is most common.

## Lava Flows

Movies and TV shows depict lava flows as a great danger to humans, but only rarely do they cause fatalities. The reason is that most lava flows do not move very rapidly, and because they are fluid they follow low areas. So once a flow begins advancing, determining the path it will take is fairly easy, and anyone in areas likely to be affected can be evacuated. From April 1990 to January 1991, lava flows covered Kalapana, Hawaii, and in the process destroyed 180 homes, highways, and points of archaeological interest. Civil Defense authorities working with geologists at the Hawaiian Volcano Observatory made important decisions regarding evacuations and road closures, and as a result there were no injuries or fatalities.

Even low-viscosity flows, such as those in Hawaii, do not move very quickly. They can move much more rapidly, though, when their margins cool to form a channel, and especially when insulated on all sides as in a **lava tube** (●**FIGURE 5.4A**), where geologists have recorded speeds of more than 50 km/hr. A lava tube forms within a flow when its margins and upper surface solidify, thereby forming a conduit through which lava can move swiftly and for great distances. When an eruption ceases, the tube drains, leaving

an empty tunnel-like structure (●**FIGURE 5.4B**). Part of the roof of a lava tube may collapse to form a *skylight* through which an active flow can be observed, or access can be gained to an inactive lava tube.

Geologists in Hawaii characterized basalt lava flows as *pahoehoe* or *aa*, although these terms are now also used elsewhere. **Pahoehoe** (pronounced *pay-hoy-hoy*) has a smooth, ropy surface much like taffy (●**FIGURE 5.5A**). The surface of an **aa** (pronounced *ah-ah*) flow consists of rough, jagged, angular blocks and fragments (●**FIGURE 5.5B**). Pahoehoe flows are hotter and thinner than aa flows; indeed, aa flows are viscous enough to break into blocks and move forward as a wall of rubble. It is worth noting that a pahoehoe flow may change along its length to aa as its viscosity increases, partly because it cools, but aa does not change to pahoehoe.

Much of the igneous rock in the upper oceanic crust is a distinctive type made up of bulbous masses of basalt that resemble pillows, hence the name **pillow lava**. Geologists realized long ago that pillow lava (●**FIGURE 5.6**) forms when lava is rapidly chilled underwater, but its formation was not observed until 1971. Divers near Hawaii saw pillows form when a blob of lava broke through the crust of an underwater lava flow and cooled very quickly, forming a pillow- or loaf-shaped structure with a glassy exterior. The remaining

---

**lava tube**  A tunnel beneath the solidified surface of a lava flow through which lava moves; also, the hollow space left when the lava within a tube drains away.

**pahoehoe**  A type of lava flow with a smooth, ropy surface.

**aa**  Lava flow with a surface of rough, angular blocks and fragments.

**pillow lava**  Bulbous masses of basalt, resembling pillows, formed when lava is rapidly chilled under water.

a)

b)

Photo by J.B. Judd/USGS

Photo provided by Hawai'i Volcanoes National Park

● **FIGURE 5.4 Lava Tubes** Lava tubes consisting of hollow spaces beneath the surfaces of lava flows are common in many areas. (a) An active lava tube in Hawaii. Part of the tube's roof has collapsed, forming a skylight. (b) A lava tube in Hawaii after the lava has drained out.

a)

J.D. Griggs, USGS

b)

Robert Tilling/USGS

● **FIGURE 5.5 Pahoehoe and aa Lava Flows** Pahoehoe and aa were named for lava flows in Hawaii, but the same kinds of flows are found in many other areas. (a) An excellent example of the taffylike appearance of pahoehoe. (b) An aa lava flow advances over an older pahoehoe flow. Notice the rubbly nature of the aa flow.

**Critical Thinking Question** What controls whether a lava flow is aa or pahoehoe?

fluid lava inside then broke through the crust of an already-formed pillow and formed another pillow, and so on.

Mafic lava flows, and some intermediate ones, as well as some rocks in dikes, sills, and volcanic necks, show a pattern of columns bounded by fractures, or what geologists call **columnar joints**. For columnar joints to form, a lava flow must cease moving, then cool and contract, thereby setting up forces that cause fractures, known as *joints*, to open. On

a lava flow's surface, the fractures are commonly polygonal (often six-sided) cracks that extend downward, thus outlining columns with their long axes perpendicular to the cooling surface (●**FIGURE 5.7**).

_____

**columnar joints** Columns in igneous rocks bounded by fractures that formed when lava or magma cooled and contracted.

● **FIGURE 5.6 Pillow Lava** Much of the upper part of the oceanic crust is composed of pillow lava that formed when lava erupted underwater. (a) Pillow lava on the seafloor in the Pacific Ocean about 240 km west of Oregon that formed about five years before this photo was taken. (b) Ancient pillow lava, now on land in Kenai Fjords National Park, Alaska.

● **FIGURE 5.7 Columnar Jointing** Columnar jointing is seen mostly in mafic lava flows and related plutonic rocks. (a) Columnar jointing at Reynisfjara Beach in Iceland. The photographer standing on one of the joints provides scale for the size of the columns. (b) Surface view of columns at Devil's Postpile National Monument, California. The straight lines and polished surface resulted from abrasion by a sediment-laden glacier that moved over its surface. Also, note the polygonal shape of the columns.

## Pyroclastic Materials

In addition to lava flows, erupting volcanoes eject pyroclastic materials (●**FIGURE 5.8**), especially **volcanic ash**, a designation for pyroclastic particles that measure less than 2.0 mm. In some cases, ash is ejected into the atmosphere and settles to the surface as an *ash fall*. In contrast to an ash fall, an *ash flow* is a cloud of ash and gas that flows along or close to the land surface. Ash flows can move faster than 100 km per hour, and some cover vast areas.

In populated areas adjacent to volcanoes, ash falls and ash flows pose serious problems, and volcanic ash in the atmosphere is a hazard to aviation. The most serious plane incident occurred in 1989, when ash from Redoubt volcano in Alaska caused all four jet engines to fail on KLM Flight 867. The plane, carrying 231 passengers, nearly crashed when it fell more than 3 km before the crew could restart the engines. The plane landed safely in Anchorage, Alaska, but it required $80 million in repairs. A more recent, but less grave, event occurred in 2016 when ash from Pavlof Volcano in Alaska disrupted flights in the area.

In addition to volcanic ash, volcanoes erupt *lapilli*, pyroclastic materials that measure from 2 mm to 64 mm, and *blocks* and *bombs*, both larger than 64 mm (Figure 5.8). Bombs have a twisted, streamlined shape, which indicates that they were erupted as globs of magma that cooled and solidified during their flight through the air. Blocks, in contrast, are angular pieces of rock ripped from a volcanic conduit or pieces of a solidified crust of a lava flow. Because of their size, lapilli, bombs, and blocks are confined to the immediate area of an eruption.

**volcanic ash** Pyroclastic materials that measure less than 2 mm.

Sue Monroe

● **FIGURE 5.8 Pyroclastic Materials** Pyroclastic materials are all particles ejected from volcanoes, especially during explosive eruptions. The volcanic bomb is elongate, because it was molten when it descended through the air. The lapilli were collected at a small volcano in Oregon, whereas the ash came from the 1980 eruption of Mount St. Helens in Washington.

# TYPES OF VOLCANOES

**LO7** Define a crater and caldera

**LO8** Define the four types of volcanoes

Simply put, a volcano is a hill or mountain that forms around a vent where lava, pyroclastic materials, and gases erupt. Although volcanoes vary in size and shape, all have a conduit or conduits leading to a magma chamber beneath the surface. Vulcan, the Roman deity of fire, was the inspiration for calling these mountains volcanoes, and because of their danger and obvious connection to Earth's interior, they have been held in awe by many cultures.

Most volcanoes have a circular depression known as a **crater** at their summit, or on their flanks, that forms by explosions or collapse. Most craters are less than 1 km across, whereas much larger rimmed depressions on volcanoes are **calderas**. In fact, some volcanoes have a summit crater within a caldera. Calderas are huge structures that form following voluminous eruptions, during which part of a magma chamber drains and the mountain's summit collapses into the vacated space below (see GEO-FOCUS). An excellent example is mis-named Crater Lake in Oregon (●**FIGURE 5.9**). Crater Lake is actually a steep-rimmed caldera that formed about 7,700 years ago in the manner just described; it is more than 1,200 m deep and measures 9.7 by 6.5 km. As impressive as Crater Lake is, it is not nearly as large as some other calderas, such as the Toba caldera in Sumatra, which is 100 km long and 30 km wide.

Geologists recognize several major types of volcanoes, but one must realize that each volcano is unique in its history of eruptions and development. For instance, the frequency of eruptions varies considerably; the Hawaiian volcanoes and Mount Etna on Sicily have erupted repeatedly, whereas Pinatubo in the Philippines erupted in 1991 for the first time in 600 years. Furthermore, some volcanoes are complex and have the characteristics of more than one type of volcano.

## Shield Volcanoes

Volcanoes that look much like the outer surface of a shield lying on the ground with the convex side up are **shield volcanoes** (●**FIGURE 5.10**). They are composed almost entirely of mafic lava flows that have low viscosity, so the flows spread out and form thin layers that slope only 2 to 10 degrees. Erupting shield volcanoes, sometimes called *Hawaiian-type eruptions*, are rather quiet compared to eruptions of many other volcanoes, particularly those at convergent plate boundaries. Magma that rises to the surface and issues as lava flows pose little danger to humans, although incandescent lava may be forcefully ejected as lava fountains, some up to 400 m high, when magma reaches the surface and gases expand.

Although eruptions of shield volcanoes tend to be rather quiet, some of the Hawaiian volcanoes have, on occasion, produced sizable explosions when groundwater instantly vaporized as it came into contact with magma. One such explosion in 1790 killed about 80 warriors in a party headed by Chief Keoua, who was leading them across the summit of Kilauea volcano.

Kilauea volcano is impressive, because it has been erupting continuously since January 3, 1983, making it the longest recorded eruption. During these 35 years, more than 4.0 km³ of molten rock has flowed out at the surface, much of it reaching the sea and forming 3.25 km² of new land on the island of Hawaii. Unfortunately, lava flows and *lahars* from Kilauea have caused considerable property damage. During its most recent eruption in 2018, nearly 2,000 residents had to be evacuated from the Leilani Estates subdivision and nearby areas, with 700 houses destroyed by lava flows. In addition, lava flows reaching the Pacific Ocean are responsible for the thick clouds produced.

Shield volcanoes are most common in the ocean basins, but some are also present on the continents—in East Africa, for instance. The island of Hawaii is made up of five huge shield volcanoes, two of which, Kilauea and Mauna Loa, are active much of the time. Mauna Loa is nearly 100 km across its base and stands more than 9.5 km above the surrounding seafloor; it has a volume estimated at 50,000 km³, making it the world's largest volcano.

## Cinder Cones

Small, steep-sided **cinder cones** made up of particles resembling cinders form when pyroclastic materials accumulate around a vent from which they erupted (●**FIGURE 5.11**). Cinder cones are small, rarely exceeding 400 m high, with

**crater** An oval to circular depression at the summit of a volcano resulting from the eruption of lava, pyroclastic materials, and gases.

**caldera** A large, steep-sided, oval to circular depression usually formed when a volcano's summit collapses into an underlying partially drained magma chamber.

**shield volcano** A dome-shaped volcano with a low, rounded profile built up mostly by overlapping basalt lava flows.

**cinder cone** A small, steep-sided volcano composed of pyroclastic materials resembling cinders that accumulate around a vent.

# GEO-FOCUS

## The Bronze Age Eruption of Santorini

Crater Lake in Oregon, which formed approximately 7,700 years ago, is the best-known caldera in the United States (Figure 5.9), but many others are equally impressive. One that formed recently, geologically speaking, resulted from a Bronze Age eruption of Santorini, an event that figured importantly in Mediterranean history (●**FIGURE 1A**). Actually, Santorini consists of five islands in that part of the Mediterranean called the Aegean Sea. The islands have a total area of 76 km², all of which owe their present configuration to a colossal volcanic eruption that took place about 3,600 years ago (estimates range from 1650 B.C. to 1596 B.C.). Indeed, the eruption was responsible for the present islands, the origin of the huge caldera, and it probably accounted for, or at least contributed to, the demise of the Minoan culture on Crete. Furthermore, some authorities think that the disappearance of much of the

original island during this eruption was the basis for Plato's story of Atlantis.

As you approach Santorini from the sea, the impression is snow-covered cliffs in the distance. On closer inspection, though, the "snow" is actually closely spaced white buildings that cover much of the higher parts of the largest island (●**FIGURE 1B**).

Perhaps the most impressive features of Santorini are the near vertical cliffs rising as much as 350 m from the sea. Actually, these cliffs are the walls of a caldera that measures about 6 by 12 km and is as much as 400 m deep. The caldera-forming eruption, known as the "Minoan eruption," spewed a tremendous amount of pumice and volcanic ash that buried the island.

The two small islands within the caldera, where volcanic activity continues, appeared above sea level in 197 B.C., and since then have grown to their present size. The most recent activity took place in

1950 on the larger of the two islands (Figure 1A).

Santorini volcano began forming two million years ago, and during the past 400,000 years, it has erupted at least 100 times, each eruption adding new layers to the island, making it larger.

Today, approximately 15,500 people live on the islands, and we know from archaeological evidence that several tens of thousands of people resided there before the Minoan eruption when Santorini was larger. However, a year or so before the catastrophic eruption, a devasting earthquake occurred and many people left the island.

The fact that the island's residents escaped is indicated by the lack of human and animal skeletons in the ruins of the civilization, the only exception being one pig skeleton. In fact, archaeological excavations, many still in progress, show that the people had time to collect their valuables and tools before evacuating the island. Their destination, however, remains a mystery.

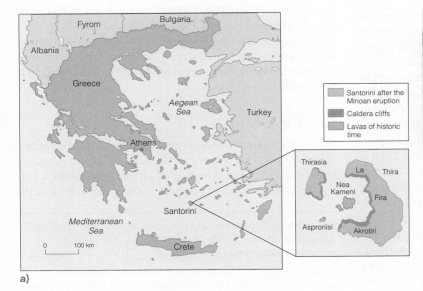

a)

James S. Monroe

b)

● **FIGURE 1** The Minoan eruption Santorini that took place between 1650 and 1596 B.C. formed a large caldera. It might have contributed to the demise of the Minoan culture on Crete and may have been the basis for Plato's account of the sinking of Atlantis. (a) Map showing Santorini and nearby areas in the Aegean Sea, which is part of the northwestern Mediterranean Sea. (b) These 350-m-high cliffs are part of the caldera wall just west of Fira.

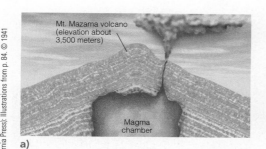

From Howell Williams, *Crater Lake: The Story of its Origin* (Berkeley, California, University of California Press): Illustrations from p. 84. © 1941 Regents of the University of California, © renewed 1969.

a)

b)

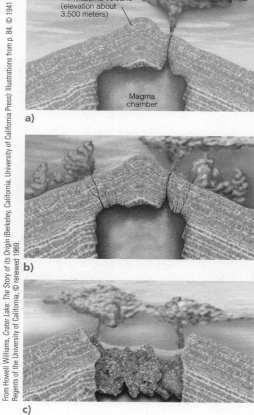

c)

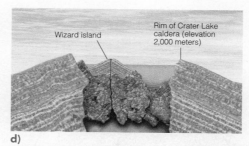

d)

e)

James S. Monroe

● **FIGURE 5.9 The Origin of Crater Lake, Oregon** Remember, Crater Lake is actually a caldera that formed when a volcano's summit collapsed into a partly drained magma chamber. (a) Eruption begins as huge quantities of ash are ejected from the volcano. (b) The collapse of the summit into the partially drained magma chamber forms a huge caldera. (c) The eruption continues as more ash and pumice are ejected into the air and pyroclastic flows move down the flanks of the mountain. (d) Postcaldera eruptions partly cover the caldera floor and the small cinder cone called Wizard Island forms. (e) View from the rim of Crater Lake showing Wizard Island. The lake is 594 m deep, making it the second deepest in North America.

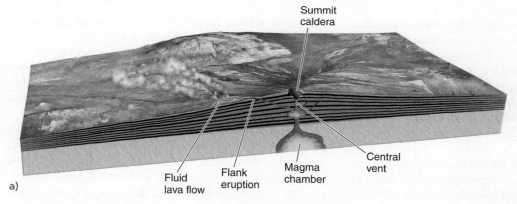

a)

b)

James S. Monroe

● **FIGURE 5.10 Shield Volcanoes** (a) Shield volcanoes consist of numerous thin basalt lava flows that build up mountains with slopes rarely exceeding 10 degrees. (b) Profile of Mauna Loa in Hawaii. Mauna Loa is one of five huge shield volcanoes that make up the island of Hawaii.

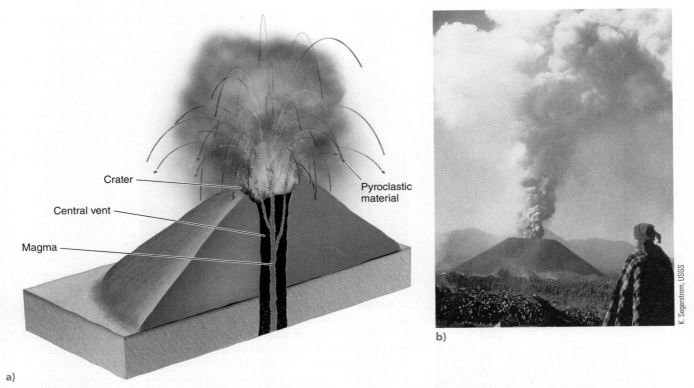

Crater

Central vent

Magma

Pyroclastic material

a)

b)

K. Segerstrom, USGS

● **FIGURE 5.11  Cinder Cones** (a) Cinder cones are composed of pyroclastic materials that accumulate around a vent from which they erupted. They are small, usually symmetrical, and steep-sided. (b) This 400-m-high cinder cone named Paricutín formed in a short time in Mexico in 1943, when pyroclastic materials began to erupt in a farmer's field. Lava flows from the volcano covered two nearby villages, but all activity ceased by 1952.

slope angles up to 33 degrees, depending on the angle that can be maintained by the angular pyroclastic materials. Many have a large, bowl-shaped crater, and if they issue any lava flows, they usually break through the base or lower flanks of the volcano. Although all cinder cones are conical, their symmetry varies from almost perfectly symmetrical to those that formed when prevailing winds caused pyroclastic materials to build up higher on the downwind side of the vent.

Many cinder cones form on the flanks, or within the calderas of larger volcanoes and represent the final stages of activity, particularly in areas of basaltic volcanism. Wizard Island in Crater Lake, Oregon, is a small cinder cone that formed after the summit of Mount Mazama collapsed to form a caldera (Figure 5.9). Cinder cones are common in the southern Rocky Mountain states, particularly New Mexico and Arizona, and many others are in California, Oregon, Washington, and Hawaii.

## Composite Volcanoes (Stratovolcanoes)

Pyroclastic layers, as well as lava flows, both of intermediate composition, are found in **composite volcanoes**, which are also called **stratovolcanoes** (●**FIGURE 5.12A**). As the lava flows cool, they typically form andesite; recall that intermediate lava flows are more viscous than mafic ones. Geologists

use the term **lahar** for volcanic mudflows, which are also common on composite volcanoes. A lahar may form when rain falls on unconsolidated pyroclastic materials and creates a muddy slurry that moves downslope. On November 13, 1985, a minor eruption of Nevado del Ruiz in Colombia melted snow and ice on the volcano, causing lahars that killed 23,000 people.

Composite volcanoes differ from shield volcanoes and cinder cones in composition and their overall shape. Remember that shield volcanoes have very low slopes, whereas cinder cones are small, steep-sided, conical volcanoes. In contrast, composite volcanoes are steep-sided near their summits, perhaps as much as 30 degrees, but the slope decreases toward the base, where it may be no more than 5 degrees. Mayon Volcano in the Philippines is one of the most nearly symmetrical composite volcanoes anywhere and has erupted 48 times during the last 400 years, most recently in 2018. Another example of a nearly symmetrical composite volcano is Mount St. Helens, Washington, which, when it began erupting on May 6, 1980, killed 63 people and devastated 600 km² of forest (●**FIGURE 5.12B**).

**composite volcano (stratovolcano)** A volcano composed of lava flows and pyroclastic layers, typically of intermediate composition, and mudflows.

**lahar** A mudflow composed of pyroclastic materials such as ash.

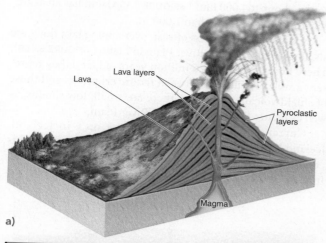

**● FIGURE 5.12 Composite Volcanoes** (a) Composite volcanoes, also called stratovolcanoes, are composed mostly of lava flows and pyroclastic materials of intermediate composition, although mudflows (lahars) are also common. (b) Mount St. Helens in Washington State as it appeared from the east in 1978, prior to its explosive eruption in 1980.

When most people think of volcanoes, they picture the graceful profiles of composite volcanoes, which are the typical large volcanoes found on the continents and island arcs. And some of these volcanoes are indeed large: Mount Shasta in northern California is composed of about 350 km³ of material and measures 20 km across its base. Remember, though, that Mauna Loa in Hawaii has an estimated volume of 50,00 km³.

## Lava Domes

Less common volcanoes are **lava domes**, also known as *volcanic domes* and *plug domes*. These structures are steep-sided, bulbous mountains that form when viscous felsic magma, and occasionally intermediate magma, is forced toward the surface (●FIGURE 5.13). Because felsic magma is so viscous, it moves upward very slowly and only when the pressure from below is great.

Lava dome eruptions are some of the most violent and destructive. In 1902, viscous magma accumulated beneath the summit of Mount Pelée on the island of Martinique. Eventually, the pressure increased until the side of the mountain

blew out in a tremendous explosion, ejecting a mobile, dense cloud of pyroclastic materials and a glowing cloud of gases and dust called a **nuée ardente** (French for "glowing cloud"). The pyroclastic flow followed a valley to the sea, but the nuée ardente jumped a ridge and engulfed the city of St. Pierre.

A tremendous blast hit St. Pierre, leveling buildings and hurling boulders, trees, and pieces of masonry down the streets (●FIGURE 5.14). Accompanying the blast was a swirling cloud of incandescent ash and gases with an internal temperature of 700°C that incinerated everything in its path.

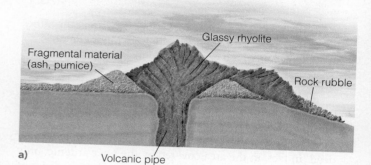

**● FIGURE 5.13 Lava Domes** Lava domes are bulbous masses of viscous magma that are emplaced in the craters of composite volcanoes, or stand alone as irregularly shaped volcanic mountains flanked by debris shed from the dome. (a) Diagram of a mass of viscous magma forming a lava dome. (b) This photograph shows Lassen Peak in Lassen Volcanic National Park in California. The dark masses of rock cooled from viscous magma. Lassen Peak erupted from 1914 to 1917.

**Critical Thinking Question** Why do you think lava domes are so dangerous?

**lava dome** A bulbous, steep-sided volcano formed by viscous magma moving upward through a volcanic conduit.

**nuée ardente** A fast-moving, dense cloud of hot pyroclastic materials and gases ejected from a volcano.

● **FIGURE 5.14 Nuée Ardente** St. Pierre, Martinique, after it was destroyed by a nuée ardente from Mount Pelée in 1902. Only two of the city's 28,000 inhabitants survived.

The nuée ardente passed through St. Pierre in two or three minutes, only to be followed by a firestorm as combustible materials burned and casks of rum exploded. But by then, most of the 28,000 residents of the city were already dead. In fact, in the area covered by the nuée ardente, only two people survived!* One survivor was on the outer edge of the nuée ardente, but even there, he was terribly burned, and his family and neighbors were all killed. The other survivor, a stevedore incarcerated the night before for disorderly conduct, was in a windowless cell partly below ground level. He remained in his cell badly burned for four days after the eruption until rescue workers heard his cries for help.

# OTHER VOLCANIC LANDFORMS

**LO9** Define fissure eruptions, basalt plateaus, and pyroclastic sheet deposits

During *fissure eruptions*, fluid lava pours out and simply builds up rather flat-lying areas, whereas huge explosive eruptions might yield *pyroclastic sheet deposits*, which, as their name implies, have a sheetlike geometry. In both cases volcanoes fail to develop.

## Fissure Eruptions and Basalt Plateaus

Rather than erupting from central vents, the lava flows comprising **basalt plateaus** issue from long cracks or fissures during **fissure eruptions**. The lava is so fluid (has such low viscosity) that it spreads out and covers vast areas. A good example is the Columbia River basalt in eastern Washington and parts of Oregon and Idaho. This huge accumulation of 17-million- to 6-million-year-old overlapping lava flows

covers about 164,000 km$^2$ (●**FIGURE 5.15A**) and has an aggregate thickness of more than 1,000 m.

Similar accumulations of vast, overlapping lava flows are also found in the Snake River Plain in Idaho (●**FIGURE 5.15B**). These flows are 5.0 to 1.6 million years old, and they represent a style of eruption between fissure eruptions and those of shield volcanoes. In fact, there are small, low shields, as well as fissure flows, in the Snake River Plain.

Currently, fissure eruptions occur only in Iceland. Iceland has several volcanoes, but the bulk of the island is composed of basalt lava flows that issued from fissures. Indeed, about half of the lava that erupted during historic

a)

b)

● **FIGURE 5.15 Basalt Plateaus** Basalt plateaus are vast areas of overlapping lava flows that issued from long fissures. Fissure eruptions take place today in Iceland; however, in the past, they formed basalt plateaus in various areas. (a) Lava flows of the Columbia River basalt in Washington State. (b) Basalt lava flows of the Snake River Plain at Malad Gorge State Park, Idaho.

---

**basalt plateau** A plateau built up by horizontal or nearly horizontal overlapping lava flows that erupted from fissures.

**fissure eruption** A volcanic eruption in which lava or pyroclastic materials issue from a long, narrow fissure (crack) or group of fissures.

---

*Although reports commonly claim that only two people survived the eruption, at least 69 and possibly as many as 111 people survived beyond the extreme margins of the nuée ardente and on ships in the harbor. Many, however, were badly injured.

time in Iceland came from two fissure eruptions, one in A.D. 930 and the other in 1783. The 1783 eruption from Laki fissure, which is more than 30 km long, accounted for lava that covered 560 km² and, in one place, filled a valley to a depth of about 200 m.

## Pyroclastic Sheet Deposits

Geologists have long been aware of vast areas covered by felsic volcanic rocks a few meters to hundreds of meters thick. Based on observations of historic pyroclastic flows, such as the nuée ardente erupted by Mount Pelée in 1902, it seems that these ancient rocks originated as pyroclastic flows—hence the name **pyroclastic sheet deposits**. They cover far greater areas than any observed during historic time, however, and apparently erupted from long fissures rather than from a central vent. The pyroclastic materials of many of these flows were so hot that they fused together to form *welded tuff*.

Geologists now think that major pyroclastic flows issue from fissures formed during the origin of calderas. For example, the Bishop Tuff of eastern California erupted shortly before the formation of the Long Valley caldera. Interestingly, earthquake activity in the Long Valley caldera and nearby areas beginning in 1978 may indicate that magma is moving upward beneath part of the caldera. Thus, the possibility of future eruptions in that area cannot be discounted.

# DISTRIBUTION OF VOLCANOES

**LO10** Explain why volcanoes are found in well-defined belts rather than being randomly distributed

**LO11** Define the circum-Pacific belt and the Mediterranean belt

**LO12** Discuss the volcanic activity of Lassen Peak and Mount St. Helens in the Cascade Range

Most of the world's active volcanoes are in well-defined zones or belts rather than being randomly distributed. The **circum-Pacific belt**, popularly called the Ring of Fire, has more than 60% of all active volcanoes. It includes volcanoes in the Andes of South America; the volcanoes of Central America, Mexico, and the Cascade Range of North America; as well as the Alaskan volcanoes and those in Japan, the Philippines, Indonesia, and New Zealand (●**FIGURE 5.16**).

**pyroclastic sheet deposit** A vast, sheetlike deposit of felsic pyroclastic materials erupted from fissures.

**circum-Pacific belt** A zone of seismic and volcanic activity and mountain building that nearly encircles the Pacific Ocean basin.

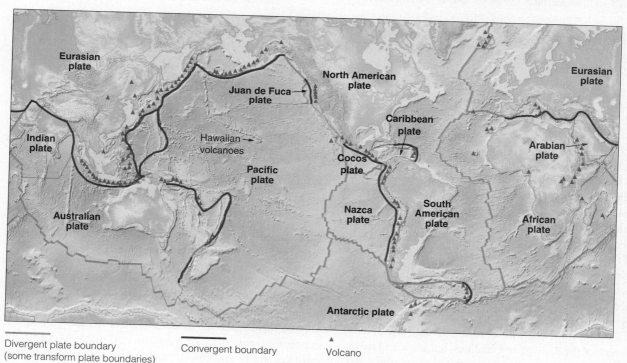

Divergent plate boundary (some transform plate boundaries)    Convergent boundary    ▲ Volcano

●**FIGURE 5.16 Volcanoes of the World** Most volcanoes are at or near convergent and divergent plate boundaries. The two major volcano belts are the circum-Pacific belt, commonly known as the Ring of Fire, with about 60% of all active volcanoes, and the Mediterranean belt, with 20% of active volcanoes. Most of the rest lie near mid-oceanic ridges.

**Critical Thinking Question** Why is magma at divergent plate boundaries mafic, whereas magma at convergent plate boundaries is intermediate or felsic?

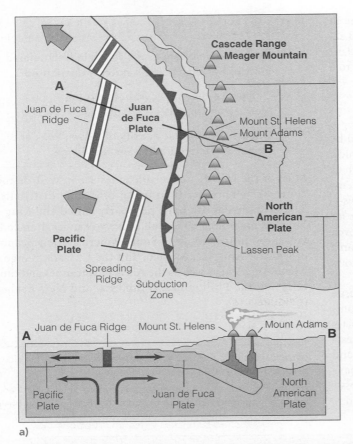

a)

b)

● **FIGURE 5.17 The Cascade Range of the Pacific Northwest** (a) Plate tectonic setting for the Pacific Northwest. Subduction of the Juan de Fuca plate accounts for ongoing volcanism in the region. (b) Lassen Peak in California erupted from 1914 to 1917. This eruption took place in 1915.

The second area of active volcanism is the **Mediterranean belt** (Figure 5.16). About 20% of all active volcanism takes place in this belt, where the famous Italian volcanoes such as Mounts Etna and Vesuvius and the Greek volcano Santorini (see GEO-FOCUS) are found.

The **Cascade Range** (●**FIGURE 5.17A**) stretches from Lassen Peak in northern California north through Oregon and Washington into British Columbia, Canada. Most of the large volcanoes in the range are composite volcanoes, but Lassen Peak in California is the world's largest lava dome. It erupted from 1914 to 1917, but has since been quiet except for ongoing hydrothermal activity (●**FIGURE 5.17B**).

What was once a nearly symmetrical composite volcano changed markedly on May 6, 1980, when Mount St. Helens in Washington (Figure 5.12b) erupted explosively, killing 63 people and leveling some 600 km² of forest. A huge lateral blast caused much of the damage and fatalities, but snow and ice on the volcano melted, and pyroclastic materials displaced water in lakes and rivers, resulting in lahars and extensive flooding.

Mount St. Helens's renewed activity, beginning in late September 2004, has resulted in dome growth and small steam and ash explosions. Scientists at the Cascades Volcano Observatory in Vancouver, Washington, issued a low-level alert for an eruption and continue to monitor the volcano. However, in January 2008, after 40 months of activity, it ceased erupting.

# PLATE TECTONICS, VOLCANOES, AND PLUTONS

**LO13**  Explain the volcanic activity that occurs along divergent plate boundaries

**LO14**  Explain the volcanic activity that occurs along convergent plate boundaries

**LO15**  Discuss how volcanism can occur within a plate, rather than along its boundaries

In Chapter 4, we discussed the origin and evolution of magma and concluded that (1) mafic magma is generated below spreading ridges and (2) intermediate and felsic magma form where an oceanic plate is subducted beneath another oceanic plate or a continental plate. Accordingly, most volcanism and emplacement of plutons take place at or near divergent and convergent plate boundaries.

**Mediterranean belt** A zone of seismic and volcanic activity extending through the Mediterranean region of southern Europe and eastward to Indonesia.

**Cascade Range** A mountain range with several active volcanoes in northern California, Oregon, Washington, and southern British Columbia, Canada.

## Igneous Activity at Divergent Plate Boundaries

Much of the mafic magma that originates at spreading ridges is emplaced as vertical dikes and gabbro plutons, thus composing the lower part of the oceanic crust. However, some rises to the surface and issues forth as submarine lava flows and pillow lava (Figure 5.6), which constitutes the upper part of the oceanic crust. Much of this volcanism goes undetected, but researchers in submersibles have seen the results of recent eruptions.

Mafic lava is very fluid, allowing gases to escape easily, and at great depth in the oceans, the water pressure is so great that explosive volcanism is prevented. In short, pyroclastic materials are rare to absent unless a volcanic center builds up above sea level. Even if this occurs, however, the mafic magma is so fluid that it forms the gently sloping layers found on shield volcanoes.

Excellent examples of divergent plate boundary volcanism are found along the Mid-Atlantic Ridge, particularly where it rises above sea level as in Iceland. The East Pacific Rise and the Indian Ridge are areas of similar volcanism. A divergent plate boundary is also present in Africa as the East African Rift system, which is well known for its volcanoes (see Figure 2.18).

## Igneous Activity at Convergent Plate Boundaries

Most of the large active volcanoes in both the circum-Pacific and Mediterranean belts are composite volcanoes near the leading edges of overriding plates along convergent plate boundaries (Figure 5.16). The overriding plate, with its chain of volcanoes, may be oceanic, as in the case of the Aleutian Islands, or it may be continental as is, for instance, the South American plate with its chain of volcanoes along its western margin.

As we noted, these volcanoes at convergent plate boundaries consist mostly of lava flows and pyroclastic materials of intermediate to felsic composition. Remember that when mafic oceanic crust partially melts, some of the magma generated is emplaced near plate boundaries as plutons, and some is erupted to build up composite volcanoes. More viscous magmas, usually of felsic composition, are emplaced as lava domes, thus accounting for the explosive eruptions that typically occur at convergent plate boundaries.

Good examples of volcanism at convergent plate boundaries are the explosive eruptions of Mount Pinatubo and Mayon volcano in the Philippines; both are near a plate boundary beneath which an oceanic plate is subducted. Mount St. Helens, Washington, is similarly situated, but it is on a continental rather than an oceanic plate.

## Intraplate Volcanism

Mauna Loa and Kilauea on the island of Hawaii and Loihi just 32 km to the south are within the interior of a rigid plate located far from any divergent or convergent plate boundary (Figure 5.16). The magma is derived from the upper mantle, as it is at spreading ridges, and accordingly is mafic, so it builds up shield volcanoes. Loihi is particularly interesting,

because it represents an early stage in the origin of a new Hawaiian island. It is a submarine volcano that rises more than 3,000 m above the adjacent seafloor, but its summit is still about 940 m below sea level.

Even though the Hawaiian volcanoes are not at or near a spreading ridge or a subduction zone, their evolution is nevertheless related to plate movements. Notice in Figure 2.9 that the ages of the rocks that comprise the Hawaiian Islands increase toward the northwest. Kauai formed 5.6 to 3.8 million years ago, whereas Hawaii began forming less than 1 million years ago, and Loihi started to develop even more recently. The islands have formed in succession as the Pacific plate moves continuously over a hot spot that is now beneath Hawaii and just to the south at Loihi.

# VOLCANIC HAZARDS, VOLCANO MONITORING, AND FORECASTING ERUPTIONS

LO16   Define what constitutes a volcanic hazard

LO17   Define the volcanic explosivity index (VEI)

LO18   Discuss the various ways volcanoes are monitored to forecast potential volcanic eruptions

Undoubtedly you suspect that living near an active volcano poses some risk, and of course this assessment is correct. But what exactly are volcanic hazards, is there any way to anticipate eruptions, and what can we do to minimize the dangers of eruptions? We have already mentioned that lava flows, with few exceptions, pose little threat to humans, although they may destroy property. Lava flows, nuée ardentes, and volcanic gases are threats during an eruption (●FIGURE 5.18); however, lahars and landslides may take place even when no eruption has occurred for a long time. Certainly, the most vulnerable areas in the United States are Alaska, Hawaii, California, Oregon, and Washington, but some other parts of the West might also experience renewed volcanism.

## How Large Is an Eruption, and How Long Do Eruptions Last?

The most widely used indication of the size of a volcanic eruption is the **volcanic explosivity index (VEI)** (●FIGURE 5.19). The VEI ranges from 0 (gentle) to 8 (cataclysmic) and is based on several aspects of an eruption, such as the volume of material explosively ejected and the height of the eruption plume. However, the volume of lava, fatalities, and property damage are not considered. For instance, the 1985 eruption of Nevado del Ruiz in Colombia killed

**volcanic explosivity index (VEI)** A semiquantitative scale for the size of a volcanic eruption based on evaluation of criteria such as the volume of material explosively erupted and the height of the eruption cloud.

● **FIGURE 5.18 Volcanic Hazards** A volcanic hazard is any manifestation of volcanism that poses a threat, including lava flows and, more importantly, volcanic gas, ash, and lahars. (a) This 2002 lava flow in Goma, Democratic Republic of Congo, killed 47 people, mostly by causing gasoline storage tanks to explode. (b) This sign at Mammoth Mountain volcano in California warns of the potential danger of $CO_2$ gas, which has killed 170 acres of trees. (c) When Mount Pinatubo in the Philippines erupted on June 15, 1991, this huge cloud of ash and steam formed over the volcano.

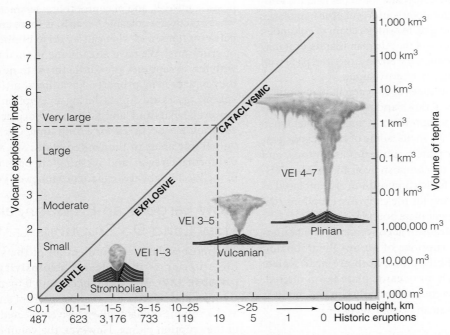

● **FIGURE 5.19 The Volcanic Explosivity Index (VEI)** In this example, an eruption with a VEI of 5 has an eruption cloud up to 25 km high and ejects at least 1 km³ of *tephra*, a collective term for all pyroclastic materials. Geologists characterize eruptions as Hawaiian (nonexplosive), Strombolian, Vulcanian, or Plinian.

**Critical Thinking Question** Why are the number of fatalities and property damage not used in assigning a VEI value to an eruption?

23,000 people, yet had a VEI value of only 3. In contrast, the huge eruption (VEI = 6) of Novarupta in Alaska in 1912 caused no fatalities or injuries. Since A.D. 1500, only the 1815 eruption of Tambora had a value of 7; it was both large and deadly. Of the several eruptions of Eyjafjallajökull in Iceland during April 2010, the largest one had a VEI of no more than 4, but because of its location and the fact that it erupted beneath glacial ice, it caused disruption of air traffic over the North Atlantic.

The duration of eruptions varies considerably. Fully 42% of about 3,300 historic eruptions lasted less than one month. About 33% erupted for one to six months, but some 16 volcanoes have been active somewhat continuously for more than 20 years. Stromboli and Mount Etna in Italy and Erta Ale in Ethiopia are good examples. For some explosive volcanoes, the time from the onset of their eruptions to the climactic event is weeks or months.

A case in point is the explosive eruption of Mount St. Helens in May, 1980, which occurred two months after eruptive activity began. Unfortunately, many volcanoes give little or no warning of such large-scale events. Of 252 explosive eruptions, 42% erupted most violently during their first day of activity.

## Is It Possible to Forecast Eruptions?

Only a few of Earth's potentially dangerous volcanoes are monitored, including some in Japan, Italy, Russia, New Zealand, and the United States. Volcano monitoring involves recording and analyzing physical and chemical changes at volcanoes (●FIGURE 5.20). Tiltmeters detect changes in the slopes of a volcano as it inflates when magma rises beneath it, and a geodimeter uses a laser beam to measure horizontal distances, which change as a volcano inflates. Geologists also monitor gas emissions, changes in groundwater level and temperature, hot springs activity, and changes in the local magnetic and electrical fields. Even the accumulating snow and ice, if any, are evaluated to anticipate hazards from floods if an eruption takes place. In fact, advances in technology are providing scientists with more efficient and accurate monitoring, including collecting and transmitting precise real-time data that can be used to help predict possible eruptions.

Of critical importance in volcano monitoring and warning of an imminent eruption is the detection of **volcanic tremor**, continuous ground motion that lasts for minutes to hours as opposed to the sudden, sharp jolts produced by most earthquakes. Volcanic tremor, also known as *harmonic tremor*, indicates that magma is moving beneath the surface.

To more fully anticipate the future activity of a volcano, its eruptive history must be known. Accordingly, geologists study the record of past eruptions preserved in rocks. Detailed studies before 1980 indicated that Mount St. Helens, Washington, had erupted explosively 14 or 15 times during the last 4,500 years, so geologists concluded that it was one of the most likely Cascade Range volcanoes to erupt again. In fact, maps they prepared showing areas in which damage from an eruption could be expected were helpful in determining which areas should have restricted access and evacuations once an eruption did take place.

---

**volcanic tremor** Ground motion lasting from minutes to hours, resulting from magma moving beneath the surface, as opposed to the sudden jolts produced by most earthquakes.

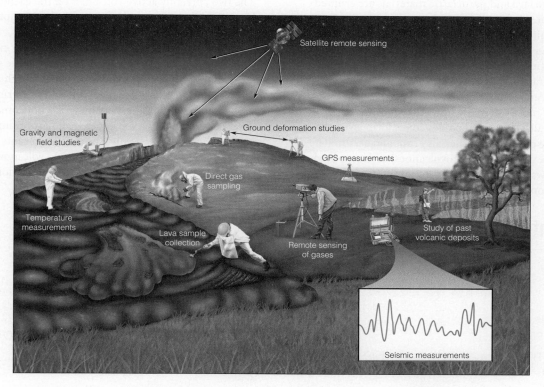

Satellite remote sensing

Ground deformation studies

Gravity and magnetic field studies

GPS measurements

Direct gas sampling

Temperature measurements

Lava sample collection

Remote sensing of gases

Study of past volcanic deposits

Seismic measurements

● **FIGURE 5.20 Volcanic Monitoring** Some important techniques used to monitor volcanoes.

Geologists successfully gave timely warnings of impending eruptions of Mount St. Helens in Washington and Mount Pinatubo in the Philippines, but in both cases, the climactic eruptions were preceded by eruptive activity of lesser intensity. In some cases, however, the warning signs are much more subtle and difficult to interpret. Numerous small earthquakes and other warning signs indicated to United States Geological Survey (USGS) geologists that magma was moving beneath the surface of the Long Valley caldera in eastern California, so in 1987, they issued a low-level warning, and then nothing happened.

Volcanic activity in the Long Valley caldera occurred as recently as 250 years ago, and there is every reason to think that it will occur again. Unfortunately, the local populace was largely unaware of the geologic history of the region, the USGS did a poor job of communicating its concerns, and premature news releases caused more concern than was justified. In any case, local residents were outraged, because the warnings caused a decrease in tourism (Mammoth Mountain on the margins of the caldera is the second-largest ski area in the country) and property values plummeted. Monitoring continues in the Long Valley caldera, and the signs of renewed volcanism, including (1) earthquake swarms, (2) trees being killed by carbon dioxide gas apparently emanating from magma (●FIGURE 5.18B), and (3) hot spring activity, cannot be ignored. In April 2006, three members of a ski patrol were killed by carbon dioxide gas that accumulated in a low area.

# Key Concepts Review

- Volcanism encompasses those processes by which magma rises to the surface as lava flows, and pyroclastic materials and gases are released into the atmosphere.
- Gases make up only a few percent by weight of magma. Most is water vapor, but sulfur gases may have far-reaching climatic effects, and carbon dioxide is dangerous.
- Aa lava flows have surfaces of jagged, angular blocks, whereas the surfaces of pahoehoe flows have a smooth, ropy surface.
- Additional features of lava flows include lava tubes and columnar joints. Lava erupted under water typically forms bulbous masses known as pillow lava.
- Volcanoes occur in various shapes and sizes, but all form where lava and pyroclastic materials are erupted from a vent.
- The summits of volcanoes have either a crater or a much larger caldera. Calderas form following voluminous eruptions, and the volcanic peak collapses into a partially drained magma chamber.
- Shield volcanoes have low, rounded profiles and are composed mostly of mafic flows that cool and form basalt. Small, steep-sided cinder cones form around a vent where pyroclastic materials erupt and accumulate. Composite volcanoes (stratovolcanoes) are composed of lava flows and pyroclastic materials of intermediate composition and volcanic mudflows.

- Viscous bulbous masses of lava, mostly of felsic composition, form lava domes, which are dangerous because they erupt explosively.
- Fluid mafic lava from fissure eruptions spreads over large areas to form basalt plateaus.
- Pyroclastic sheet deposits result from huge eruptions of ash and other pyroclastic materials, particularly when calderas form.
- Approximately 80% of all volcanic eruptions take place in the circum-Pacific and Mediterranean belts, mostly along convergent plate boundaries. Most of the rest of the eruptions occur along mid-oceanic ridges or their extensions onto land.
- The two currently active volcanoes on the island of Hawaii, and one just to the south, lie above a hot spot over which the Pacific plate moves.
- Geologists have devised a volcanic explosivity index (VEI) to give a semiquantitative measure of the size of an eruption. Volume of material erupted and the height of the eruption plume are criteria used to determine the VEI; fatalities and property damage are not considered.
- To effectively monitor volcanoes, geologists evaluate several physical and chemical aspects of volcanic regions. Of particular importance in monitoring volcanoes and forecasting eruptions is detecting volcanic tremor and determining the eruptive history of a volcano.

# Important Terms

| | | |
|---|---|---|
| aa  92 | crater  95 | pillow lava  92 |
| basalt plateau  100 | fissure eruption  100 | pyroclastic sheet deposit  101 |
| caldera  95 | lahar  98 | shield volcano  95 |
| Cascade Range  102 | lava dome  99 | volcanic ash  94 |
| cinder cone  95 | lava tube  92 | volcanic explosivity index (VEI)  103 |
| circum-Pacific belt  101 | Mediterranean belt  102 | volcanic tremor  105 |
| columnar joints  93 | nuée ardente  99 | volcanism  91 |
| composite volcano (stratovolcano)  98 | pahoehoe  92 | volcano  91 |

# Review Questions

1. Which one of the following statements is correct?
   a. Most volcanism takes place in continental interiors.
   b. Lava flows on composite volcanoes are predominantly intermediate in composition.
   c. Volcanism in the Cascade Range takes place along a divergent plate boundary.
   d. The VEI rating for an eruption depends on the number of fatalities and injuries.
   e. Earth's Moon has many active volcanoes.

2. The shaking that occurs when magma moves beneath the surface is called
   a. columnar vibrations.
   b. basalt accumulation.
   c. volcanic tremor.
   d. cratering.
   e. fissure eruption.

3. An incandescent cloud of gas and particles erupted by a volcano is a
   a. spatter cone.
   b. lapilli.
   c. pressure ridge.
   d. nuée ardent.
   e. welded tuff.

4. Water-saturated flows of volcanic debris are called _____ and are common on _____ volcanoes.
   a. fissure eruption/cinder cone
   b. pillow lava/submarine
   c. lahars/composite
   d. lava flows/dome
   e. volcanic bombs/shield

5. The most common gas emitted by volcanoes is
   a. hydrogen sulfide.
   b. fluorine.
   c. methane.
   d. oxygen.
   e. water vapor.

6. What kinds of data do geologists evaluate when they monitor volcanoes and warn of impending eruptions?

7. Why are eruptions of mafic magma rather quiet, whereas those of felsic magma are commonly explosive?

8. What criteria do geologists use to assign a volcanic explosivity index (VEI) value to an eruption?

9. Suppose you find rock exposures on land made up of pillow lava overlain by deep-sea sedimentary rocks. Where and how did the pillow lava form, and what type of rock would you expect to find beneath the pillow lava?

# Creative Thinking Visual Question

Identify the type of volcano and the kind of lava flow shown in this image (●**FIGURE 1**).

● **FIGURE 1** Volcanic features on Medicine Lake Volcano in California.

Sue Monroe

# 6

# WEATHERING, SOIL, AND SEDIMENTARY ROCKS

Weathering and erosion has produced this badlands topography in Badlands National Park, South Dakota. The steep slopes, angular ridges, and peaks are typical of badlands topography. The exposed rock formations were originally deposited as sand in stream channels and mud on floodplains.

# INTRODUCTION

All rocks at or near Earth's surface—as well as rocklike substances such as pavement and concrete in sidewalks, bridges, and foundations—decay and crumble with age. In short, they experience **weathering**, defined as the physical breakdown and chemical alteration of Earth materials as they are exposed to the atmosphere, hydrosphere, and biosphere. Actually, weathering is a group of physical and chemical processes that alter Earth materials so that they are more nearly in equilibrium with a new set of environmental conditions. For instance, many igneous and metamorphic rocks form within Earth's crust where pressure and temperature are high or no oxygen or water is present. These same rocks at the surface, though, are exposed to lower temperature and pressure, water, the atmosphere, and the activities of organisms (●FIGURE 6.1A). Thus, interactions of Earth materials with the hydrosphere, atmosphere, and biosphere bring about changes as they break down physically (*disintegrate*) and change chemically (*decompose*).

When **parent material**, that is, rocks and minerals exposed to weathering, breaks down into smaller pieces (●FIGURE 6.1B), or perhaps dissolves, some of its constituent minerals are altered or dissolved. This weathered material may accumulate and be further modified to form *soil*. Much of it, however, is removed by **erosion**, which is the wearing away of soil and rock by geologic agents such as running water. This eroded material is transported elsewhere by running water, wind, glaciers, and marine currents, and is eventually deposited as *sediment*, the raw material for *sedimentary rocks*.

Earth's crust is composed mostly of *crystalline rock*, a term that refers loosely to metamorphic and igneous rocks, except those made up of pyroclastic materials. Nevertheless, sediment and sedimentary rocks are by far the most common materials in surface exposures and in the shallow subsurface, even though they make up perhaps only 5% of the crust. They cover approximately two-thirds of the continents and most of the seafloor, except spreading ridges. All rocks are important in deciphering Earth history, but sedimentary rocks have a special place in this endeavor, because they preserve evidence of surface processes responsible for them, as well as most fossils, the only evidence of prehistoric life.

# HOW ARE EARTH MATERIALS ALTERED?

LO1   Explain how earth materials are altered

LO2   Define mechanical weathering and chemical weathering

LO3   Explain how mechanical weathering and chemical weathering act in tandem to weather rocks

Weathering occurs at or near the surface, but the rocks it acts upon are not structurally and compositionally homogeneous throughout, which accounts for **differential weathering**. This is weathering that takes place at different rates even in the same area, so it commonly results in irregular surfaces. Differential weathering and *differential erosion*—that is, variable rates of erosion—combine to yield some unusual and even bizarre features, such as *hoodoos*, *spires*, and *arches* (see the Chapter Opening photo).

Geologists characterize weathering as *mechanical* and *chemical*, both of which proceed simultaneously on parent

---

**weathering**   The physical breakdown and chemical alteration of rocks and minerals at or near Earth's surface.

**parent material**   The material that is chemically and mechanically weathered to yield sediment and soil.

**erosion**   The removal of weathered materials from their source area by running water, wind, glaciers, and waves.

**differential weathering**   Weathering that occurs at different rates on rocks, thereby yielding an uneven surface.

a)

b)

●**FIGURE 6.1 Weathering of Granite**  (a) This exposure of granite has been so thoroughly altered by weathering that only spherical masses of the original rock are visible. (b) Close-up view of weathered granite. Mechanical weathering has predominated, so the particles are mostly small pieces of granite and minerals including quartz, feldspars, and biotite.

Sue Monroe

material, as well as on materials in transport and those deposited as sediment. In short, all surface or near-surface materials weather, although one type of weathering may predominate depending on such variables as climate and rock type.

## Mechanical Weathering

**Mechanical weathering** takes place when physical forces break Earth materials into smaller pieces that retain the composition of the parent material. Granite, for instance, might yield smaller pieces of granite or individual grains of quartz, potassium feldspars, plagioclase feldspars, and biotite (Figure 6.1b).

**Frost action,** involving water repeatedly freezing and thawing in the cracks and pores of rocks, is particularly effective where temperatures commonly fluctuate above and below freezing. Frost action is so effective because water expands by about 9% when it freezes, thus exerting great force on the walls of a crack, widening and extending it by *frost wedging* (●**FIGURE 6.2A**). Repeated freezing and thawing dislodge angular pieces of rock from the parent material that tumble downslope and accumulate as **talus** (●**FIGURE 6.2B**).

Some rocks form at great depth and are stable under tremendous pressure. Granite crystallizes far below the surface, so that when it is uplifted and eroded, its contained energy is released by outward expansion, a phenomenon known as **pressure release**. The outward expansion results in the origin of fractures called *sheet joints* that more or less parallel the exposed rock surface (●**FIGURE 6.3A**). Sheet joint-bounded slabs of rock slip or slide off the parent rock (●**FIGURE 6.3B**), leaving large, rounded masses known as **exfoliation domes** (●**FIGURE 6.3C**).

That solid rock expands and produces fractures might be counterintuitive but is a well-known phenomenon. In deep mines, masses of rock detach from the sides of the excavation, often explosively. These *rock bursts* and less violent *popping* pose a danger to mine workers, and in South Africa, they are responsible for approximately 20 deaths per year.

During **thermal expansion and contraction**, the volume of rocks changes as they heat up and then cool down. The temperature may vary as much as 30°C a day in a desert, and because rocks conduct heat poorly, they heat and expand on the outside more than the inside. Even dark minerals absorb heat faster than light-colored ones, so differential expansion takes place between minerals. Surface expansion might generate enough stress to cause fracturing, but experiments in which rocks are heated and cooled repeatedly to simulate years of such activity indicate that thermal expansion and contraction are of minor importance in mechanical weathering.

The formation of salt crystals exerts enough force to widen cracks and dislodge particles in porous, granular rocks such as sandstone. And even in rocks with an interlocking mosaic of crystals, such as granite, **salt crystal growth** pries loose individual minerals. It takes place mostly in hot, arid regions, but also probably affects rocks in some coastal areas.

---

**mechanical weathering**　Disaggregation of rocks by physical processes that yields smaller pieces that retain the composition of the parent material.

**frost action**　The disaggregation of rocks by repeated freezing and thawing of water in cracks and crevasses.

**talus**　Accumulation of coarse, angular rock fragments at the base of a slope.

**pressure release**　A mechanical weathering process in which rocks that formed under pressure expand on being exposed at the surface.

**exfoliation dome**　A large, rounded dome of rock resulting when concentric layers of rock are stripped from the surface of a rock mass.

**thermal expansion and contraction**　A type of mechanical weathering in which the volume of rocks changes in response to heating and cooling.

**salt crystal growth**　A mechanical weathering process in which salt crystals growing in cracks and pores disaggregate rocks.

a)

Talus

James S. Monroe

b)

●**FIGURE 6.2 Frost Wedging** (a) Frost wedging takes place when water seeps into cracks and expands as it freezes. Angular pieces of rock are pried loose by repeated freezing and thawing. (b) Frost wedging and other mechanical weathering processes produced this talus accumulation at the base of this bluff along the Henrys Fork of the Snake River in Idaho.

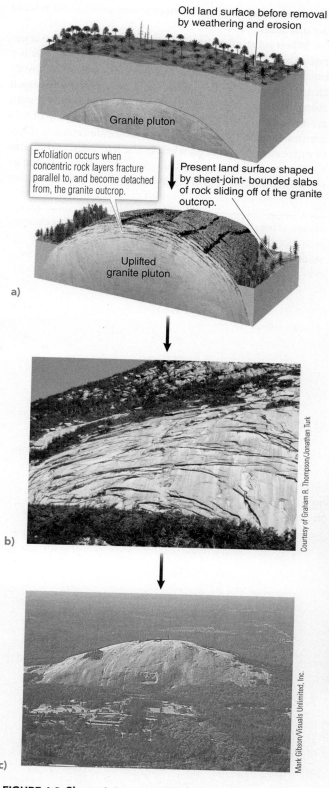

Old land surface before removal by weathering and erosion

Granite pluton

Exfoliation occurs when concentric rock layers fracture parallel to, and become detached from, the granite outcrop.

Present land surface shaped by sheet-joint- bounded slabs of rock sliding off of the granite outcrop.

Uplifted granite pluton

a)

b)

c)

Courtesy of Graham R. Thompson/Jonathan Turk

Mark Gibson/Visuals Unlimited, Inc.

● **FIGURE 6.3 Sheet Joints and Exfoliation Domes** (a) Formation of exfoliation dome. (b) Downwardly inclined sheet joint-bounded slabs have started moving down-slope toward the road below. (c) Stone Mountain in Georgia is a large exfoliation dome.

**Critical Thinking Question** Why do you think these sheet joints in part (b) might be a problem for highway maintenance crews?

Animals and plants also participate in the mechanical alteration of rocks. Burrowing animals, such as worms, reptiles, rodents, termites, and ants, constantly mix soil and sediment particles and bring material from depth to the surface, where further weathering occurs. The roots of plants, especially large bushes and trees, wedge themselves into cracks in rocks and further widen them (●**FIGURE 6.4A**).

James S. Monroe

a)

James S. Monroe

b)

● **FIGURE 6.4 Organisms and Weathering** (a) These trees in the Black Hills of South Dakota contribute to mechanical weathering as they grow in cracks in the rocks, thereby breaking the parent material into smaller pieces. (b) The orange and gray masses on this rock at Grimes Point Archaeological Site in Nevada are lichens, which are composite organisms composed of fungi and algae. Lichens derive their nutrients from the rock and thus contribute to chemical weathering.

**Critical Thinking Question** Does vegetation, especially trees, have a detrimental effect on rocklike substances such as foundations and sidewalks?

## Chemical Weathering

In contrast to mechanical weathering, **chemical weathering** changes the composition of parent materials by chemical alteration. For example, several clay minerals (sheet silicates) form by the chemical and structural alteration of other minerals, such as potassium feldspars and plagioclase feldspars, both of which are framework silicates. Other minerals completely decompose during chemical weathering, but some chemically stable minerals are simply liberated from the parent material.

Important agents of chemical weathering include atmospheric gases, especially oxygen, water, and acids (see GEO-FOCUS). Organisms also play an important role. Rocks with lichens (composite organisms made up of fungi and algae) on their surfaces undergo more rapid chemical alteration than do lichen-free rocks (●**FIGURE 6.4B**). In addition, plants remove ions from soil water and reduce the chemical stability of soil minerals, and plant roots release organic acids.

When **solution** takes place, the ions of a substance separate in a liquid, and the solid substance dissolves. Water is a remarkable solvent, because its molecules have an asymmetric shape, consisting of one oxygen atom with two hydrogen atoms arranged so that the angle between the two hydrogen atoms is about 104 degrees (●**FIGURE 6.5**). Because of this asymmetry, the oxygen end of the molecule retains a slight negative electrical charge, whereas the hydrogen end retains a slight positive charge. When a soluble substance such as the mineral halite (NaCl) comes into contact with a water

---

**chemical weathering**   The decomposition of rocks by chemical alteration of parent material.

**solution**   A reaction in which the ions of a substance become dissociated in a liquid and the solid substance dissolves.

---

# GEO-FOCUS

## Industrialization and Acid Rain

One result of industrialization is atmospheric pollution, which causes smog, possible disruption of the ozone layer, global warming, and acid rain. Acidity, a measure of hydrogen ion concentration, is measured on the pH scale (●**FIGURE 1**). A pH value of 7 is neutral, whereas acidic conditions correspond to values less than 7, and values greater than 7 denote alkaline, or basic, conditions. Normal rain has a pH value of about 5.6, making it slightly acidic, but acid rain has a pH of less than 5.0. In addition, some areas experience acid snow and even acid fog with a pH as low as 1.7.

Several natural processes, including volcanism and decaying vegetation, release gases into the atmosphere that contribute to acid rain. Human activities, however, are its main source. The exhaust from automobiles and other motor vehicles releases nitrogen oxides, which, when combined with water vapor in the atmosphere, produces nitric acid ($HNO_3$). The burning of coal (which contains sulfur) by

power companies and other industries releases sulfur, which is oxidized and forms sulfur dioxide ($SO_2$). As sulfur dioxide rises into the atmosphere, it reacts with oxygen and water droplets to form sulfuric acid ($H_2SO_4$), the main component of acid rain (●**FIGURE 2**).

Acid rain was first recognized in Sweden in 1872, but it was not until 1961 that it became an environmental concern when scientists realized that acid rain is corrosive and irritating, kills vegetation, and has a detrimental effect on surface waters. Since then, the effects of

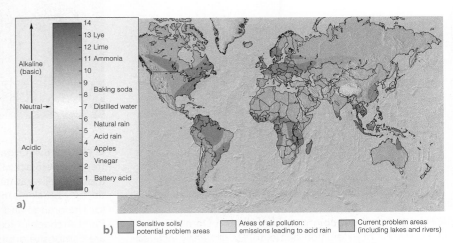

●**FIGURE 1  Acid Rain** (a) Values less than 7 on the pH scale indicate acidic conditions, whereas those greater than 7 are alkaline. The pH scale is a logarithmic scale, so a decrease of one unit is a 10-fold increase in acidity. (b) Areas where acid rain is now a problem, and areas where the problem may develop.

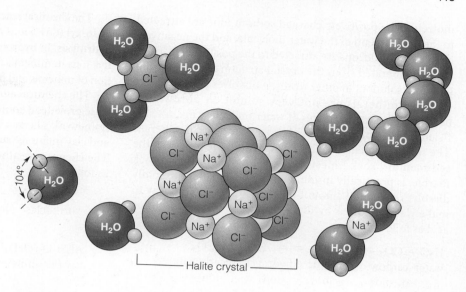

● **FIGURE 6.5 The Solution of Halite** The asymmetric arrangement of hydrogen atoms causes a water molecule to have a slight positive electrical charge at its hydrogen end and a slight negative charge at its oxygen end. The mineral halite (NaCl) goes into solution because sodium atoms are attracted to the oxygen end of a water molecule, whereas chlorine atoms are attracted to the hydrogen end of the molecule.

acid rain have been readily apparent in Europe (especially in eastern Europe), and the eastern part of North America.

The areas affected by acid rain are not limited to where industrial sources are located, but invariably lie downwind from plants that emit sulfur gases. These winds can blow chemicals into the atmosphere for hundreds and thousands of kilometers from their source, before they are deposited.

The effects of acid rain can vary, depending on the topography, soil, amount of rain or snow, and the underlying geology. For example, in areas underlain by limestone or alkaline soils, acid rain tends to be neutralized, but granite has little or no modifying effect. Small lakes lose their ability to neutralize acid rain and become more and more acidic until various organisms disappear, and in some cases all life-forms eventually die.

Acid rain also causes increased chemical weathering of limestone and marble and, to a lesser degree, sandstone. The effects are especially evident on buildings, monuments, and tombstones, as in Gettysburg National Military Park in Pennsylvania.

What can be done about the problem of acid rain? Since the 1980s, emissions of sulfur dioxide and nitrogen oxide, the two main components in acid rain, have sharply decreased. This is due, in part, to legislative mandates, federal and state government programs, and switching to less polluting fossil fuels, such as natural gas. In addition, the increased use of renewable energy, such as solar and wind power in electricity-generating plants, continues to help reduce acid rain levels.

Acid rain, like global warming, is a multifaceted problem that knows no national boundaries. It is just one of many environmental issues, like climate change, that are interconnected and are part of a dynamic and ever-changing Earth.

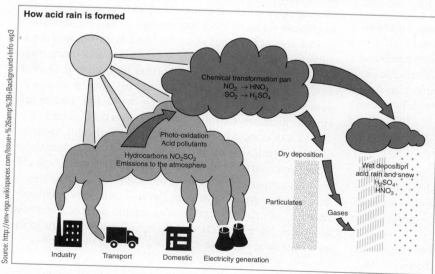

Source: http://env-ngo.wikispaces.com/Issue+-%26amp%3B+Background+Info-wp3

● **FIGURE 2** Sources of sulfur dioxide ($SO_2$) and its reaction to form sulfuric acid ($H_2SO_4$). About half of the acidic gases and particles fall to Earth in dry deposition, whereas wet deposition applies to acid rain, snow, or fog.

molecule, the positively charged sodium ions are attracted to the negative end of the water molecule, and the negatively charged chloride ions are attracted to the positive end. Thus, ions are liberated from the crystal structure, and the solid goes into solution; in other words, it dissolves (Figure 6.5).

Most minerals are not very soluble in pure water, because the attractive forces of water molecules are not sufficient to overcome the forces between particles in minerals. The mineral calcite ($CaCO_3$), the major constituent of the sedimentary rock limestone and the metamorphic rock marble, is practically insoluble in pure water, but it rapidly dissolves if a small amount of acid is present. One way to make water acidic is by dissociating the ions of carbonic acid as follows:

$$H_2O + CO_2 \rightleftharpoons H_2CO_3 \rightleftharpoons H^+ + HCO_3^-$$
water carbon    carbonic    hydrogen bicarbonate
dioxide    acid    ion    ion

According to this chemical equation, water and carbon dioxide combine to form *carbonic acid*, a small amount of which dissociates to yield hydrogen and bicarbonate ions. The concentration of hydrogen ions determines the acidity of a solution: the more hydrogen ions present, the stronger the acid.

The atmosphere is mostly nitrogen and oxygen, but approximately 0.03% is carbon dioxide, causing rain to be slightly acidic. Decaying organic matter and the respiration of organisms produce carbon dioxide in soils, so groundwater is usually slightly acidic. Climate also affects the acidity, however, with arid regions tending to have alkaline groundwater (that is, it has a low concentration of hydrogen ions). Whatever the source of carbon dioxide, once an acidic solution is present, calcite rapidly dissolves according to this reaction:

$$CaCO_3 + H_2O + CO_2 \rightleftharpoons Ca^{++} + 2HCO_3^-$$
calcite    water carbon    calcium    bicarbonate
dioxide    ion    ion

The term **oxidation** has a variety of meanings for chemists, but in chemical weathering, it refers to reactions with oxygen to form an *oxide* (one or more metallic elements combined with oxygen) or, if water is present, a *hydroxide* (a metallic element or radical combined with $OH^-$). For example, iron rusts when it combines with oxygen to form the iron oxide hematite:

$$4Fe + 3O_2 \rightarrow 2Fe_2O_3$$
iron    oxygen    iron oxide
(hematite)

Atmospheric oxygen is abundantly available for oxidation reactions, but oxidation is a slow process unless water is present. Thus, most oxidation is carried out by oxygen dissolved in water.

Oxidation is important in the alteration of ferromagnesian silicates such as olivine, pyroxenes, amphiboles, and biotite. Iron in these minerals combines with oxygen to form the reddish iron oxide hematite ($Fe_2O_3$) or the yellowish or brown hydroxide limonite [$FeO(OH) \cdot nH_2O$]. The yellow, brown, and red colors of many soils and sedimentary rocks are caused by the presence of small amounts of hematite or limonite.

The chemical reaction between the hydrogen ($H^+$) ions and hydroxyl ($OH^-$) ions of water and a mineral's ions is known as **hydrolysis**. In hydrolysis, hydrogen ions actually replace positive ions in minerals. Such replacement changes the composition of minerals and liberates iron that then may be oxidized.

The chemical alteration of the potassium feldspar orthoclase provides a good example of hydrolysis. All feldspars are framework silicates, but when altered, they yield soluble salts and clay minerals, such as kaolinite, which are sheet silicates. The chemical weathering of orthoclase by hydrolysis occurs as follows:

$$2KAlSi_3O_8 + 2H^+ + 2HCO_3^- + H_2O \rightarrow$$
orthoclase hydrogen bicarbonate    water
ion    ion

$$Al_2Si_2O_5(OH)_4 + 2K^+ + 2HCO_3^- + 4SiO_2$$
clay (kaolinite)    potassium    bicarbonate    silica
ion    ion

In this reaction, hydrogen ions attack the ions in the orthoclase structure, and some liberated ions are incorporated into a developing clay mineral. The potassium and bicarbonate ions go into solution and combine to form a soluble salt. On the right side of the equation is excess silica that would not fit into the crystal structure of the clay mineral.

**THE RATE OF CHEMICAL WEATHERING** Chemical weathering operates on the surfaces of particles, so that it alters rocks and minerals from the outside inward, but the rate at which it proceeds depends on several factors. One is simply the presence or absence of fractures, because fluids seep along fractures and thus weathering is more intense here (Figure 6.1a).

Because chemical weathering affects particle surfaces, the greater the surface area, the more effective the weathering. It is important to realize that small particles have larger surface areas compared to their volume than do large particles. Notice in ●**FIGURE 6.6** that a block measuring 1 m on a side has a total surface area of 6 $m^2$, but when the block is broken into pieces measuring 0.5 m on a side, the total surface area increases to 12 $m^2$. And, if these pieces are further reduced to 0.25 m on a side, the total surface area increases to 24 $m^2$. It should be pointed out, however, that although the surface area in this example increases, the total volume remains the same at 1 $m^3$.

We can conclude that mechanical weathering contributes to chemical weathering by yielding smaller particles with greater surface area compared to their volume. Actually, your own experiences with particle size verify our contention about surface area and volume. Because of its very small particle size, powdered sugar gives an intense burst of sweetness as the tiny pieces dissolve rapidly, but otherwise it is the same as the granular sugar we use on our cereal or in our coffee.

---

**oxidation** The reaction of oxygen with other atoms to form oxides or, if water is present, hydroxides.

**hydrolysis** The chemical reaction between hydrogen ($H^+$) ions and hydroxyl ($OH^-$) ions of water and a mineral's ions.

● **FIGURE 6.6 Particle Size and Chemical Weathering** (a) As a rock is divided into smaller particles, its surface area increases, but its volume of 1 m³ remains the same. The surface area is 6 m². (b) The surface area is 12 m². (c) The surface area is 24 m², but the volume remains the same at 1 m³. Small particles have more surface area in relation to their volume than do large particles.

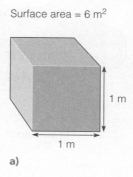
Surface area = 6 m²
1 m, 1 m
a)

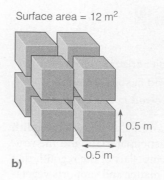
Surface area = 12 m²
0.5 m, 0.5 m
b)

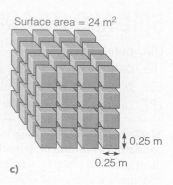
Surface area = 24 m²
0.25 m, 0.25 m
c)

It is not surprising that chemical weathering is more effective in the tropics than in arid and arctic regions, because temperatures and rainfall are high, evaporation rates are low, and vegetation and animal life are much more abundant. Consequently, the effects of weathering extend to greater depths, perhaps several tens of meters, but they extend only centimeters to a few meters deep in arid and arctic regions.

Parent material is another control on the rate of chemical weathering, because some rocks are more resistant to chemical alteration than others. The metamorphic rock quartzite is an extremely stable substance that alters slowly compared to most other rocks. In contrast, basalt, which contains large amounts of calcium-rich plagioclase and pyroxene minerals, decomposes rapidly, because these minerals are chemically unstable. In fact, the stability of common minerals is just the opposite of their order of crystallization in Bowen's reaction series: The minerals that form last in this series are more stable, whereas those that form early are easily altered, because they are most out of equilibrium with their conditions of formation (see Figure 4.3).

In **spheroidal weathering**, a stone, even one that is rectangular to begin with, weathers to form a more spherical shape, because that is the most stable shape it can assume. The reason is that on a rectangular stone, the corners are attacked by weathering from three sides, and the edges are attacked from two sides, but the flat surfaces weather more or less uniformly (●**FIGURE 6.7**). Consequently, the corners and edges are altered more rapidly, the material sloughs off, a more spherical shape develops, and all surfaces weather at the same rate.

**spheroidal weathering** A type of chemical weathering in which corners and sharp edges of rocks weather more rapidly than flat surfaces, thus yielding spherical shapes.

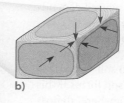
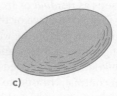
a) b) c)

● **FIGURE 6.7 Spheroidal Weathering** (a) The rectangular blocks outlined by fractures are attacked by chemical weathering processes. (b) Corners and edges weather most rapidly. (c) When the blocks are weathered so that they are nearly spherical, their surfaces weather evenly and no further change in shape takes place. (d) An exposure of granite showing spheroidal weathering in California.

James S. Monroe
d)

# SOIL AND ITS ORIGIN

**LO4** Define soil

**LO5** Define each of the five soil horizons

**LO6** Discuss the factors involved in soil formation

**LO7** Discuss the processes involved in soil degradation

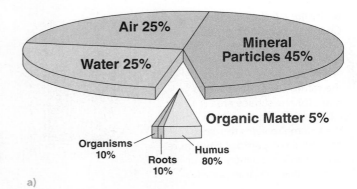

a)

Much of Earth's land surface is covered by a layer of **regolith** consisting of sediment, pyroclastic materials, and the residue formed in place by weathering. Part of the regolith that contains air, water, and organic matter and supports vegetation is **soil**. Obviously, plants grow in soil, from which they receive most of their nutrients and water, and, of course, many land-dwelling animals depend directly or indirectly on soils for their nutrients.

A good soil for farming and gardening is made up of about 45% solid particles derived by weathering of parent material, and most of the rest of its volume consists of voids filled with air or water or both (●**FIGURE 6.8A**). Another important constituent of soil is *humus*, which is carbon that forms by bacterial decay of organic matter and is very resistant to further decay. Even fertile soils may have as little as 5% humus, but it is essential as a source of plant nutrients, and it enhances a soil's capacity to retain moisture.

The solid particles in soils may be sand- and silt-sized mineral grains such as quartz, feldspars, and others, which hold soil particles apart and allow oxygen and water to circulate more freely. Clay minerals are also important, because they supply nutrients to plants and aid in soil-water retention. Should excess clay be present, though, a soil drains poorly, and is sticky when wet, and hard when dry.

We characterize soils as *residual* or *transported* depending on whether they formed in place or the materials composing them were transported from the weathering site. For example, if a body of granite weathers, and the residue that accumulates over the granite is converted to soil, the soil so formed is residual. In contrast, transported soil forms if this same weathering residue is transported elsewhere, deposited, and then converted to soil.

## The Soil Profile

Observed in vertical cross section, soil is made up of distinct layers, or **soil horizons**, that differ in texture, structure, composition, and color (●**FIGURE 6.8B**). From the surface downward, the soil horizons are designed O, A, E, B, and C, although the boundaries between horizons are transitional, and in some cases horizon E is not present.

Horizon O is only a few centimeters thick and is composed of organic matter (Figure 6.8b). Plant remains in various states of decomposition are clearly visible in the upper part of this horizon, but its lower part consists of humus. Indeed, the upper and lower parts of horizon O are sometimes referred to as O1 and O2, respectively.

| | |
|---|---|
| O | Loose leaves and organic debris |
| | Partly decomposed organic debris |
| A | Topsoil; dark in color; rich in organic matter |
| E | Zone of intense leaching or eluviation |
| B | Subsoil, zone of accumulation |
| | Transition to C |
| C | Partly weathered parent material |
| | Parent material |

b)

●**FIGURE 6.8 The Composition of Soil and Soil Horizons** (a) Soils are made up mostly of minerals and rock fragments derived by weathering, air, water, and organic matter. Most of the organic matter is humus. (b) The soil horizons in a fully developed soil. Horizon O is only a few centimeters thick, but it has been exaggerated here to show some detail.

**regolith** The layer of unconsolidated rock and mineral fragments and soil that covers most of the land surface.

**soil** Regolith consisting of weathered materials, water, air, and humus that can support vegetation.

**soil horizon** A distinct soil layer that differs from other soil layers in texture, structure, composition, and color.

Horizon A, also called *topsoil*, has more organic matter than the horizons below and is characterized by intense biological activity because plant roots, fungi, bacteria, and worms are abundant (Figure 6.8b). In fact, the earthy aroma of a freshly plowed soil comes from threadlike soil bacteria. In soils developed over a long time, horizon A is composed mostly of clays and chemically stable minerals such as quartz. Because soil formation starts at the surface and works downward, horizon A has been altered longest and is the most changed from the parent material than the horizons below.

Below horizon A in some soils is a pale layer with little carbon from which most of the small particles have been removed. This horizon E, as it is called, is present in older, more mature soils, and results from *eluviation*, a process of leaching of minerals by downward-moving soil water. Some of this material is then deposited in the horizon below.

Horizon B, or *subsoil*, has fewer organisms and less organic matter than horizon A (Figure 6.8b). This horizon is also called the *zone of accumulation* because soluble materials leached from above accumulate as irregular masses. If horizon A is eroded, leaving horizon B exposed, plants do not grow as well, and if it is clayey, it is harder when dry and stickier when wet than the other soil horizons.

Partially altered bedrock grading down into unaltered bedrock with little organic matter characterizes horizon C (Figure 6.8b). In the horizons above C, the parent material has been so thoroughly altered that it is no longer recognizable, but in horizon C minerals and rock fragments of parent material are easy to identify.

Soils are subdivided, classified, and mapped on the basis of the development and composition of the various soil horizons. The Soil Survey Division of the Natural Resources Conservation Service (the NRCS) divides soils into 12 soil orders, and further subdivides them into small groupings. The 12 soil orders are based on the interaction of such features and processes as parent material, vegetation, and climate. It is beyond the scope of this chapter to go into detail about the 12 soil orders, but information about them is available from the NRCS, a branch of the United States Department of Agriculture.

## Factors That Control Soil Formation

Complex interactions between several factors account for soil type, thickness, and fertility, but climate is the single most important influence (●**FIGURE 6.9**). For example, soils that form in rather humid regions have most of their soluble minerals leached out and horizon A may be gray, but more commonly, it is black because of abundant organic matter.

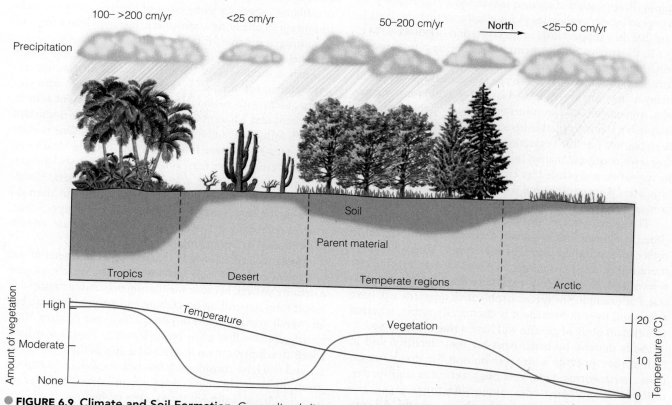

●**FIGURE 6.9  Climate and Soil Formation** Generalized diagram showing soil formation as a function of the relationship between climate and vegetation, which alter parent material over time.

**Critical Thinking Question** Why does soil extend to much greater depths in the tropics compared to deserts?

a)    b)

● **FIGURE 6.10  Alkali Soil and Laterite**  (a) An alkali soil near Fallon, Nevada. The white material is sodium carbonate or potassium carbonate. Notice that only a few hardy plants grow in this soil. (b) Laterite, shown here in Madagascar, is a deep red soil that forms in response to intense chemical weathering in the tropics.

**Critical Thinking Question**  Laterite supports lush vegetation, but is not very good for agriculture. Why?

Soils that form in semi-arid to arid regions have much less organic matter and more unstable minerals in horizon A because so little water is available to leach them out. However, soluble minerals such as calcite ($CaCO_3$) are taken into solution and precipitated in horizon B as irregular masses of *caliche*. Precipitation of sodium salts in some arid-region soils where soil water evaporates yields *alkali soils* that are so alkaline that they support little or no vegetation (●**FIGURE 6.10A**).

In the tropics, where chemical weathering is intense and leaching of most minerals is complete, a soil termed **laterite** forms (●**FIGURE 6.10B**). These red soils extend to depths of many meters and are composed largely of aluminum hydroxides, iron oxides, and clay minerals. Laterite supports lush vegetation but is not very fertile because most plant nutrients have been leached out; the vegetation depends mostly on the surface layer of organic matter. In fact, when laterite is cleared of its vegetation and planted for crops, it can sustain farming for only a few years until the soil is depleted; in which case, native farmers simply clear another area and repeat the process.

The same rock type can yield different soils in different climatic regimes, and in the same climatic regime, the same soils can develop on different rock types. Thus, climate is more important than parent material in determining the type of soil. Nevertheless, rock type does exert some control. For example, the metamorphic rock quartzite will have a thin soil over it, because it is chemically stable, whereas an adjacent body of granite will have a much deeper soil.

Soils depend on organisms for their fertility, and in return they provide a suitable habitat for many organisms. Earthworms, ants, sow bugs, termites, centipedes, millipedes, and nematodes, along with fungi, algae, and single-celled organisms, make their homes in soil. All contribute to soil formation and provide humus when they die and decompose by bacterial action.

Much of the humus in soils comes from grasses or leaf litter that microorganisms decompose to obtain food. In so doing, they break down organic compounds in plants and release nutrients back into the soil. In addition, organic acids from decaying soil organisms are important in further weathering of parent materials and soil particles.

Burrowing animals constantly churn and mix soils, and their burrows provide avenues for gases and water. Soil organisms, especially some types of bacteria, are extremely important in changing atmospheric nitrogen into a form of soil nitrogen suitable for use by plants.

The difference in elevation between high and low points in a region is called *relief*. Because climate is such an important factor in soil formation, and climate changes with elevation, areas with considerable relief have different soils in mountains and adjacent lowlands. *Slope*, another important control, influences soil formation in two ways. One is *slope angle*; steep slopes have little or no soil, because weathered materials erode faster than soil-forming processes operate. The other factor is *slope direction*. In the Northern Hemisphere, north-facing slopes receive less sunlight than do south-facing slopes and have cooler internal temperatures, support different vegetation, and, if in a cold climate, remain snow covered or frozen longer.

How much time is needed to develop a centimeter of soil or a fully developed soil a meter or so deep? We cannot give a definite answer, because weathering proceeds at vastly different rates depending on climate and parent material, but an overall average might be about 2.5 cm per century. Nevertheless, a lava flow a few centuries old in Hawaii may have a well-developed soil on it, whereas a flow of the same age in Iceland will have considerably less soil. As you might expect, given the same climatic conditions, soil develops faster on unconsolidated sediment than it does on bedrock.

**laterite**  A red soil, rich in iron or aluminum, or both, resulting from intense chemical weathering in the tropics.

## Soil Degradation

From the human perspective, soil forms so slowly that it is a nonrenewable resource. So any soil losses that exceed the rate of soil formation are viewed with alarm. Likewise, any reduction in soil fertility or production is a cause for concern. Any process that removes soil or makes it less productive is defined as **soil degradation**, a serious problem that includes erosion, chemical deterioration, and physical changes.

Erosion is an ongoing natural process, but it is usually slow enough for soil formation to keep pace. Unfortunately, some human practices add to the problem. Removing natural vegetation by plowing, overgrazing, overexploitation for firewood, and deforestation all contribute to erosion by wind and running water. The Dust Bowl that developed in several U.S. Great Plains states during the 1930s is a poignant example of just how effective wind erosion is on soil that has been pulverized and exposed by plowing (●FIGURE 6.11A).

Wind has caused considerable soil erosion in some areas, but running water is much more powerful. Some soil is removed by *sheet erosion*, which involves the removal of thin layers of soil more or less evenly over a broad, sloping surface. *Rill erosion*, in contrast, takes place when running water scours small, troughlike channels. Channels shallow enough to be eliminated by plowing are *rills*, but those too deep (about 30 cm) to be plowed over are *gullies* (●FIGURE 6.11B, C). Where gullying is extensive, croplands can no longer be tilled and must be abandoned.

Soil undergoes chemical deterioration when its nutrients are depleted and its productivity decreases. Loss of soil nutrients is most notable in many of the populous developing nations, where soils are overused to maintain high levels of agricultural productivity. Chemical deterioration is also caused by insufficient use of fertilizers and by clearing soils of their natural vegetation.

Other types of chemical deterioration are pollution and *salinization*, which occurs when the concentration of salts increases in a soil, making it unfit for agriculture. Improper disposal of domestic and industrial wastes, oil and chemical spills, and the concentration of insecticides and pesticides in soils all cause pollution.

Soil deteriorates physically when it is compacted by the weight of heavy machinery and livestock, especially cattle. Compacted soils are more costly to plow, and plants have a more difficult time emerging from them. Furthermore, water does not readily infiltrate, so more runoff occurs, which in turn accelerates the rate of water erosion.

Problems experienced in the past have stimulated the development of methods to minimize soil erosion on agricultural lands. Crop rotation, contour plowing, and strip cropping, as well as the construction of terraces have all proved helpful. So has *no-till planting*, in which the residue from the harvested crop is left on the ground to protect the surface from the ravages of wind and water.

**soil degradation** Any process leading to a loss of soil productivity; may involve erosion, chemical pollution, or compaction.

**●FIGURE 6.11 Soil Degradation Resulting from Erosion**
(a) The Dust Bowl of the 1930s was a time of drought and dust storms resulting in wind erosion of soil in Kansas, Oklahoma, Texas, New Mexico, and Colorado. This huge dust storm approaching Stratford, Texas, was photographed in 1935.
(b) Rill erosion in a field in Michigan during a rainstorm. The rill was later plowed over. (c) This large gully cuts across a farmer's field in Glenn County, California. Notice the cattle in the field for scale.

# WEATHERING AND RESOURCES

**LO8** **Discuss the relationship between weathering and various natural resources**

Soils are certainly one of our most precious natural resources. Indeed, if it were not for soils, food production on Earth would be vastly different and capable of supporting far fewer people. Other aspects of soils are also important economically. We discussed the origin of laterite in response to intense chemical weathering in the tropics and we noted further that laterite is not very productive (Figure 6.10b). If the parent material is rich in aluminum, however, the ore of aluminum, called *bauxite*, accumulates in horizon B. Although some bauxite is found in Arkansas, Alabama, and Georgia, the United States currently imports more than 50% of its aluminum needs.

Bauxite and other accumulations of valuable minerals form by the selective removal of soluble substances during chemical weathering and are known as *residual concentrations*. Bauxite is certainly a good example, but other deposits that formed in a similar manner are those rich in iron, manganese, clays, nickel, phosphate, tin, diamonds, and gold. Some of the sedimentary iron deposits in the Lake Superior region of the United States and Canada were enriched by chemical weathering when soluble parts of the deposits were carried away. Some kaolinite deposits in the southern United States formed when chemical weathering altered feldspars in pegmatites or as residual concentrations of clay-rich limestones and dolostones.

# SEDIMENT AND SEDIMENTARY ROCKS

**LO9** **Define a sedimentary rock**
**LO10** **Discuss the process of lithification**

Weathering is fundamental to the origin of **sediment** and **sedimentary rocks**, and so are *erosion* and *deposition*—that is, the movement of sediment from the weathering site by processes such as running water, wind, and glaciers, and its accumulation in some area (●**FIGURE 6.12**). One important criterion for classifying detrital sediment is particle size.

**sediment** Loose aggregate of solids derived by weathering from preexisting rocks, or solids precipitated from solution by inorganic chemical processes or extracted from solution by organisms.

**sedimentary rock** Any rock composed of sediment, such as limestone and sandstone.

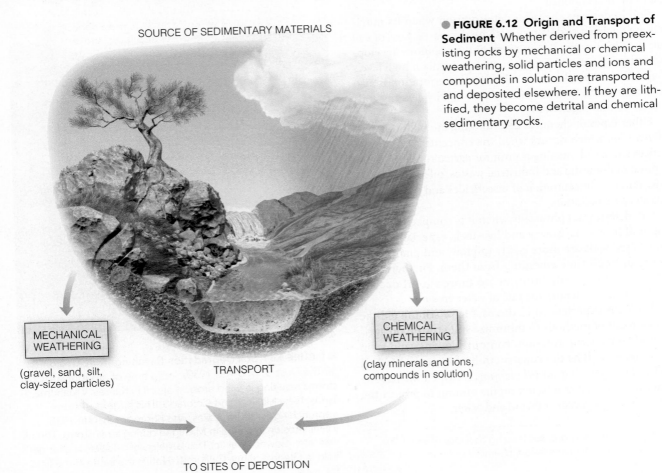

SOURCE OF SEDIMENTARY MATERIALS

MECHANICAL WEATHERING

(gravel, sand, silt, clay-sized particles)

TRANSPORT

CHEMICAL WEATHERING

(clay minerals and ions, compounds in solution)

TO SITES OF DEPOSITION

● **FIGURE 6.12 Origin and Transport of Sediment** Whether derived from preexisting rocks by mechanical or chemical weathering, solid particles and ions and compounds in solution are transported and deposited elsewhere. If they are lithified, they become detrital and chemical sedimentary rocks.

Particles described as *gravel* measure more than 2 mm, and *sand* measures between 1/16 and 2 mm. *Silt* applies to any particle from 1/256 to 1/16 mm. None of these designations imply anything about composition; most gravel is made up of rock fragments—that is, small pieces of granite, basalt, or any other type of rock—but most sand and silt grains are single minerals. Particles smaller than 1/256 mm are termed *clay*, but clay has two meanings. One is size designation, but the term also refers to certain sheet silicates called *clay minerals*. However, most clay minerals are also clay sized. The common name for mixtures of silt and clay are called *mud*.

Remember that *detrital sediment* is solid particles derived from other rocks. *Chemical sediment*, in contrast, comes from chemicals that were derived from other rocks that were then extracted from solution by inorganic chemical processes, such as evaporating seawater, or by the activities of organisms. Snails, oysters, corals, and some plants make their skeletons of minerals, especially aragonite or calcite ($CaCO_3$) or silica ($SiO_2$). In any case, minerals form that may be converted into sedimentary rock.

All detrital sediment is transported some distance from its source, but chemical sediment forms in the area where it is deposited. During transport of detrital sediment, sand and gravel particles collide, and *abrasion* wears away the sharp corners and edges, a process called *rounding* (●FIGURE 6.13A). Transport also results in *sorting*, which refers to the size distribution of particles in sediment or sedimentary rocks, which are poorly sorted if a wide range of size is present and well sorted if all the particles are about the same size (●FIGURE 6.13B). Rounding and sorting may seem unimportant, but both influence how readily groundwater, petroleum, and gas move through sediments

and sedimentary rocks, which is essential to our efforts to recover these materials. They are also useful for determining how sediment deposition occurred, a topic covered more fully in a later section.

Regardless of how detrital sediment is transported or how chemical sediment forms, it is eventually deposited in some geographic area known as a **depositional environment** (●FIGURE 6.14). Deposition may take place in a stream channel or on its floodplain, in a lake, on a beach, or on the deep seafloor where physical and biological processes impart distinctive characteristics to the accumulating sediment. The three broad depositional settings geologists recognize are *continental* (on the land), *transitional* (on or near seashores), and *marine* (in the seas), each with several specific depositional environments (Figure 6.14).

## How Does Sediment Become Sedimentary Rock?

Mud in lakes and sand and gravel in stream channels or on beaches are good examples of sediment. To convert these aggregates of particles into sedimentary rocks requires **lithification** by compaction, cementation, or both (Figure 6.14).

To illustrate the relative importance of compaction and cementation, consider detrital deposits of mud and sand. In both cases, the sediment consists of solid particles and

---

**depositional environment** Any site such as a floodplain or beach where physical, biologic, and chemical processes yield a distinctive kind of sedimentary deposit.

**lithification** The process of converting sediment into sedimentary rock by compaction and cementation.

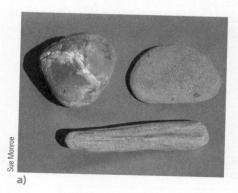

a)

c)

b)

●FIGURE 6.13 **Rounding and Sorting in Sediments** (a) Beginning students mistake rounding to mean ball-shaped or spherical. These three stones are all rounded, but only the one at the upper left is spherical. (b) Deposit of very poorly sorted, but well-rounded gravel. (c) Angular, poorly sorted gravel. Note the quarter for scale.

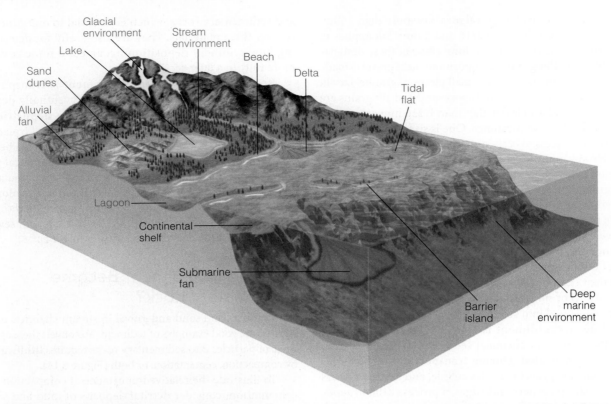

● **FIGURE 6.14 Depositional Environments** Continental environments are shown in red type. The environments along the shoreline, shown in blue type, are transitional between continental and marine. The others, shown in black type, are marine environments.

*pore spaces*, the voids between particles. These deposits are subjected to **compaction** from their own weight and the weight of any additional sediment deposited on top of them, thereby reducing the amount of pore space and the volume of the deposit. Our hypothetical mud deposit may have 80% water-filled pore space, but after compaction, its volume is reduced by as much as 40% (●**FIGURE 6.15**). The sand deposit, with as much as 50% pore space, is also compacted, but far less than the mud deposit, so that the grains fit more tightly together (Figure 6.15).

Compaction alone is sufficient for lithification of mud, but for sand and gravel, **cementation,** involving the precipitation of minerals in pore spaces, is also necessary. The two most common chemical cements are calcium carbonate ($CaCO_3$) and silicon dioxide ($SiO_2$), but iron oxide and hydroxide cement, such as hematite ($Fe_2O_3$) and limonite [$FeO(OH) \cdot nH_2O$], are found in some sedimentary rocks. Recall that calcium carbonate readily dissolves in water that contains a small amount of carbonic acid, and chemical weathering of feldspars and other minerals yields silica in solution. Cementation takes place when minerals precipitate in the pore spaces of sediment from circulating water, thereby binding the loose particles together. Iron oxide and hydroxide cements account for the red, yellow, and brown sedimentary rocks found in many areas (see the Chapter Opening photo).

By far the most common chemical sediments are calcium carbonate mud and sand- and gravel-sized accumulations of calcium carbonate grains, such as shells and shell fragments. Compaction and cementation also take place in these sediments, converting them into limestone, but compaction is less effective, because cementation takes place soon after deposition. In any case, the cement is calcium carbonate derived by partial solution of some of the particles in the deposit.

# TYPES OF SEDIMENTARY ROCKS

**LO11 Name and define each of the different types of sedimentary rocks**

Thus far, we have considered the origin of sediment, its transport, deposition, and lithification. We now turn to the types of sedimentary rocks and how they are classified. The two broad classes or types of sedimentary rocks are *detrital* and *chemical*, although the latter has a subcategory known as *biochemical* (●**TABLE 6.1**).

**compaction** Reduction in the volume of a sedimentary deposit that results from its own weight and the weight of any additional sediment deposited on top of it.

**cementation** The process whereby minerals crystallize in the pore spaces of sediment and bind the loose particles together.

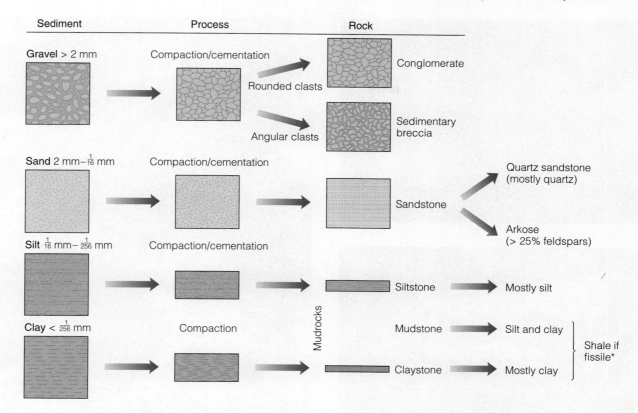

*Fissile refers to rocks that split along closely spaced planes.

● **FIGURE 6.15 Lithification and Classification of Detrital Sedimentary Rocks** Notice that little compaction takes place in gravel and sand.

| TABLE 6.1 | Classification of Chemical and Biochemical Sedimentary Rocks |
|---|---|

**Chemical Sedimentary Rocks**

| Texture | Composition | Rock Name | |
|---|---|---|---|
| Varies | Calcite ($CaCO_3$) | Limestone | Carbonate rocks |
| Varies | Dolomite [$CaMg(CO_3)_2$] | Dolostone | |
| Crystalline | Gypsum ($CaSO_4 \cdot 2H_2O$) | Rock gypsum | Evaporites |
| Crystalline | Halite (NaCl) | Rock salt | |

**Biochemical Sedimentary Rocks**

| Clastic | Calcite ($CaCO_3$) shells | Limestone (various types, such as chalk and coquina) |
|---|---|---|
| Usually crystalline | Altered microscopic shells of $SiO_2$ | Chert (various color varieties) |
| Amorphous | Carbon from altered land plants | Coal (lignite, bituminous, anthracite) |

## Detrital Sedimentary Rocks

**Detrital sedimentary rocks** are made up of solid particles such as gravel, sand, silt, and clay, and all have a *clastic texture*, meaning they are composed of particles or fragments known as *clasts*. The several varieties of detrital rocks are classified by the size of their constituent particles, although composition is used to modify some rock names.

Both *conglomerate* and *sedimentary breccia* are composed of gravel-sized particles (**FIGURES 6.15** and ●**6.16A, B**), but conglomerate has rounded gravel, whereas sedimentary breccia has angular gravel. Conglomerate is common, but sedimentary breccia is rare, because gravel becomes rounded very quickly during transport. Considerable energy is needed to transport gravel, so conglomerate is usually found in high-energy environments such as stream channels and beaches.

*Sand* is a size designation for particles between 1/16 mm and 2 mm, so any mineral or rock fragment can be in *sandstone*. Geologists recognize varieties of sandstone based on mineral content (Figures 6.15 and 6.16a). *Quartz sandstone* is

**detrital sedimentary rock** Sedimentary rock composed of the solid particles (detritus) of preexisting rocks.

● **FIGURE 6.16 Detrital Sedimentary Rocks** (a) Sandstone overlain by conglomerate at Lillooet, British Columbia, Canada. (b) Sedimentary breccia in Death Valley, California. Notice the angular gravel-sized particles. The largest clast is about 12 cm across. (c) Exposure of shale in Tennessee.

the most common and, as the name implies, is made up mostly of quartz sand. Another variety of sandstone, called *arkose*, contains at least 25% feldspar minerals. Sandstone is found in many depositional environments, including stream channels, sand dunes, beaches, barrier islands, deltas, and the continental shelf.

*Mudrock* is a general term that includes all detrital sedimentary rocks composed of silt- and clay-sized particles (Figure 6.15). Varieties include *siltstone* (mostly silt-sized particles), *mudstone* (a mixture of silt and clay), and *claystone* (primarily clay-sized particles). Some mudstones and claystones are designated *shale* if they are fissile, meaning that they break along closely spaced parallel planes (●**FIGURE 6.16C**). Even weak currents transport silt- and clay-sized particles, and deposition takes place only where currents and fluid turbulence are minimal, as in the quiet offshore waters of lakes or in lagoons.

## Chemical and Biochemical Sedimentary Rocks

During chemical weathering, several compounds and ions go into solution and provide the raw materials for chemical and biochemical sedimentary rocks. For example, seawater contains silica ($SiO_2$), calcium (Ca), carbonate ($CO_3$), sulfate ($SO_4$), potassium (K), sodium (Na), and chlorine (Cl), and many other substances that, under certain conditions, are extracted from the water to form minerals that make up **chemical sedimentary rocks**. Organisms play an important role in the origin of some of these rocks, which are designated **biochemical sedimentary rocks**. Some chemical sedimentary rocks have a crystalline texture, meaning that they are made up of a mosaic of interlocking mineral crystals as in rock salt. Others have a clastic texture as in some limestones composed of fragmented shells.

By far the most common chemical sedimentary rocks are the **carbonate rocks**, so-called because they contain the carbonate radical $(CO_3)^{-2}$. Several rocks meet this

**chemical sedimentary rock** Sedimentary rock composed of minerals that were dissolved during chemical weathering and later precipitated from seawater, more rarely lake water, or extracted from solution by organisms.

**biochemical sedimentary rock** Any sedimentary rock produced by the chemical activities of organisms.

**carbonate rock** Any rock, such as limestone and dolostone, composed mostly of carbonate minerals.

criterion but only two are common; limestone, composed of calcite ($CaCO_3$), and dolostone, composed of dolomite [$CaMg(CO_3)_2$] (Table 6.1). The origin of limestone is fairly straightforward. Recall that calcite in the presence of acidic groundwater rapidly goes into solution, but the chemical reaction leading to solution is reversible, so under some conditions calcite can precipitate from solution. Thus, some limestone forms by inorganic chemical precipitation from seawater, more rarely lake water.

Most limestone is biochemical because organisms are so important in its origin. Indeed, the skeletons of sea-dwelling animals are common in many varieties of limestone (●**FIGURE 6.17A**). *Coquina* is a type of limestone composed almost entirely of fragmented seashells (●**FIGURE 6.17B**), and *chalk* is a soft type of limestone that consists of microscopic shells. One distinctive variety of limestone contains small spherical grains called *ooids* that have a small nucleus around which concentric layers of calcite precipitate. Lithified deposits of ooids form *oolitic limestone* (●**FIGURE 6.17C**).

Dolostone is similar to limestone, but it forms mostly by the alteration of limestone when magnesium replaces some of the calcium in calcite, thereby converting it to dolomite. This may take place in a lagoon where evaporation of seawater takes place, enriching the remaining seawater in magnesium that permeates limestone and brings about a chemical change.

Most of you are aware that seawater is salty—that is, it contains sodium and chlorine and several other compounds and ions in solution. If you were to take a glass of seawater and let it evaporate completely, you would find a layer of minerals on the bottom of the glass. Obviously evaporation is involved in the origin of these minerals and their corresponding rocks, which are collectively called **evaporites** (Table 6.1). The most familiar evaporites are

---

**evaporite** Any sedimentary rock, such as rock salt, formed by inorganic chemical precipitation of minerals from evaporating water.

Sue Monroe

**a)**

Sue Monroe

**b)**

Stan Celestian/Glendale Community College

Sue Monroe

**c)**

● **FIGURE 6.17 Chemical Sedimentary Rocks—Limestone** (a) Limestone with numerous fossil shells is called fossiliferous limestone. (b) Coquina is limestone composed entirely of broken shells. (c) This oolitic limestone is made up partly of ooids (see inset), which are rather spherical grains of calcium carbonate.

**Critical Thinking Question** Coquina is composed of fragmented seashells, so would you expect it to accumulate on a Florida beach, on the deep seafloor, or on a river's floodplain? Explain.

*rock salt*, composed of the mineral halite (NaCl), and *rock gypsum*, made up of the mineral gypsum (CaSO$_4$·2H$_2$O) (●**FIGURE 6.18A, B**). Compared with sandstone, mudrocks, and limestone, evaporites are not very common, and yet they are significant deposits in areas such as Michigan, Ohio, New York, and the Gulf Coast region.

*Chert* is a hard rock consisting of microscopic crystals of silica (SiO$_2$) (Table 6.1, ●**FIGURE 6.18C**). Perhaps you have heard of *flint*, which is simply chert, colored black by inclusions of organic matter, or *jasper*, which is red or brown chert due to its iron oxide content. Because chert is hard and lacks cleavage, it can be shaped to form sharp cutting edges for tools, spear points, and arrowheads. Some chert is found as irregular masses in other rocks, especially in limestone, whereas other chert is in distinct layers of *bedded chert* made up of tiny shells of silica-secreting organisms and thus is biochemical (Figure 6.18c).

We previously mentioned that *coal* consists of altered organic matter but is nevertheless a biochemical

sedimentary rock (Table 6.1, ●**FIGURE 6.18D**). It forms from vegetation that accumulates in bogs and swamps where the water is deficient in oxygen. Bacteria that decompose vegetation can live without oxygen, but their wastes must be oxidized, and because little or no oxygen is present, wastes build up and kill the bacteria. Decomposition ceases and the vegetation forms organic muck, which if buried and compressed becomes *peat*, the first step in forming coal.

Where peat is abundant, as in Ireland and Scotland, it is used for fuel, but if it is altered further by deeper burial, and especially if it is also heated, it becomes dull black coal called *lignite*. During the change from peat to lignite, the volatile elements are driven off, increasing the amount of carbon; peat has about 50% carbon, whereas about 70% is present in lignite. Further changes yield *bituminous coal*, with about 80% carbon, which is dense, black, and so thoroughly altered that plant remains are rarely seen. The highest grade of coal is *anthracite*, a metamorphic type of coal with up to 98% carbon.

● **FIGURE 6.18 Evaporites, Chert, and Coal** (a) This cylindrical core of rock salt was taken from an oil well in Michigan. (b) Rock gypsum. When deeply buried, gypsum (CaSO$_4$·2H$_2$O) loses its water and is converted to anhydrite (CaSO$_4$). (c) Bedded chert exposed in Marin County, California. Most of the layers are about 5 cm thick. (d) Bituminous coal is the most common type of coal used for fuel.

# SEDIMENTARY FACIES

## L012    Define a marine transgression and marine regression

Long ago, geologists realized that when they traced layers of sediment or sedimentary rock laterally, they saw that the layers changed in composition, texture, or both. They concluded that these changes resulted from the simultaneous operation of different processes in adjacent depositional environments. For example, sand may be deposited in a high-energy nearshore marine environment, whereas mud and carbonate sediments accumulate simultaneously in the laterally adjacent low-energy offshore environments (●**FIGURE 6.19**). Deposition in each environment produces **sedimentary facies,** bodies of sediment that possess distinctive physical, chemical, and biological attributes.

Many sedimentary rocks in the interiors of continents show clear evidence of deposition in marine environments.

The rock layers in Figure 6.19 (*left*), for example, consist of a sandstone facies that was deposited in a nearshore marine environment overlain by shale and limestone facies deposited in offshore environments. Geologists explain this vertical sequence of facies by deposition occurring during a time when sea level rose with respect to the continents. As sea level rises, the shoreline moves inland, giving rise to a **marine transgression** (Figure 6.19), and the depositional environments parallel to the shoreline migrate landward. As a result, offshore facies are superimposed over nearshore facies, thus accounting for the vertical succession of sedimentary facies.

---

**sedimentary facies** Any aspect of a sedimentary rock unit that makes it recognizably different from adjacent sedimentary rocks of the same or approximately the same age.

**marine transgression** The invasion of a coastal area or a continent by the sea, resulting from a rise in sea level or subsidence of the land.

---

**Three Stages of Marine Transgression**

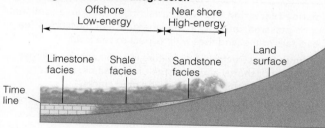

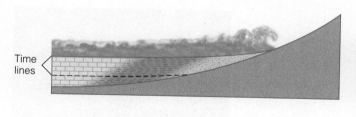

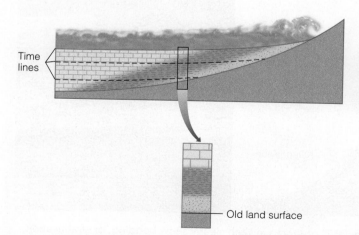

**Three Stages of Marine Regression**

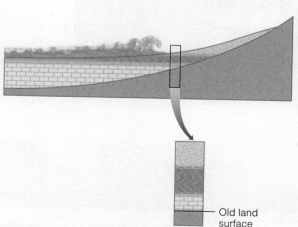

●**FIGURE 6.19  Marine Transgressions and Regressions** Notice that the sandstone, shale, and limestone facies are deposited simultaneously in adjacent environments. Also note the vertical succession of facies that result from marine transgressions and regressions.

Even though the nearshore environment is long and narrow at any particular time, deposition takes place continuously as the environment migrates landward. The sand deposit may be tens to hundreds of meters thick but has horizontal dimensions of length and width measured in hundreds of kilometers.

The opposite of a marine transgression is a **marine regression** (Figure 6.19). If sea level falls with respect to a continent, the shoreline and environments that parallel the shoreline move seaward. The vertical sequence produced by a marine regression has facies of the nearshore environment superposed over facies of offshore environments.

# READING THE STORY PRESERVED IN SEDIMENTARY ROCKS

LO13  **Explain how each of the following sedimentary structures form: cross-bedding, graded bedding, ripple marks, and mud cracks**

LO14  **Define a fossil**

LO15  **Discuss how geologists determine an ancient environment of deposition**

No one was present when ancient sediments were deposited, so geologists must evaluate those aspects of sedimentary rocks that allow them to make inferences about the original depositional environment. Sedimentary textures such as sorting and rounding can give clues to depositional processes. The sand in windblown dunes tends to be well sorted and well rounded, whereas poor sorting is typical of glacial deposits. The geometry, or three-dimensional shape, is another important aspect of sedimentary rock bodies.

Marine transgressions and regressions yield sediment bodies with a blanket or sheetlike geometry, but sand deposits in stream channels are long and narrow and are described as having a *shoestring* geometry. Sedimentary textures and geometry alone are usually insufficient to determine depositional environment, but when considered with other sedimentary rock properties, especially *sedimentary structures* and *fossils*, geologists can reliably determine the history of a deposit.

## Sedimentary Structures

Physical and biological processes operating in depositional environments are responsible for features known as **sedimentary structures**. One of the most common consists of distinct layers known as **strata** or **beds** (●**FIGURE 6.20A**),

---

**marine regression**  The withdrawal of the sea from a continent or coastal area, resulting in the emergence of the land as sea level falls or the land rises with respect to sea level.

**sedimentary structure**  Any feature in sedimentary rock that formed at or shortly after the time of deposition, such as cross-bedding, animal burrows, and mud cracks.

**strata (singular, stratum)**  Refers to layering in sedimentary rocks.

**bed**  An individual layer of rock, especially sediment or sedimentary rock.

a)

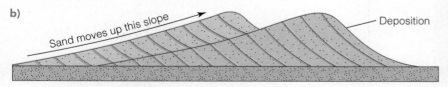

Wind or current direction

b)

Sand moves up this slope — Deposition

c)

● **FIGURE 6.20 Bedding (Stratification) and Cross-Bedding**  (a) These rocks in Utah clearly show bedding, or stratification. (b) Origin of cross-bedding by deposition on the sloping surface of a desert dune. Cross-bedding is also common in dunelike structures in stream and river channels. (c) Horizontal bedding and cross-bedding in sandstone at Wisconsin Dells, Wisconsin. About 10 m of strata are visible in this image.

with individual layers of less than a millimeter up to many meters thick. These strata or beds are separated from one another by surfaces above and below in which the rocks differ in composition, texture, color, or a combination of features.

Many sedimentary rocks have **cross-bedding**, in which layers are arranged at an angle to the surface on which they were deposited (●**FIGURE 6.20B, C**). Cross-beds are found in many depositional environments, such as sand dunes in deserts and along shorelines, as well as in stream-channel deposits. Invariably, cross-beds result from transport and deposition by wind or water currents, and the cross-beds are inclined downward in the same direction that the current flowed. Thus, ancient deposits with cross-beds inclined down toward the south, for example, indicate that the currents responsible for them flowed from north to south.

Some individual sedimentary rock layers show an upward decrease in grain size, termed **graded bedding**, mostly formed by turbidity current deposition. A *turbidity current* is an underwater flow of sediment and water with a greater density than sediment-free water. Because of its greater density, a turbidity current flows downslope until it reaches the relatively flat seafloor, or lake floor, where it slows and begins depositing large particles followed by progressively smaller ones (●**FIGURE 6.21**).

The surfaces that separate layers in sand deposits commonly have **ripple marks,** small ridges with intervening troughs, giving them a corrugated appearance. Some ripple marks are asymmetrical in cross section, with a gentle slope on one side and a steep slope on the other. Currents that flow in one direction, as in stream channels, generate these so-called *current ripple marks* (●**FIGURE 6.22A, B**). Because the steep slope of these ripples is on the downstream side, they are good indications of ancient current directions. In contrast, *wave-formed ripple marks* tend to be symmetrical in cross section and, as their name implies, are generated by the to-and-fro motion of waves (●**FIGURE 6.22C, D**).

When clay-rich sediment dries, it shrinks and develops intersecting fractures called **mud cracks** (●**FIGURE 6.23**).

Mud cracks in ancient sedimentary rocks indicate that the sediment was deposited in an environment where periodic drying took place, such as on a river floodplain, near a lakeshore, or where muddy deposits are exposed along seacoasts at low tide.

## Fossils: Remains and Traces of Ancient Life

**Fossils**, the remains or traces of once-living organisms, are interesting as evidence of prehistoric life (●**FIGURE 6.24**) and are also important for determining depositional environments. The actual remains of organisms are known as *body fossils*, whereas any indication of organic activity, such as tracks and trails, are *trace fossils*. Most people are familiar with fossils of dinosaurs and some other land-dwelling animals but are unaware that fossils of invertebrates—animals lacking a segmented vertebral column, such as corals, clams, snails, and a variety of microorganisms—are much more useful because they are so common.

It is true that the remains of land-dwelling creatures and plants can be washed into marine environments, but most are preserved in rocks deposited on land or perhaps transitional environments such as deltas. In contrast, fossils of corals tell us that the rocks in which they are preserved were deposited in the ocean.

Microfossils are particularly useful for environmental interpretations because hundreds or even thousands are

---

**cross-bedding** A type of bedding in which layers are deposited at an angle to the surface on which they accumulate, as in sand dunes.

**graded bedding** A sedimentary layer that shows a decrease in grain size from bottom to top.

**ripple mark** A wavelike (undulating) structure produced in granular sediment, especially sand, by unidirectional wind and water currents or by oscillating wave currents.

**mud crack** A crack in clay-rich sediment that forms in response to drying and shrinkage.

**fossil** The remains or traces of once-living organisms.

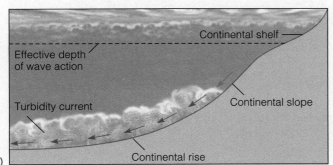

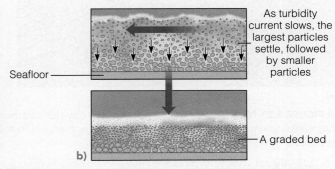

**● FIGURE 6.21 Turbidity Currents and the Origin of Graded Bedding** (a) A turbidity current flows downslope along the seafloor (or a lake bottom) because it is denser than sediment-free water. (b) The flow slows and deposits progressively smaller particles, thus forming a graded bed.

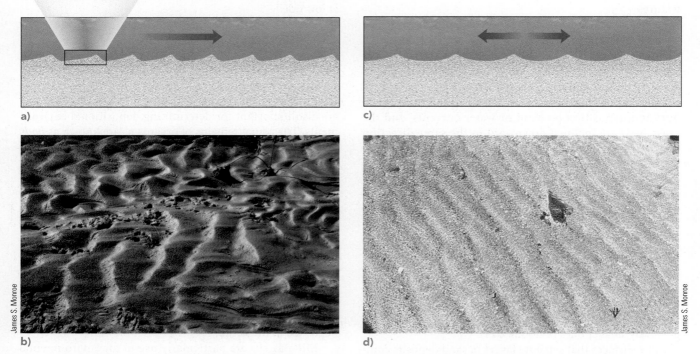

**● FIGURE 6.22  Current and Wave-Formed Ripple Marks**  (a) Current ripple marks form in response to currents that flow in one direction, as in a stream. The enlargement shows the cross-beds in an individual ripple. (b) Current ripple marks that formed in a stream channel. Flow was from right to left. (c) The to-and-fro motion of waves in shallow water yield wave-formed ripple marks. (d) Wave-formed ripple marks in sand in shallow seawater.

**● FIGURE 6.23  Mud Cracks**  Mud cracks form in clay-rich sediments when they dry and contract. (a) Mud cracks in a present-day environment. (b) Ancient mud cracks in Glacier National Park in Montana. Note that the cracks have been filled in with sediment.

**Critical Thinking Question**  What inferences can you make about the environment in which the mud in part (a) was deposited?

a)

b)

●**FIGURE 6.24 Fossils** (a) Skeleton of the dinosaur *Allosaurus*. (b) Shells of the marine animals called ammonites which are members of the class Cephalopoda, that includes squids and octopuses.

recovered from small rock samples. When drilling for oil, geologists recover small rock chips called *well cuttings* that may contain numerous fossils of tiny organisms. These fossils are routinely used to determine depositional environments, and to match up or correlate rocks of the same relative age in different areas. In addition, fossils provide some of the evidence for organic evolution.

## Determining the Environment of Deposition

What kinds of evidence would allow you to determine how a layer of sandstone was deposited? Certainly you would consider texture—that is, rounding and sorting—and also the kinds of sedimentary structures and fossils, if any. You might also compare features in the sandstone with those seen in sand deposits forming today. But are you justified in using present-day processes and deposits to infer what happened when no human observers were present?

Actually, you are familiar with the reasoning used to interpret events you did not witness. Skid marks on a street, broken glass, and a damaged power pole almost certainly indicate that a vehicle hit the pole. If you see a badly burned, shattered tree in the forest you could conclude that a bomb damaged it, but in the absence of any bomb fragments or residue, you would, no doubt, decide that the tree was hit by lightning. Geologists use exactly the same kind of reasoning—that is, their understanding of natural processes—when they evaluate evidence preserved in sedimentary rocks; when the rocks formed

is irrelevant. Geologists simply rely on the principle of *uniformitarianism* in making these interpretations (see Chapter 1).

So what about that sandstone we mentioned at the beginning of this section? Suppose that it has symmetrical ripples and fossils of marine-dwelling organisms, in which case you would no doubt conclude that it was deposited in a shallow marine environment. If, on the other hand, in addition to symmetrical ripples it contained dinosaur and land-plant fossils, you would probably conclude that it was deposited near a lakeshore. Many other features in sedimentary rocks are used in a similar fashion. Ooids (Figure 6.17c) form today in shallow marine environments with vigorous currents, and we have every reason to think that ancient ones formed in the same way. Glacial deposits are typically poorly sorted, show little stratification, and have other features that indicate glacial transport and deposition.

The Navajo Sandstone in the southwestern United States is made up of well-sorted sand measuring 0.2 to 0.5 mm in diameter; it has large cross-beds and shows the footprints of land-dwelling animals (●**FIGURE 6.25A**). Long ago, geologists concluded that the Navajo Sandstone formed as ancient sand dunes. In fact, the inclination of the cross-beds indicates that the wind blew mostly from the northeast.

Evidence from the sequence of rocks exposed in the lower part of the Grand Canyon in ●**FIGURE 6.25B** indicates that they were deposited in shallow seas during a marine transgression as shown in Figure 6.19 (*left*).

● **FIGURE 6.25 Ancient Sedimentary Rocks and Their Interpretation** (a) The Jurassic-age Navajo Sandstone in Zion National Park in Utah is a wind-blown dune deposit. Vertical fractures intersect cross-beds, hence the name Checkerboard Mesa for this rock exposure. (b) View of three formations in the Grand Canyon of Arizona. These rocks were deposited during a marine transgression. Compare with the vertical sequence of rocks in Figure 6.19 (left side).

# IMPORTANT RESOURCES IN SEDIMENTARY ROCKS

**LO16**   Discuss the various resources found in sedimentary rocks

Sand and gravel are essential to the construction industry; pure clay deposits are used for ceramics, and limestone is used in the manufacture of cement and in blast furnaces, where iron ore is refined to make steel. Evaporites are the source of table salt, as well as chemical compounds, and rock gypsum is used to manufacture wallboard. Phosphate-bearing sedimentary rock is used in fertilizers and animal-feed supplements.

Some valuable sedimentary deposits are found in streams and on beaches where minerals were concentrated during transport and deposition. These *placer deposits* are surface accumulations resulting from the separation and concentration of materials of greater density from those of lesser density. Much of the gold recovered during the initial stages of the California gold rush (1849–1853) was mined from placer deposits.

## Coal

Historically, most coal mined in the United States has been bituminous coal from the Appalachian region that formed in coastal swamps during the Pennsylvanian Period (299–323

million years ago). Huge lignite and subbituminous coal deposits in the western United States are becoming increasingly important. During 2016, about 728.4 million metric tons of coal was mined in this country, more than 70% of it from mines in Wyoming, West Virginia, Illinois, and Kentucky.

Anthracite coal is especially desirable because it burns more efficiently than other types of coal. Unfortunately, it is the least common variety, so most coal used for heating buildings and generating electricity is bituminous (Figure 6.18d). *Coke*, a hard, gray substance consisting of the fused ash of bituminous coal, is used in blast furnaces where steel is produced. Synthetic oil and gas and several other products are also made from bituminous coal and lignite.

## Petroleum and Natural Gas

Petroleum and natural gas are *hydrocarbons*, meaning that they are composed of hydrogen and carbon. The remains of microscopic organisms settle to the seafloor, or lake floor in some cases, where little oxygen is present to decompose them. If buried beneath layers of sediment, they are heated and transformed into petroleum and natural gas. The rock in which hydrocarbons form is *source rock*, but for them to accumulate in economic quantities, they must migrate from the source rock into some kind of *reservoir rock*. Finally, the reservoir rock must have an overlying, nearly impervious *cap rock*; otherwise, the hydrocarbons would eventually reach the surface and

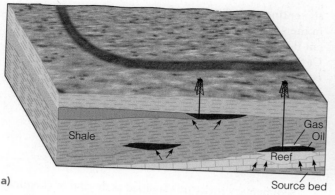

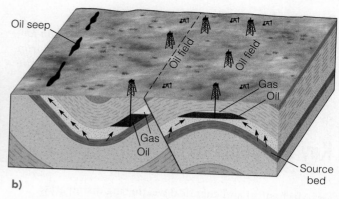

● **FIGURE 6.26 Oil, Natural Gas, and Banded Iron Formation** The arrows in parts (a) and (b) indicate the direction of migration. (a) Two examples of stratigraphic traps; one in sand within shale and the other in a buried reef. (b) Two examples of structural traps: one formed by folding and the other by faulting. (c) This banded iron formation at Ishpeming, Michigan, consists of alternating layers of red chert and silver-colored iron minerals.

*James S. Monroe*

escape (●**FIGURE 6.26A, B**). Effective reservoir rocks must have appreciable pore space and good *permeability*, the capacity to transmit fluids; otherwise, hydrocarbons cannot be extracted from them in reasonable quantities.

New technology, however, has resulted in the ability to extract gas and oil from organic-rich shales, which are impermeable. This technique, called *hydraulic fracturing*, has revolutionized the petroleum industry and turned the United States into a global energy exporter (see Chapter 13 GEO-FOCUS).

Many hydrocarbon reservoirs consist of nearshore marine sandstones with nearby fine-grained, organic-rich source rocks. These are called *stratigraphic traps*, because they owe their existence to variations in the strata (Figure 6.26a). Indeed, some of the oil in the Persian Gulf region and Michigan is trapped in ancient reefs that are also good stratigraphic traps. *Structural traps* result when rocks are deformed by folding, fracturing, or both. In sedimentary rocks that have been deformed into a series of folds, hydrocarbons migrate to the high parts of these structures (Figure 6.26b).

## Uranium

Most of the uranium used in nuclear reactors in North America comes from the complex potassium-, uranium-, vanadium-bearing mineral *carnotite*, which is found in some sedimentary rocks. Some uranium is also derived from *uraninite* ($UO_2$), a uranium oxide in granitic rocks and hydrothermal veins. Uraninite is easily oxidized and dissolved in groundwater, transported elsewhere, and chemically reduced and precipitated in the presence of organic matter.

The richest uranium ores in the United States are widespread in the Colorado Plateau area of Colorado and adjoining parts of Wyoming, Utah, Arizona, and New Mexico. These ores, consisting of fairly pure masses and encrustations of carnotite, are associated with plant remains in sandstones that formed in ancient stream channels. Although most of these ores are associated with fragmentary plant remains, some petrified trees also contain large quantities of uranium.

Large reserves of low-grade uranium ore are likewise found in the Chattanooga Shale. The uranium is finely disseminated in this black, organic-rich mudrock that underlies large parts of several states, including Illinois, Indiana, Ohio, Kentucky, and Tennessee. The top four producers of uranium in the world are Kazakhstan, Canada, Australia, and Niger.

## Banded Iron Formation

The chemical sedimentary rock known as *banded iron formation* consists of alternating thin layers of chert and iron minerals, mostly the iron oxides hematite and magnetite (●**FIGURE 6.26C**). Banded iron formations are present on all continents and account for most of the iron ore mined in the world today. Vast banded iron formations are present in the Lake Superior region of the United States and Canada, as well as Australia, Brazil, and several other countries.

# Key Concepts Review

- Mechanical and chemical weathering disintegrate and decompose parent material so that it is more nearly in equilibrium with new physical and chemical conditions.
- The products of weathering include rock fragments and minerals liberated from parent material as well as soluble compounds and ions in solution.
- Weathering yields materials that may become soil or sedimentary rock.
- Mechanical weathering processes include frost action, pressure release, thermal expansion and contraction, salt crystal growth, and the activities of organisms. The particles yielded retain the composition of the parent material.
- Chemical weathering by solution, hydrolysis, and oxidation results in a chemical change in parent material and proceeds most rapidly in hot, wet environments.
- Mechanical weathering contributes to chemical weathering by breaking parent material into smaller pieces, thereby exposing more surface area.
- Soils possess horizons designated, in descending order, as O, A, E, B, and C, which differ from one another in texture, composition, structure, and color.
- The important factors controlling soil formation are climate, parent material, organic activity, relief and slope, and time.
- Soils in humid regions are darker and more fertile than those of semiarid regions. Laterite is soil that forms in the tropics where chemical weathering is intense.
- Soil degradation results from erosion as well as from physical and chemical deterioration. Human activities, such as construction, agriculture, deforestation, waste disposal, and chemical spills, all contribute to soil degradation.
- Lithification is the process of converting sediment into sedimentary rock by compaction, cementation, or both. Silica and calcium carbonate are the most common chemical cements.

- Sedimentary rocks are classified as detrital or chemical.
- Detrital sedimentary rocks consist of particles (gravel, sand, silt, and clay) derived from preexisting rocks.
- Chemical sedimentary rocks are derived from substances in solution by inorganic chemical processes or the activities of organisms. A subcategory called biochemical sedimentary rocks is also recognized.
- Carbonate rocks contain minerals with the carbonate radical $(CO_3)^{-2}$ as in limestone and dolostone.
- Evaporites include rock salt and rock gypsum, both of which form by inorganic precipitation of minerals from evaporating water.
- Coal is a type of biochemical sedimentary rock composed of the altered remains of land plants.
- Sedimentary facies are bodies of sediment or sedimentary rock that are recognizably different from adjacent sediments or rocks.
- Vertical sequences of rocks with offshore facies overlying nearshore facies form when sea level rises with respect to the land, causing a marine transgression. A rise in the land relative to sea level causes a marine regression, which results in nearshore facies overlying offshore facies.
- Sedimentary structures, such as bedding, crossbedding, and ripple marks, help geologists determine ancient current directions and depositional environments.
- Fossils provide the only record of prehistoric life and are useful for correlation and environmental interpretations.
- Depositional environments of ancient sedimentary rocks are determined by studying all aspects of the rocks and making comparisons with present-day sediments deposited by known processes.
- Many sediments and sedimentary rocks, including sand, gravel, evaporates, coal, petroleum and natural gas, uranium, and banded iron formations, are important natural resources.

# Important Terms

bed   128
biochemical sedimentary rock   124
carbonate rock   124
cementation   122
chemical sedimentary rock   124
chemical weathering   112
compaction   122
cross-bedding   129
depositional environment   121
detrital sedimentary rock   123
differential weathering   109
erosion   109
evaporite   125
exfoliation dome   110
fossil   129

frost action   110
graded bedding   129
hydrolysis   114
laterite   118
lithification   121
marine regression   128
marine transgression   127
mechanical weathering   110
mud crack   129
oxidation   114
parent material   109
pressure release   110
regolith   116
ripple mark   129
salt crystal growth   110

sediment   120
sedimentary facies   127
sedimentary rock   120
sedimentary structure   128
soil   116
soil degradation   119
soil horizon   116
solution   112
spheroidal weathering   115
strata (singular, stratum)   128
talus   110
thermal expansion and
contraction   110
weathering   109

# Review Questions

1. The unconsolidated sediment and pyroclastic materials as well as the debris derived by weathering that cover much of the land surface is called
   a. soil degradation.
   b. regolith.
   c. caliche.
   d. exfoliation.
   e. humus.

2. Cross-bedding preserved in sedimentary rocks is a good indication of
   a. the intensity of organic activity.
   b. depositional environment.
   c. sedimentary facies.
   d. ancient current directions.
   e. a marine regression.

3. The loose, angular debris derived by weathering that accumulates at the base of a slope is called
   a. chert.
   b. humus.
   c. laterite.
   d. regolith.
   e. talus.

4. Graded bedding commonly results from deposition by
   a. turbidity currents.
   b. wind.
   c. landslides.
   d. glaciers.
   e. rivers.

5. A vertical sequence of sedimentary rocks in which marine offshore facies overlie nearshore facies results from
   a. gravitational settling.
   b. deposition in a river system.
   c. a marine transgression.
   d. turbidity current deposition.
   e. differential weathering.

6. Describe how deposits of mud and sand are lithified.

7. How do parent material, particle size, and climate control the rate of chemical weathering?

8. How does mechanical weathering differ from, and contribute to, chemical weathering?

9. How is it possible to determine how ancient sedimentary rocks were deposited given that no human observers were present to witness their deposition?

# Creative Thinking Visual Question

●**FIGURE 1** shows an exposure of granite. Explain which weathering phenomena accounts for its present surface expression.

Courtesy of Frank Hanna

● **FIGURE 1** Exposure of granite modified by weathering.

saiko3p/Shutterstock.com

The Taj Mahal in India is constructed mostly of marble.
In addition to its main use as
a building material, marble
was used throughout the
structure in artworks and
intricately carved marble
flowers. It took more than
20,000 workers and 17 years
to build the Taj Mahal from
1631 to 1648.

# METAMORPHISM
# AND METAMORPHIC
# ROCKS

**7**

# INTRODUCTION

Marble is a remarkable stone that has a variety of uses. Formed from limestone or dolostone by the metamorphic processes of heat and pressure, marble comes in a variety of colors and textures. It has been used by sculptors and architects for many centuries in statuary and monuments. For example, the marble statue *Aphrodite of Melos*, also known as *Venus de Milo*, is one of the most recognizable works of art in the world. And the Peace Monument at Pennsylvania Avenue on the west side of the Capitol in Washington, DC, is constructed from white marble from Carrara, Italy, a locality famous for its marble. Marble is also used as a facing and main stone in many buildings and various other structures. The Taj Mahal in India is constructed mostly of Makrana marble quarried from hills just southwest of Jaipur in Rajasthan (see the Chapter Opening photo). Moreover, marble is used in floor tiling and other ornamental and structural uses. In addition, ground marble is used in toothpaste and it is a source of lime in agricultural fertilizers.

**Metamorphic rocks** (from the Greek *meta*, "change," and *morpho*, "shape") like marble are the third major group of rocks we will be examining. They result from the transformation of other rocks by processes that typically occur beneath Earth's surface (see Figure 1.14). During **metamorphism**, rocks are subjected to sufficient heat, pressure, and fluid activity to change their mineral composition, texture, or both, thus forming new rocks that usually do not look anything like the original rock before it was metamorphosed. These transformations take place below the melting temperature of the rock; otherwise, an igneous rock would result.

A useful analogy for metamorphism is baking a cake. Just like a metamorphic rock, the resulting cake depends on the ingredients, their proportions, how they are mixed together, how much water or milk is added, and the temperature and length of time used for baking the cake.

Except for marble and slate, most people are not familiar with metamorphic rocks. Students frequently ask us why is it important to study metamorphic rocks and processes? The answer is always, "Just look around you."

A large portion of Earth's continental crust is composed of metamorphic and igneous rocks. Together, they form the crystalline basement rocks underlying the sedimentary rocks of a continent's surface. Some of the oldest known rocks, dated at about 4.0 billion years (●**FIGURE 7.1**), are metamorphic, which means that they formed from even older rocks! In addition, many metamorphic minerals and rocks, such as garnets, talc, asbestos, marble, and slate, are also economically important and useful.

# THE AGENTS OF METAMORPHISM

**L01**  Define the three principal agents of metamorphism

The three principal agents of metamorphism are *heat*, *pressure*, and *fluid activity*. Time is also important to the metamorphic process, because chemical reactions proceed at

● **FIGURE 7.1  Acasta Gneiss**  This metamorphic rock, found in Canada, is estimated to be about 4.0 billion years old, making it one of the oldest known rocks on Earth. Gneiss is a foliated metamorphic rock.

**Critical Thinking Question**  If the Acasta Gneiss (a metamorphic rock) is one of the oldest known rocks on Earth, why does Earth have to be older than 4 billion years?

different rates and thus require different amounts of time to complete. Reactions involving silicate compounds are particularly slow, and because most metamorphic rocks are composed of silicate minerals, it is thought that metamorphism is a very slow geologic process.

During metamorphism, the original rock, which was in equilibrium with its environment—meaning that it was chemically and physically stable under those conditions—undergoes changes to achieve equilibrium with its new environment. These changes may result in the formation of new minerals, a change in the texture of the rock, or both. In some instances, the change is minor, and features of the original rock can still be recognized. In other cases, the rock changes so much that the identity of the original rock can be determined only with great difficulty, if at all.

**MINDTAP**
From Cengage

Animation: Exploring the Agents of Metamorphism, Part 1

## Heat

**Heat** is an important agent of metamorphism, because it increases the rate of chemical reactions that may produce

---

**metamorphic rock**  Any rock that has been changed from its original condition by heat, pressure, and the chemical activity of fluids, as in marble and slate.

**metamorphism**  The phenomenon of changing rocks subjected to heat, pressure, and fluids so that they are in equilibrium with a new set of environmental conditions. Metamorphism takes place in the solid state.

**heat**  An agent of metamorphism.

minerals different from those in the original rock. Heat may come from lava, magma, or as a result of deep burial in the crust due to subduction along a convergent plate boundary.

When rocks are intruded by bodies of magma, they are subjected to intense heat that affects the surrounding rock. The most intense heating usually occurs adjacent to the magma body and gradually decreases with distance from the intrusion. The zone of metamorphosed rocks that forms in the country rock adjacent to an intrusive igneous body is usually distinct and easy to recognize.

It is known that temperature increases with depth. Some rocks that form at the surface may be transported to great depths by subduction along a convergent plate boundary and thus subjected to increasing temperature and pressure. During subduction, some minerals may be transformed into other minerals that are more stable under the higher temperature and pressure conditions.

## Pressure

During burial, rocks are subjected to increasingly greater pressure, just as you feel greater pressure the deeper you dive into a body of water. Whereas the pressure you feel is known as *hydrostatic pressure*, because it comes from the water surrounding you, rocks undergo **lithostatic pressure**, which means that the *stress* (force per unit area) on a rock in Earth's crust is the same in all directions (●**FIGURE 7.2A**). A similar situation occurs when an object is immersed in water. For example, the deeper a cup composed of Styrofoam™ is submerged in the ocean, the smaller it gets, because pressure increases with depth and is exerted on the cup equally in all directions, thereby compressing the cup (●**FIGURE 7.2B**).

Along with lithostatic pressure resulting from burial, rocks may also experience **differential pressure** (●**FIGURE 7.3**). In this case, the stresses are not equal in all directions, but are stronger from some directions than from others. Differential pressures typically occur when two plates collide, thus producing distinctive metamorphic textures and features.

## Fluid Activity

In almost every region of metamorphism, water and carbon dioxide ($CO_2$) are present in varying amounts along mineral grain boundaries or in the pore spaces of rocks. These fluids, which may contain ions in solution, enhance metamorphism by increasing the rate of chemical reactions. Under dry conditions, most minerals react very slowly, but when even small amounts of fluid are introduced, reaction rates increase. This is mainly because ions can move readily through the fluid, and thus enhance chemical reactions and the formation of new minerals.

The following reaction provides a good example of how new minerals can be formed by **fluid activity**. Seawater

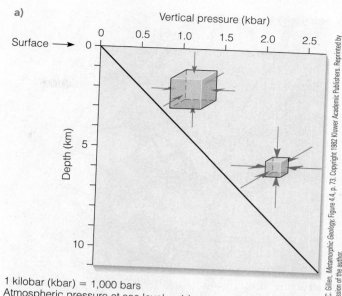

a)

1 kilobar (kbar) = 1,000 bars
Atmospheric pressure at sea level = 1 bar

*From C. Gillen, Metamorphic Geology, Figure 4.4, p. 73. Copyright 1982 Kluwer Academic Publishers. Reprinted by permission of the author.*

b)

*David J. and Jane M. Matty*

● **FIGURE 7.2 Lithostatic Pressure** (a) Lithostatic pressure is applied equally in all directions in Earth's crust due to the weight of overlying rocks. Thus, pressure increases with depth, as indicated by the sloping black line. (b) A similar situation occurs when 200-ml cups composed of Styrofoam™ are lowered to ocean depths of approximately 750 m and 1,500 m. Increased water pressure is exerted in all directions on the cups, and they consequently decrease in volume while maintaining their general shape.

moving through hot basaltic rock in the oceanic crust transforms olivine into the metamorphic mineral serpentine:

$$2Mg_2SiO_4 + 2H_2O \rightarrow Mg_3Si_2O_5(OH)_4 + MgO$$

olivine          water          serpentine          carried away in solution

**lithostatic pressure** Pressure exerted on rocks by the weight of overlying rocks.

**differential pressure** Pressure that is not applied equally to all sides of a rock body.

**fluid activity** An agent of metamorphism in which water and carbon dioxide promote metamorphism by increasing the rate of chemical reactions.

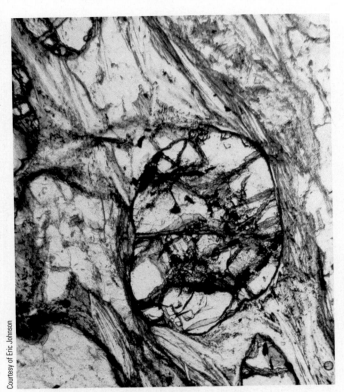

● **FIGURE 7.3 Differential Pressure** Differential pressure results from stress that is unequally applied to an object. Rotated garnets are a good example of the effects of differential pressure applied to a rock during metamorphism. In this example from a schist in northeast Sardinia, stress was applied in opposite directions on the left and right side of the garnet (*center*), causing it to rotate.

The chemically active fluids important in the metamorphic process come primarily from three sources: (1) water trapped in the pore spaces of sedimentary rocks as they form, (2) the volatile fluid within magma, and (3) the dehydration of water-bearing minerals such as gypsum ($CaSO_4 \cdot 2H_2O$) and some clays.

# THE THREE TYPES OF METAMORPHISM

**L02**  Define the three types of metamorphism

**L03**  Discuss why the boundary between them is not always distinct

**L04**  Define regional metamorphism

**L05**  Define index minerals

**L06**  Discuss how the presence of index minerals allows geologists to determine metamorphic grade

Geologists recognize three major types of metamorphism: (1) *contact (thermal) metamorphism*, in which magmatic heat and fluids act to produce change; (2) *dynamic metamorphism*, which is principally the result of high differential

pressures associated with intense deformation; and (3) *regional metamorphism*, which occurs within a large area and is associated with major mountain-building episodes. Even though we will discuss each type of metamorphism separately, the boundary between them is not always distinct and depends largely on which of the three metamorphic agents was dominant.

## Contact Metamorphism

**Contact (thermal) metamorphism** takes place when a body of magma alters the surrounding country rock. At shallow depths, intruding magma raises the temperature of the surrounding rock, causing thermal alteration. Furthermore, the release of hot fluids into the country rock by the cooling intrusion can aid in the formation of new minerals.

Important factors in contact metamorphism are the initial temperature, the size of the intrusion, the fluid content of the magma, and the nature of the country rock. Sometimes all three of these factors are involved. The initial temperature of an intrusion depends, in part, on its composition; mafic magmas are hotter than felsic magmas (see Chapter 4) and hence have a greater thermal effect on the rocks surrounding them. The size of the intrusion is also important. In the case of small intrusions, such as dikes and sills, usually only those rocks in immediate contact with the intrusion are affected. Because large intrusions, such as batholiths, take a long time to cool, the increased temperature in the surrounding rock may last long enough for a larger area to be affected.

The area of metamorphism surrounding an intrusion is an **aureole**, and the boundary between an intrusion and its aureole may be either sharp or transitional (●**FIGURE 7.4**). Metamorphic aureoles vary in width depending on the size, temperature, and composition of the intruding magma, as well as the mineralogy of the surrounding country rock. Aureoles range from a few centimeters wide, bordering small dikes and sills, to several hundred meters or even several kilometers wide around large plutons.

The degree of metamorphic change within an aureole generally decreases with distance from the intrusion, reflecting the decrease in temperature from the original heat source. The region, or zone, closest to the intrusion, and hence subject to the highest temperatures, commonly contains high-temperature metamorphic minerals (that is, minerals in equilibrium with the higher-temperature environment) such as sillimanite. The outer zones, that is, those farthest from the intrusion, are typically characterized by lower-temperature metamorphic minerals such as chlorite, talc, and epidote.

Contact metamorphism can result not only from igneous intrusions but also from lava flows, either along mid-oceanic

**contact (thermal) metamorphism** Metamorphism of country rock adjacent to a pluton and beneath a lava flow.

**aureole** A zone surrounding a pluton in which contact metamorphism took place.

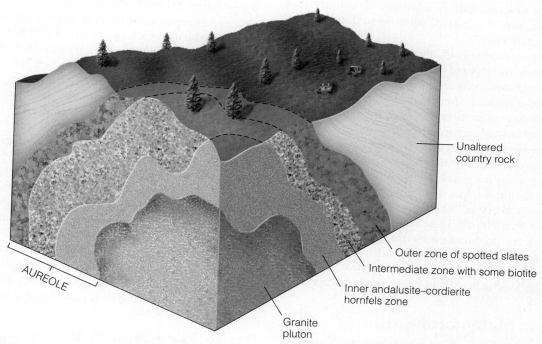

**● FIGURE 7.4  Metamorphic Aureole**  A metamorphic aureole, the area surrounding an intrusion, consists of zones that reflect the degree of metamorphism. The metamorphic aureole associated with this idealized granite pluton contains three zones of mineral assemblages reflecting the decrease in temperature with distance from the intrusion. An inner andalusite–cordierite hornfels zone forms adjacent to the pluton and is reflective of the high temperatures near the intrusion. This is followed by an intermediate zone of extensive recrystallization in which some biotite develops, and farthest from the intrusion is an outer zone characterized by spotted slates.

**● FIGURE 7.5  Contact Metamorphism from a Lava Flow**  A highly weathered basaltic lava flow near Susanville, California, has altered an underlying rhyolitic volcanic ash by contact metamorphism. The red zone below the lava flow has been baked by the heat of the lava when it flowed over the ash layer. The lava flow displays spheroidal weathering, a type of weathering common in fractured rocks (see Chapter 6).

ridges or from lava flowing over land and thermally altering the underlying rocks (**●FIGURE 7.5**). Whereas recognizing a recent lava flow and the resulting contact metamorphism of the rocks below is easy, it is less obvious whether an igneous body is intrusive or extrusive in a rock outcrop where sedimentary rocks occur above and below the igneous body. Recognizing which sedimentary rock units have been metamorphosed enables geologists to determine whether the

igneous body is intrusive (such as a sill or dike) or extrusive (lava flow). Such a determination is critical in reconstructing the geologic history of an area and may have important economic implications as well.

Fluids also play an important role in contact metamorphism. Magma is usually wet and contains hot, chemically active fluids that may emanate into the surrounding rock. These fluids can react with the rock and aid in the formation of new minerals. In addition, the country rock may contain pore fluids that, when heated by magma, increase reaction rates.

Because heat and fluids are the primary agents of contact metamorphism, two types of contact metamorphic rocks are generally recognized: those resulting from baking of country rock and those altered by hot solutions. Many of the rocks that result from contact metamorphism have the texture of porcelain; that is, they are hard and fine-grained. This is particularly true for rocks with a high clay content, such as shale. Such texture results because the clay minerals in the rock are baked, just as a clay pot is baked when fired in a kiln.

During the final stages of cooling, when an intruding magma begins to crystallize, large amounts of hot, watery solutions are often released. These solutions may react with the country rock and produce new metamorphic minerals. This process, which usually occurs near Earth's surface, is called *hydrothermal alteration* (from the Greek *hydro*, "water," and *therme*, "heat") and may result in valuable mineral deposits.

## Dynamic Metamorphism

Most **dynamic metamorphism** is associated with *faults* (fractures along which movement has occurred) or fault zones, where rocks are subjected to high levels of differential pressure. The metamorphic rocks that result from pure dynamic metamorphism are called *mylonites*, and typically they are restricted to narrow zones adjacent to faults. Mylonites are hard, dense, fine-grained rocks, many of which are characterized by thin laminations (●**FIGURE 7.6**). Tectonic settings where mylonites occur include the Moine Thrust Zone in northwest Scotland, the Adirondack Highlands in New York, and portions of the San Andreas Fault in California (see Chapter 2).

● **FIGURE 7.6 Mylonite** An outcrop of mylonite from the Adirondack Highlands, New York. Mylonites result from dynamic metamorphism, where rocks are subjected to high levels of differential pressure. Note the thin laminations (closely spaced layers) that are characteristic of many mylonites.

Courtesy of Eric Johnson

## Regional Metamorphism

Most metamorphic rocks result from **regional metamorphism**, which occurs over a large area and is usually caused by tremendous temperatures, pressures, and deformation, all occurring together within the deeper portions of the crust. Regional metamorphism is most obvious along convergent plate boundaries, where rocks are intensely deformed and recrystallized during convergence and subduction. Within these metamorphic rocks there is usually a gradation of metamorphic intensity from areas that were subjected to the most intense pressures or the highest temperatures, or both, to areas of lower pressures and temperatures. Such a gradation in metamorphism can be recognized by the metamorphic minerals that are present.

Regional metamorphism is not, however, confined only to convergent margins. It can also occur in areas where plates diverge, although usually at much shallower depths because of the high geothermal gradient associated with these areas.

## Index Minerals and Metamorphic Grade

From field studies and laboratory experiments, certain minerals are known to form only within specific temperature and pressure ranges. Such minerals are known as **index minerals**, because their presence allows geologists to recognize low-, intermediate-, and high-grade metamorphism (●**FIGURE 7.7**).

*Metamorphic grade* is a term that generally characterizes the degree to which a rock has undergone metamorphic change (Figure 7.7). Although the boundaries between the different metamorphic grades are not sharp, the distinction is nonetheless useful for communicating in a general way the degree to which rocks have been metamorphosed. The presence of index minerals thus helps determine metamorphic grade. For example, when a clay-rich rock such as shale undergoes regional metamorphism, the mineral chlorite first begins to crystallize under relatively low temperatures of about 200°C. Its presence in these rocks thus indicates low-grade metamorphism. If temperatures and pressures continue to increase, new minerals form to replace chlorite, because they are more stable under those new conditions. Thus, there is a progression in the appearance of new minerals from chlorite—whose presence indicates low-grade metamorphism—to biotite and garnet, which are good index minerals for intermediate-grade metamorphism, and then to sillimanite, whose

---

**dynamic metamorphism** Metamorphism in fault zones where rocks are subjected to high differential pressure.

**regional metamorphism** Metamorphism that occurs over a large area, resulting from high temperatures, tremendous pressure, and the chemical activity of fluids within the crust.

**index mineral** A mineral that forms within specific temperature and pressure ranges during metamorphism.

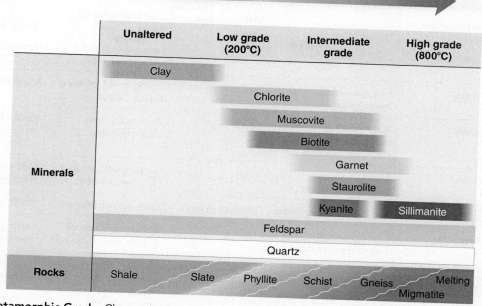

● **FIGURE 7.7  Metamorphic Grade** Change in mineral assemblage and rock type with increasing metamorphism of shale. When a clay-rich rock such as shale is subjected to increasing metamorphism, new minerals form, as shown by the colored bars. The progressive appearance of certain minerals, known as index minerals, allows geologists to recognize low-, inter- mediate-, and high-grade metamorphism.

presence indicates high-grade metamorphism and temperatures exceeding 500°C (Figure 7.7).

Different rock compositions develop different sets of index minerals. For example, clay-rich rocks such as shale will develop the index minerals shown in Figure 7.7, whereas a sandy dolomite will produce a different set of index minerals as metamorphism progresses, because it has a different mineral composition than shale. Thus, a particular set of index minerals will form based on the original composition of the parent rock undergoing metamorphism.

# HOW ARE METAMORPHIC ROCKS CLASSIFIED?

**L07** Discuss the criteria used for classifying metamorphic rocks

**L08** Define each of the common metamorphic rocks

For purposes of classification, metamorphic rocks are commonly divided into two groups: those exhibiting a *foliated texture* (from the Latin *folium*, "leaf") and those with a *nonfoliated texture* (●TABLE 7.1).

## Foliated Metamorphic Rocks

Rocks subjected to heat and differential pressure during metamorphism typically have minerals arranged in a parallel fashion, giving them a **foliated texture** (●FIGURE 7.8). The size and shape of the mineral grains determine whether the foliation is fine or coarse. Low-grade metamorphic rocks, such as slate, have a finely foliated texture in which the mineral grains are so small that they cannot be distinguished without magnification. High-grade foliated rocks, such as gneiss, are coarse-grained, such that the individual grains can easily be seen with the unaided eye. Foliated metamorphic rocks can be arranged in order of increasingly coarse grain size and perfection of foliation.

*Slate* is a very fine-grained foliated metamorphic rock that commonly exhibits *slaty cleavage* (●FIGURE 7.9). It results from regional metamorphism of shale or, more rarely, volcanic ash. Because it can easily be split along cleavage planes into flat pieces, slate is an excellent rock for roofing and floor tiles, billiard and pool table tops, and blackboards. The different colors of slate are caused by minute amounts of graphite (black), iron oxide (red and purple), and chlorite (green).

*Phyllite* is similar in composition to slate but is coarser grained. The minerals, however, are still too small to be identified without magnification. Phyllite can be distinguished from slate by its glossy or lustrous sheen (●FIGURE 7.10) and represents an intermediate grain size between slate and schist.

**foliated texture** A texture in metamorphic rocks in which platy and elongated minerals are aligned in a parallel fashion.

**TABLE 7.1** Classification of Common Metamorphic Rocks

| Texture | Metamorphic Rock | Typical Minerals | Metamorphic Grade | Characteristics Of Rocks | Parent Rock |
|---------|------------------|------------------|-------------------|--------------------------|-------------|
| Foliated | Slate | Clays, micas, chlorite | Low | Fine-grained, splits easily into flat pieces | Mudrocks, volcanic ash |
| | Phyllite | Fine-grained quartz, micas, chlorite | Low to medium | Fine-grained, glossy or lustrous sheen | Mudrocks |
| | Schist | Micas, chlorite, quartz, talc, hornblende, garnet, staurolite, graphite | Low to high | Distinct foliation, minerals visible | Mudrocks, carbonates, mafic igneous rocks |
| | Gneiss | Quartz, feldspars, hornblende, micas | High | Segregated light and dark bands visible | Mudrocks, sandstones, felsic igneous rocks |
| | Amphibolite | Hornblende, plagioclase | Medium to high | Dark, weakly foliated | Mafic igneous rocks |
| | Migmatite | Quartz, feldspars, hornblende, micas | High | Streaks or lenses of granite intermixed with gneiss | Felsic igneous rocks mixed with sedimentary rocks |
| Nonfoliated | Marble | Calcite, dolomite | Low to high | Interlocking grains of calcite or dolomite, reacts with HCl | Limestone or dolostone |
| | Quartzite | Quartz | Medium to high | Interlocking quartz grains, hard, dense | Quartz sandstone |
| | Greenstone | Chlorite, epidote, hornblende | Low to high | Fine-grained, green | Mafic igneous rocks |
| | Hornfels | Micas, garnets, andalusite, cordierite, quartz | Low to medium | Fine-grained, equidimensional grains, hard, dense | Mudrocks |
| | Anthracite | Carbon | High | Black, lustrous, subconcoidal fracture | Coal |

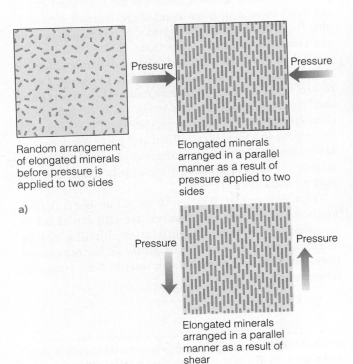

a)

Random arrangement of elongated minerals before pressure is applied to two sides

Elongated minerals arranged in a parallel manner as a result of pressure applied to two sides

Elongated minerals arranged in a parallel manner as a result of shear

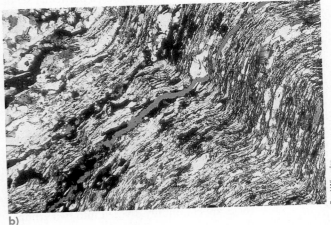

Reed Wicander

b)

● **FIGURE 7.8  Foliated Texture** (a) When rocks are subjected to differential pressure, the mineral grains are typically arranged in a parallel manner, producing a foliated texture. (b) Photomicrograph of a metamorphic rock with a foliated texture showing the parallel arrangement of mineral grains.

● **FIGURE 7.9 Slate** (a) Hand specimen of red slate. (b) Slate roof of Chalet Enzian, Switzerland.

● **FIGURE 7.10 Phyllite** Hand specimen of phyllite. Note the lustrous sheen, as well as the bedding (upper left to lower right) at an angle to the cleavage of the specimen.

**Critical Thinking Question** How do you distinguish between bedding in a sedimentary rock and cleavage in a metamorphic rock?

*Schist* is most commonly produced by regional metamorphism. The type of schist formed depends on the intensity of metamorphism and the character of the original rock (●**FIGURE 7.11**). Metamorphism of many rock types can yield schist, but most schist appears to have formed from clay-rich sedimentary rocks (Table 7.1).

All schists contain more than 50% platy and elongated minerals, all of which are large enough to be clearly visible. Their mineral composition imparts a *schistosity*, or *schistose foliation*, to the rock that usually produces a wavy type of parting when split. Schistosity is common in low- to high-grade metamorphic environments, and each type of schist is known by its most conspicuous mineral or minerals, such as mica schist, chlorite schist, or garnet-mica schist (Figure 7.11).

*Gneiss* is a high-grade metamorphic rock that is streaked, having segregated bands of light and dark minerals. Gneisses are composed of granular minerals such as quartz, feldspar, or both, with lesser percentages of platy or elongated minerals such as micas or amphiboles (●**FIGURE 7.12**). Quartz and feldspar are the principal minerals of the light-colored mineral

● **FIGURE 7.11 Schist** (a) Almandine garnet crystals in a mica schist. (b) Mica schist, Manhattan, New York.

bands, whereas biotite and hornblende make up the dark-colored mineral bands. Gneiss typically breaks in an irregular manner, much like coarsely crystalline nonfoliated rocks.

Reed Wicander

● **FIGURE 7.12  Gneiss**  Gneiss is characterized by segregated bands of light and dark minerals. This folded gneiss is exposed at Wawa, Ontario, Canada.

Most gneiss probably results from recrystallization of clay-rich sedimentary rocks during regional metamorphism (Table 7.1). Gneiss also can form from igneous rocks such as granite or older metamorphic rocks.

Another fairly common foliated metamorphic rock is *amphibolite.* A dark rock, it is composed mainly of hornblende and plagioclase. The alignment of the hornblende crystals produces a slightly foliated texture. Many amphibolites result from intermediate- to high-grade metamorphism of basalt and ferromagnesian-rich mafic rocks.

In some areas of regional metamorphism, exposures of "mixed rocks" called *migmatites,* having both igneous and high-grade metamorphic characteristics, are present

(●**FIGURE 7.13**). Migmatites are thought to result from the extremely high temperatures produced during metamorphism. However, part of the problem in determining the origin of migmatites is explaining how the granitic component formed. According to one model, the granitic magma formed in place by the partial melting of rock during intense metamorphism. Such an origin is possible provided that the host rocks contained quartz and feldspars and that water was present.

Others argue that the characteristic layering or wavy appearance of migmatites arises by the redistribution of minerals during recrystallization in the solid state—that is, through purely metamorphic processes.

## Nonfoliated Metamorphic Rocks

In some metamorphic rocks, the mineral grains do not show a discernable preferred orientation. Instead, these rocks consist of a mosaic of roughly equidimensional minerals and are characterized as having a **nonfoliated texture** (●**FIGURE 7.14**). Most nonfoliated metamorphic rocks result from contact or regional metamorphism of rocks with no platy or elongate minerals. Frequently, the only indication that a granular rock has been metamorphosed is the large grain size resulting from recrystallization.

Nonfoliated metamorphic rocks are generally of two types: those composed of mainly one mineral—for example, marble or quartzite—and those in which the different

---

**nonfoliated texture**  A metamorphic texture in which there is no discernable preferred orientation of minerals.

Copyright and Photograph by Dr. Parvinder S. Sethi

● **FIGURE 7.13  Migmatite**  A migmatite boulder in Rocky Mountain National Park, near Estes Park, Colorado. Migmatites consist of high-grade metamorphic rock intermixed with streaks or lenses of granite.

● **FIGURE 7.14 Nonfoliated Texture** Nonfoliated textures are characterized by a mosaic of roughly equidimensional minerals, as in this photomicrograph of marble.

mineral grains are too small to be seen without magnification, such as greenstone and hornfels.

*Marble* is a well-known metamorphic rock composed predominantly of calcite or dolomite; its grain size ranges from fine to coarsely granular. Marble results from either contact or regional metamorphism of limestones or dolostones (●**FIGURE 7.15** and Table 7.1). Pure marble is snowy white or bluish; however, many color varieties exist because of the presence of mineral impurities in the original sedimentary rock.

*Quartzite* is a hard, compact rock typically formed from quartz sandstone under intermediate- to high-grade metamorphic conditions during contact or regional metamorphism (●**FIGURE 7.16**). Because recrystallization is so complete, metamorphic quartzite is of uniform strength and therefore usually breaks across the component quartz grains rather than around them when it is struck. Pure quartzite is white; however, iron and other impurities commonly impart a pinkish-red or other color to it. Quartzite is commonly used as foundation material for road and railway beds.

The name *greenstone* is applied to any compact, dark green, altered, mafic igneous rock that formed under low- to high-grade metamorphic conditions. The green color results from the presence of chlorite, epidote, and hornblende.

*Hornfels* is a common, fine-grained, nonfoliated metamorphic rock resulting from contact metamorphism and composed of various equidimensional mineral grains. The composition of hornfels depends directly on the composition of the original rock, and many compositional varieties are known. Most hornfels, however, are apparently derived from contact metamorphism of clay-rich sedimentary rocks or impure dolostones.

*Anthracite* is a black, lustrous, hard coal that contains a high percentage of fixed carbon and a low percentage of volatile matter and is highly valued by people who burn coal

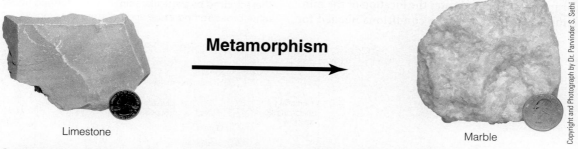

Limestone        **Metamorphism**        Marble

● **FIGURE 7.15 Marble** Metamorphism of the sedimentary rock limestone or dolostone yields marble.

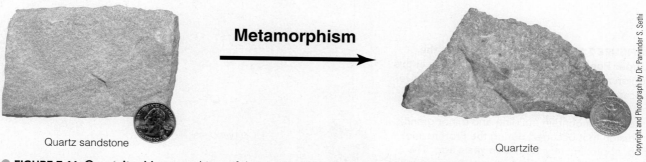

Quartz sandstone        **Metamorphism**        Quartzite

● **FIGURE 7.16 Quartzite** Metamorphism of the sedimentary rock quartz sandstone yields quartzite.

for heating and power. Anthracite usually forms from the metamorphism of lower-grade coals by heat and pressure, and many geologists consider it to be a metamorphic rock (Table 7.1).

# METAMORPHIC ZONES AND FACIES

**LO9   Explain the difference between metamorphic zones and metamorphic facies**

While mapping the 400- to 440-million-year-old Dalradian schists of Scotland in the late 1800s, George Barrow and other British geologists made the first systematic study of metamorphic zones. Here, clay-rich sedimentary rocks had been subjected to regional metamorphism, and the resulting metamorphic rocks were divided into different zones based on the presence of distinctive silicate mineral assemblages. These mineral assemblages, each recognized by the presence of one or more index minerals, indicate different degrees of metamorphism. The index minerals that Barrow and his associates chose to represent increasing metamorphic intensity were chlorite, biotite, garnet, staurolite, kyanite, and sillimanite (Figure 7.7), which we now know all result from the recrystallization of clay-rich sedimentary rocks.

The successive appearance of metamorphic index minerals indicates gradually increasing or decreasing intensity of metamorphism. Going from lower- to higher-grade metamorphic zones, the first appearance of a particular index mineral indicates the location of the minimum temperature and pressure conditions needed for

the formation of that mineral. When the locations of the first appearances of that index mineral are connected on a map, the result is a line of equal metamorphic intensity, or an *isograd*. The region between two adjacent isograds makes up a single **metamorphic zone**—a belt of rocks displaying the same general degree of metamorphism. By mapping adjoining metamorphic zones, geologists can reconstruct metamorphic conditions throughout an entire area (●FIGURE 7.17).

Not long after Barrow and his coworkers completed their work, geologists in Norway and Finland came up with a different method of mapping metamorphism that was more useful than the metamorphic zone approach. A **metamorphic facies** is defined as a group of metamorphic rocks characterized by particular mineral assemblages formed under broadly similar temperature and pressure conditions (●FIGURE 7.18). Each facies is named after its most characteristic rock or mineral. For example, the green metamorphic mineral chlorite, which forms under relatively low temperatures and pressures, yields rocks belonging to the *greenschist facies*. Under increasingly higher temperatures and pressures, mineral assemblages indicative of the *amphibolite* and *granulite facies* develop.

Although usually applied to areas where the original rocks were clay-rich, the concept of metamorphic facies can also be used with modification in other situations. It cannot, however,

---

**metamorphic zone** The region between lines of equal metamorphic intensity known as isograds.

**metamorphic facies** A group of metamorphic rocks characterized by particular minerals that formed under the same broad temperature and pressure conditions.

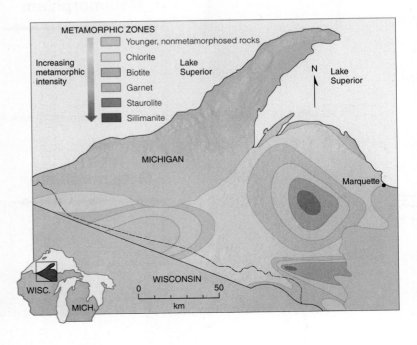

● **FIGURE 7.17 Metamorphic Zones in the Upper Peninsula of Michigan** The zones in this region are based on the presence of distinctive silicate mineral assemblages resulting from the metamorphism of sedimentary rocks during an interval of mountain building and minor granitic intrusion that occurred during the Proterozoic Eon, approximately 1.5 billion years ago. The lines separating the different metamorphic zones are isograds.

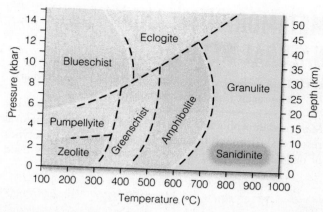

**FIGURE 7.18 Metamorphic Facies and Their Associated Temperature–Pressure Conditions** A temperature–pressure diagram showing under what conditions various metamorphic facies occur. A metamorphic facies is characterized by a particular mineral assemblage that formed under the same broad temperature–pressure conditions. Each facies is named after its most characteristic rock or mineral.

**Critical Thinking Question** Go to a point that is represented by 200°C and 2 kbar of pressure. What metamorphic facies is represented by those conditions? If the pressure is raised to 12 kbar, what facies is represented by the new conditions? What change in depth of burial is required to effect the pressure change from 2 to 12 kbar?

be used in areas where the original rocks were pure quartz sandstones or pure limestones or dolostones. Such rocks, regardless of the imposed temperature and pressure conditions, will yield only quartzites and marbles, respectively. In such cases, all one can say is that "metamorphism happened."

# PLATE TECTONICS AND METAMORPHISM

**LO10** Discuss the relationship between plate tectonics and metamorphism

Although metamorphism is associated with all three types of plate boundaries, it is most common along convergent plate margins. Metamorphic rocks form at convergent plate boundaries because temperature and pressure increase there as a result of plate collisions.

**FIGURE 7.19** illustrates the various metamorphic facies produced along a typical oceanic–continental convergent plate boundary. When an oceanic plate collides with a continental plate, tremendous pressure is generated as the oceanic plate is subducted. Because rock is a poor heat conductor, the cold descending oceanic plate heats slowly, and metamorphism is caused mostly by increasing pressure with depth. Metamorphism in such an environment produces rocks typical of the blueschist facies (low temperature, high pressure). Geologists use the presence of blueschist facies rocks as evidence of ancient subduction zones.

As subduction along the oceanic–continental convergent plate boundary continues, both temperature and pressure increase with depth and yield high-grade metamorphic rocks.

Eventually, the descending plate begins to melt and generate magma that moves upward. This rising magma may alter the surrounding rock by contact metamorphism, producing migmatites in the deeper portions of the crust and hornfels at shallower depths. High temperatures and low to medium pressures characterize such an environment.

Although metamorphism is most common along convergent plate margins, many divergent plate boundaries are

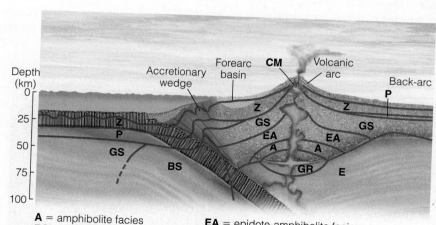

**A** = amphibolite facies
**BS** = blueschist facies
**CM** = contact metamorphic zone shown in green
**E** = eclogite facies
**EA** = epidote-amphibolite facies
**GR** = granulite facies
**GS** = greenschist facies
**P** = prehnite-pumpellyite facies
**Z** = zeolite facies

**FIGURE 7.19 Relationship of Facies to Major Tectonic Features at an Oceanic–Continental Convergent Plate Boundary**

characterized by contact metamorphism. Rising magma at mid-oceanic ridges heats the adjacent rocks, producing contact metamorphic minerals and textures. In addition, fluids emanating from the rising magma—and its reaction with seawater—very commonly produce metal-bearing hydrothermal solutions that may precipitate minerals of economic value, such as the copper ores of Cyprus.

# METAMORPHISM AND NATURAL RESOURCES

**LO11    Explain how various types of natural resources are the result from metamorphism**

Many metamorphic minerals and rocks are valuable natural resources. Although these resources include various types of ore deposits, the two most familiar and widely used

# GEO-FOCUS

## Asbestos: Good or Bad?

Asbestos (from the Latin, meaning "unquenchable") is a general term applied to any silicate mineral that easily separates into flexible fibers. The combination of such features as fire resistance and flexibility makes asbestos an important industrial material of considerable value.

Asbestos is divided into two broad groups: serpentine asbestos and amphibole asbestos. *Chrysotite* is the fibrous form of serpentine asbestos (Figure 7.20); it is the most valuable type and constitutes the bulk of all commercial asbestos. Its strong, silky fibers are easily spun and can withstand temperatures as high as 2,750°C. Among the varieties of amphibole asbestos, *crocidolite* is the most common. Crocidolite is a long, coarse, spinning fiber that is stronger but more brittle than chrysotile and less resistant to heat.

In 1986, Congress passed the Asbestos Hazard Emergency Response Act that required all schools be inspected for asbestos and take appropriate abatement action. That law and current policy of the U.S. Environmental Protection Agency (EPA) mandates that all forms of asbestos are treated as identical hazards. However, numerous studies indicate that only the amphibole forms (crocidolite) constitute a known health hazard because their long, straight, thin fibers can penetrate the lungs and stay there. Thus, crocidolite, and not chrysotile, is overwhelmingly responsible for asbestos-related lung cancer (●FIGURE 1). Because approximately 95% of the asbestos in place in the United States is chrysotile, and the cost of removing asbestos from buildings where it has been

installed can cost billions of dollars, many people question whether the dangers from asbestos are exaggerated.

Unless the material containing the asbestos is disturbed, asbestos does not shed fibers and thus does not contribute to airborne asbestos that can be inhaled. Furthermore, improper removal of asbestos can lead to contamination, resulting in far higher concentrations of airborne asbestos fibers than if the asbestos had been sealed and left in place.

The problem of asbestos contamination is a good example of how geology affects our lives and why we should have a basic knowledge of science before making decisions that could have broad economic and societal impacts.

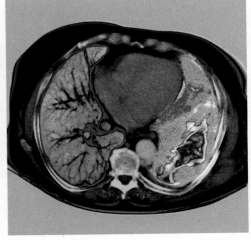

● **FIGURE 1 Lung Cancer** Computerized tomography scan of an axial section through the chest of a patient with a mesothelioma cancer (light red). It is surrounding and constricting the lung at right (pink). The other lung (dark blue) has a healthy pleura (dark red). The spine (lower center, light blue), the descending aorta (green), and the heart (dark green between lungs) are also seen. Mesothelioma is a malignant cancer of the pleura, the membrane lining the chest cavity and lungs. It is usually caused by asbestos exposure. It often reaches a large size, as here, before diagnosis, and prognosis is then poor.

SPL/Science Source

metamorphic rocks are marble (Figure 7.15) and slate (Figure 7.9), which have been used for centuries in a variety of ways.

Many ore deposits result from contact metamorphism during which hot, ion-rich fluids migrate from igneous intrusions into the surrounding rock, thereby producing rich ore deposits. The most common sulfide ore minerals associated with contact metamorphism are bornite and chalcopyrite (copper), galena (lead) (Figure 3.12b), pyrite (iron) (Figure 3.13a), and sphalerite (zinc); two common iron oxide ore minerals are hematite and magnetite. Tin and tungsten are also important ores associated with contact metamorphism.

Other economically important metamorphic minerals include asbestos (see GEO-Focus), used for insulation and fireproofing in buildings and building materials (●FIGURE 7.20), talc for talcum powder, graphite for pencils and dry lubricants, and garnets and corundum, which are used as abrasives or gemstones, depending on their quality. In addition, andalusite, kyanite, and sillimanite, which all have the same chemical composition, but differ in crystal

● **FIGURE 7.20 Specimen of Chrysotile Asbestos from Arizona** Chrysotile is the fibrous form of serpentine asbestos and the most commonly used in buildings and other structures.

structure, are used in manufacturing high-temperature porcelains and temperature-resistant materials for such products as sparkplugs and furnace linings.

# Key Concepts Review

- Metamorphic rocks result from the transformation of other rocks, usually beneath Earth's surface, as a consequence of one, or a combination, of three agents: heat, pressure, and fluid activity.
- Heat for metamorphism comes from intrusive magmas, extrusive lava flows, or deep burial. Pressure is either lithostatic (uniformly applied stress) or differential (stress unequally applied from different directions). Fluids trapped in sedimentary rocks or emanating from intruding magmas can enhance chemical changes and the formation of new minerals.
- The three major types of metamorphism are contact, dynamic, and regional.
- Contact metamorphism results when magma or lava alters the surrounding country rock.
- Dynamic metamorphism is associated with fault zones where rocks are subjected to high differential pressure.
- Most metamorphic rocks result from regional metamorphism, which occurs over a large area and is usually caused by tremendous temperatures, pressures, and deformation within the deeper portions of the crust.
- Metamorphic grade generally characterizes the degree to which a rock has undergone metamorphic change.
- Index minerals—minerals that form only within specific temperature and pressure ranges—allow geologists to recognize low-, intermediate-, and high-grade metamorphism.
- Metamorphic rocks are primarily classified according to their texture. In a foliated texture, platy and elongate

minerals have a preferred orientation. A nonfoliated texture does not exhibit any discernible preferred orientation of the mineral grains.
- Foliated metamorphic rocks can be arranged in order of increasing grain size, perfection of their foliation, or both. Slate is fine-grained, followed by (in increasingly larger grain size) phyllite and schist; gneiss displays segregated bands of minerals. Amphibolite is another fairly common foliated metamorphic rock. Migmatites have both igneous and high-grade metamorphic characteristics.
- Marble, quartzite, greenstone, hornfels, and anthracite are common nonfoliated metamorphic rocks.
- Metamorphic zones are based on index minerals and are areas of rock that all have similar grades of metamorphism; that is, they have all experienced the same intensity of metamorphism.
- A metamorphic facies is a group of metamorphic rocks whose minerals are formed under a particular range of temperatures and pressures. Each facies is named after its most characteristic rock or mineral.
- Metamorphism occurs along all three types of plate boundaries, but it is most common at convergent plate margins.
- Many metamorphic rocks and minerals, such as marble, slate, graphite, talc, and asbestos, are valuable natural resources. In addition, many ore deposits are the result of metamorphism and include copper, tin, tungsten, lead, iron, and zinc.

## Important Terms

aureole   140
contact (thermal) metamorphism   140
differential pressure   139
dynamic metamorphism   142
fluid activity   139

foliated texture   143
heat   138
index mineral   142
lithostatic pressure   139
metamorphic facies   148

metamorphic rock   138
metamorphic zone   148
metamorphism   138
nonfoliated texture   146
regional metamorphism   142

## Review Questions

1. From which of the following rock groups can metamorphic rocks form?
   a. volcanic
   b. sedimentary
   c. plutonic
   d. metamorphic
   e. all of these

2. Which is the correct metamorphic sequence of increasingly coarser grain size?
   a. gneiss → schist → phyllite → slate
   b. phyllite → slate → schist → gneiss
   c. schist → slate → gneiss → phyllite
   d. slate → phyllite → schist → gneiss
   e. slate → schist → phyllite → gneiss

3. Metamorphism is most common along what type of plate boundary?
   a. divergent
   b. transform
   c. lithospheric
   d. aseismic
   e. convergent

4. Which of the following are the three agents of metamorphism?
   a. gravity, fluid activity, pressure
   b. heat, pressure, gravity
   c. gravity, heat, fluid activity
   d. heat, pressure, fluid activity
   e. none of these

5. Concentric zones surrounding an igneous intrusion and characterized by distinctive mineral assemblages are
   a. aureoles.
   b. hydrothermal regions.
   c. regional facies.
   d. thermodynamic rings.
   e. metamorphic zones.

6. How do metamorphic rocks record the influence of differential pressure in their structures and mineral textures?

7. Why is it important for people to know something about metamorphism, metamorphic rocks, and how they form?

8. If plate tectonic movement did not exist, could there be metamorphism?

9. Describe the two types of metamorphic texture, and discuss how they are produced.

## Creative Thinking Visual Question

Foliated metamorphic rocks are characterized by having their minerals arranged in a parallel fashion, such that they can be split along foliation planes (Figure 7.8). In this photo (●**FIGURE 1**) of 500-million- to 1-billion-year-old schists and gneisses of the Blue Ridge Belt in southwestern Virginia, the foliation planes are dipping in the same direction as the slope of the hillside. What problem does this present in terms of potential landslides along the roadway? Hint: Note the several large rocks along the base of the hillside.

● **FIGURE 1** Schists and gneisses in the Blue Ridge Belt, Virginia.

U.S. Navy

The 7.0-magnitude earthquake that struck the island nation of Haiti on January 12, 2010, destroyed its capital city, Port-au-Prince, and devastated the surrounding areas, killing more than 223,000 people.

# EARTHQUAKES AND EARTH'S INTERIOR

8

# INTRODUCTION

## L01 Define an earthquake

In the afternoon of January 12, 2010, a 7.0-magnitude earthquake struck the island nation of Haiti. According to official estimates, 223,000 people died, at least 300,000 were injured, and more than 285,000 residences and businesses were destroyed or severely damaged. Widespread devastation occurred in the capital of Port-au-Prince and elsewhere throughout the region, exacerbated by an almost total collapse of the vital infrastructure needed to respond to such a disaster, including medical, transportation, and communication systems.

A little more than a year later, on March 11, 2011, a 9.0-magnitude earthquake and tsunami struck Japan, causing more than 20,000 deaths and also resulting in tremendous destruction, including severe damage to a nuclear power plant in the northeastern part of the island. Neither of these earthquakes is the first, nor will they be the last major devastating earthquakes in these regions of the world.

Earthquakes, along with volcanic eruptions, are manifestations of Earth's dynamic and active makeup. As one of nature's most frightening and destructive phenomena, earthquakes have always aroused feelings of fear and have been the subject of myths and legends. What makes an earthquake so frightening is that when it begins, there is no way to tell how long it will last or how violent it will be. Approximately 13 million people have died in earthquakes during the past 4,000 years, with about three million of these deaths occurring during the last century (●**TABLE 8.1**). Geologists define an **earthquake** as the shaking or trembling of the ground caused by the sudden release of energy, usually as a result of *faulting*, which involves the displacement of rocks along fractures (we discuss the different types of faults in Chapter 9). After an earthquake, continuing adjustments along a fault may generate a series of earthquakes known as *aftershocks*. Most aftershocks are smaller than the main shock, but they can still cause considerable damage to already weakened structures.

Why should you study earthquakes? The obvious reason is that they are destructive and cause many deaths and injuries to people living in earthquake-prone areas. Earthquakes also affect the economies of many countries in terms of cleanup costs, lost jobs, and lost business revenues. From a purely personal standpoint, you should be interested in earthquakes because you may be caught in one someday. Even if you don't live in an area that is subject to earthquakes, you may travel to places where the threat of earthquakes exists, and you should know what to do if you experience one. Such knowledge may help you avoid serious injury or even death.

**TABLE 8.1** Some Significant Earthquakes

| Year | Location | Magnitude (Estimated before 1935) | Deaths (Estimated) |
|------|----------|-----------------------------------|--------------------|
| 1556 | China (Shanxi Province) | 8.0 | 1,000,000 |
| 1755 | Portugal (Lisbon) | 8.6 | 70,000 |
| 1906 | U.S.A. (San Francisco, California) | 8.3 | 3,000 |
| 1923 | Japan (Tokyo) | 8.3 | 143,000 |
| 1960 | Chile | 9.5 | 5,700 |
| 1964 | U.S.A (Anchorage, Alaska) | 9.2 | 131 |
| 1976 | China (Tangshan) | 8.0 | 242,000 |
| 1985 | Mexico (Mexico City) | 8.1 | 9,500 |
| 1988 | Armenia | 6.9 | 25,000 |
| 1990 | Iran | 7.3 | 50,000 |
| 1993 | India | 6.4 | 30,000 |
| 1995 | Japan (Kobe) | 7.2 | >6,000 |
| 1999 | Turkey | 7.4 | 17,000 |
| 2001 | India | 7.9 | >14,000 |
| 2003 | Iran | 6.6 | 43,000 |
| 2004 | Indonesia | 9.0 | >230,000 |
| 2005 | Pakistan | 7.6 | >86,000 |
| 2006 | Indonesia | 6.3 | >6,200 |
| 2008 | China (Sichuan Province) | 7.9 | >69,000 |
| 2010 | Haiti | 7.0 | 223,000 |
| 2010 | Chile | 8.8 | 525 |
| 2011 | Japan | 9.0 | >20,000 |
| 2012 | Iran | 6.4 | >300 |
| 2017 | Mexico | 7.1 | >200 |

# ELASTIC REBOUND THEORY

## L02 Explain the elastic rebound theory

Based on studies conducted after the 1906 San Francisco earthquake, H. F. Reid of The Johns Hopkins University proposed the **elastic rebound theory** to explain how energy is

**earthquake** Vibrations caused by the sudden release of energy, usually as a result of displacement of rocks along faults.

**elastic rebound theory** An explanation for the sudden release of energy that causes earthquakes when deformed rocks fracture and rebound to their original undeformed condition.

released during earthquakes. Reid studied three sets of measurements taken across a portion of the San Andreas Fault that had broken during the 1906 earthquake. The measurements revealed that points on opposite sides of the fault had moved 3.2 m during the 50-year period prior to breakage in 1906, with the west side moving northward.

According to Reid, rocks on opposite sides of the San Andreas Fault had been storing energy and bending slightly for at least 50 years before the 1906 earthquake. Any straight line, such as a fence or road that crossed the San Andreas Fault, was gradually bent, because rocks on one side of the fault moved relative to rocks on the other side. Eventually, the strength of the rocks was exceeded, and they then fractured. After fracturing, the rocks on both sides of the fault rebounded, or "snapped back," to their former undeformed shape, and the energy stored was released as earthquake waves radiating outward from the break (●FIGURE 8.1A). Additional field and laboratory studies conducted by Reid and others have confirmed that elastic rebound is the mechanism by which energy is released during earthquakes.

A useful analogy is that of bending a long, straight stick over your knee. As the stick bends, it deforms and eventually reaches the point at which it breaks. When this happens, the two pieces of the original stick snap back into their original straight position. Likewise, rocks subjected to intense forces bend until they break, and they then return to their original position, releasing energy in the process.

a)

Fault

Fence

Original position

Deformation

Rupture and release of energy

Rocks rebound to original undeformed shape

b)

Copyright and Photograph by Dr. Parvinder S. Sethi

● **FIGURE 8.1 The Elastic Rebound Theory** (a) According to the elastic rebound theory, rocks experiencing deformation store energy and bend. When the internal strength of the rocks is exceeded, they rupture, releasing their accumulated energy, and "snap back" or rebound to their former undeformed shape. This sudden release of energy is what causes an earthquake. (b) During the 1906 San Francisco earthquake, this fence in Marin County was displaced by almost 5 m. Whereas many people would see just a broken fence, a geologist sees that the fence has moved, or been displaced, and would look for evidence of a fault. A geologist would also notice that the ground has been displaced toward the right side, relative to his or her view. Regardless of what side of the fence you stand on, you must look to the right to see the other part of the fence. Try it!

# SEISMOLOGY

**L03**   Define seismology

**L04**   Explain how a seismograph works and what it records

**L05**   Define an earthquake's focus

**L06**   Define an earthquake's epicenter

**L07**   Explain the relationship between an earthquake's focus and plate boundaries

**Seismology**, the study of earthquakes, emerged as a true science during the 1880s with the development of **seismographs**, instruments that detect, record, and measure the vibrations produced by an earthquake (●**FIGURE 8.2**). The record made by a seismograph is called a *seismogram*. Modern seismographs have electronic sensors and record movements precisely using computers, rather than simply relying on the drum strip charts commonly used on older seismographs (●**FIGURE 8.2A**).

When an earthquake occurs, energy in the form of *seismic waves* radiates out from the point of release (●**FIGURE 8.3**). These waves are somewhat analogous to the ripples that move out concentrically from the point where a stone is thrown into a pond. Unlike waves on a pond, however, seismic waves move outward in all directions from their source.

Earthquakes take place because rocks are capable of storing energy, but their strength is limited, so if enough force is present, they rupture and thus release their stored energy.

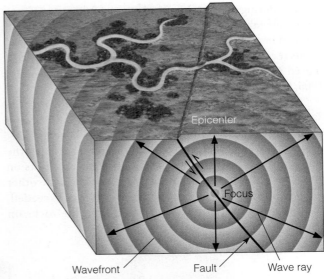

● **FIGURE 8.3 The Focus and Epicenter of an Earthquake** The focus of an earthquake is the location where the rupture begins and energy is released. The place on the surface vertically above the focus is the epicenter. Seismic wavefronts move out in all directions from their source, the focus of an earthquake.

**Critical Thinking Question** Why isn't the epicenter located where the fault emerges at Earth's surface?

**seismology** The study of earthquakes.

**seismograph** An instrument that detects, records, and measures the various waves produced by earthquakes.

● **FIGURE 8.2 Seismographs** (a) Seismographs record ground motion during an earthquake. The record produced is a seismogram. The seismograph shown here records earthquakes on a strip of paper attached to a rotating drum. (b) A horizontal-motion seismograph. Because of its inertia, the suspended mass that contains the marker remains stationary while the rest of the structure moves along with the ground during an earthquake. As long as the length of the arm is not parallel to the direction of ground movement, the marker will record the earthquake waves on the rotating drum. This seismograph would record waves from west or east, but to record waves from the north or south, another seismograph at right angles to this one is needed. (c) A vertical-motion seismograph. This seismograph operates on the same principle as the horizontal-motion instrument and records vertical ground movement.

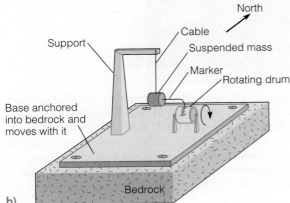

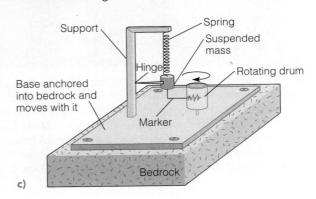

In other words, most earthquakes result when movement occurs along faults, most of which are related, at least indirectly, to plate movements. Once rupturing begins, it moves along the fault at several kilometers per second for as long as conditions for failure exist. The longer the fault along which movement occurs, the more time it takes for the stored energy to be released, and therefore the longer the ground will shake. During some very large earthquakes, the ground might shake for 3 minutes, a seemingly brief time, but interminable if you are experiencing the earthquake firsthand!

## The Focus and Epicenter of an Earthquake

The location within Earth's lithosphere where rupturing begins—that is, the point at which energy is first released—is an earthquake's **focus**, or *hypocenter*. What we usually hear in news reports, however, is the location of the **epicenter**, the point on Earth's surface directly above the focus (Figure 8.3).

Seismologists recognize three categories of earthquakes based on focal depth. *Shallow-focus* earthquakes have focal depths of less than 70 km from the surface, whereas those with foci between 70 and 300 km are *intermediate-focus*, and the foci

of those characterized as *deep-focus* are more than 300 km deep. However, earthquakes are not evenly distributed among these three categories. Approximately 90% of all earthquake foci are at depths of less than 100 km, whereas only about 3% of all earthquakes are deep-focus. Shallow-focus earthquakes are, with few exceptions, the most destructive, because the energy they release has little time to dissipate before reaching the surface.

A definite relationship exists between earthquake foci and plate boundaries. Earthquakes generated along divergent or transform plate boundaries are invariably shallow-focus, whereas many shallow-focus earthquakes and nearly all intermediate- and deep-focus earthquakes occur along convergent margins (●FIGURE 8.4). Furthermore, a pattern emerges when the focal depths of earthquakes near island arcs and their adjacent ocean trenches are plotted. Notice in ●FIGURE 8.5 that the focal depth increases beneath the Tonga Trench in a narrow, well-defined zone that dips

---

**focus** The site within Earth where an earthquake originates and energy is released.

**epicenter** The point on Earth's surface directly above the focus of an earthquake.

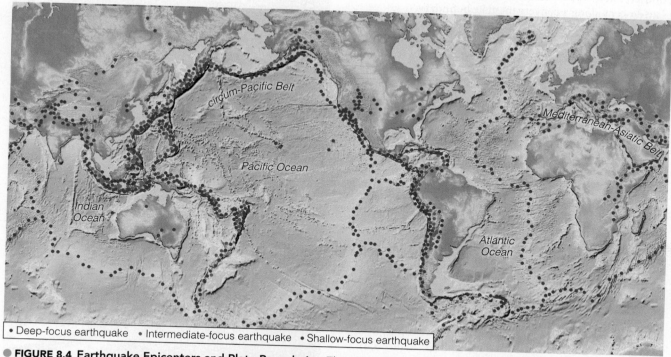

• Deep-focus earthquake    • Intermediate-focus earthquake    • Shallow-focus earthquake

● **FIGURE 8.4 Earthquake Epicenters and Plate Boundaries** This map of earthquake epicenters shows that most earthquakes occur within seismic zones that correspond closely to plate boundaries. Approximately 80% of earthquakes occur within the circum-Pacific belt, 15% within the Mediterranean–Asiatic belt, and the remaining 5% within plate interiors and along oceanic spreading ridges. The dots represent earthquake epicenters and are divided into shallow-, intermediate-, and deep-focus earthquakes. Along with shallow-focus earthquakes, nearly all intermediate- and deep-focus earthquakes occur along convergent plate boundaries.

**Critical Thinking Question** Why are nearly all intermediate- and deep-focus earthquakes associated with convergent plate boundaries?

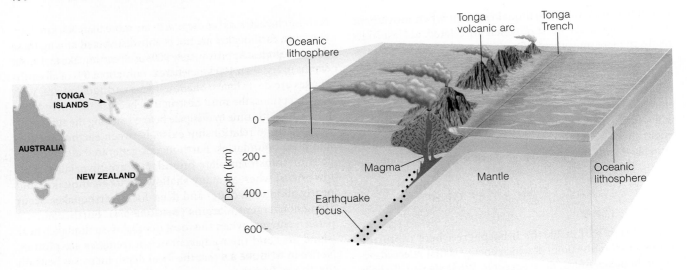

● **FIGURE 8.5  Benioff Zones** Focal depth increases in a well-defined zone that dips approximately 45 degrees beneath the Tonga volcanic arc in the South Pacific. Dipping seismic zones are called *Benioff* or *Benioff–Wadati zones*.

approximately 45 degrees. Dipping seismic zones, called *Benioff* or *Benioff–Wadati zones*, are common to convergent plate boundaries where one plate is subducted beneath another. Such dipping seismic zones indicate the angle of plate descent along a convergent plate boundary.

# WHERE DO EARTHQUAKES OCCUR, AND HOW OFTEN?

**LO8**  Name the two seismic belts where 95% of earthquakes take place

**LO9**  Explain why earthquakes occur in these two seismic belts

**LO10**  Identify where the remaining 5% of earthquakes occur

**LO11**  Discuss what geologists think is the possible cause of these earthquakes

No place on Earth is immune to earthquakes, but almost 95% take place in seismic belts corresponding to plate boundaries where plates converge, diverge, and slide past each other. The relationship between plate margins and the distribution of earthquakes is readily apparent when the locations of earthquake epicenters are superimposed on a map showing the boundaries of Earth's plates (Figure 8.4).

The majority of all earthquakes (approximately 80%) occur in the *circum-Pacific belt*, a zone of seismic activity nearly encircling the Pacific Ocean basin. Most of these earthquakes result from convergence along plate margins, as in the case of the 2011 Japanese earthquake. The earthquakes along the North American Pacific Coast, especially in California, are also in this belt, but here, plates slide past one another rather than converge. The January 17, 1994, Northridge earthquake (●**FIGURE 8.6**) occurred along this plate boundary.

● **FIGURE 8.6 Earthquake Damage in the Circum-Pacific Belt** One of the several elevated freeway spans that collapsed during the January, 1994 Northridge, California, earthquake in which 61 people died.

The second major seismic belt, accounting for 15% of all earthquakes, is the *Mediterranean–Asiatic belt*. This belt extends westward from Indonesia through the Himalayas, across Iran and Turkey, and westward through the Mediterranean region of Europe. The 2004 earthquake in Indonesia that killed more than 230,000 people is an example of the destructive earthquakes that strike this region (Table 8.1).

The remaining 5% of earthquakes occur mostly in the interiors of plates and along oceanic spreading-ridge systems. Most of these earthquakes are not strong, although several major intraplate earthquakes of note have occurred. For example, the 1811 and 1812 earthquakes near New Madrid, Missouri, killed approximately 20 people and nearly destroyed the town. So strong were these earthquakes that they were felt from the Rocky Mountains to the Atlantic Ocean and from the Canadian border to the Gulf of Mexico. Within the immediate area, numerous buildings were destroyed and forests were flattened. The land sank several meters in some areas, causing flooding, and reportedly the Mississippi River reversed its flow during the shaking and changed its course slightly.

The cause of intraplate earthquakes is not well understood, but geologists think that they arise from localized stresses caused by the compression that most plates experience along their margins. A useful analogy is moving a house. Regardless of how careful the movers are, moving something so large without its internal parts shifting slightly is impossible. Similarly, plates are not likely to move without some internal stresses that occasionally cause earthquakes.

It is estimated that there are 500,000 detectable earthquakes annually, most of which are too small to be felt. These earthquakes result from the energy released as continual

adjustments take place between the various plates. On average, however, more than 10,000 earthquakes per year can be felt and can cause various amounts of damage, depending on how strong they are and where they occur.

# SEISMIC WAVES

**LO12**   Define the two types of body waves
**LO13**   Define the two types of surface waves

Many people have experienced an earthquake, but most are probably unaware that the shaking they feel and the damage to structures are caused by the arrival of *seismic waves*, a general term encompassing all waves generated by an earthquake. When movement on a fault takes place, energy is released in the form of two kinds of seismic waves that radiate outward in all directions from an earthquake's focus. *Body waves*, so called because they travel through the solid body of Earth, are somewhat like sound waves, and *surface waves*, which travel along the ground surface, are analogous to undulations or waves on water surfaces.

## Body Waves

An earthquake generates two types of body waves: P-waves and S-waves (●**FIGURE 8.7**). **P-waves**, or *primary waves*, are

**P-wave** A compressional, or push-pull, wave; the fastest seismic wave and one that can travel through solids, liquids, and gases; also called a primary wave.

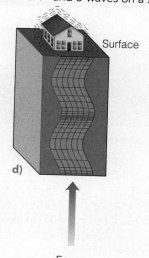

● **FIGURE 8.7 Primary and Secondary Seismic Body Waves** Body waves travel through Earth. (a) Undisturbed material for reference. (b) Primary waves (P-waves) compress and expand material in the same direction that they travel. (c) Secondary waves (S-waves) move material perpendicular to the direction of wave movement. (d) The effect of P- and S-waves on a surface structure.

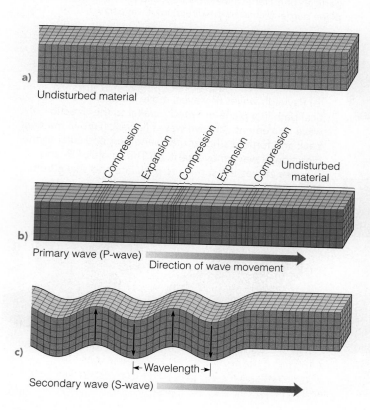

a) Undisturbed material

Compression  Expansion  Compression  Expansion  Compression  Undisturbed material

b) Primary wave (P-wave)
Direction of wave movement

c)
|←Wavelength→|
Secondary wave (S-wave)

Surface

d)

Focus

the fastest seismic waves and can travel through solids, liquids, and gases. P-waves are *compressional*, or *push-pull*, *waves* and are similar to sound waves in that they move material forward and backward along a line in the same direction that the waves themselves are moving (●**FIGURE 8.7B**). Thus, the material through which a P-wave travels is expanded and compressed as the waves move through it and returns to its original size and shape after the wave passes by.

**S-waves**, or *secondary waves,* are somewhat slower than P-waves and can travel only through solids. S-waves are *shear waves* because they move the material perpendicular to the direction of travel, thereby producing shear stresses in the material they move through (●**FIGURE 8.7C**). Because liquids (as well as gases) are not rigid, they have no shear strength, and S-waves cannot be transmitted through them.

The velocities of P- and S-waves are determined by the density and elasticity of the materials through which they travel. For example, seismic waves travel more slowly through rocks of greater density but more rapidly through rocks with greater elasticity. *Elasticity* is a property of solids, such as rocks, and means that once they have been deformed by an applied force, they return to their original shape when the force is no longer present. Because P-wave velocity is greater than S-wave velocity in all materials, P-waves always arrive at seismic stations first.

## Surface Waves

Surface waves travel along the surface of the ground or just below it and are slower than body waves (●**FIGURE 8.8**). Unlike the sharp jolting and shaking that body waves cause, surface waves generally produce a rolling or swaying motion, much like the experience of being on a boat.

Seismologists recognize several types of surface waves. The two most important are Rayleigh waves and Love waves, named after the British scientists who discovered them, Lord Rayleigh and A. E. H. Love. **Rayleigh waves (R-waves)** are generally the slower of the two and behave like water waves in that they move forward while the individual particles of material move in an elliptical path within a vertical plane oriented in the direction of wave movement (●**FIGURE 8.8B**). The motion of a **Love wave (L-wave)** is similar to that of an S-wave, but the individual particles of the material move only back and forth in a horizontal plane perpendicular to the direction of wave travel (●**FIGURE 8.8C**).

# LOCATING AN EARTHQUAKE

**LO14    Explain how an earthquake's epicenter is located**

We mentioned that news articles commonly report an earthquake's epicenter, but just how is the location of an epicenter determined? Once again, geologists rely on the study of seismic waves. We know that P-waves travel faster than S-waves,

---

**S-wave**  A shear wave that moves material perpendicular to the direction of travel, thereby producing shear stresses in the material it moves through; also known as a secondary wave; S-waves travel only through solids.

**Rayleigh wave (R-wave)**  A surface wave in which individual particles of material move in an elliptical path within a vertical plane oriented in the direction of wave movement.

**Love wave (L-wave)**  A surface wave in which the individual particles of material move only back and forth in a horizontal plane perpendicular to the direction of wave travel.

● **FIGURE 8.8 Rayleigh and Love Seismic Surface Waves**  Surface waves travel along Earth's surface or just below it. (a) Undisturbed material for reference. (b) Rayleigh waves (R-waves) move material in an elliptical path in a plane oriented parallel to the direction of wave movement. (c) Love waves (L-waves) move material back and forth in a horizontal plane perpendicular to the direction of wave movement. (d) The arrival of R- and L-waves causes the surface to undulate and shake from side to side.

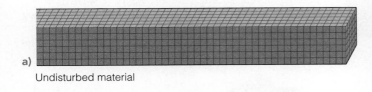
a)
Undisturbed material

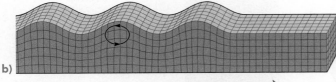
b)
Rayleigh wave (R-wave)

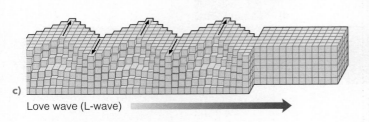
c)
Love wave (L-wave)

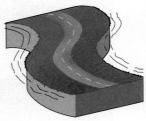
d)    Rayleigh wave                                    Love wave

nearly twice as fast in all substances, so P-waves arrive at a seismograph station first, followed some time later by S-waves. Both P- and S-waves travel directly from the focus to the seismograph station through Earth's interior, but L- and R-waves arrive last, because they are the slowest, and they also travel the longest route along the surface (●**FIGURE 8.9A**). However, only the P- and S-waves need concern us here, because they are the ones important in finding an epicenter.

Seismologists, who are geologists that study seismology, have accumulated a tremendous amount of data over the years and now know the average speeds of P- and S-waves for any specific distance from their source. These P- and S-wave travel times are published in *time–distance graphs* that illustrate the difference between the arrival times of the two waves as a function of the distance between a seismograph and an earthquake's focus (●**FIGURE 8.9B**). That is, the farther the waves travel, the greater the *P–S time interval*, which is simply the time difference between the arrivals of P- and S-waves (Figure 8.9a, b).

If the P–S time intervals are known from at least three seismograph stations, then the epicenter of any earthquake can be determined (●**FIGURE 8.10**). Here is how it works: Subtracting the arrival time of the first P-wave from the arrival time of the first S-wave gives the P–S time interval for each seismic station. Each of these time intervals is plotted on a time–distance graph, and a line is drawn straight down to the distance axis of the graph, thus giving the distance from the focus to each seismic station (Figure 8.9b). Next, a circle whose radius equals the distance shown on the time–distance graph from each of the seismic stations is drawn on a map (Figure 8.10). The intersection of the three circles is the location of the earthquake's epicenter. It should be obvious from Figure 8.10 that P–S time intervals from at least three seismic stations are needed. If only one were used, the epicenter could be at any location on the circle drawn around that station; using two stations would give two possible locations for the epicenter.

Determining the focal depth of an earthquake is much more difficult and considerably less precise than finding its

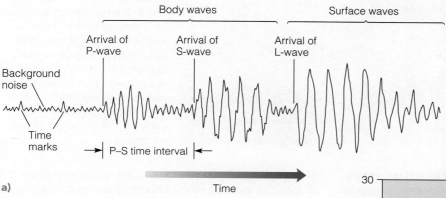

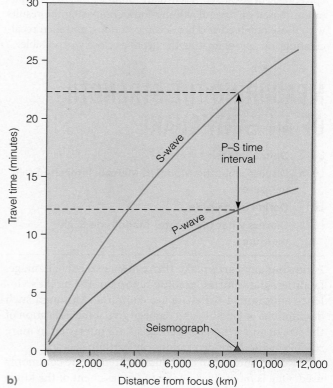

● **FIGURE 8.9 Determining the Distance from an Earthquake**
(a) A schematic seismogram showing the arrival order and pattern produced by P-, S-, and L-waves. When an earthquake occurs, body and surface waves radiate out from the focus at the same time. Because P-waves are the fastest, they arrive at a seismograph first, followed by S-waves, and then by surface waves, which are the slowest waves. The difference between the arrival times of the P- and S-waves is the *P–S time interval*; it is a function of the distance of the seismograph station from the focus. (b) A time–distance graph showing the average travel times for P- and S-waves. The farther away a seismograph station is from the focus of an earthquake, the longer the interval between the arrival of the P- and S-waves, and hence the greater the distance between the P- and S-wave curves on the time–distance graph as indicated by the P–S time interval. For example, let's assume the difference in arrival times between the P- and S-waves is 10 minutes (P–S time interval). Using the Travel time (minutes) scale, measure how long 10 minutes is (P–S time interval), and move that distance between the S-wave curve and the P-wave curve until the line touches both curves as shown. Then draw a line straight down to the Distance from focus (km) scale. That number is the distance the seismograph is from the earthquake's focus. In this example, the distance is almost 9,000 km.

Sue Monroe

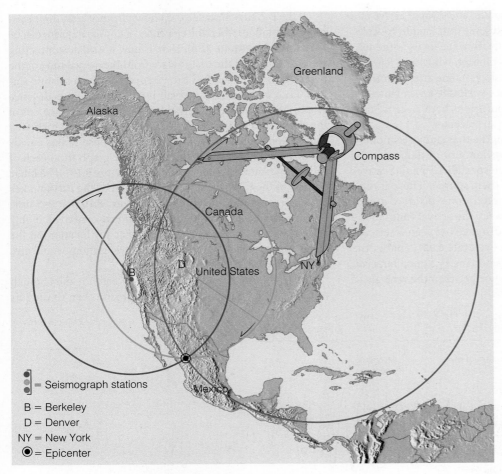

● **FIGURE 8.10 Determining the Epicenter of an Earthquake** Three seismograph stations are needed to locate the epicenter of an earthquake. The P–S time interval is plotted on a time–distance graph for each seismograph station to determine the distance that station is from the epicenter. A circle with that radius is drawn from each station, and the intersection of the three circles is the epicenter of the earthquake.

epicenter. The focal depth is usually found by making computations based on several assumptions, comparing the results with those obtained at other seismic stations, and then recalculating and approximating the depth as closely as possible.

# MEASURING THE STRENGTH OF AN EARTHQUAKE

LO15    Define intensity

LO16    Discuss what the Modified Mercalli Intensity Scale measures

LO17    Define magnitude

LO18    Discuss what the Richter Magnitude Scale measures

Following any earthquake that causes extensive damage, fatalities, and injuries, graphic reports of the quake's violence and human suffering are common. Although such descriptions of fatalities and damage give some indication of the size of an earthquake, geologists are interested in more reliable methods for determining an earthquake's size.

Two measures of an earthquake's strength are commonly used. One is *intensity*, a qualitative assessment of the kinds of damage done by an earthquake. The other, *magnitude*, is a quantitative measure of the amount of energy released by an earthquake. Each method provides important information that can be used to prepare for future earthquakes.

## Intensity

**Intensity** is a subjective or qualitative measure of the kind of damage done by an earthquake as well as people's reaction to it. Since the mid-19th century, geologists have used intensity as a rough approximation of the size and strength of an earthquake. The most common intensity scale used in the United States is the **Modified Mercalli Intensity Scale**, which has values ranging from I to XII (●**TABLE 8.2**).

Intensity maps can be constructed for regions hit by earthquakes by dividing the affected region into various intensity zones. The intensity value given for each zone is the maximum intensity that the earthquake produced for that zone. Even though intensity maps are not precise, because of the subjective nature of the measurements, they do provide

---

**intensity** The subjective measure of the kind of damage done by an earthquake, as well as people's reaction to it.

**Modified Mercalli Intensity Scale** A scale with values from I to XII used to characterize earthquakes based on damage.

**TABLE 8.2** Modified Mercalli Intensity Scale

I   Not felt except by a very few under especially favorable circumstances.

II   Felt by only a few people at rest, especially on upper floors of buildings.

III   Felt quite noticeably indoors, especially on upper floors of buildings, but many people do not recognize it as an earthquake. Standing automobiles may rock slightly.

IV   During the day felt indoors by many, outdoors by few. At night some awakened. Sensation like heavy truck striking building, standing automobiles rocked noticeably.

V   Felt by nearly everyone, many awakened. Some dishes, windows, etc. broken, a few instances of cracked plaster. Disturbance of trees, poles, and other tall objects sometimes noticed.

VI   Felt by all, many frightened and run outdoors. Some heavy furniture moved. A few instances of fallen plaster or damaged chimneys. Damage slight.

VII   Everybody runs outdoors. Damage negligible in buildings of good design and construction; slight to moderate in well-built ordinary structures; considerable in poorly built or badly designed structures; some chimneys broken. Noticed by people driving automobiles.

VIII   Damage slight in specially designed structures; considerable in normally constructed buildings with possible partial collapse; great in poorly built structures. Fall of chimneys, monuments, walls. Heavy furniture overturned. Sand and mud ejected in small amounts.

IX   Damage considerable in specially designed structures. Buildings shifted off foundations. Ground noticeably cracked. Underground pipes broken.

X   Some well-built wooden structures destroyed; most masonry and frame structures with foundations destroyed; ground badly cracked. Rails bent. Landsides considerable from river banks and steep slopes. Water splashed over river banks.

XI   Few, if any (masonry) structures remain standing. Bridges destroyed. Broad fissures in ground. Underground pipelines completely out of service.

XII   Damage total. Waves seen on ground surface. Objects thrown upward into the air.

*Source:* U.S. Geological Survey.

geologists with a rough approximation of the location of the earthquake, the kind and extent of the damage done, and the effects of local geology on different types of building construction. Because intensity is a measure of the kind of damage done by an earthquake, insurance companies still classify earthquakes on the basis of intensity.

Generally, a large earthquake will produce higher intensity values than a small earthquake, but many other factors besides the amount of energy released by an earthquake also affect its intensity. These include distance from the epicenter, focal depth of the earthquake, population density, geology of the area, type of building construction employed, and duration of shaking.

## Magnitude

If earthquakes are to be compared quantitatively, we must use a scale that measures the amount of energy released and is independent of intensity. Charles F. Richter, a seismologist at the California Institute of Technology, developed such a scale in 1935. The **Richter Magnitude Scale** measures earthquake **magnitude**, which is the total amount of energy released by an earthquake at its source. It is an open-ended scale with values beginning at zero. The largest magnitude recorded was a 9.5-magnitude earthquake in Chile on May 22, 1960 (Table 8.1).

Scientists determine the magnitude of an earthquake by measuring the amplitude of the largest seismic wave as recorded on a seismogram (●**FIGURE 8.11**). To avoid large

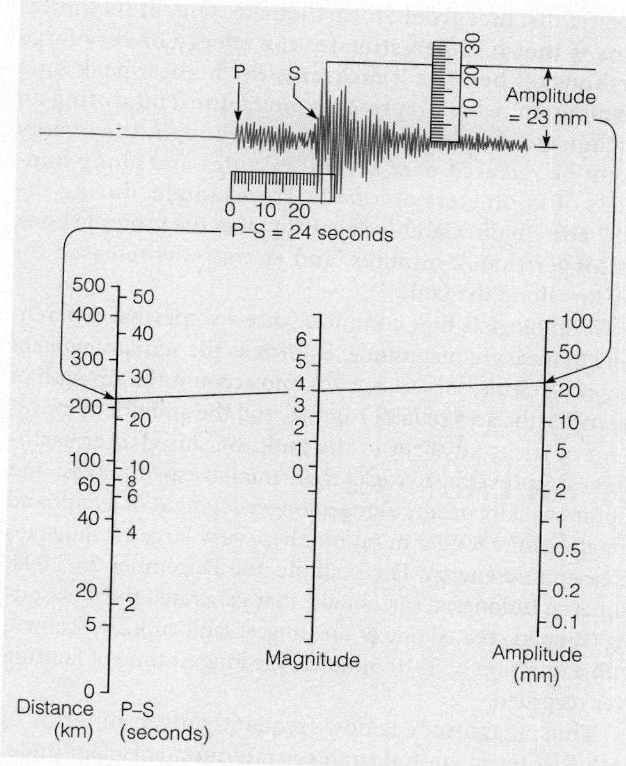

● **FIGURE 8.11 Richter Magnitude Scale** The Richter Magnitude Scale measures the total amount of energy released by an earthquake at its source. The magnitude is determined by measuring the maximum amplitude of the largest seismic wave and marking it on the right-hand scale. The difference between the arrival times of the P- and S-waves (recorded in seconds) is marked on the left-hand scale. When a line is drawn between the two points, the magnitude of the earthquake is the point at which the line crosses the center scale.

**Richter Magnitude Scale** An open-ended scale that measures the amount of energy released during an earthquake.

**magnitude** The total amount of energy released by an earthquake at its source.

numbers, Richter used a conventional base-10 logarithmic scale to convert the amplitude of the largest recorded seismic wave to a numerical magnitude value. Therefore, each whole-number increase in magnitude represents a 10-fold increase in wave amplitude. For example, the amplitude of the largest seismic wave for an earthquake of magnitude 6 is 10 times that produced by an earthquake of magnitude 5, 100 times as large as a 4-magnitude earthquake, and 1,000 times that of an earthquake of magnitude 3 ($10 \times 10 \times 10 = 1,000$).

A common misconception about the size of earthquakes is that an increase of one unit on the Richter Magnitude Scale—a 7 versus a 6, for instance—means a 10-fold increase in size. It is true that each whole-number increase in magnitude represents a 10-fold increase in the wave amplitude, but each magnitude increase of one unit corresponds to a roughly 30-fold increase in the amount of energy released (actually, it is 31.5, but 30 is close enough for our purposes). This means, for example, that the 2011 Japanese earthquake, with a magnitude of 9.0, released nearly 900 times more energy than the 2010 Haiti earthquake, with a magnitude 7.0 ($30 \times 30 = 900$)!

The Richter Magnitude Scale was devised to measure earthquake waves on a particular seismograph and at a specific distance from an earthquake. One of its limitations is that it underestimates the energy of very large earthquakes because it measures the highest peak on a seismogram, which represents only an instant during an earthquake. For large earthquakes, though, the energy might be released over several minutes and along hundreds of kilometers of a fault. For example, during the 1857 Fort Tejon, California earthquake, the ground shook for longer than 2 minutes, and energy was released for 360 km along the fault.

Seismologists now commonly use a somewhat different scale to measure magnitude. Known as the *seismic-moment magnitude scale*, this scale takes into account the strength of the rocks, the area of fault rupture, and the amount of movement of rocks adjacent to the fault. Because larger earthquakes rupture more rocks than smaller earthquakes, and rupture usually occurs along a longer segment of a fault and therefore for a longer duration, these very large earthquakes release more energy. For example, the December 26, 2004, Sumatra, Indonesia, earthquake that generated the devastating tsunami created one of the longest fault ruptures known, with a duration of 8–10 minutes, the longest time of faulting ever recorded.

Thus, magnitude is now frequently given in terms of both Richter magnitude and seismic-moment magnitude. As an example, the 2004 Sumatra, Indonesia, earthquake is given a Richter magnitude of 9.0 and a seismic-moment magnitude of 9.1–9.3. Because the Richter Magnitude Scale is most commonly used in the news, we will use that scale here.

# THE DESTRUCTIVE EFFECTS OF EARTHQUAKES

LO19   Discuss the factors that determine the destructiveness of an earthquake

LO20   List the four major destructive effects caused by earthquakes

LO21   Discuss what the outcomes are for each type of destructive event

The number of deaths and injuries, as well as the amount of property damage resulting from an earthquake, depends on several factors. Generally speaking, earthquakes that occur during working and school hours in densely populated urban areas are the most destructive and cause the most fatalities and injuries. However, magnitude, duration of shaking, distance from the epicenter, geology of the affected region, and the type of structures are also important considerations. Given these variables, it should not be surprising that a comparatively small earthquake can have disastrous effects whereas a much larger one might go largely unnoticed, except perhaps by seismologists.

The destructive effects of earthquakes include ground shaking, fire, seismic sea waves, and landslides, as well as panic, disruption of vital services, and psychological shock.

## Ground Shaking

Ground shaking, the most obvious and immediate effect of an earthquake, varies depending on the earthquake's magnitude, distance from the epicenter, and type of underlying materials in the area—unconsolidated sediment or fill versus bedrock, for instance. Certainly, ground shaking is terrifying, and it may be violent enough for fissures to open in the ground. Nevertheless, contrary to popular myth, fissures do not swallow up people and buildings and then close on them. And although California will no doubt have big earthquakes in the future, rocks cannot store enough energy to displace a landmass as large as California into the Pacific Ocean as some alarmists claim.

The effects of ground shaking, such as collapsing buildings, falling building facades and window glass, and toppling monuments and statues, cause more damage and result in more loss of life and injuries than any other earthquake hazard (see GEO-FOCUS). Structures built on solid bedrock generally suffer less damage than those built on poorly consolidated material such as water-saturated sediments or artificial fill (●**FIGURE 8.12**).

Structures built on poorly consolidated or water-saturated material are subjected to ground shaking of longer duration and greater S-wave amplitude than structures built on bedrock (Figure 8.12). In addition, fill and water-saturated sediments tend to liquefy, or behave as a fluid, a process known

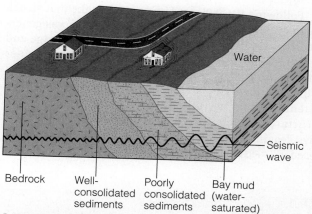

● **FIGURE 8.12 Relationship between Seismic Wave Amplitude and Underlying Geology** The amplitude and duration of seismic waves generally increase as the waves pass from bedrock to poorly consolidated or water-saturated material. Thus, structures built on weaker material typically suffer greater damage than similar structures built on bedrock, because the shaking lasts longer.

as *liquefaction*. When shaken, the individual grains lose cohesion, and the ground flows. Two dramatic and classic examples of damage resulting from liquefaction are Niigata, Japan, and Turnagain Heights, Alaska. In Niigata, Japan, large apartment buildings were tipped to their sides after the water-saturated soil of the hillside collapsed (●**FIGURE 8.13**). In Turnagain Heights, Alaska, many homes were destroyed when the Bootlegger Cover Clay lost all of its strength when it was shaken by the 1964 earthquake (see Figure 10.17).

Besides the magnitude of an earthquake and the underlying geology, the material used and the type of construction also affect the amount of damage done. Adobe and

● **FIGURE 8.13 Liquefaction** The effects of ground shaking on water-saturated soil are dramatically illustrated by the collapse of these buildings in Niigata, Japan, during a 1964 earthquake. The buildings, which were designed to be earthquake resistant, fell over on their sides intact when the ground below them underwent liquefaction.

● **FIGURE 8.14 Ground Shaking** Many buildings collapsed or were severely damaged as a result of ground shaking during the August 17, 1999 Turkey earthquake, which killed more than 17,000 people.

mud-walled structures are the weakest and almost always collapse during an earthquake. Unreinforced brick structures and poorly built concrete structures are also particularly susceptible to collapse, such as was the case in the 1999 Turkey earthquake in which an estimated 17,000 people died (●**FIGURE 8.14**). The 1976 earthquake in Tangshan, China, which killed more than 242,000 people, completely leveled the city, because hardly any structures were built to resist seismic forces. In fact, most had unreinforced brick walls, which have no flexibility, and consequently they collapsed during the shaking.

## Fire

In many earthquakes, particularly in urban areas, fire is a major hazard. Nearly 90% of the damage done in the 1906 San Francisco earthquake was caused by fire. The shaking severed many of the electrical and gas lines, which touched off flames and started fires throughout the city. Because the earthquake ruptured water mains, there was no effective way to fight the fires that raged out of control for three days, destroying much of the city.

Eighty-three years later, during the 1989 Loma Prieta earthquake, a fire broke out in the Marina district of San Francisco. This time, however, the fire was contained within a small area because San Francisco had a system of valves throughout its water and gas pipeline system so that lines could be isolated from breaks.

During the September 1, 1923, earthquake in Japan, fires destroyed 71% of the houses in Tokyo and practically all of the houses in Yokohama. In all, 576,262 houses were destroyed by fire, and 143,000 people died, many as a result of fire fanned by unusually high winds, which in some cases created fire tornadoes up to 100 m high (●**FIGURE 8.15**).

## Tsunami: Killer Waves

On December 26, 2004, a 9.0-magnitude earthquake struck 160 km off the west coast of northern Sumatra, Indonesia, generating the deadliest tsunami in history. Within hours, walls of water as high as 10.5 m pounded the coasts of Indonesia, Sri Lanka, India, Thailand, Somalia, Myanmar, Malaysia, and the Maldives, killing more than 230,000 people and causing billions of dollars in damage.

Following the 9.0-magnitude earthquake that struck Japan on March 11, 2011, a massive tsunami was generated that resulted in tremendous property damage and loss of life along the coastline of Japan. Within minutes after the earthquake, walls of water, some as high as 37 m, inundated low-lying areas along the Japanese coast and extended as far as 10 km inland. Boats, vehicles, and structures were destroyed or swept aside as if they were toys, leaving a wake of destruction as the waters finally receded (●FIGURE 8.16).

# GEO-FOCUS

## Designing and Building Earthquake-Resistant Structures

One way to reduce property damage, injuries, and loss of life is to design and build structures that are as earthquake resistant as possible. Many things can be done to improve the safety of current structures and of new buildings.

To design earthquake-resistant structures, engineers must understand the dynamics and mechanics of earthquakes, including the type and duration of the ground motion and how rapidly the ground accelerates during an earthquake. An understanding of the area's geology is also important because certain ground materials such as water-saturated sediments

or landfill can lose their strength and cohesiveness during an earthquake (see Figure 8.13). Finally, engineers must be aware of how different structures behave under different earthquake conditions.

With the level of technology currently available, a well-designed,

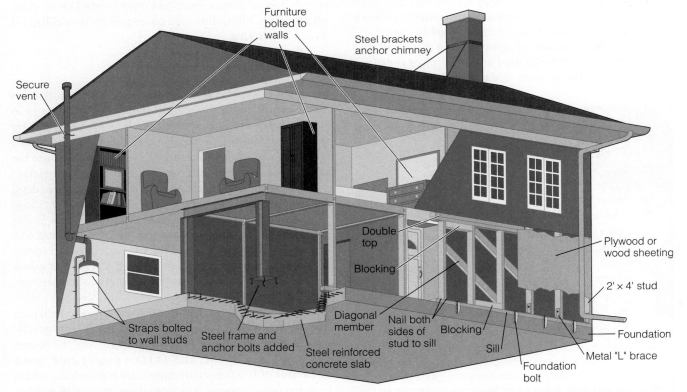

● **FIGURE 1** Some of useful things a homeowner can do to reduce damage to a building because of ground shaking during an earthquake. Notice that the structure must be solidly attached to its foundation and bracing the walls helps prevent damage from horizontal motion.

More recently, on September 28, 2018, a 7.5-magnitude earthquake struck Indonesia's Sulawawesi Island, causing a tsunami that swept through Palu Bay, destroying Palu, a city of 380,000 people. Almost immediately following the earthquake, eyewitnesses reported water being sucked out to sea and then surging back through the long, narrow bay, producing waves up to 7 m high.

In addition to the tsunami, the earthquake also triggered widespread liquefaction of soils, in which structures, people, and livestock were literally swallowed up by the unstable mud produced by the earthquake's shaking of the ground.

In all, more than 1,400 people died and more than 70,000 were left homeless from this tragedy.

And if that wasn't enough, five days later, a volcano erupted 941 km to the south, but fortunately, it did not cause any additional damage to the area. The earthquake, tsunami, and volcanic eruption in this region serve to remind us of the tremendous forces unleashed as a result of Earth's dynamic activity taking place along convergent plate boundaries.

All three of the aforementioned earthquakes generated what is popularly called a "tidal wave," but is more correctly

properly constructed building should be able to withstand small, short-duration earthquakes of less than magnitude 5.5 with little or no damage. In moderate earthquakes (magnitude 5.5–7.0), the damage suffered should not be serious and should be repairable. In a major earthquake of greater than magnitude 7.0, the building should not collapse, although it may later have to be demolished.

Many factors enter into the design of an earthquake-resistant structure, but the most important is that the building be tied together; that is, the foundation, walls, floors, and roof should all be joined together to create a structure that can withstand both horizontal and vertical shaking (●FIGURE 1). Almost all the structural failures resulting from earthquake ground movement occur at weak connections, where the various parts of a structure are not securely tied together. Buildings with open or unsupported first stories are particularly susceptible to damage. Some reinforcement must be done, or collapse is a distinct possibility.

Damage to high-rise structures can be minimized or prevented by using diagonal steel beams to help prevent swaying. In addition, tall buildings in earthquake-prone areas are now commonly placed on layered steel and rubber structures and devices similar

to shock absorbers that help decrease the amount of sway.

What about structures built many years ago? Just as in new buildings, the most important thing that can be done to increase the stability and safety of older structures is to tie together the different components of each building. Although such modifications are expensive, they are usually cheaper than having to replace a building that was destroyed by an earthquake.

Another problem related to both new and older buildings is that even with strict building codes, damage will still result if there is not adherence to construction standards for structures in earthquake-prone areas. Such was the case in the Sichuan, China, earthquake in 2008, in which more than 69,000 people died, and almost 80% of the building in the area were destroyed, many because of shoddy construction (●FIGURE 2).

● **FIGURE 2** During the 2008 Sichuan, China earthquake, structures of masonry construction, coupled with little or no reinforcing, collapsed, leading to tremendous damage and loss of life.

● **FIGURE 8.15** Ryounkaku Tower is the only structure still standing in Asakusa Park, Tokyo, following the 8.3-magnitude earthquake that struck Japan in 1923. Approximately 38,000 people were killed when a fire whirl swept through the former Army Clothing Depot site, where they had evacuated.

● **FIGURE 8.16 Japanese Tsunami** Aerial view of Minamisanriku, Miyagi Prefecture in northeastern Japan, showing houses clogged with debris and a large boat that had been swept inland as a result of the tsunami.

termed a *seismic sea wave* or **tsunami**, a Japanese term meaning "harbor wave." The term *tidal wave* nevertheless persists in popular literature and some news accounts, but

these waves are not caused by, or related to, tides. Indeed, tsunami are destructive sea waves generated when the sea floor undergoes sudden, vertical movements. Many result from submarine earthquakes, but volcanoes at sea or submarine landslides can also cause them. For example, the 1883 eruption of Krakatau between Java and Sumatra generated a large sea wave that killed 36,000 on nearby islands.

Once a tsunami is generated, it can travel across an entire ocean and cause devastation far from its source. In the open sea, tsunami travel at several hundred kilometers per hour and commonly go unnoticed as they pass beneath ships because they are usually less than 1 m high and the distance between wave crests is typically hundreds of kilometers. When they enter shallow water, however, the waves slow down and water piles up to heights anywhere from a meter or two to many meters high. The 1946 tsunami that struck Hilo, Hawaii, was 16.5 m high, and the one that struck Japan in 2011 was reported to have reached more than twice that height in some areas. In any case, the tremendous energy possessed by a tsunami is concentrated on a shoreline when it hits, either as a large breaking wave or, in some cases, what appears to be a very rapidly rising tide.

One of nature's warning signs of an approaching tsunami is a sudden withdrawal of the sea from a coastal region. In fact, the sea might withdraw so far that it cannot be seen and the seafloor is laid bare over a huge area. On more than one occasion, people have rushed out to inspect exposed reefs or to collect fish and shells only to be swept away when the tsunami arrived.

Following the tragic 1946 tsunami that struck Hilo, Hawaii, the U.S. Coast and Geodetic Survey established a Pacific Tsunami Early Warning System in Ewa Beach, Hawaii. This system combines seismographs and instruments that detect earthquake-generated sea waves. Whenever a strong earthquake takes place anywhere within the Pacific basin, its location is determined, and instruments are checked to see whether a tsunami has been generated. If it has, a warning is sent out to evacuate people from low-lying areas that may be affected.

Unfortunately, no such warning system exists for the Indian Ocean. If one had been in place, it is possible that the death toll from the December 26, 2004, tsunami would have been significantly lower.

## Ground Failure

Earthquake-triggered landslides are particularly dangerous in mountainous regions and have been responsible for tremendous amounts of damage and many deaths. The 1959 Hebgen Lake earthquake in Madison Canyon, Montana, for example, caused a huge rock slide (●**FIGURE 8.17**), and the 1970 Peru earthquake caused an avalanche that destroyed the town of Yungay and killed an estimated 66,000 people. Most of the 100,000 deaths from the 1920 earthquake in Gansu, China, resulted when cliffs composed of loess (wind-deposited silt) collapsed.

**tsunami** A large sea wave that is usually produced by an earthquake, but can also result from submarine landslides and volcanic eruptions.

a)

b)

Source of landslide
Landslide deposit

● **FIGURE 8.17 Ground Failure** On August 17, 1959, an earthquake with a Richter magnitude of 7.3 shook southwestern Montana and a large area in adjacent states. (a) The fault scarp in this image was produced when the block in the background moved up several meters compared to the one in the foreground. (b) The earthquake also triggered a landslide (visible in the distance) that blocked the Madison River in Montana and created Earthquake Lake (foreground). The slide entombed approximately 26 people in a campground at the very bottom of the valley.

# EARTHQUAKE PREDICTION

**LO22** Discuss the factors involved predicting the likelihood of possible future earthquakes

A successful prediction must include a time frame for the occurrence of an earthquake, its location, and its strength. Despite the tremendous amount of information geologists have gathered about the cause of earthquakes, successful predictions, especially short-range ones that would provide warning of an impending earthquake, are still rare. Nevertheless, if reliable predictions can be made, they can greatly reduce the number of deaths and injuries.

From an analysis of historic records and the distribution of known faults, geologists construct *seismic risk maps* that indicate the likelihood and potential severity of future earthquakes based on the intensity of past earthquakes. An international effort by scientists from several countries resulted in the publication of the first Global Seismic Hazard Assessment Map in December 1999 (●**FIGURE 8.18**). Although such

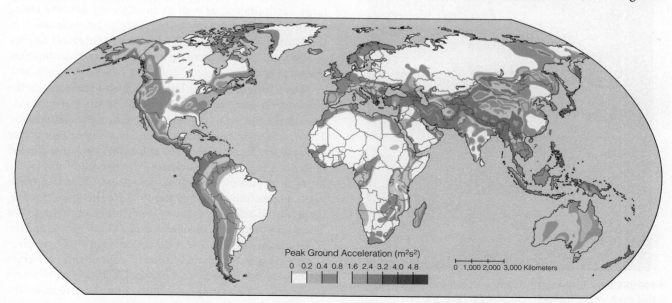

Peak Ground Acceleration (m²s²)
0  0.2  0.4  0.8  1.6  2.4  3.2  4.0  4.8

0  1,000 2,000 3,000 Kilometers

● **FIGURE 8.18 Global Seismic Hazard Assessment Map** The Global Seismic Hazard Assessment Program published this seismic hazard map showing peak ground accelerations. The values are based on a 90% probability that the indicated horizontal ground acceleration during an earthquake is not likely to be exceeded in 50 years. The higher the number, the greater the hazard. As expected, the greatest seismic risks are in the circum-Pacific belt and the Mediterranean–Asiatic belt.

maps cannot be used to predict when an earthquake will take place in any particular area, they are useful in anticipating future earthquakes and helping people plan and prepare for them.

Another method geologists use to determine the possibility of a future seismic event is known as *paleoseismology*, which, as its name implies, is the study of prehistoric earthquakes. As more people move into seismically active areas, it is important to know the frequency of past earthquakes in that region. In this way, prudent decisions can be made about what precautions need to be taken in developing an area, and how stringent the building codes for a region should be.

A typical technique used in paleoseismic studies is excavating a trench across active faults in the area to be studied, and dating the sediments disturbed by prehistoric earthquakes (●FIGURE 8.19). In this manner, geologists can determine the frequency of past earthquakes and when the last earthquake occurred, thus providing a basis for estimating the probability, and potential severity, of future earthquakes.

An interesting case in point concerns an ancient earthquake in what is now Seattle, Washington. Data from a variety of sources indicates that a shallow-focus earthquake of at least magnitude 7 occurred in the Seattle area approximately 1,000 years ago. If history and events of the geologic past are any guide, it is likely that another large earthquake will strike the Seattle area in the foreseeable future. When this will occur cannot yet be predicted, but it would be wise to plan for such an eventuality.

Based on studies conducted during the past several decades, it has been shown that most earthquakes are preceded by both short-term and long-term changes within Earth. Such changes are called *precursors* and may be useful in earthquake prediction.

One long-range prediction technique used in seismically active areas involves plotting the location of major earthquakes and their aftershocks to detect areas that have had major earthquakes in the past but are currently inactive. Such regions are said to be *locked* and not releasing energy. Nevertheless, pressure is continuing to accumulate in these regions because of plate motions, making these *seismic gaps* prime locations for future earthquakes. Several seismic gaps along the San Andreas Fault have the potential for future major earthquakes (●FIGURE 8.20). A major earthquake that damaged Mexico City in 1985 occurred along a seismic gap in the convergence zone along the west coast of Mexico.

Other earthquake precursors that may be useful in making short-term predictions include slight changes in elevation and tilting of the land surface, fluctuations in the water level in wells, changes in Earth's magnetic field, and the electrical resistance of the ground.

## Earthquake Prediction Programs

Currently, only a handful of nations, including the United States, have government-sponsored earthquake prediction programs. These programs include laboratory and field studies of rock behavior before, during, and after large earthquakes, as well as monitoring activity along major active faults. Most earthquake prediction work in the United States is done by the U.S. Geological Survey (USGS) and involves research into all aspects of earthquake-related phenomena.

The Chinese have perhaps the most ambitious earthquake prediction program in the world, which is understandable considering their long history of destructive earthquakes. Their earthquake prediction program was initiated soon after two large earthquakes occurred at Xingtai (300 km southwest of Beijing) in 1966, and includes extensive study and monitoring of all possible earthquake precursors. Chinese seismologists successfully predicted the 1975 Haicheng earthquake, but unfortunately, they failed to predict the devastating 1976 Tangshan earthquake that killed at least 242,000 people, and the 2008 Sichuan earthquake that killed more than 69,000 people.

Progress is being made toward dependable, accurate earthquake predictions, and studies are underway to assess public reactions to long-, medium-, and short-term earthquake warnings. However, unless short-term warnings are actually followed by an earthquake, most people will probably ignore the warnings, as they frequently do now for hurricanes, tornadoes, and tsunami.

● **FIGURE 8.19** Geologists examining a trench across an active fault in California to determine possible seismic hazards. Excavating trenches is a common method used by geologists to gather information about ancient earthquakes in a region, and to help assess the potential for future earthquakes and the damage that they might cause.

© John Karachewski Geoscapes Photography

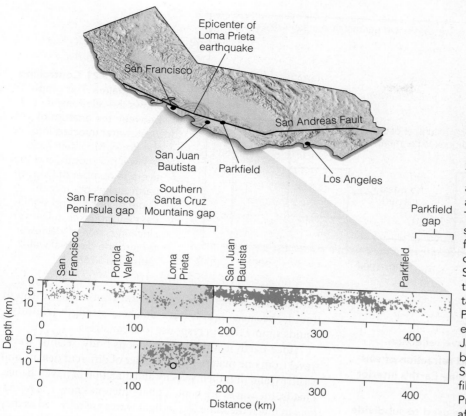

● **FIGURE 8.20 Earthquake Precursors**
Seismic gaps are one type of earthquake precursor that can indicate a potential earthquake in the future. These seismic gaps are regions along a fault that are locked, meaning they are not moving and releasing energy. Three gaps are evident in this cross section along the San Andreas Fault from north of San Francisco to south of Parkfield. The first is between San Francisco and Portola Valley, the second near Loma Prieta Mountain, and the third is southeast of Parkfield. The top section shows the epicenters of earthquakes between January 1969 and July 1989. The bottom section shows the southern Santa Cruz Mountains gap after it was filled by the October 17, 1989, Loma Prieta earthquake (*open circle*) and its aftershocks.

# EARTHQUAKE CONTROL

**LO23** Discuss the viability and possibility of earthquake control

Although reliable earthquake prediction is still in the future, can anything be done to control, or at least partly control, earthquakes? Because of the tremendous energy involved, it seems unlikely that humans will ever be able to prevent earthquakes. However, it may be possible to gradually release the energy stored in rocks, thus decreasing the probability of a large earthquake and extensive damage.

During the early- to mid-1960s, Denver, Colorado, experienced numerous small earthquakes, which was surprising because Denver had not been prone to earthquakes in the past. Geologist David M. Evans suggested that the earthquakes were directly related to the injection of contaminated wastewater into a 3,674-m-deep disposal well at the Rocky Mountain Arsenal, northeast of Denver. A USGS study concluded that the pumping of waste fluids into the fractured rocks beneath the disposal well decreased the friction on opposite sides of the fractures—in effect lubricating them so that movement occurred, thus causing the earthquakes that Denver experienced.

Interestingly, a high degree of correlation was found when comparing the number of earthquakes in Denver against the average amount of contaminated fluids injected into the disposal well per month (●**FIGURE 8.21**). Furthermore, when no waste fluids were injected, earthquake activity decreased dramatically.

In a situation similar to Denver, the number of earthquakes in Texas, Oklahoma, and the Dakotas has dramatically increased. Because of the thousands of gas wells drilled recently using hydraulic fracturing in which large amounts of high-pressure water mix are injected into the wells to increase pore space and open up preexisting fractures and thus allowing gas to flow into the wells, there has been an increase in earthquakes in these areas. This has led many people to conclude that there is a link between hydraulic fracturing and the onset of small earthquakes. Although there is currently no clear consensus as to whether there is a cause-and-effect relationship between the two events, geologists and seismologists are continuing studies to determine if there is a connection between them.

Based on the results of the Denver study, and the possible link between the injection of high pressure fluids into wells, some geologists have proposed that fluids be pumped into the locked segments, or seismic gaps, of active faults to cause small- to moderate-sized earthquakes. They think that this would relieve the pressure on the fault and prevent a major earthquake from occurring.

Although this plan is intriguing, it has many potential problems. For instance, there is no guarantee that only a small earthquake might result. Instead, a major earthquake might occur, causing tremendous property damage and loss of life, especially in a densely populated area. Who would be responsible? Certainly, a great deal more research is needed before such an experiment is performed, even in an area of low population density.

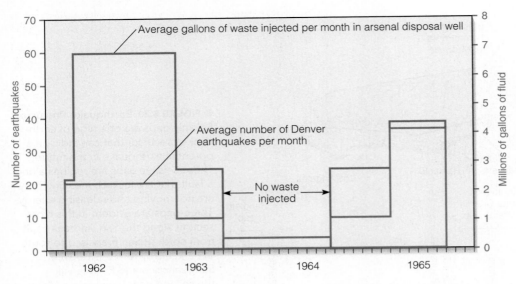

● **FIGURE 8.21 Controlling Earthquakes** This graph shows the relationship between the amount of wastewater injected into the Rocky Mountain Arsenal well per month, and the average number of Denver earthquakes per month. There have been no significant earthquakes in Denver since injection of wastewater into the disposal well ceased in 1966.

# EARTH'S INTERIOR

**LO24**   Define wave refraction and wave reflection

**LO25**   Discuss how the reflection and refraction of seismic waves are used to determine Earth's interior

**LO26**   Define a discontinuity

**LO27**   Explain how discontinuities are used to subdivide Earth's interior

During most of historic time, Earth's interior was perceived as an underground world of vast caverns, heat, and sulfur gases populated by demons. By the 1860s, scientists knew the average density of Earth and that pressure and temperature increased with depth. And even though Earth's interior is hidden from direct observation, scientists now have a reasonably good idea of its internal structure and composition.

Earth is generally depicted as consisting of concentric layers that differ in composition and density, separated from adjacent layers by rather distinct boundaries (●**FIGURE 8.22**). Recall that the outermost layer, or *crust*, is Earth's thin skin. Below the crust and extending about halfway to Earth's center is the *mantle*, which comprises more than 80% of Earth's volume. The central part of Earth consists of a *core*, which is divided into a solid inner portion and a liquid outer part (Figure 8.21).

The behavior and travel times of P- and S-waves provide geologists with information about Earth's internal structure. Seismic waves travel outward as wavefronts from their source areas, although it is most convenient to depict them as *wave rays*, which are lines showing the direction of movement of small parts of wavefronts (Figure 8.3).

As we noted earlier, P- and S-wave velocity is determined by the density and elasticity of the materials they travel through, both of which increase with depth. Wave velocity is slowed by increasing density but increases in materials with greater elasticity. Because elasticity increases with depth faster than density, a general increase in seismic wave velocity takes place as the waves penetrate to greater depths. P-waves travel faster than S-waves under all circumstances,

but unlike P-waves, S-waves are not transmitted through liquids, because liquids have no shear strength (rigidity); liquids simply flow in response to shear stress.

Since Earth is not a homogeneous body, seismic waves travel from one material into another of different density and elasticity, and thus their velocity and direction of travel change. That is, the waves are bent, a phenomenon known as *refraction*, in much the same way as light waves are refracted as they pass from air into more-dense water. Because seismic waves pass through materials of differing density and elasticity, they are continually refracted, so that their paths are curved. Wave rays travel in a straight line only when their direction of travel is perpendicular to a boundary (●**FIGURE 8.23**).

In addition to refraction, seismic rays are *reflected*, much as light is reflected from a mirror. When seismic rays encounter a boundary separating materials of different density or elasticity, some of a wave's energy is reflected back to the surface (Figure 8.23). If we know the wave velocity and the time required for the wave to travel from its source to the boundary and back to the surface, we can calculate the depth of the reflecting boundary. Such information is useful in determining not only Earth's internal structure but also the depths of sedimentary rocks that may contain petroleum. Seismic reflection is a common tool used in petroleum exploration.

Although changes in seismic wave velocity occur continuously with depth, P-wave velocity increases suddenly at the base of the crust and decreases abruptly at a depth of approximately 2,900 km (●**FIGURE 8.24**). These marked changes in seismic wave velocity indicate a boundary called a **discontinuity**, across which a significant change in Earth materials or their properties occurs. Discontinuities are the basis for subdividing Earth's interior into concentric layers.

---

**discontinuity** A boundary across which seismic wave velocity or direction of travel changes abruptly, such as the mantle–core boundary.

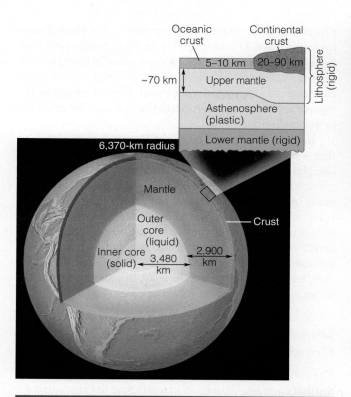

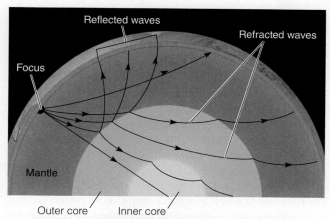

● **FIGURE 8.23 Refraction and Reflection of Seismic Waves** Refraction and reflection of P-waves as they encounter boundaries separating materials of different density or elasticity. Notice that the only wave ray not refracted is the one perpendicular to boundaries.

**Critical Thinking Question** If Earth was a homogeneous body, how would seismic waves behave as they moved through Earth?

### Earth's Composition and Density

| | Composition | Density (g/cm³) |
|---|---|---|
| Continental crust | Average composition of granodiorite | ≈2.7 |
| Oceanic crust | Upper part basalt, lower part gabbro | ≈3.0 |
| Mantle | Peridotite (made up of ferromagnesian silicates) | 3.3–5.7 |
| Outer core | Iron with perhaps 12% sulfur, silicon, oxygen, nickel, and potassium | 9.9–12.2 |
| Inner core | Iron with 10–20% nickel | 12.6–13.0 |
| Earth | | 5.5 |

● **FIGURE 8.22 Earth's Internal Structure** The inset shows Earth's outer part in more detail. The asthenosphere is solid but behaves plastically and flows.

# THE CORE

**LO28** Explain how the P- and S-wave shadow zones are used to define the configuration of Earth's core and the boundary between it and the overlying mantle

**LO29** Discuss how the density and composition of the core is determined

In 1906, R. D. Oldham of the Geological Survey of India realized that seismic waves arrived later than expected at seismic stations more than 130 degrees from an earthquake focus. He postulated that Earth has a core that transmits seismic waves more slowly than shallower Earth

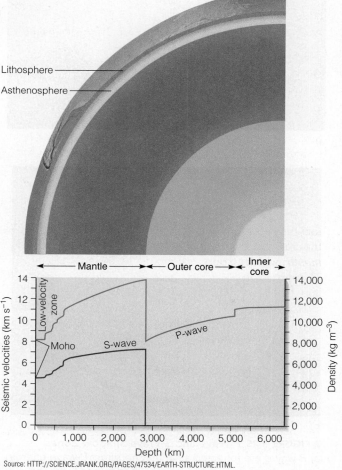

Source: HTTP://SCIENCE.JRANK.ORG/PAGES/47534/EARTH-STRUCTURE.HTML.

● **FIGURE 8.24 Seismic Wave Velocities** Profiles showing seismic wave velocities versus depth. Several discontinuities are shown, across which seismic wave velocities change rapidly.

materials. We now know that P-wave velocity decreases markedly at a depth of 2,900 km, which indicates an important discontinuity now recognized as the core–mantle boundary (Figure 8.24).

Because of the sudden decrease in P-wave velocity at the core–mantle boundary, P-waves are refracted in the core. Consequently, little P-wave energy reaches the surface in the area between 103 degrees and 143 degrees from an earthquake focus (●FIGURE 8.25A). This **P-wave shadow zone**, as it is called, is an area in which little P-wave energy is recorded by seismographs.

The P-wave shadow zone is not a perfect shadow zone because some weak P-wave energy is recorded within it. Scientists proposed several hypotheses to account for

this observation, but all were rejected by the Danish seismologist Inge Lehmann, who in 1936 postulated that the core is not entirely liquid as previously thought. She proposed that seismic wave reflection from a solid inner core accounts for the arrival of weak P-wave energy in the P-wave shadow zone, a proposal that was quickly accepted by seismologists.

In 1926, the British physicist Harold Jeffreys realized that S-waves were not simply slowed by the core, but were completely blocked by it. So, besides a P-wave shadow zone, a much larger and more complete **S-wave shadow zone** also exists (●FIGURE 8.25B). At locations greater than 103 degrees from an earthquake focus, no S-waves are recorded, which indicates that S-waves cannot be transmitted through the core. S-waves will not pass through a liquid, so it seems that the outer core must be liquid or behave as a liquid. The inner core, however, is thought to be solid, because P-wave velocity increases at the base of the outer core.

## Density and Composition of the Core

The core constitutes 16.4% of Earth's volume and nearly one-third of its mass. Geologists can estimate the core's density and composition by using seismic evidence and laboratory experiments. Furthermore, meteorites, which are thought to represent remnants of the material from which the solar system formed, are used to make estimates of density and composition. From these studies, the density of the outer core has been determined to vary from 9.9 to 12.2 g/cm$^3$. At Earth's center, the pressure is equivalent to approximately 3.5 million times normal atmospheric pressure.

The core cannot be composed of minerals common at the surface, because, even under the tremendous pressures at great depth, they would still not be dense enough to yield an average density of 5.5 g/cm$^3$ for Earth. Both the outer and inner cores are thought to be composed largely of iron, but pure iron is too dense to be the sole constituent of the outer core. It must be "diluted" with elements of lesser density. Laboratory experiments and comparisons with iron meteorites indicate that perhaps 12% of the outer core consists of sulfur and possibly some silicon, oxygen, nickel, and potassium (Figure 8.22).

In contrast, pure iron is not dense enough to account for the estimated density of the inner core, so perhaps 10%–20% of the inner core consists of nickel. These metals form an iron–nickel alloy thought to be sufficiently dense under the pressure at that depth to account for the density of the inner core.

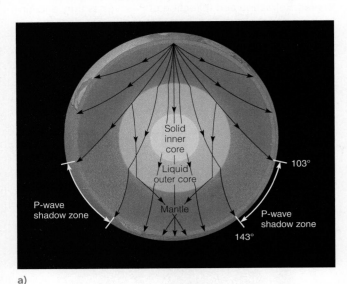

a)

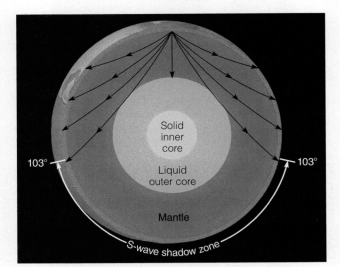

b)

● **FIGURE 8.25 P-Wave and S-Wave Shadow Zones**
(a) P-waves are refracted such that no direct P-wave energy reaches the surface in the P-wave shadow zone. (b) The presence of an S-wave shadow zone indicates that S-waves are being blocked within Earth.

**P-wave shadow zone** An area between 103 and 143 degrees from an earthquake's focus where little P-wave energy is recorded by seismographs.

**S-wave shadow zone** Those areas more than 103 degrees from an earthquake's focus where no S-waves are recorded.

# EARTH'S MANTLE

LO30   Define the Mohorovičić discontinuity

LO31   Explain how the mantle's structure, density, and composition have been determined

Another significant discovery about Earth's interior was made in 1909 when the Yugoslavian seismologist Andrija Mohorovičić detected a seismic discontinuity at a depth of about 30 km. While studying the arrival times of seismic waves from Balkan (part of southeastern Europe) earthquakes, Mohorovičić noticed that seismic stations a few hundred kilometers from an earthquake's epicenter were recording two distinct sets of P- and S-waves.

From his observations, Mohorovičić concluded that a sharp boundary separates rocks with different properties at a depth of about 30 km. He postulated that P-waves below this boundary travel at 8 km per second, whereas those above the boundary travel at 6.75 km per second. When an earthquake occurs, some waves travel directly from the focus to a seismic station, whereas others travel through the deeper layer, and some of their energy is refracted back to the surface (●FIGURE 8.26). The waves traveling through the deeper layer (the mantle) travel farther to a seismic station, but they do so more rapidly and arrive before those that travel more slowly in the shallower layer.

The boundary identified by Mohorovičić separates the crust from the mantle and is now called the **Mohorovičić discontinuity**, or simply the Moho. It is present everywhere except beneath spreading ridges. However, its depth varies: beneath continents, it ranges from 20 to 90 km, with an average of 35 km; beneath the sea floor, it is 5 to 10 km deep.

## The Mantle's Structure, Density, and Composition

Although seismic wave velocity in the mantle increases with depth, several discontinuities exist. Between depths of 100 and 250 km, both P- and S-wave velocities decrease markedly. This 100- to 250-km-deep layer is the *low-velocity zone*, which corresponds closely to the *asthenosphere* (Figure 8.24), a layer in which the rocks are close to their melting point and are less elastic, accounting for the observed decrease in seismic wave velocity. The asthenosphere is an important zone because it is where most magma is generated, especially under the ocean basins. Furthermore, it lacks strength, flows plastically, and is thought to be the layer over which the plates of the outer, rigid *lithosphere* move.

Other discontinuities are also present at deeper levels within the mantle. But unlike those between the crust and mantle or between the mantle and core, these probably represent structural changes in minerals rather than compositional changes. In other words, geologists think that the mantle is composed of the same material throughout, but the structural states of minerals such as olivine change with depth.

Although the mantle's density, which varies from 3.3 to 5.7 g/cm³, can be inferred rather accurately from seismic waves, its composition is less certain. The igneous rock *peridotite* (see Figure 4.10), containing mostly ferromagnesian silicates, is considered the most likely component of the upper mantle. Laboratory experiments indicate that it possesses physical properties that account for the mantle's density and observed rates of seismic wave transmissions. Peridotite also forms the lower parts of igneous rock sequences thought to be fragments of the oceanic crust and upper mantle emplaced on land.

**Mohorovičić discontinuity**  The boundary between Earth's crust and mantle.

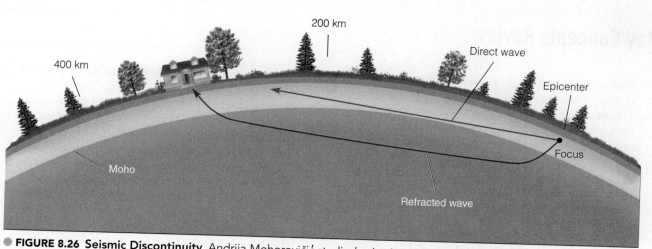

● **FIGURE 8.26 Seismic Discontinuity** Andrija Mohorovičić studied seismic waves and detected a seismic discontinuity at a depth of about 30 km. The deeper, faster seismic waves arrive at seismic stations first, even though they travel farther. This discontinuity, now known as the Moho, is between the crust and mantle.

# EARTH'S INTERNAL HEAT

**LO32** Define Earth's geothermal gradient

**LO33** Discuss why it is thought that much of Earth's internal heat is generated by radioactive decay

During the 19th century, scientists realized that the temperature in deep mines increases with depth, and this same trend has been observed in deep drill holes. This temperature increase with depth, or **geothermal gradient**, is approximately 25°C/km near the surface. In areas of active or recently active volcanism, the geothermal gradient is greater than in adjacent nonvolcanic areas, and temperature rises faster beneath spreading ridges than elsewhere beneath the seafloor.

Much of Earth's internal heat is generated by radioactive decay, especially the decay of isotopes of uranium and thorium and, to a lesser degree, potassium-40. When these isotopes decay, they emit energetic particles and gamma rays that heat surrounding rocks. Because rock is such a poor conductor of heat, it takes little radioactive decay to build up considerable heat, given enough time.

Unfortunately, the geothermal gradient is not useful for estimating temperatures at great depth. If we were simply to extrapolate from the surface downward, the temperature at 100 km would be so high that, despite the great pressure, all known rocks would melt. Current estimates of the temperature at the base of the crust are 800°C to 1,200°C. The latter figure seems to be an upper limit; if it were any higher, melting would be expected. Furthermore, fragments of mantle rock, thought to have come from depths of 100–300 km, appear to have reached equilibrium at these depths at a temperature of approximately 1,200°C. At the core–mantle boundary, the temperature is probably between 2,500°C and 5,000°C; the wide range of values indicates the uncertainties of such estimates. If these figures are reasonably accurate, the geothermal gradient in the mantle is only about 1°C/km.

# EARTH'S CRUST

**LO34** Discuss the composition, density, and thickness of continental crust

**LO35** Discuss the composition, density, and thickness of oceanic crust

Our main concern in the latter part of this chapter is Earth's interior; however, to be complete, we must briefly discuss the crust, which, along with the upper mantle, constitutes the lithosphere.

Continental crust is complex, consisting of all rock types, but it is usually described as "granitic," meaning that its overall composition is similar to that of granitic rocks. With the exception of metal-rich rocks such as iron ore deposits, most rocks of the continental crust have densities between 2.5 g/cm$^3$ and 3.0 g/cm$^3$, with the average density of the crust being about 2.7 g/cm$^3$. P-wave velocity in continental crust is approximately 6.75 km/sec, but at the base of the crust, P-wave velocity abruptly increases to about 8 km/sec. Continental crust averages 35 km thick, but its thickness varies from 20 km to 90 km. Beneath mountain ranges such as the Rocky Mountains, the Alps in Europe, and the Himalayas in Asia, continental crust is much thicker than it is in adjacent areas. In contrast, continental crust is much thinner than average beneath the Rift Valleys of East Africa and in a large area called the Basin and Range Province in the western United States and northern Mexico. The crust in these areas has been stretched and thinned in what appear to be the initial stages of rifting (see Chapter 2).

In contrast to continental crust, oceanic crust is simpler, consisting of gabbro in its lower part and overlain by basalt. It is thinnest, about 5 km, at spreading ridges, and nowhere is it thicker than 10 km. Its average density of 3.0 g/cm$^3$ accounts for the fact that it transmits P-waves at approximately 7 km/sec. In fact, this P-wave velocity is what one would expect if oceanic crust is composed of basalt and gabbro.

---

**geothermal gradient** Earth's temperature increase with depth; it averages 25°C/km near the surface but varies from area to area.

# Key Concepts Review

- Earthquakes are vibrations caused by the sudden release of energy, usually along a fault.
- The elastic rebound theory is an explanation for how energy is released during earthquakes. As rocks on opposite sides of a fault are subjected to force, they accumulate energy and slowly deform until their internal strength is exceeded. At that time, a sudden movement occurs along the fault, releasing the accumulated energy, and the rocks snap back to their original, undeformed shape.
- Seismology is the study of earthquakes. Earthquakes are recorded on seismographs, and the record of an earthquake is a seismogram.
- An earthquake's focus is the location where rupture within Earth's lithosphere occurs and energy is released.

- The epicenter is the point on Earth's surface directly above the focus.
- Approximately 80% of all earthquakes occur in the circum-Pacific belt, 15% within the Mediterranean–Asiatic belt, and the remaining 5% mostly in the interior of the plates and along oceanic spreading ridges.
- The two types of body waves are P-waves (primary waves), which are compressional (expanding and compressing) and the fastest seismic waves, traveling through all material, and S-waves (secondary waves), which are shear (moving material perpendicular to the direction of travel) and slower than P-waves and can travel only through solids.
- There are several types of surface waves, but the two most important are Rayleigh waves (R-waves) and

Love waves (L-waves), which move along or just below Earth's surface.

- An earthquake's epicenter is determined using a time–distance graph of the P- and S-waves to calculate how far away a seismic station is from an earthquake. The greater the difference in arrival times between the two waves, the farther away the seismic station is from the earthquake. Three seismic stations are needed to locate the epicenter.
- Intensity is a subjective, or qualitative, measure of the kind of damage done by an earthquake. It is expressed in values from I to XII in the Modified Mercalli Intensity Scale.
- The Richter Magnitude Scale measures an earthquake's magnitude, which is the total amount of energy released by an earthquake at its source. It is an open-ended scale with values beginning at 1. Each increase in magnitude number represents about a 30-fold increase in energy released.
- The seismic-moment magnitude scale more accurately measures the total energy released by very large earthquakes.
- The destructive effects of earthquakes include ground shaking, fire, tsunami, landslides, and disruption of vital services.
- Seismic risk maps help geologists determine the likelihood and potential severity of future earthquakes based on the intensity of past earthquakes.
- Paleoseismology is the study of prehistoric earthquakes and can be used to estimate the probability and potential severity of future earthquakes.
- Earthquake precursors are changes preceding an earthquake and include seismic gaps, changes in surface elevations, and fluctuations of water levels in wells.
- A variety of earthquake research programs are under way in various countries. Studies indicate that most people would probably not heed a short-term earthquake warning.
- Although it is unlikely that earthquakes can ever be prevented, it might be possible to release small amounts of the energy stored in rocks and thus avoid a large, devastating earthquake.
- Various studies indicate that Earth has an outer layer of oceanic and continental crust below which lies a rocky mantle and an iron-rich core with a solid inner part and a liquid outer part.
- Density and elasticity of Earth materials determine the velocity of seismic waves. Seismic waves are refracted when their direction of travel changes. Wave reflection occurs at boundaries across which the properties of rocks change.
- Geologists use the behavior of P- and S-waves and the presence of the P- and S-wave shadow zones to estimate the density and composition of Earth's interior, as well as to estimate the size and depth of the core and mantle.
- Earth's inner core is probably made up of iron and nickel, whereas the outer core is mostly iron with 10%–20% other substances.
- Peridotite, an igneous rock composed mostly of ferromagnesian silicates, is the most likely rock of Earth's mantle.
- Oceanic crust is composed of basalt and gabbro, whereas continental crust has an overall composition similar to that of granite. The Moho is the boundary between the crust and the mantle.
- The geothermal gradient of 25°C/km cannot continue to great depths; within the mantle and core, it is probably about 1°C/km. The temperature at Earth's center is estimated to be 6,500°C.

# Important Terms

| | | |
|---|---|---|
| discontinuity 172 | Love wave (L-wave) 160 | Richter Magnitude Scale 163 |
| earthquake 154 | magnitude 163 | seismograph 156 |
| elastic rebound theory 154 | Modified Mercalli Intensity Scale 162 | seismology 156 |
| epicenter 157 | Mohorovičić discontinuity 175 | S-wave 160 |
| focus 157 | P-wave 159 | S-wave shadow zone 174 |
| geothermal gradient 176 | P-wave shadow zone 174 | tsunami 168 |
| intensity 162 | Rayleigh wave (R-wave) 160 | |

# Review Questions

1. Regions along a major fault that are locked and not releasing energy are prime sites for future earthquakes and are known as
   a. epicenters.
   b. foci.
   c. hypocenters.
   d. discontinuities.
   e. seismic gaps.

2. The minimum number of seismographs needed to determine an earthquake's epicenter is
   a. 1.
   b. 2.
   c. 3.
   d. 4.
   e. 5.

3. The majority of all earthquakes take place in the
   a. spreading-ridge zone.
   b. Mediterranean–Asiatic belt.
   c. continental interiors.
   d. circum-Pacific belt.
   e. circum-Atlantic belt.

4. How much more energy is released by a 5.0-magnitude earthquake than a 2.0-magnitude earthquake?
   a. 4
   b. 810,000
   c. 27,000
   d. 90
   e. 2,500,000

5. The seismic discontinuity at the base of the crust is known as the
   a. transition zone.
   b. magnetic reflection point.
   c. low-velocity zone.
   d. Moho.
   e. high-velocity zone.

6. Refer to the graph in Figure 8.11. A seismograph in Berkeley, California, recorded the arrival time of an earthquake's P-waves as 6:59:54 p.m. and the S-waves as 7:00:02 p.m. The maximum amplitude of the S-waves as recorded on the seismogram was 75 mm. What was the magnitude of the earthquake, and how far away from Berkeley did it occur?

7. Why do scientists think that the inner core is solid and the outer core is liquid? What is the core composed of, and what is the evidence for this conclusion?

8. Why do insurance companies use the qualitative Modified Mercalli Intensity Scale instead of the quantitative Richter Magnitude Scale in classifying earthquakes?

9. From the arrival times of P- and S-waves shown in the accompanying chart, and the graph in Figure 8.9b, calculate how far away from each seismograph station the earthquake occurred. How would you determine the epicenter of this earthquake?

|  | Arrival Time of P-Wave | Arrival Time of S-Wave |
|---|---|---|
| Station A | 2:59:03 p.m. | 3:04:03 p.m. |
| Station B | 2:51:16 p.m. | 3:01:16 p.m. |
| Station C | 2:48:25 p.m. | 2:55:55 p.m. |

# Creative Thinking Visual Question

At 3:02 a.m., on August 17, 1999, violent shaking from a 7.4-magnitude earthquake awakened millions of people in Turkey. Unfortunately for many, their houses or apartment buildings collapsed, causing an estimated 17,000 deaths, at least 50,000 injuries, and leaving more than 150,000 buildings moderately to heavily damaged.

Like California, Turkey is situated in an earthquake-prone area. However, there are typically many more deaths and much greater destruction from earthquakes in Turkey than from earthquakes of similar size along the San Andreas Fault. Why is this so? What could some of the factors be that lead to so much more damage in Turkey than in California, even when the earthquakes are the same magnitudes? To answer this question, look at ●**FIGURE 1A** and **B** and think about how the types of construction, population density, building codes, the type of fault movement, and other factors generally lead to a greater number of deaths and more destruction in Turkey than in California.

a)

b)

●**FIGURE 1 The Izmit, Turkey, Earthquake of August 17, 1999** (a) Some of the numerous severely damaged structures in Izmit, Turkey. (b) Damage ranged from minimal to complete in this area of Turkey, following the 7.4-magnitude earthquake on August 17, 1999.

9

View of the Karakorum Range, Pakistan from the Goro II camp in the center of Baltoro glacier. The Karakorum Range is part of the Himalayan mountain system and lies on the borders of Pakistan, India, and China. It has more peaks higher than 8,000 m than any other mountain range, including K2, which at 8,611 m is the second-highest peak on Earth.

# DEFORMATION, MOUNTAIN BUILDING, AND THE CONTINENTS

179

# INTRODUCTION

"Solid as a rock" implies permanence and durability, but you already know that when rocks weather they disaggregate and decompose (see Chapter 6), and they behave very differently during metamorphism (see Chapter 7). Under the tremendous pressure and high temperature at several kilometers below the surface, rock layers yield to forces and actually crumple or fold, yet remain solid. And at shallower depths they may fracture, or fold and fracture. In any case, dynamic forces within Earth cause **deformation**, a general term for all changes in the shape or volume of rocks.

The fact that dynamic forces continue to operate within Earth is obvious from seismic activity, volcanism, plate movements, and the evolution of mountains in South America, Asia, and elsewhere. In short, Earth is an active planet with several processes driven by internal heat, especially plate movements; much of Earth's seismic activity, volcanism, deformation, and mountain building take place at convergent plate boundaries. The other terrestrial planets and Earth's moon, with the possible exception of Venus, show little evidence of ongoing deformation, volcanism, and so on.

The origin of Earth's truly large mountain ranges on the continents involves tremendous deformation—usually accompanied by emplacement of plutons, volcanism, and metamorphism—at convergent plate boundaries. And, in some cases, this activity continues even now (see the Chapter Opening photo). Thus, deformation and mountain building are closely related topics, and accordingly we consider both in this chapter.

The past and continuing evolution of continents involves not only deformation at continental margins but also additions of new material to existing continents, a phenomenon known as *continental accretion* (see Chapter 17). North America, for instance, has not always had its present shape and area. Indeed, it began evolving during the Archean Eon (4.6 to 2.5 billion years ago) as new material was added to the continent at deformation belts along its margins.

Much of this chapter is devoted to a review of *geologic structures*, such as folded and fractured rock layers resulting from deformation; their descriptive terminology; and the forces responsible for them. There are several practical reasons to study deformation and mountain building. For one thing, deformed rock layers provide a record of the kinds

---

**deformation**  A general term for any change in shape or volume, or both, of rocks in response to stress; involves folding and fracturing.

---

# GEO-FOCUS

## Engineering and Geology

On March 12, 1928, the St. Francis Dam in Southern California failed, resulting in a flood wave about 43 m high that killed at least 450 people; 179 victims of the flood were never found. Geology had been a consideration in large-scale projects for many years previously, but following this catastrophic event, geologic input became standard practice in ventures such as building dams. So, as you might expect, engineering geology is a specialty that applies the concepts of geology to engineering projects, which may involve slope stability studies, and studies for acceptable locations for dams, power plants, highways, tunnels, canals, and structures designed to protect riverbank and seashore communities.

A good example is the concern prior to building the Mackinac (pronounced "mack-in-aw") Bridge, a huge suspension bridge that connects the Upper and Lower Peninsulas of Michigan (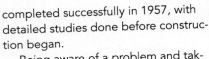**FIGURE 1**). Geologists and engineers were aware that some of the rock in the area, called Mackinac Breccia, was a collapse breccia, or rubble that formed when caverns collapsed. The concern was whether the breccia or any uncollapsed caverns beneath the area would support the weight of the huge piers and abutments for the bridge. The project was completed successfully in 1957, with detailed studies done before construction began.

Being aware of a problem and taking remedial action sometimes comes too late. For instance, engineers were aware that the Santa Monica Freeway in the Los Angeles area would likely be damaged during an earthquake and retrofitting was scheduled for February 1994. Unfortunately, the

● **FIGURE 1** Before the Mackinac Bridge, that connects the Lower and Upper Peninsulas of Michigan, could be built, geologists and engineers had to determine whether the Mackinac Breccia could support such a large structure.

and intensities of forces that operated during the past. Thus, interpretations of these structures allow us to satisfy our curiosity about Earth history, and, in addition, such studies are essential in engineering endeavors such as choosing sites for dams, bridges, and nuclear power plants, especially if they are in areas of ongoing deformation (see GEO-FOCUS). Also, many aspects of mining and exploration for petroleum and natural gas rely on correctly identifying geologic structures.

# ROCK DEFORMATION: HOW DOES IT OCCUR?

**LO1**  Define rock deformation

**LO2**  Define stress and strain

**LO3**  Define the three varieties of stress

**LO4**  Define the three types of strain

We defined the term *deformation* as any change in the volume or shape of rocks, but our reference to rocks means rock layers as depicted in many images in this text, not isolated stones you might find in your driveway or on a streambed. The type of rock is irrelevant, although layered rocks show the effects of deformation most clearly. In any case, rock layers may be crumpled into folds or fractured as a result of **stress**, which results from force applied to a given area of rock. The rock's internal strength resists stress, but if the stress is great enough, the rock undergoes **strain**, which is simply deformation caused by stress. The terminology is a little confusing at first, but the following discussion and reference to ●**FIGURE 9.1** will help you understand these terms.

## Stress and Strain

Remember that stress is the force applied to a given area of rock, usually expressed in kilograms per square centimeter ($kg/cm^2$). For example, the stress, or force, exerted by a person walking on an ice-covered pond is a function of the person's weight and the area beneath her or his feet. The ice's internal strength resists the stress unless the stress is too great, in which case the ice may bend or crack as it is

---

**stress**  The force per unit area applied to a material such as rock.

**strain**  Deformation caused by stress.

---

Northridge earthquake struck on January 17, 1994, and part of the freeway collapsed. Of course, this and other similar events provide important information that can be incorporated into engineering practice to make freeways, buildings, and bridges safer during earthquakes.

Along rugged seacoasts and in mountainous areas, highways are notoriously unstable and slump or slide from hillsides (●FIGURE 2). When this happens, engineering geologists are consulted for their recommendations for stabilizing slopes. They may suggest building retaining walls or drainage systems to keep the slopes dry, planting vegetation, or, in some cases, simply rerouting the highway if it is too costly to maintain in its present position.

A recent example of the problems associated with building roads along rugged coastlines occurred in early 2017, when El Niño rains triggered mud and rock slides along Highway 1

in Big Sur, about 240 km south of San Francisco, California. Although it was initially thought that the highway might not be able to be repaired, the California Department of Transportation was able to clear the debris and rebuild that portion of the highway that had been destroyed. During the 14 months it took to complete the project, revenues from local businesses that depend on tourism in and around the Big Sur area, fell as much as 90%!

Remember that geologic hazards, which include flooding, landslides,

earthquakes, and volcanic eruptions, to name just a few, account for thousands of fatalities and billions of dollars in property damage every year. Whereas the incidence of hazards has not increased, fatalities and damages have grown because more and more people live in disaster-prone area. And although we cannot eliminate geologic hazards, we can better understand them, and, at the very least, decrease the amount of damage and human suffering.

Dr. Marli Miller/Visuals Unlimited/Getty Images

●**FIGURE 2**  The Devil's Slide area along the Pacific Coast of California. The thin line on the hillside is California Highway 1, which is periodically closed due to slides. Not only are the rocks here weak and susceptible to sliding and slumping but the San Andreas Fault is nearby. The state built a tunnel, which was completed in March 2013, to bypass this area.

**● FIGURE 9.1 Stress and Strain Exerted on an Ice-Covered Pond** (a) The woman weighs 65 kg (6,500 g). Her weight is imparted to the ice through her feet, which have a contact area of 120 cm². The stress she exerts on the ice (6,500/120 cm²) is 54 g/cm². This is sufficient stress to cause the ice to crack. (b) To avoid plunging into the freezing water, the woman lies out flat, thereby decreasing the stress she exerts on the ice. Her weight remains the same, but her contact area with the ice is 3,150 cm², so the stress is only about 2 g/cm² (6,500 g/3,150 cm²), which is well below the threshold needed to crack the ice.

strained (deformed) (Figure 9.1). To avoid breaking through the ice, the person may lie down; this does not reduce the person's weight on the ice, but it does distribute their weight over a larger area, thus reducing the stress per unit area.

Although stress is force per unit area, it comes in three varieties: *compression, tension,* and *shear*. In **compression**, materials are squeezed, or compressed, by forces directed toward one another along the same line, as when you squeeze a rubber ball in your hand. Rock layers in compression tend to be shortened in the direction of stress by either folding or fracturing (●**FIGURE 9.2A**). **Tension** results from forces acting along the same line, but in opposite directions. Tension tends to lengthen rocks or pull them apart (●**FIGURE 9.2B**). Incidentally, rocks are much stronger in compression than they are in tension. In **shear stress**, forces act parallel to one another, but in opposite directions, resulting in deformation by displacement along closely spaced planes (●**FIGURE 9.2C**).

## Types of Strain

Geologists characterize strain as **elastic strain** if deformed rocks return to their original shape when the deforming forces are relaxed. In Figure 9.1, the ice on the pond may bend under a person's weight but return to its original shape once the person leaves. As you might expect, rocks are not very elastic, but Earth's crust behaves elastically when loaded by glacial ice and depressed into the mantle.

As stress is applied, rocks respond first by elastic strain, but when strained beyond their elastic limit, they undergo **plastic strain** as when they yield by folding, or they behave like brittle solids and **fracture**. In either folding or fracturing, the strain is permanent; that is, the rocks do not recover their original shape or volume even if the stress is removed.

Whether strain is elastic, plastic, or produces fractures depends on the kind of stress applied, pressure and temperature, rock type, and the length of time rocks are subjected to stress. A small stress applied over a long period, as on a mantelpiece supported only at its ends, will cause the

---

**compression** Stress resulting when rocks are squeezed by external forces directed toward one another.

**tension** A type of stress in which forces act in opposite directions but along the same line, thus tending to stretch an object.

**shear stress** The result of forces acting parallel to one another but in opposite directions; results in deformation by displacement of adjacent layers along closely spaced planes.

**elastic strain** A type of deformation in which the material returns to its original shape when stress is relaxed.

**plastic strain** Permanent deformation of a solid with no failure by fracturing.

**fracture** A break in rock resulting from intense applied pressure.

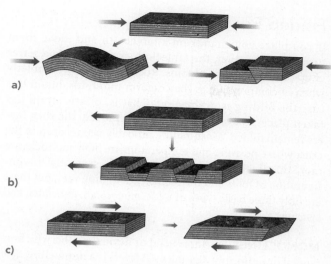

**FIGURE 9.2 Stress and Possible Types of Resulting Deformation** (a) Compression causes shortening of rock layers by folding or faulting. (b) Tension lengthens rock layers and causes faulting. (c) Shear stress causes deformation by displacement along closely spaced planes.

rock to sag; that is, the rock deforms plastically. By contrast, a large stress applied rapidly to the same object, as when struck by a hammer, results in fracture. Rock type is important, because not all rocks have the same internal strength and thus respond to stress differently. Some rocks are *ductile*, whereas others are *brittle*, depending on the amount of plastic strain they exhibit. Brittle rocks show little or no plastic strain before they fracture, but ductile rocks exhibit a great deal.

Many rocks show the effects of plastic strain that must have taken place deep within the crust. At or near the surface, rocks commonly behave like brittle solids and fracture, but at depth, they more often yield by plastic deformation, and thus become more ductile with increasing pressure and temperature.

# STRIKE AND DIP: THE ORIENTATION OF DEFORMED ROCK LAYERS

**LO5** Define strike and dip

**LO6** Explain how strike and dip are used to determine the orientation of rock layers

One concept in geology is the *principle of original horizontality*, meaning that sediments accumulate in horizontal or nearly horizontal layers (see Figure 16.3). Thus, if we observe steeply inclined sedimentary rocks, we are justified in inferring that they were deposited nearly horizontally, lithified, and then tilted into their present position. Rock layers deformed by folding, faulting, or both are no longer in their original position, so geologists use *strike* and *dip* to describe their orientation with respect to a horizontal plane.

By definition, **strike** is the direction of a line formed by the intersection of a horizontal plane and an inclined plane. The surfaces of the rock layers in **FIGURE 9.3** are good examples of inclined planes, whereas the land surface is a horizontal plane. The direction of the line formed at the intersection of these planes is the strike of the rock layers. The strike line's orientation is determined by using a compass to measure its angle with respect to north. **Dip** is a measure of an inclined plane's deviation from horizontal, so it must be measured at right angles to strike direction (Figure 9.3).

Geologic maps showing the age, aerial distribution, and geologic structures of rocks use a special symbol to indicate strike and dip. A long line oriented in the appropriate direction indicates strike, and a short line perpendicular to the

**strike** The direction of a line formed by the intersection of an inclined plane and a horizontal plane.

**dip** A measure of the maximum angular deviation of an inclined plane from horizontal.

**FIGURE 9.3 Strike and Dip of Deformed Rock Layers** We can infer that the sedimentary rocks in this illustration were deposited horizontally or nearly so, lithified, and then deformed. To describe their orientation, geologists use the terms *strike* and *dip*. Strike is the line formed by the intersection of a horizontal plane with an inclined plane. Dip is the maximum angular deviation of the inclined plane from horizontal. Notice the symbol that shows strike and dip.

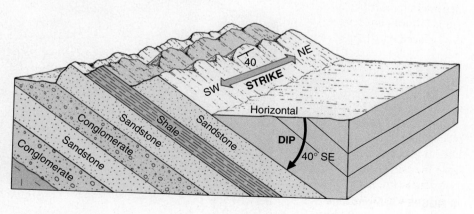

strike line shows the direction of dip (Figure 9.3). Adjacent to the strike and dip symbol is a number corresponding to the dip angle. The usefulness of strike and dip symbols will become apparent in the sections on folds and faults.

# DEFORMATION AND GEOLOGIC STRUCTURES

**LO7**  Define a geologic structure

**LO8**  List the three basic types of folds

**LO9**  Explain how each fold structure forms

**LO10**  Define joints

**LO11**  Define a fault

**LO12**  Define the terms hanging wall block and footwall block as they relate to a fault

**LO13**  Discuss the types of dip-slip faults

**LO14**  Define a strike-slip fault

Remember that *deformation* and its synonym *strain* refer to changes in the shape or volume of rocks. During deformation, rocks might be crumpled into folds, or they might be fractured, or perhaps folded and fractured. Any of these features resulting from deformation is referred to as a **geologic structure**. Geologic structures are found almost everywhere rocks are exposed, and many are detected far below the surface by drilling and several geophysical techniques.

## Folded Rock Layers

If you place your hands on a tablecloth and move them toward one another, the tablecloth crumples into a series of up- and down-arched folds. Rock layers behave similarly when in compression as they deform into **folds**, but in this case the folding is permanent. That is, plastic strain has taken place, so even if the stress is removed, the rock layers remain folded. Most folding probably occurs deep in the crust where pressure and temperature are high and rocks are more ductile than they are at or near the surface. The configuration of folds and the intensity of folding vary, but there are only three basic types of folds: *monoclines, anticlines,* and *synclines.*

**MONOCLINES**  A simple bend or flexure in otherwise horizontal or uniformly dipping rock layers is a **monocline**. The large monocline in ●**FIGURE 9.4** formed when the Bighorn Mountains in Wyoming rose vertically along a fracture. The fracture did not penetrate to the surface, and as uplift of the mountains proceeded, the near-surface rocks were bent such

---

**geologic structure**  Any feature in rocks that results from deformation, such as folds, joints, and faults.

**fold**  A type of geologic structure in which planar features in rock layers such as bedding and foliation have been bent.

**monocline**  A bend or flexure in otherwise horizontal or uniformly dipping rock layers.

Monocline

●**FIGURE 9.4 Monocline**
A monocline in the Bighorn Mountains in Wyoming.

Sue Monroe

● **FIGURE 9.5 Anticlines and Synclines** Folded rocks in the Calico Mountains of southeastern California. Compression was responsible for these folds, which are, from left to right, a syncline, an anticline, and a syncline.

that they now appear to be draped over the margin of the uplifted block. In a manner of speaking, a monocline is simply one-half of an anticline or syncline.

**ANTICLINES AND SYNCLINES** An **anticline** is an up-arched or convex upward fold with the oldest rock layers in its core, whereas a **syncline** is a down-arched or concave downward fold in which the youngest rock layers are in its core (●**FIGURE 9.5**). Anticlines and synclines also have an axial plane connecting the points of maximum curvature of each folded layer (●**FIGURE 9.6**); the axial plane divides folds into halves, each half being a *limb*. Because folds are most often found in a series of anticlines alternating with synclines, an anticline and adjacent syncline share a limb.

Folds are commonly exposed to view in areas of deep erosion, but even where eroded, strike and dip and the relative ages of the folded rock layers easily distinguish anticlines from synclines. Notice in ●**FIGURE 9.7** that in the surface view of the anticline, each limb dips outward, or away from the center of the fold, and the oldest exposed rocks are in the fold's core. In an eroded syncline, though, each limb dips inward toward the fold's center, where the youngest exposed rocks are found.

The folds described so far are *upright*, meaning that their axial planes are vertical, and both fold limbs dip at the same angle (Figure 9.7). In many folds, the axial plane is not vertical, the limbs dip at different angles, and the folds are characterized as *inclined* (●**FIGURE 9.8A**). If both limbs dip in the same direction, the fold is *overturned*. That is, one limb has been rotated more than 90 degrees from its original position so that it is now upside down (●**FIGURE 9.8B**). In some

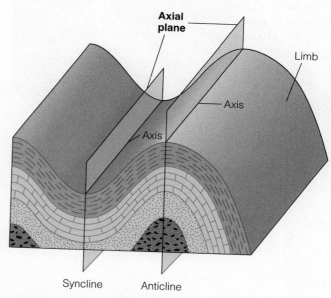

● **FIGURE 9.6 Syncline and Anticline Axial Planes** Syncline and anticline showing the axial plane, axis, and fold limbs.

**anticline** A convex upward fold in which the oldest exposed rocks coincide with the fold axis and all strata dip away from the axis.

**syncline** A down-arched fold in which the youngest exposed rocks coincide with the fold axis and all strata dip toward the axis.

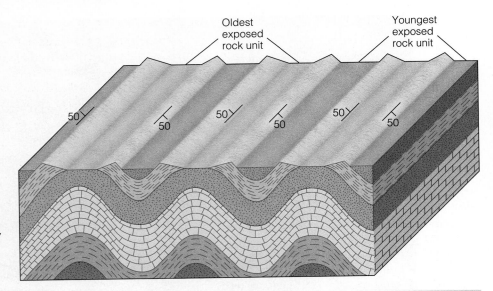

● **FIGURE 9.7 Eroded Anticlines and Synclines** Geologists identify eroded anticlines and synclines by strike and dip and the relative ages of the folded rock layers.

**Critical Thinking Question** Show on this illustration the direction of the forces that caused deformation.

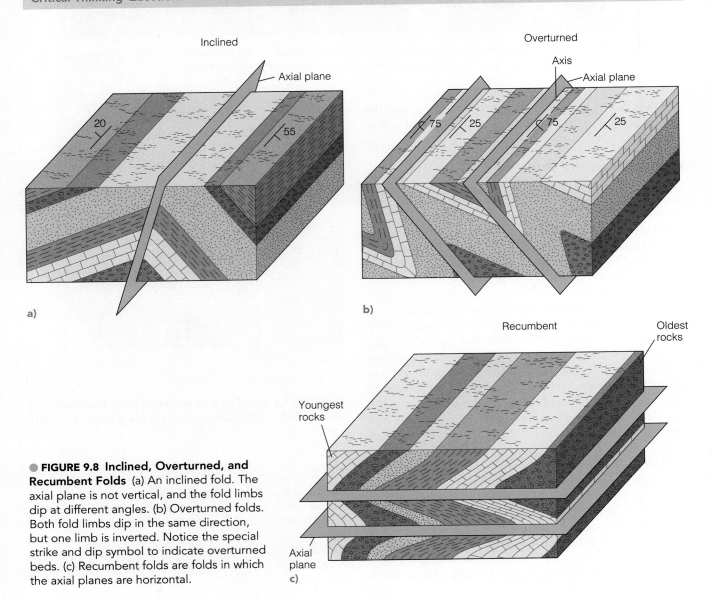

● **FIGURE 9.8 Inclined, Overturned, and Recumbent Folds** (a) An inclined fold. The axial plane is not vertical, and the fold limbs dip at different angles. (b) Overturned folds. Both fold limbs dip in the same direction, but one limb is inverted. Notice the special strike and dip symbol to indicate overturned beds. (c) Recumbent folds are folds in which the axial planes are horizontal.

areas, deformation has been so intense that axial planes of folds are now horizontal, giving rise to what geologists call *recumbent folds* (●**FIGURE 9.8C**). Overturned and recumbent folds are particularly common in mountains resulting from compression at convergent plate boundaries.

**PLUNGING FOLDS** In some folds, the fold axis—a line formed by the intersection of the axial plane with the folded layers—is horizontal and the folds are characterized as *nonplunging* (Figure 9.7). Much more commonly, though, fold axes are inclined so that they appear to plunge beneath adjacent rocks, and the folds are said to be *plunging* (●**FIGURE 9.9**).

It might seem that with this additional complication, differentiating plunging anticlines from plunging synclines would be much more difficult. However, you can use exactly the same criteria that you used for nonplunging folds. Therefore, all rock layers dip away from the fold axis in plunging anticlines and toward the axis in plunging

synclines. The oldest exposed rocks are in the core of an eroded plunging anticline, whereas the youngest exposed rock layers are found in the core of an eroded plunging syncline (●**FIGURE 9.9B**).

In Chapter 6, we noted that anticlines form one type of structural trap in which petroleum and natural gas might accumulate. As a matter of fact, most of the world's petroleum production comes from anticlines, although other geologic structures and stratigraphic traps are also important.

**DOMES AND BASINS** Anticlines and synclines are elongated structures, meaning that their length greatly exceeds their width. In contrast, folds that are nearly equidimensional (i.e., circular) are *domes* and *basins*. In a **dome**, all

---

**dome** A rather circular geologic structure in which all rock layers dip away from a central point and the oldest exposed rocks are at the dome's center.

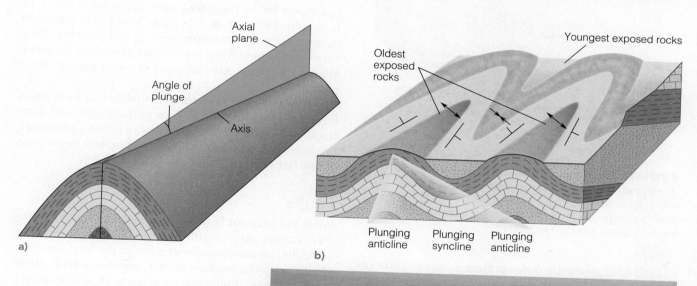

●**FIGURE 9.9 Plunging Folds** (a) A plunging fold. (b) Surface and cross-sectional views of plunging folds. The long arrow is the geologic symbol for a plunging fold; it shows the direction of plunge. (c) The Sheep Mountain Anticline in Wyoming You can tell that this is an anticline by the strike and dip symbols. The long line shows the axis of the anticline and its direction of plunge.

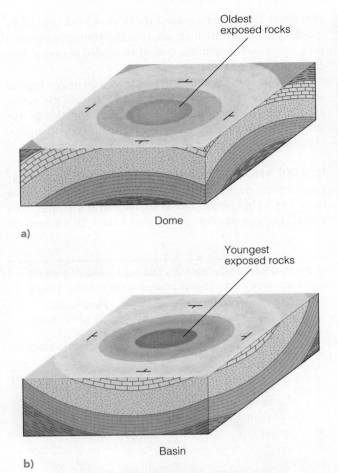

a)

Dome

b)

Basin

● **FIGURE 9.10 Domes and Basins** (a) Notice that in a dome, the oldest exposed rocks are in the center and all rocks dip outward from a central point. (b) In a basin, the youngest exposed rocks are in the center and all rocks dip inward toward a central point.

of the folded strata dip outward from a central point (as opposed to outward from a line as in an anticline), and the oldest exposed rocks are at the center of the fold (●**FIGURE 9.10A**). In contrast, a **basin** has all strata dipping inward toward a central point, and the youngest exposed rocks are at the fold's center (●**FIGURE 9.10B**).

Unfortunately, the terms *dome* and *basin* are also used to distinguish high and low areas of Earth's surface, but domes and basins as defined here do not necessarily correspond with mountains or valleys. In some of the following discussions, we will use these terms in other contexts, but we will try to be clear when we refer to surface elevations as opposed to geologic structures.

## Joints

**Joints** are fractures along which no movement has taken place parallel with the fracture surface (●**FIGURE 9.11**), although they may open up; that is, joints may show movement perpendicular to the fracture. Remember that rocks near the surface are brittle and therefore

Anna Stowe Landscapes UK/Alamy Stock Photo

● **FIGURE 9.11 Joints** Fractures along which no movement has taken place parallel with the fracture surface are called joints. These joints are exposed in the bed rock along the beach at Nash Point in Wales.

commonly fail by fracturing when subjected to stress. In fact, almost all near-surface rocks have joints that form in response to compression, tension, and shearing. They vary from minute fractures to those extending for many kilometers and are often arranged in two or perhaps three prominent sets.

We already discussed columnar joints that form when lava or magma in some shallow plutons cools and contracts (see Figure 5.7). Another type of jointing previously discussed is sheet joints that form in response to pressure release (see Figure 6.3b).

## Faults

Joints and faults are both fractures, but on joints no movement occurs parallel with the fracture surface, whereas on **faults**, blocks on opposite sides of the fracture move parallel with the fracture surface, which is a **fault plane** (●**FIGURE 9.12A**). Faults that penetrate to the surface might show a *fault scarp*, a bluff or cliff formed by vertical movement (●**FIGURE 9.12B**). In some cases, the fault plane is scratched and polished, but in others the movement of the blocks on opposite sides of the fault grind and pulverize the rock into a zone of *fault breccia*.

Notice in Figure 9.12a that the rocks overlying the fault make up the **hanging wall block**, whereas those beneath

---

**basin** An oval to circular fold in which all strata dip inward toward a central point and the youngest exposed strata are in the center.

**joint** A fracture along which no movement has occurred or where movement is perpendicular to the fracture surface.

**fault** A fracture along which rocks on opposite sides of the fracture have moved parallel with the fracture surface.

**fault plane** A fault surface that is more or less planar.

**hanging wall block** The block of rock that overlies a fault plane.

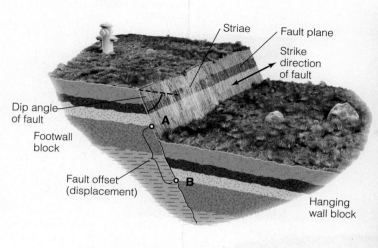

Striae — Fault plane

Strike
direction
of fault

Dip angle
of fault

Footwall
block

Fault offset
(displacement)

Hanging
wall block

a)

Fault
plane

David J. Matty

b)

● **FIGURE 9.12 Faults** Fractures along which movement has occurred parallel to the fracture surface are called faults. (a) Terms used to describe the orientation of a fault plane. Striae are scratch marks that form when one block slides past another. You can measure offset or displacement on a fault wherever the truncated end of one feature (point A) can be related to its equivalent along the fault (B). (b) A polished, scratched fault plane and fault scarp near Klamath Falls, Oregon.

the fault constitute the **footwall block**. You can recognize these two blocks on any fault with the exception of one that is vertical—that is, one that dips at 90 degrees. In order to identify some faults, you must identify these two blocks and determine the direction of relative movement of the blocks. By *relative movement*, we mean which block appears to have moved up or down the fault plane. For example, in Figure 9.12a, the footwall block may have moved up, the hanging wall may have moved down, or both blocks may have moved. Nevertheless, the hanging wall block appears to have moved down relative to the footwall block.

Remember that geologists use *strike* and *dip* to define the orientation of dipping rock layers. Fault planes are also dipping planes, so this same concept applies to them. In fact, two types of faults are recognized based on whether the blocks on opposite sides of the fault moved parallel to the direction of dip (dip-slip faults) or along the direction of strike (strike-slip faults).

## Dip-Slip Faults

All movement on **dip-slip faults** takes place parallel with the fault's dip; that is, movement is vertical, either up or down the fault plane. In ●**FIGURE 9.13A**, for example, the hanging wall block moved down relative to the footwall block, giving rise to a **normal fault** (●**FIGURE 9.14A**). In contrast, **reverse faults** are those on which the hanging wall block moves up relative to the footwall block (●**FIGURES 9.13B** and **9.14B**). In ●**FIGURE 9.13C**, the hanging wall block has also moved up relative to the footwall block, but the fault dips at less than 45 degrees, so it is a special type of reverse fault known as a **thrust fault**.

Numerous normal faults are present along one or both sides of mountain ranges in the Basin and Range Province of the western United States where the crust is being stretched and thinned. The Sierra Nevada at the western margin of the Basin and Range is bounded by normal faults, and the range has risen along these faults so that it now stands more than 3,000 m above the lowlands to the east.

Large-scale examples of both reverse and thrust faults are found in mountain ranges that formed at convergent plate margins, where one would expect compression (Figures 9.13b, c).

## Strike-Slip Faults

**Strike-slip faults**, resulting from shear stresses, show horizontal movement with blocks on opposite sides of the fault sliding past one another (●**FIGURE 9.13D**). In other words, all movement is in the direction of the fault plane's strike. Several large strike-slip faults are known, but the best studied

**footwall block** The block of rock that lies beneath a fault plane.

**dip-slip fault** A fault on which all movement is parallel with the dip of the fault plane.

**normal fault** A dip-slip fault on which the hanging wall block has moved downward relative to the footwall block.

**reverse fault** A dip-slip fault on which the hanging wall block has moved upward relative to the footwall block.

**thrust fault** A type of reverse fault in which a fault plane dips less than 45 degrees.

**strike-slip fault** A fault involving horizontal movement of blocks of rock on opposite sides of a fault plane.

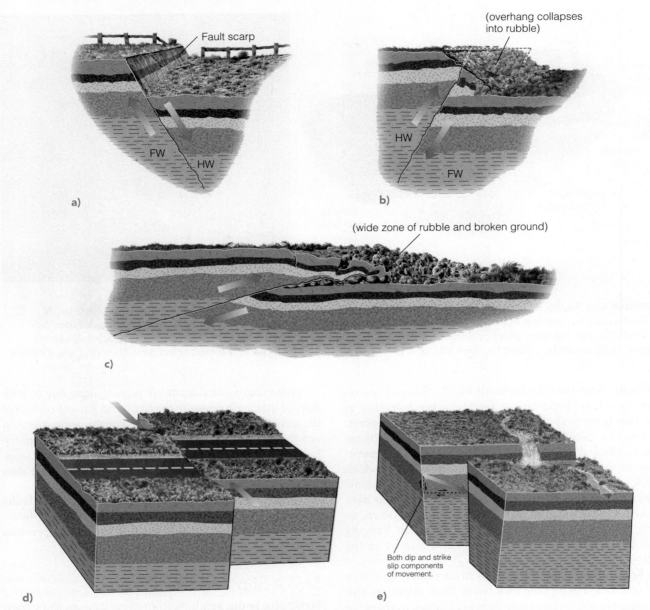

● **FIGURE 9.13 Types of Faults** (a) Normal fault—hanging wall block (HW) moves down relative to the footwall block (FW). (b) Reverse fault—hanging wall block moves up relative to the footwall block. (c) A thrust is a type of reverse fault with a fault plane dipping at less than 45 degrees. (d) Left-lateral strike-slip fault. (e) An oblique-slip fault involves a combination of dip-slip and strike-slip movements.

is the San Andreas Fault, which cuts through coastal California, where numerous earthquakes have occurred, including the tragic 1906 San Francisco earthquake. Recall from Chapter 2 that the San Andreas Fault is called a *transform fault* in plate tectonics terminology.

Strike-slip faults are characterized as right-lateral or left-lateral, depending on the apparent direction of offset. In Figure 9.13d, for example, observers looking at the block on the opposite side of the fault from their location notice that it appears to have moved to the left. Accordingly, this is a *left-lateral strike-slip fault*. If it had been a *right-lateral strike-slip fault*, the block across the fault from the observers would appear to have moved to the right.

## Oblique-Slip Faults

The movement on most faults is primarily dip-slip or strike-slip, but on **oblique-slip faults**, both types of movement take place. Strike-slip movement might be accompanied by a component of dip-slip, giving rise to a combined movement that includes left-lateral and reverse, or right-lateral and normal, or any other combination (●**FIGURE 9.13E**).

---

**oblique-slip fault** A fault showing both dip-slip and strike-slip movement.

James S. Monroe

a)

James S. Monroe

b)

● **FIGURE 9.14  Dip-Slip Faults**  (a) A normal fault at Mt. Carmel Junction, Utah. The hanging wall block (on the right) has moved down about 2 m relative to the footwall block. (b) A small reverse fault. Notice that the hanging wall block (on the right) has moved up relative to the footwall block.

# DEFORMATION AND THE ORIGIN OF MOUNTAINS

LO15  Define an orogeny

LO16  Discuss the relationship between plate tectonics and mountain building

LO17  Discuss the features produced by orogenic activity occurring at oceanic–oceanic plate boundaries

LO18  Discuss the features produced by orogenic activity occurring at oceanic–continental plate boundaries

LO19  Discuss the features produced by orogenic activity occurring at continental–continental plate boundaries

LO20  Discuss the relationship between terranes and mountain systems

*Mountain* is a designation for any area of land that stands significantly higher, at least 300 m, than the surrounding country and has a restricted summit area. Some mountains are single, isolated peaks, but most are parts of linear associations of peaks and ridges known as *mountain ranges* that are related in age and origin. A *mountain system* is a complex linear zone of deformation and crustal thickening that consists of several or many mountain ranges. The Teton Range in Wyoming is one of many ranges in the Rocky Mountains. Mountains form in several ways, but the truly large mountains on continents result mostly from compression-induced deformation at convergent plate boundaries (see Chapter 2).

Animation: Continent-Continent Collision

## Mountain Building

*Block-faulting* is one way that mountains form, which is caused by movement on normal faults in response to tension, so that one or more blocks are elevated relative to adjacent blocks (●**FIGURE 9.15**). A classic example is the Basin and Range Province, which is centered on Nevada but extends into adjacent states, and on into Mexico. Differential

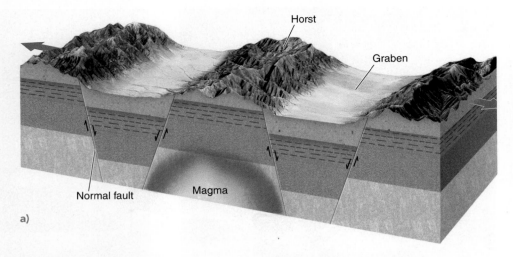

a)

b)

● **FIGURE 9.15 Block Faulting and the Origin of Horsts and Grabens** (a) Movement on parallel normal faults accounts for many of the mountain ranges in the Basin and Range Province of the western United States. (b) The Stillwater Range in Nevada is a horst bounded by normal faults.

movement on faults has produced uplifted blocks called *horsts* and down-dropped blocks called *grabens*. Erosion of the horsts has produced mountainous topography.

Volcanic outpourings form chains of volcanic mountains such as the Hawaiian Islands, where a plate moves over a hot spot (see Figure 2.9). Some mountains, such as the Cascade Range of the Pacific Northwest, are made up almost entirely of volcanic rocks, and the mid-ocean ridges are also mountains. However, most mountains on land are composed of all rocks types and show clear evidence of deformation by compression.

## Plate Tectonics and Mountain Building

Any theory that accounts for mountain building, or what geologists call **orogeny**, must adequately explain the characteristics of mountain ranges, such as their geometry and location; they tend to be long and narrow and at or near plate margins. Mountains also show intense deformation, especially compression-induced overturned and recumbent folds, as well as reverse and thrust faults. Furthermore, granitic plutons and regional metamorphism characterize the interiors or cores of mountain ranges. Another feature is sedimentary rocks that are now far above sea level but that were deposited in shallow and deep marine environments.

Deformation and associated activities at convergent plate boundaries are certainly important processes in mountain building. They account for a mountain system's location and geometry, as well as complex geologic structures, plutons, and metamorphism. Yet, the present-day topographic expression of mountains is also related to surface processes, such as mass wasting (gravity-driven processes including landslides), glaciers, and running water. In other words, erosion also plays an important role in the evolution of mountains.

**orogeny** An episode of mountain building involving deformation, usually accompanied by igneous activity, metamorphism, and crustal thickening.

## OROGENIES AT OCEANIC–OCEANIC PLATE BOUNDARIES

Deformation, igneous activity, and the origin of a volcanic island arc characterize orogenies that take place where oceanic lithosphere is subducted beneath oceanic lithosphere (see Figure 2.19a). Sediments derived from the island arc are deposited in an adjacent oceanic trench and then deformed and scraped off against the landward side of the trench (●FIGURE 9.16). These deformed sediments are part of a subduction complex, or an *accretionary wedge*, of intricately folded rocks cut by numerous thrust faults resulting from compression. In addition, orogenies in this setting are characterized by low-temperature, high-pressure metamorphism of the blueschist facies (see Figure 7.18).

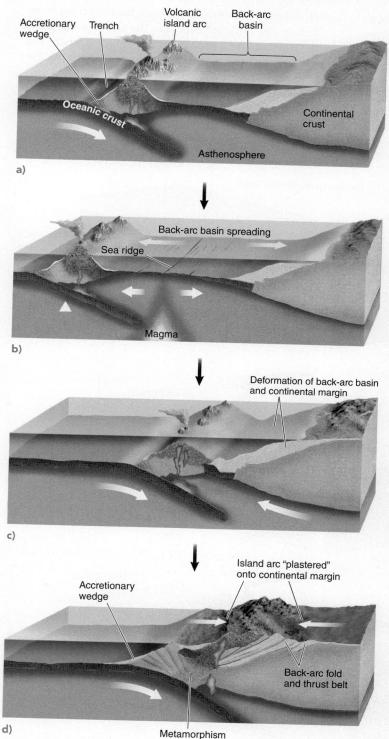

●FIGURE 9.16 **Orogeny and the Origin of a Volcanic Island Arc at an Oceanic–Oceanic Plate Boundary** (a) Subduction of an oceanic plate and the origin of a volcanic island arc and a back-arc basin. (b) Continued subduction and back-arc spreading. (c) Back-arc basin begins to close, resulting in deformation of back-arc basin and continental margin deposits. (d) Thrusting of back-arc sediments onto the adjacent continent and suturing of the island arc to the continent.

Deformation caused largely by the emplacement of plutons also takes place in the island arc system, where many rocks show evidence of high-temperature, low-pressure metamorphism. The overall effect of an island arc orogeny is the origin of two, more-or-less, parallel orogenic belts consisting of a landward volcanic island arc underlain by batholiths, and a seaward belt of deformed trench rocks (Figure 9.16). The Japanese Islands are a good example.

In the area between an island arc and its nearby continent, the back-arc basin, volcanic rocks and sediments derived from the island arc and the adjacent continent are also deformed as the plates continue to converge. The sediments are intensely folded and displaced toward the continent along low-angle thrust faults. Eventually, the entire island arc complex is fused to the edge of the continent, and the back-arc basin sediments are thrust onto the continent, forming a thick stack of thrust sheets (Figure 9.16).

## OROGENIES AT OCEANIC–CONTINENTAL PLATE BOUNDARIES

The Andes Mountains in South America are the best example of continuing orogeny at an oceanic–continental plate boundary (see Figure 2.19b). Among the ranges of the Andes are the highest mountain peaks in the Americas and many active volcanoes. Furthermore, the west coast of South America is an extremely active segment of the circum-Pacific earthquake belt, and one of Earth's great oceanic trench systems, the Peru–Chile Trench, lies just off the coast.

Before 200 million years ago, the western margin of South America was a broad continental shelf where sediments accumulated much as they do now along the East Coast of North America. However, when Pangaea split apart along what is now the Mid-Atlantic Ridge, the South American plate moved westward. As a consequence, the oceanic lithosphere west of South America began subducting beneath the continent (●FIGURE 9.17). Subduction resulted

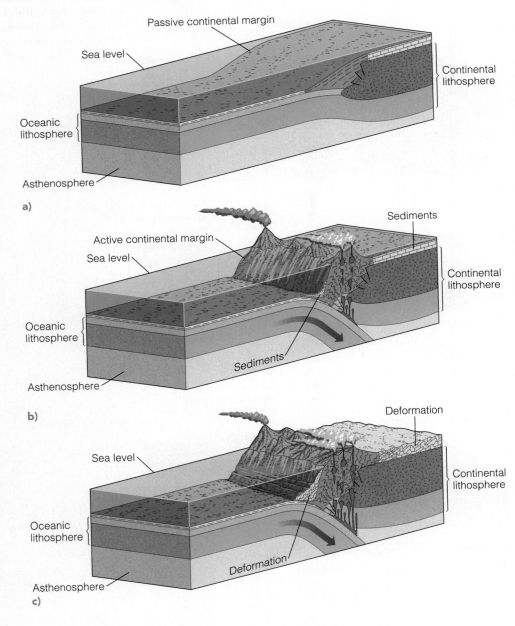

● **FIGURE 9.17 The Andes Mountains in South America** (a) Prior to 200 million years ago, the western margin of South America was a passive continental margin. (b) Orogeny began when this area became an active continental margin as the South American plate moved to the west and collided with oceanic lithosphere. (c) Continued deformation, plutonism, and volcanism.

in partial melting of the descending plate, which produced the andesitic volcanic arc of composite volcanoes, and the West Coast became an active continental margin. Felsic magmas, mostly of granitic composition, were emplaced as large plutons beneath the arc (Figure 9.17).

As a result of the events just described, the Andes Mountains consist of a central core of granitic rocks capped by andesitic volcanoes. To the west of this central core along the coast are the deformed rocks of the accretionary wedge. And, to the east of the central core are intensely folded sedimentary rocks that were thrust eastward onto the continent (Figure 9.17). Present-day subduction, volcanism, and seismicity along South America's west coast indicate that the Andes Mountains are still forming.

**OROGENIES AT CONTINENTAL–CONTINENTAL PLATE BOUNDARIES** The best example of an orogeny along a continental–continental plate boundary is the Himalayas of Asia (see Figure 2.19c). The Himalayas began forming when India collided with Asia about 40 to 50 million years ago. Before that time, India was far south of Asia and separated from it by an ocean basin (●FIGURE 9.18A). As the Indian plate moved northward, a subduction zone formed along the southern margin of Asia where oceanic lithosphere was consumed. Magma rose to form a volcanic arc, and large granite plutons were emplaced into what is now Tibet. At this stage, the activity along Asia's southern margin was similar to what is now occurring along the west coast of South America.

The ocean separating India from Asia continued to close, and India eventually collided with Asia (Figure 9.18a). As a result, two continental plates became welded, or sutured, together. Thus, the Himalayas are now within a continent rather than along a continental margin. The leading margin of India was pushed beneath Asia, causing crustal thickening, thrusting, and uplift. Sedimentary rocks that had been deposited in the sea south of Asia were thrust northward, and two major thrust faults carried rocks of Asian origin onto the Indian plate. Rocks deposited in the shallow seas along India's northern margin now form the higher parts of the Himalayas (●FIGURE 9.18B). Since its collision with Asia, India has been thrust horizontally about 2,000 km beneath Asia and now moves north at several centimeters per year.

Other mountain systems also formed as a result of collisions between two continental plates. The Urals in Russia and the Appalachians of North America were formed by such collisions. In addition, the Arabian plate is now colliding with Asia along the Zagros Mountains of Iran.

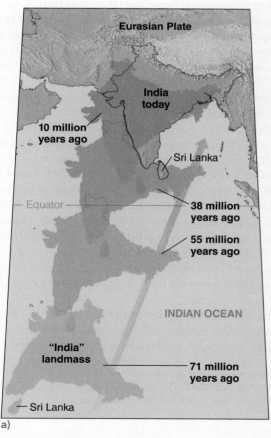

a)

b)

● **FIGURE 9.18 Orogeny at a Continental–Continental Plate Boundary and the Origin of the Himalayas of Asia**
(a) During its long journey north, India moved 15 to 20 cm per year; however, beginning 40 to 50 million years ago, its rate of movement decreased markedly as it collided with the Eurasian plate. (b) The Karakoram Range seen here from Karimaba, Pakistan, is within the Himalayan orogen. The range lies on the border of Pakistan, China, and India.

# Terranes and the Origin of Mountains

In the preceding section, we discussed orogenies along convergent plate boundaries that result in adding material to a continent, a process termed **continental accretion**. Much of the material added to continental margins is eroded older continental crust, but some plutonic and volcanic rocks are new additions. During the 1970s and 1980s, however, geologists discovered that parts of many mountain systems are also made up of small, accreted lithospheric blocks that clearly originated elsewhere. These **terranes,**[*] as they are called, are fragments of seamounts, island arcs, and small pieces of continents that were carried on oceanic plates that collided with continental plates, thus adding them to the continental margins.

**MINDTAP**
From Cengage

Animation: Continental Accretion

# EARTH'S CRUST

LO21   **Define the principle of isostasy**
LO22   **Define isostatic rebound**

Continental crust stands higher than oceanic crust, but why should this be so? Why do mountains stand higher than surrounding areas? To answer these questions, we must examine Earth's crust in more detail. You already know that continental crust is granitic in composition, with an overall density of 2.7 g/cm$^3$, whereas oceanic crust is made up of basalt and gabbro, and its density is 3.0 g/cm$^3$ (see Figure 8.22). In most places, continental crust is about 35 km thick, except beneath mountain systems, where it is much thicker. The oceanic crust, in contrast, varies from only 5 to 10 km thick. So, these differences in composition, as well as variations in crustal thickness, account for why mountains stand high and why continents stand higher than ocean basins.

## Floating Continents?

How is it possible for a solid (continental crust) to float in another solid (the mantle)? Floating brings to mind a ship at sea or a block of wood in water; however, continents do not behave in this manner. Or do they? Actually, they do float, in a manner of speaking, but a complete answer requires more discussion on the concept of gravity and on the *principle of isostasy*.

Isaac Newton formulated the law of universal gravitation in which the force of gravity ($F$) between two masses ($m_1$ and $m_2$) is directly proportional to the products of their masses and inversely proportional to the square of the distance between their centers of mass. This means that an

attractive force exists between any two objects, and the magnitude of that force varies depending on the masses of the objects, and the distance between their centers.

Gravitational attraction would be the same everywhere on the surface if Earth were perfectly spherical, homogeneous throughout, and did not rotate. But, because Earth varies in all of these aspects, the force of gravity varies from area to area.

## Principle of Isostasy

Geologists realized long ago that mountains are not simply piles of materials on Earth's surface, and in 1865, George Airy proposed that, in addition to projecting high above sea level, mountains also project far below the surface and thus have a low-density root (●FIGURE 9.19). In effect, he was saying that the thicker crust of mountains float on denser rock at depth, with their excess mass above sea level compensated for by low-density material at depth. Another explanation was proposed by J. H. Pratt, who thought that mountains were high because they were composed of rocks of lower density than those in adjacent regions.

Actually, both Airy and Pratt were correct, because there are places where density or thickness accounts for differences in the level of the crust. For example, Pratt's hypothesis was confirmed because (1) continental crust is thicker and less dense than oceanic crust and thus stands high, and (2) the mid-oceanic ridges stand high because the crust there is hot and less dense than cooler crust elsewhere. Airy, on the other hand, was correct in his claim that the crust, continental or oceanic, "floats" on the mantle, which has a density of 3.3 g/cm$^3$ in its upper part. However, we have not yet explained what we mean by one solid floating in another solid.

This phenomenon of Earth's crust floating in the denser mantle is now known as the **principle of isostasy**, which is easy to understand by an analogy to an iceberg (●FIGURE 9.20). Ice is slightly less dense than water, so it floats. According to Archimedes' principle of buoyancy, an iceberg sinks in water until it displaces a volume of water whose weight is equal to that of the ice. When the iceberg has sunk to an equilibrium position, only about 10% of its volume is above water level. If some of the ice above water level should melt, the iceberg rises to maintain equilibrium with the same proportion of ice above and below the water.

Earth's crust is similar to the iceberg in that it sinks into the mantle to its equilibrium level. Where the crust is thickest, as beneath mountains, it sinks farther down into

---

[*]Some geologists prefer the terms *suspect terrane, exotic terrane,* or *displaced terrane.* Notice also the spelling of *terrane* as opposed to the more familiar *terrain,* the latter a geographic term indicating a particular area of land.

---

**continental accretion**  The process in which continents grow by additions of Earth materials along their margins.

**terrane**  A small lithospheric block with characteristics quite different from those of surrounding rocks. Terranes probably consist of seamounts, oceanic rises, and other seafloor features accreted to continents during orogenies.

**principle of isostasy**  The theoretical concept of Earth's crust "floating" on a dense underlying layer.

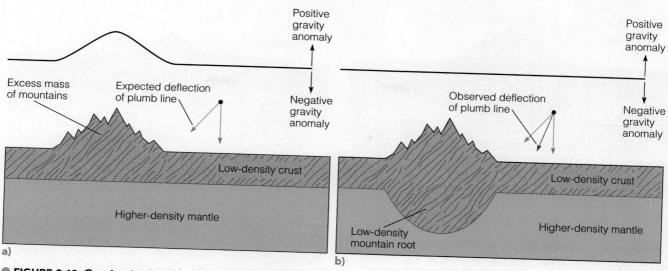

● **FIGURE 9.19 Gravity Anomalies** (a) A plumb line (a cord with a suspended weight) is normally vertical, pointing to Earth's center of gravity. Near a mountain range, the plumb line should be deflected as shown if the mountains are simply thicker, low-density material resting on denser material, and a gravity survey across the mountains would indicate a positive gravity anomaly. (b) The actual deflection of the plumb line as measured by British surveyors during a survey in India was less than expected. It was explained by postulating that the Himalayas have a low-density root. A gravity survey, in this case, would show no anomaly because the mass of the mountains above the surface is compensated for at depth by low-density material displacing denser material.

● **FIGURE 9.20 The Principle of Isostasy** An iceberg sinks to an equilibrium level with 10% of its mass above water level. The larger iceberg sinks farther below and rises higher above the water surface than the smaller one. If some of the ice above water level melts, the icebergs will rise to maintain the same proportions of ice above and below water level.

the mantle, and it also rises higher above the surface. And because continental crust is thicker and less dense than oceanic crust, it stands higher than the ocean basins. Remember, the mantle is hot, yet solid, and under tremendous pressure, so it behaves in a fluid-like manner.

Some of you might realize that crust floating on the mantle raises an apparent contradiction. In Chapter 8, we said that the mantle is a solid because it transmits S-waves,

which do not move through a fluid. But according to the principle of isostasy, the mantle behaves as a fluid. When considered in terms of the brief time required for S-waves to pass through it, the mantle is indeed solid. But when subjected to stress over long periods, it yields by flowage; thus, at this time scale, it is regarded as a viscous fluid.

## Isostatic Rebound

What happens when a ship is loaded with cargo and then later unloaded? Of course, it first sinks lower in the water and then rises, but it always finds its equilibrium position. Earth's crust responds similarly to loading and unloading, but it does so much more slowly. For example, if the crust is loaded, as when widespread glaciers accumulate, the crust sinks farther into the mantle to maintain equilibrium. The crust behaves similarly in areas where huge quantities of sediment accumulate.

If loading by glacial ice or sediment depresses Earth's crust farther into the mantle, it follows that when vast glaciers melt or where deep erosion takes place, the crust should rise back up to its equilibrium level—and in fact it does. This phenomenon, known as **isostatic rebound**, is taking place in Scandinavia, which was covered by a thick ice sheet until about 10,000 years ago; it is now rebounding at about 1 m per century. In fact, coastal cities in Scandinavia have rebounded rapidly enough that docks constructed several centuries ago are now far from shore.

**isostatic rebound** The phenomenon in which unloading of the crust causes it to rise until it attains equilibrium with the underlying upper mantle.

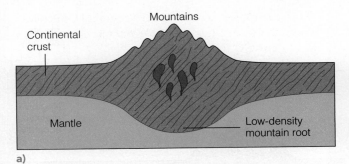

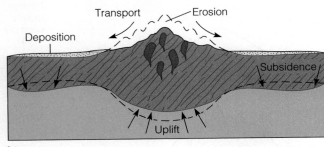

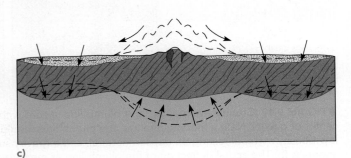

● **FIGURE 9.21  Isostatic Rebound**  A diagrammatic representation showing the isostatic response of the crust to erosion (unloading) and widespread deposition (loading). (a) The crust and mantle before erosion and deposition. (b) Erosion of the mountains and deposition in adjacent areas. Isostatic rebound begins. (c) Continuing erosion, deposition, and isostatic rebound.

Isostatic rebound has also occurred in eastern Canada, where the crust has risen as much as 100 m in the last 6,000 years.

●**FIGURE 9.21** shows the response of Earth's continental crust to loading and unloading as mountains form and evolve. Recall that during an orogeny, emplacement of plutons, metamorphism, and general thickening

of the crust accompany deformation. However, as the mountains erode, isostatic rebound takes place and the mountains rise, whereas adjacent areas of sedimentation subside (Figure 9.21). If continued long enough, the mountains will disappear and then can be detected only by the plutons and metamorphic rocks that show their former existence.

# Key Concepts Review

- Folded and fractured rocks have been deformed or strained by applied stresses.
- Stress is compression, tension, or shear. Elastic strain is not permanent, but plastic strain and fracture are, meaning that rocks do not return to their original shape or volume when the deforming forces are removed.
- Strike and dip are used to define the orientation of deformed rock layers. This same concept applies to other planar features, such as fault planes.
- Anticlines and synclines are up- and down-arched folds, respectively, and result from compressional forces. They are identified by strike and dip of the folded rocks and the relative ages of rocks in these folds.
- Domes and basins are the circular to oval equivalents of anticlines and synclines, but they are commonly much larger structures.
- The two structures that result from fracture are joints and faults. Joints show no movement parallel with the fracture surface, whereas faults do.

- On dip-slip faults, all movement is up or down the dip of the fault. If the hanging wall moves relatively down, it is a normal fault, but if the hanging wall moves relatively up, it is a reverse fault. Normal faults result from tension; reverse faults result from compression.
- In strike-slip faults, all movement is along the strike of the fault. These faults are either right lateral or left lateral, depending on the apparent direction of offset of one block relative to the other.
- Oblique-slip faults show components of both dip-slip and strike-slip movement.
- A variety of processes account for the origin of mountains. Some involve little or no deformation, but the large mountain systems on the continents are the result of deformation occurring at convergent plate boundaries.
- Subduction of an oceanic plate beneath another oceanic plate or beneath a continental plate causes an orogeny. At an oceanic–oceanic boundary, a volcanic island arc intruded by plutons forms, whereas at an

oceanic–continental boundary, a volcanic arc forms on the continental plate. In both cases, deformation and metamorphism occur.
- Some mountain systems are within continents far from a present-day plate boundary. These mountains formed when two continental plates collided and became sutured.
- Geologists now realize that orogenies also involve collisions of terranes with continents.

- Continental crust is characterized as granitic, and it is much thicker and less dense than oceanic crust, which is composed of basalt and gabbro.
- According to the principle of isostasy, Earth's crust floats in equilibrium in the denser mantle below. Continental crust stands higher than oceanic crust because it is thicker and less dense.

## Important Terms

anticline   185
basin   188
compression   182
continental accretion   196
deformation   180
dip   183
dip-slip fault   189
dome   187
elastic strain   182
fault   188
fault plane   188
fold   184

footwall block   189
fracture   182
geologic structure   184
hanging wall block   188
isostatic rebound   197
joint   188
monocline   184
normal fault   189
oblique-slip fault   190
orogeny   192
plastic strain   182
principle of isostasy   196

reverse fault   189
shear stress   182
strain   181
stress   181
strike   183
strike-slip fault   189
syncline   185
tension   182
terrane   196
thrust fault   189

## Review Questions

1. If Earth's crust is first loaded by vast glaciers and then unloaded when the glaciers melt, the crust will rise, a phenomenon called.
   a. tensional release.
   b. recumbent release.
   c. compressional release.
   d. plastic stress.
   e. isostatic rebound.

2. A fault on which the hanging wall block has moved down relative to the footwall block is a(n) _____ fault.
   a. normal
   b. oblique
   c. strike-slip
   d. reverse
   e. thrust

3. An ongoing orogeny along an oceanic–continental plate boundary is responsible for the
   a. Himalayas.
   b. Andes Mountains.
   c. Basin and Range Province.
   d. San Andreas Fault.
   e. Michigan Basin.

4. Which one of the following statements is incorrect?
   a. Anticlines are folds in which the strata dip away from the axis.
   b. Terranes result from movement along a divergent boundary.
   c. Joints are fractures along which no movement has taken place parallel with the fracture surface.
   d. Folds, joints, and faults are collectively known as geologic structures.
   e. A strike-slip fault is the result of shear stress in which movement is horizontal with blocks on opposite sides of the fault sliding past one another.

5. A type of stress in which forces acting along the same line, but in opposite directions, thus lengthing or pulling apart an object is _____
   a. compression.
   b. tension.
   c. isostatic rebound.
   d. strain.
   e. fracture.

6. Suppose that rocks along a strike-slip fault were displaced 200 km in 5 million years. What was the average rate of movement per year? Is this average likely to

represent the actual rate of displacement on the fault? Explain.

7. How would you explain stress and strain to someone unfamiliar with the concept?

8. What sequence of events takes place during an orogeny at an oceanic–continental plate boundary?

9. Discuss how time, rock type, temperature, and pressure influence rock deformation.

# Creative Thinking Visual Question

This geologic cross section (●**FIGURE 1**) shows several rock layers and a fault. Your task is to decipher the history of the area. What event took place first, second, and so on? What kind of fold is present? What kind of fault? To fully answer these questions, recall the discussions of sedimentary structures in Chapter 6.

~~ Wave-formed ripples

ᐁᐁᐁ Mud cracks

Graded bedding

Persis Sturges

● **FIGURE 1** Cross section showing deformed strata and geologic structures in an area.

# 10

# MASS WASTING

Floods and landslides in January 2011 devastated numerous towns of the Mountainous Region (Regiao Serrana), in the Brazilian state of Rio de Janeiro (Brazil) killing more than 900 people and leaving at least 14,000 people homeless. During several days of heavy rains, portions of this hillside in Nova Friburgo gave way, causing a landslide that sent mud, vegetation, and buildings sliding down its slope.

201

# INTRODUCTION

In January 2011, southeast Brazil was struck by devastating floods and landslides triggered by unusually heavy torrential rains that dumped a month's worth of rain in just a few days (see the Chapter Opening photo). The hardest hit area was the mountainous Serrana region north of Rio de Janeiro, where relentless flooding and landslides destroyed hundreds of homes, killing more than 900 people, leaving at least 14,000 people homeless, and causing the equivalent of $1.2 billion in property damage. The death toll easily surpassed that of the 1967 mudslides in Caraguatatuba, making this Brazil's worst natural disaster in more than four decades. Collapsed roads and bridges further exacerbated the situation, hampering rescue operations and emergency aid, and lack of power and telephone service contributed further to the misery of the survivors. Although the floods and landslides swept away homes of both rich and poor alike, the poorer rural areas suffered the most because many structures were built in unstable areas.

This terrible tragedy illustrates how geology affects all of our lives. The underlying causes of the mudslides in Brazil can be found anywhere in the world. In fact, *landslides* (a general term for mass movements of Earth materials) cause, on average, between 25 and 50 deaths per year in the United States, and thousands per year worldwide. Furthermore, it is estimated that landslides result in between $2 billion and $4 billion damage annually in the United States. By being able to recognize and understand how landslides occur and what the results may be, we can find ways to reduce the hazards and minimize damage in terms of both human life and property damage.

**Mass wasting** (also called *mass movement*) is defined as the downslope movement of material under the direct influence of gravity. Most types of mass wasting are aided by weathering and usually involve surficial material. The material moves at rates ranging from almost imperceptible, as in the case of creep, to extremely fast, as in a rockfall or slide. Although water can play an important role, the relentless pull of gravity is the major force behind mass wasting.

# FACTORS THAT INFLUENCE MASS WASTING

LO1  **Define mass wasting**

LO2  **Define shear strength**

LO3  **Discuss what is meant by a slope being in a state of dynamic equilibrium**

LO4  **List the seven main factors that can cause mass wasting**

LO5  **Discuss how each of these factors or processes contributes to mass wasting**

Mass wasting is an important geologic process that can occur at any time and almost any place. Although all major landslides have natural causes, many smaller ones are the result of human activity and could have been prevented or their damage minimized.

When the gravitational force acting on a slope exceeds its resisting force, slope failure (mass wasting) occurs. The resisting forces that help maintain slope stability include the slope material's strength and cohesion, the amount of internal friction between grains, and any external support of the slope (●FIGURE 10.1A). These factors collectively define a slope's **shear strength**.

Opposing a slope's shear strength is the force of gravity. Gravity operates vertically but also has a component of force acting parallel to the slope, thereby causing instability (Figure 10.1a). The steeper a slope's angle is, the greater the component of force acting parallel to the slope, and the greater the chance for mass wasting. The steepest angle that a slope can maintain without collapsing is its *angle of repose* (●FIGURE 10.1B). At this angle, the shear strength of the slope's material exactly counterbalances the force of gravity. For unconsolidated material, the angle of repose normally ranges between 25 and 40 degrees. Slopes steeper than 30 degrees usually consist of unweathered solid rock.

All slopes are in a state of *dynamic equilibrium*, which means that they are constantly adjusting to new conditions. Although we tend to view mass wasting as a disruptive and usually destructive event, it is one of the ways that a slope adjusts to new conditions. Whenever a building or road is constructed on a hillside, the equilibrium of that slope is affected. The slope must then adjust, sometimes by mass wasting, to this new set of conditions.

Many factors can cause mass wasting: a change in slope angle, weakening of material by weathering, increased water content, changes in the vegetation cover, and overloading. Although most of these processes are interrelated, we will examine them separately for ease of discussion, but we will also show how they individually and collectively affect a slope's equilibrium.

## Slope Angle

Slope angle is probably the major cause of mass wasting. Generally speaking, the steeper the slope, the less stable it is. Therefore, steep slopes are more likely to experience mass wasting than gentle ones.

A number of processes can oversteepen a slope. One of the most common is undercutting by stream or wave action (●FIGURE 10.2). This process removes the slope's base, increases the slope angle, and thereby increases the

---

**mass wasting**  The downslope movement of Earth materials under the influence of gravity.

**shear strength**  The resisting forces that help maintain a slope's stability.

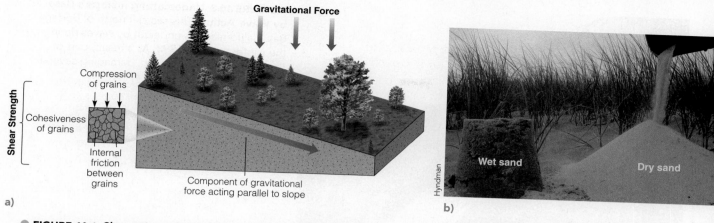

**●FIGURE 10.1  Slope Shear Strength** (a) A slope's shear strength depends on the slope material's strength and cohesion, the amount of internal friction between grains, and any external support of the slope. These factors promote slope stability. The force of gravity operates vertically but also has a component of force acting parallel to the slope. When this force, which promotes instability, exceeds a slope's shear strength, slope failure occurs. (b) The angle of repose is a function of sheer strength. Dry sand usually has an angle of repose of about 30 degrees. With wet sand, shear strength is increased, and much steeper angles of repose are possible, such as this pile that is nearly vertical.

**●FIGURE 10.2  Undercutting a Slope's Base by Stream Erosion** (a) Undercutting by stream erosion removes a slope's base, (b) which then increases the slope angle, and can lead to slope failure. (c) Undercutting by stream erosion caused slumping along this stream near Weidman, Michigan. Notice the scarp, which is the exposed surface of the underlying material following slumping.

gravitational force acting parallel to the slope. Wave action, especially during storms, often results in mass movements along the shores of oceans or large lakes (●FIGURE 10.3).

Excavations for road cuts and hillside building sites are another major cause of slope failure (●FIGURE 10.4). Grading the slope too steeply or cutting into its side increases the stress in the rock or soil until it is no longer strong enough to remain at the steeper angle, and mass movement ensues. Such action is analogous to undercutting by streams or waves, and has the same result, thus explaining why so many mountain roads are plagued by frequent mass movements.

● **FIGURE 10.3 Undercutting a Slope's Base by Wave Action** This sea cliff north of Bodega Bay, California, was undercut by waves during the winter of 1997–1998. As a result, part of the land slid into the ocean, damaging several houses.

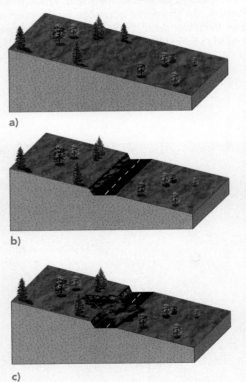

● **FIGURE 10.4 Highway Excavation and Mass Wasting** (a) Highway excavations disturb the equilibrium of a slope by (b) removing a portion of its support, as well as oversteepening it at the point of excavation, which can result in (c) landslides along the highway. (d) Construction of State Road 20 resulted in a rockfall on April 12, 2017 that completely blocked the highway east of Newhalem, Washington.

## Weathering and Climate

Mass wasting is more likely to occur in loose or poorly consolidated slope material than in bedrock. As soon as rock is exposed at Earth's surface, weathering begins to disintegrate and decompose it, reducing its shear strength and increasing its susceptibility to mass wasting. The deeper the weathering zone extends, the greater the likelihood of some type of mass movement.

Recall that some rocks are more susceptible to weathering than others and that climate plays an important role in the rate and type of weathering (see Chapter 6). In the tropics, where temperatures are high and considerable rain falls, the effects of weathering extend to depths of several tens of meters, and mass movements most commonly occur in the deep weathering zone. In arid and semiarid regions, the weathering zone is usually considerably shallower. Nevertheless, intense, localized cloudbursts can drop large quantities of water on an area in a short time. With little vegetation to absorb this water, runoff is rapid and frequently results in mudflows.

## Water Content

The amount of water in rock or soil influences slope stability. Large quantities of water from melting snow or heavy rainfall greatly increase the likelihood of slope failure. The additional weight that water adds to a slope can be enough to cause mass movement. Furthermore, water percolating through a slope's material helps to decrease friction between grains, thus contributing to a loss of cohesion. For example, slopes composed of dry clay are usually quite stable, but when wetted, they quickly lose cohesiveness and internal friction and become an unstable slurry. This occurs because clay, which can hold large quantities of water, consists of platy particles that easily slide over each other when wet. For this reason, clay beds are frequently the slippery layer along which overlying rock units slide downslope.

## Vegetation

Vegetation affects slope stability in several ways. By absorbing the water from a rainstorm, vegetation decreases water saturation of a slope's material that would otherwise lead to a loss of shear strength. Vegetation's root system also helps stabilize a slope by binding soil particles together and holding the soil to bedrock.

The removal of vegetation by either natural or human activity is a major cause of many mass movements. Summer brush and forest fires in southern California, especially those in 2017, which were the most destructive on record, with a cost of more than $13 billion in damage, frequently leave the hillsides bare of vegetation. Fall rainstorms then saturate the ground, causing mudslides that do tremendous damage and cost millions of dollars to clean up.

## Overloading

Overloading is almost always the result of human activity and typically results from the dumping, filling, or piling up of material. Under natural conditions, a material's load is carried by its grain-to-grain contacts, with the friction between the grains maintaining a slope. The additional weight created by overloading increases the water pressure within the material, which in turn decreases its shear strength, thereby weakening the slope material. If enough material is added, the slope will eventually fail, sometimes with tragic consequences.

## Geology and Slope Stability

The relationship between the topography of an area and its geology is important in determining slope stability (•FIGURE 10.5). If the rocks underlying a slope dip in the

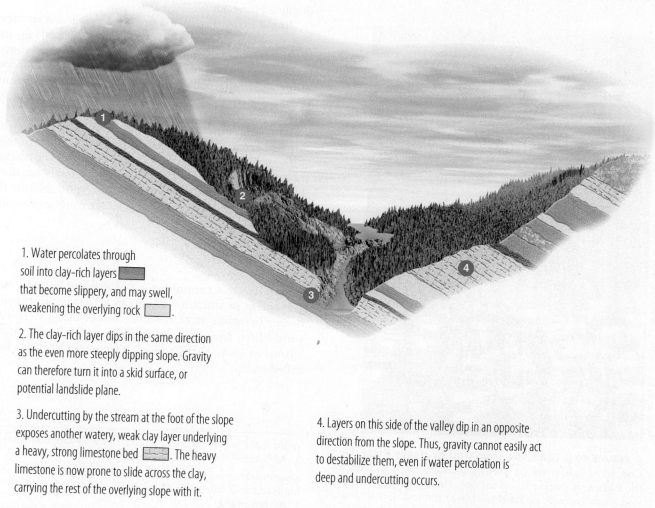

1. Water percolates through soil into clay-rich layers that become slippery, and may swell, weakening the overlying rock.

2. The clay-rich layer dips in the same direction as the even more steeply dipping slope. Gravity can therefore turn it into a skid surface, or potential landslide plane.

3. Undercutting by the stream at the foot of the slope exposes another watery, weak clay layer underlying a heavy, strong limestone bed. The heavy limestone is now prone to slide across the clay, carrying the rest of the overlying slope with it.

4. Layers on this side of the valley dip in an opposite direction from the slope. Thus, gravity cannot easily act to destabilize them, even if water percolation is deep and undercutting occurs.

● **FIGURE 10.5 Geology, Slope Stability, and Mass Wasting** Rocks dipping in the same direction as a hill's slope are particularly susceptible to mass wasting.

same direction as the slope, mass wasting is more likely to occur than if the rocks are horizontal or dip in the opposite direction. When the rocks dip in the same direction as the slope, water can percolate along the various bedding planes and decrease the cohesiveness and friction between adjacent rock units. This is particularly true when clay layers are present, because clay becomes slippery when wet.

Even if the rocks are horizontal or dip in a direction opposite to that of the slope, joints may dip in the same direction as the slope. Water migrating through them weathers the rock and expands these openings until the weight of the overlying rock causes it to fall.

## Triggering Mechanisms

The factors discussed thus far all contribute to slope instability. Most, though not all, rapid mass movements are triggered by a force that temporarily disturbs slope equilibrium. The most common triggering mechanisms are strong vibrations from earthquakes and excessive amounts of water from a winter snow melt or a heavy rainstorm (●FIGURE 10.6).

Photograph by R.L. Schuster, U.S. Geological Survey

● **FIGURE 10.6 Landslide Triggered by Heavy Rains, La Conchita, California** La Conchita, California, is located at the base of a steep-sloped terrace. Heavy rains and irrigation of an avocado orchard (visible at the top of the terrace) contributed to the landslide that destroyed nine homes in 1995. Ten years later (2005), similar factors caused another massive landslide in the same area.

# TYPES OF MASS WASTING

**LO6** List the three major criteria used to classify mass movements

**LO7** Discuss what constitutes rapid mass wasting

**LO8** Discuss what constitutes slow mass wasting

**LO9** Define a rock fall

**LO10** Define a slide

**LO11** Explain the difference between a slump and a rock slide

**LO12** Define a flow

**LO13** Explain how the seven types of flows differ from each other

Mass movements are generally classified on the basis of three major criteria (●TABLE 10.1): (1) rate of movement (rapid or slow); (2) type of movement (primarily falling, sliding, or flowing); and (3) type of material involved (rock, soil, or debris). Even though many slope failures are combinations of different materials and movements, the resulting mass movements are typically classified according to their dominant behavior.

**Rapid mass movements** involve a visible movement of material. Such movements usually occur quite suddenly, and the material moves quickly downslope. Rapid mass movements are potentially dangerous and frequently result in loss of life and property damage. Most rapid mass movements occur on relatively steep slopes and can involve rock, soil, or debris.

**Slow mass movements** advance at an imperceptible rate and are usually detectable only by the effects of their movement, such as tilted trees and power poles or cracked foundations. Although rapid mass movements are more dramatic, slow mass movements are responsible for the downslope transport of a much greater volume of weathered material.

## Falls

**Rockfalls** are a common type of extremely rapid mass movement in which rocks of any size may fall through the air (●FIGURE 10.7). Rockfalls occur along steep canyons, cliffs, and road cuts, and they build up accumulations of loose rocks and rock fragments at their base called *talus*.

**rapid mass movement** Any kind of mass wasting that involves a visible downslope displacement of material.

**slow mass movement** Mass movement that advances at an imperceptible rate and is usually detectable only by the effects of its movement.

**rockfall** A type of extremely fast mass wasting in which rocks fall through the air.

**TABLE 10.1** Classification of Mass Movements and Their Characteristics

| Type of Movement | Subdivision | Characteristics | Rate of Movement |
|---|---|---|---|
| Falls | Rockfall | Rocks of any size fall through the air from steep cliffs, canyons, and road cuts | Extremely rapid |
| Slides | Slump | Movement occurs along a curved surface of rupture; most commonly involves unconsolidated or weakly consolidated material | Extremely slow to moderate |
| | Rock slide | Movement occurs along a generally planar surface | Rapid to very rapid |
| Flows | Mudflow | Consists of at least 50% silt- and clay-sized particles and up to 30% water | Very rapid |
| | Debris flow | Contains larger-sized particles and less water than mudflows | Rapid to very rapid |
| | Earthflow | Thick, viscous, tongue-shaped mass of wet regolith | Slow to moderate |
| | Quick clays | Composed of fine silt and clay particles saturated with water; when disturbed by a sudden shock, lose their cohesiveness and flow like a liquid | Rapid to very rapid |
| | Solifluction | Water-saturated surface sediment | Slow |
| | Creep | Downslope movement of soil and rock | Extremely slow |
| Complex | | Combination of different movement types | Slow to extremely rapid |

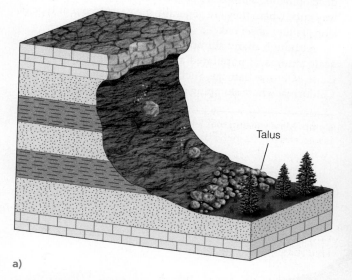

a)

b)

Copyright and Photograph by Dr. Parvinder S. Sethi

● **FIGURE 10.7  Rockfalls** (a) Rockfalls result from failure along cracks, fractures, or bedding planes in the bedrock, and are common features in areas of steep cliffs. (b) A rockfall of granite in Yosemite National Park, California.

Rockfalls result from failure along joints or bedding planes in the bedrock and are commonly triggered by natural or human undercutting of slopes or by earthquakes. Many rockfalls in cold climates are the result of frost wedging (see Figure 6.2). Chemical weathering caused by water percolating through the fissures in carbonate rocks (limestone, dolostone, and marble) is also responsible for many rockfalls.

Rockfalls range in size from small rocks falling from a cliff to massive falls involving millions of cubic meters of debris that destroy buildings, bury towns, and block highways. Rockfalls are a particularly common hazard in mountainous areas where roads have been built by blasting and grading through steep hillsides of bedrock (●**FIGURE 10.7B**). Slopes that are particularly susceptible to rockfalls are sometimes covered with wire mesh in an effort to prevent dislodged rocks from falling onto the road below (●**FIGURE 10.8**).

## Slides

A **slide** involves movement of material along one or more surfaces of failure. The type of material may be soil, rock, or a combination of the two, and it may break apart during movement or remain intact. A slide's rate of movement can vary from extremely slow to very rapid (Table 10.1).

**slide** Mass wasting involving movement of material along one or more surfaces of failure.

● **FIGURE 10.8  Minimizing Damage from Rock Falls**  Wire mesh has been used to cover this steep slope near Narvik in northern Norway. This is a common practice in mountainous areas to prevent rocks from falling on the road.

Two types of slides are generally recognized: (1) slumps, or rotational slides, in which movement occurs along a curved surface; and (2) rock or block slides, which move along a more or less planar surface.

A **slump** involves the downward movement of material along a curved surface of rupture and is characterized by the backward rotation of the slump block (●**FIGURE 10.9**). Slumps usually occur in unconsolidated or weakly consolidated material and range in size from small individual sets, such as those occurring along stream banks (Figure 10.2c), to massive, multiple sets that affect large areas and cause considerable damage.

Slumps can be caused by a variety of factors, but the most common is erosion along the base of a slope, which removes support for the overlying material. This local steepening may be caused naturally by stream erosion along its banks (Figure 10.2c) or by wave action at the base of a coastal cliff (●**FIGURE 10.10**).

Slope oversteepening can also be caused by human activity, such as the construction of highways and housing developments. Slumps are particularly prevalent along highway cuts, where they are generally the most frequent type of slope failure observed.

Although many slumps are merely a nuisance, large-scale slumps in populated areas and along highways can cause extensive damage. Such is the case in coastal southern California, where slumping and sliding have been a constant

---

**slump**  Mass wasting that takes place along a curved surface of failure and results in the backward rotation of the slump mass.

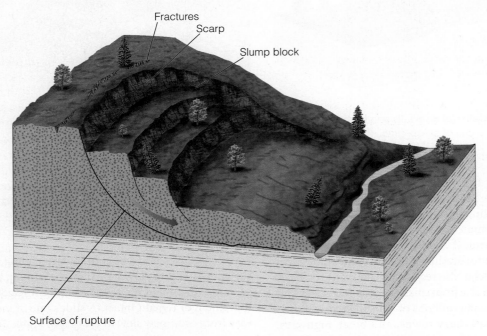

● **FIGURE 10.9  Slumping**  In a slump, material moves downward along the curved surface of a rupture, causing the slump block to rotate backward. Most slumps involve unconsolidated or weakly consolidated material, and are typically caused by erosion along the slope's base.

● **FIGURE 10.10 Slumping in the Pacific Palisades, California** Undercutting of steep sea cliffs by wave action resulted in massive slumping in the Pacific Palisades area of southern California on March 31 and April 3, 1958. Highway 1 was completely blocked. Note the heavy earth-moving equipment for scale.

problem, resulting in the destruction of many homes and the closing and relocation of numerous roads and highways (Figure 10.10) (see GEO-FOCUS).

A **rock slide**, or *block slide*, occurs when rocks move downslope along a more or less planar surface. Most rock slides take place because the local slopes and rock layers dip in the same direction (**FIGURES 10.5** and ●**10.11**), although they can also occur along fractures parallel to a slope. Rock slides are common occurrences along the southern California coast, such as Point Fermin, where seaward-dipping rocks with interbedded slippery clay layers are undercut by waves, causing numerous slides.

Farther south from Point Fermin is the town of Laguna Beach, where residents were hit by rock slides and mudslides in 1978, 1998, and 2005 (●**FIGURE 10.12**). Just as at Point Fermin, the rocks at Laguna Beach dip about 25 degrees in the same direction as the slope of the canyon walls and contain clay beds that "lubricate" the overlying rock layers, causing the rocks and the houses built on them to slide. Percolating water from heavy rains wets subsurface clayey siltstone, thus reducing its shear strength and helping to activate the slide. In addition, these slides are part of a larger ancient slide complex.

Not all rock slides are the result of rocks dipping in the same direction as a hill's slope. The rock slide at Frank, Alberta, Canada, on April 29, 1903, illustrates how nature and human activity can combine to create a situation with tragic results (●**FIGURE 10.13**).

It would appear at first glance that the coal-mining town of Frank, lying at the base of Turtle Mountain, was in no danger from a landslide (Figure 10.13). After all, many of the rocks dipped away from the mining valley. The joints in the massive limestone composing Turtle Mountain, however, dip steeply toward the valley and are essentially parallel with the slope of the mountain itself. Furthermore, Turtle Mountain is supported by weak siltstones, shales, and coal layers that underwent slow plastic deformation from the weight of the overlying massive limestone. Coal mining along the base of the valley also contributed to the stress on the rocks by removing some of the underlying support. All of these factors, as well as the frost action and chemical weathering

---

**rock slide** Rapid mass wasting in which rocks move downslope along a more or less planar surface.

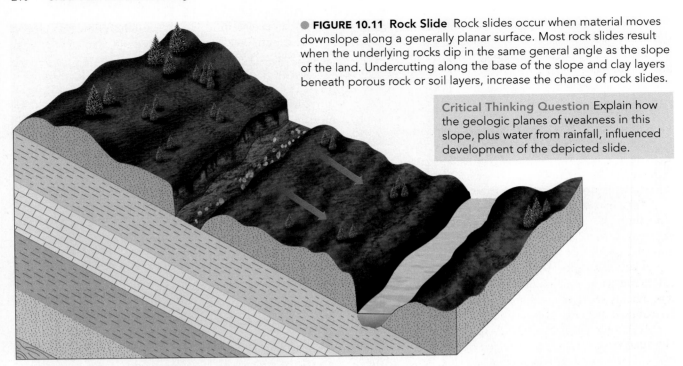

**FIGURE 10.11 Rock Slide** Rock slides occur when material moves downslope along a generally planar surface. Most rock slides result when the underlying rocks dip in the same general angle as the slope of the land. Undercutting along the base of the slope and clay layers beneath porous rock or soil layers, increase the chance of rock slides.

**Critical Thinking Question** Explain how the geologic planes of weakness in this slope, plus water from rainfall, influenced development of the depicted slide.

# GEO-FOCUS

## Southern California Landslides

Southern California is no stranger to landslides. La Conchita, Point Fermin, Pacific Palisades, and Laguna Beach are all locations that have suffered damaging mass movements during the past 50 years. Two regions in particular, La Conchita and Laguna Beach, have been in the news because of the landslides that destroyed numerous homes.

La Conchita is a small community along the coast at the base of a 100-m-high terrace, 120 km northwest of Los Angeles, California. On March 4, 1995, following a period of heavy rains, a massive 200,000 m³ slide destroyed or damaged nine homes in its path (see Figure 10.6).

Almost 10 years later, following a week of heavy rainfall which saturated the hillside and previous landslide deposits, another landslide occurred in the same area. This time 10 people were killed and 15 homes were buried under 10 m and 363,000,000 kg of mud (**FIGURE 1**).

What went wrong and why was this situation repeated? The rocks that make up the steep-sloped terrace behind La Conchita consist of soft, weak, and porous sediments that are not well lithified, and thus are easily weathered and susceptible to mass wasting. In addition, an irrigated avocado orchard sits on top of the hill, contributing to the water percolating through the porous sediments and rocks and contributing to the instability of the hillside. Add in heavy rainfall over an extended period, and you have all the ingredients for a landslide in the making. An ancient landslide area to begin with, a steep slope that has been undercut at its base by a road, well-saturated sediments decreasing the cohesion of the sediments that hold the hillside together, and continuing rains all contribute to the making of a landslide. And the potential is still there for another

**FIGURE 1** La Conchita, California, 2005. Factors similar to those in 1995 caused another massive landslide in the same area. The landslide can clearly be seen in the center of this photograph, and the scarp and the remains of the 1995 landslide are still visible on the right side, as is the avocado orchard in the foreground.

that widened the joints, finally resulted in a massive rock slide. Approximately 40 million m³ of rock slid down Turtle Mountain along joint planes, killing 70 people and partially burying the town of Frank.

## Flows

Mass movements in which material flows as a viscous fluid or displays plastic movement are termed *flows*. Their rate of movement ranges from extremely slow to extremely rapid (Table 10.1). In many cases, mass movements begin as falls, slumps, or slides, and change into flows farther downslope.

Of the major mass movement types, **mudflows** are the most fluid and move most rapidly (at speeds up to 80 km per hour). They consist of at least 50% silt- and clay-sized material combined with a significant amount of water (up to 30%). Mudflows are common in arid and semi-arid environments, where they are triggered by heavy rainstorms that quickly saturate the regolith, turning it into a raging flow of mud that engulfs everything in its path.

Mudflows can also occur in mountain regions (●**FIGURE 10.14**) and in areas covered by volcanic ash, where they can be particularly destructive (see Chapter 5). Because

mudflows are so fluid, they generally follow preexisting channels until the slope decreases or the channel widens, at which point they fan outward.

**Debris flows** are composed of larger particles than mudflows and do not contain as much water. Consequently, they are usually more viscous than mudflows, typically do not move as rapidly, and rarely are confined to preexisting channels. Debris flows can be just as damaging, though, because they can transport large objects (●**FIGURE 10.15**).

**Earthflows** move more slowly than either mudflows or debris flows. An earthflow slumps from the upper part of a hillside, leaving a scarp, and flows slowly downslope

---

**mudflow**  A flow consisting mostly of clay- and silt-sized particles and up to 30% water that moves downslope under the influence of gravity.

**debris flow**  A type of mass wasting that involves a viscous mass of soil, rock fragments, and water that moves downslope; debris flows have larger particles than mudflows and contain less water.

**earthflow**  A mass-wasting process involving the downslope movement of water-saturated soil.

---

landslide, with no guarantee it will not happen again in the near future.

Farther south in Laguna Beach, another landslide, in this case a rock slide, destroyed 18 expensive hillside homes and severely damaged approximately 20 others on June 1, 2005 (●**FIGURE 2**). Similar to what happened in 1978 (see Figure 10.12), the main triggering mechanism was probably unusually heavy winter rains, in this case the second-rainiest season on record. In this area of southern California, the rocks dip in the same direction as the slope and contain numerous clay beds interbedded with porous sandstones. Such conditions, when combined with heavy rainfall, are ideal for rock slides. It should come as no surprise that the area where both the 1978 and 2005 rock slides occurred is also part of an ancient slide complex.

Can anything be done to prevent future landslides? The short answer is probably no. Decreasing the slope, benching the hillside, and ensuring that there is sufficient drainage and a good cover of vegetation are all steps

that can minimize future mass wasting. But the sad fact is that the geologic conditions are such that future landslides are inevitable as the landscape seeks equilibrium conditions by adjusting its slope. Add in the fact that the

coastal terraces of Laguna Beach offer some of the most breathtaking views of the Pacific and people are willing to pay a premium to live here, and you have the formula for future landslides, loss of life, and property damage.

AP Images/NICK UT

●**FIGURE 2**  A rock slide on June 1, 2005, destroyed 18 expensive homes and damaged at least 20 others in Laguna Beach, California. Heavy rains, combined with unstable underlying geology, contributed to this landslide.

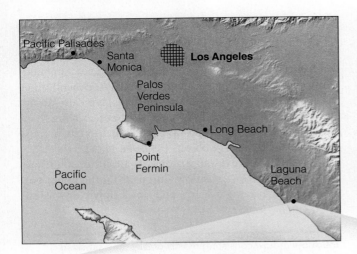

● **FIGURE 10.12 Rock Slide, Laguna Beach, California**
A combination of interbedded clay layers that become slippery when wet, rocks dipping in the same direction as the slope of the sea cliffs, and undercutting of the sea cliffs by wave action activated a rock slide at Laguna Beach, California, that destroyed numerous homes and cars on October 2, 1978. This same area was hit by another rock slide in 2005.

**Critical Thinking Question** Is there anything that could have been done after the first rock slide that might have prevented the second one 27 years later? If not, should people be allowed to build on known active slide areas?

as a thick, viscous, tongue-shaped mass of wet regolith (●**FIGURE 10.16**). Like mudflows and debris flows, earthflows can be of any size and are frequently destructive. They occur most commonly in humid climates on grassy, soil-covered slopes following heavy rains.

Some clays spontaneously liquefy and flow like water when they are disturbed. Such **quick clays** have caused serious damage and loss of lives in Sweden, Norway, eastern Canada, and Alaska. Quick clays are composed of fine silt and clay particles made by the grinding action of glaciers. Geologists think that these fine sediments were originally deposited in a marine environment, where their pore space was filled with saltwater. The ions in saltwater helped establish strong bonds between the clay particles, thus stabilizing and strengthening the clay. When the clays were subsequently uplifted above sea level, the saltwater was flushed out by fresh groundwater, reducing the effectiveness of the ionic bonds between the clay particles and thereby reducing the overall strength and cohesiveness of the clay. Consequently,

when the clay is disturbed by a sudden shock or shaking, it essentially turns to a liquid and flows.

An excellent example of the damage that can be done by quick clays occurred in the Turnagain Heights area of Anchorage, Alaska, in 1964 (●**FIGURE 10.17**). Underlying most of the Anchorage area is the Bootlegger Cove Clay, a massive clay unit of poor permeability. Because the Bootlegger Cove Clay forms a barrier that prevents groundwater from flowing through the adjacent glacial deposits to the sea, considerable hydraulic pressure builds up on the landward side of the clay. Some of this water has flushed out the saltwater in the clay and has saturated the lenses of sand and silt associated with the clay beds. When the 9.2-magnitude Good Friday earthquake struck on March 27, 1964, the shaking turned parts of the Bootlegger Cove Clay into a

---

**quick clay** A clay deposit that spontaneously liquefies and flows when disturbed.

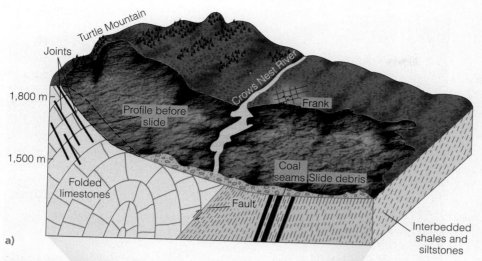

**● FIGURE 10.13 Rock Slide, Turtle Mountain, Canada** (a) The tragic Turtle Mountain rock slide that killed 70 people and partially buried the town of Frank, Alberta, Canada, on April 29, 1903, was caused by a combination of factors. These included joints that dipped in the same direction as the slope of Turtle Mountain, a fault part-way down the mountain, weak shale and siltstone beds underlying the base of the mountain, and mined-out coal seams. (b) Results of the 1903 rock slide at Frank.

quick clay and precipitated a series of massive slides in the coastal bluffs that destroyed most of the homes in the Turnagain Heights subdivision (●**FIGURE 10.17B**).

**Solifluction** is the slow downslope movement of water-saturated surface sediment. Solifluction can occur in any climate where the ground becomes saturated with water, but is most common in areas of permafrost.

**Permafrost**, ground that remains permanently frozen, covers nearly 20% of the world's land surface (●**FIGURE 10.18A**). During the warmer season, when the upper portion of the permafrost thaws, water and surface sediment form a soggy mass that flows by solifluction and produces a characteristic lobate topography (●**FIGURE 10.18B**).

---

**solifluction** Mass wasting involving the slow downslope movement of water-saturated surface materials; occurs especially at high elevations or high latitudes where the flow is underlain by frozen soil.

**permafrost** Ground that remains permanently frozen.

● **FIGURE 10.14 Mudflow, Estes Park, Colorado** Mudflows move swiftly downslope, engulfing everything in their path. Note how this mudflow in Rocky Mountain National Park has fanned out at the base of the hill. Also note the small lake adjacent to the mudflow that was formed after this mudflow created a dam across the stream.

Copyright and Photograph by Dr. Parvinder S. Sethi

● **FIGURE 10.15 Debris Flow, Ophir Creek, Nevada** A debris flow and damaged house in lower Ophir Creek, western Nevada. Note the many large boulders that are part of the debris flow. Debris flows do not contain as much water as mudflows and typically are composed of larger particles.

B. Pipkin, University of Southern California

● **FIGURE 10.16 Earthflow** (a) Earthflows form tongue-shaped masses of wet regolith that move slowly downslope. They occur most commonly in humid climates on grassy, soil-covered slopes. (b) An earthflow near Baraga, Michigan.

Scarp

a)

Scarp

b)

Reed Wicander

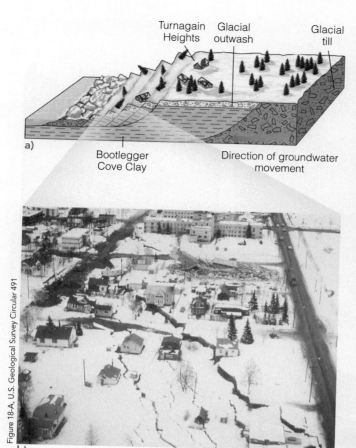

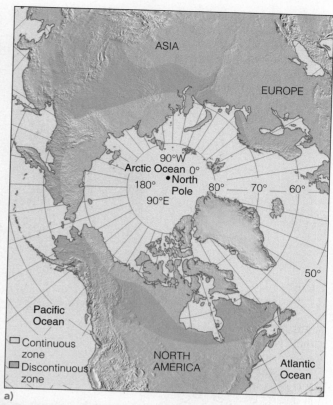

**FIGURE 10.17 Quick Clay Slide, Anchorage, Alaska**
(a) Ground shaking by the 1964 Alaska earthquake turned parts of the Bootlegger Cove Clay into a quick clay, causing numerous slides. (b) Low-altitude photograph of the Turnagain Heights subdivision of Anchorage shows some of the numerous landslide fissures that developed, as well as the extensive damage to buildings in the area. The remains of the Four Seasons apartment building can be seen in the background.

As might be expected, many problems are associated with construction in a permafrost environment. For example, when an uninsulated building is constructed directly on permafrost, heat escapes through the floor, thaws the ground below, and turns it into a soggy, unstable mush. Because the ground is no longer solid, the building settles unevenly into the ground, and numerous structural problems result (●FIGURE 10.19).

**Creep**, the slowest type of flow, is the most widespread and significant mass wasting process in terms of the total amount of material moved downslope and the monetary damage it does annually. Creep involves extremely slow downhill movement of soil or rock. Although it can occur anywhere and in any climate, it is most effective and significant as a geologic agent in humid regions.

Because the rate of movement is essentially imperceptible, we are frequently unaware of creep's existence until we notice its effects: tilted trees and power poles, broken streets

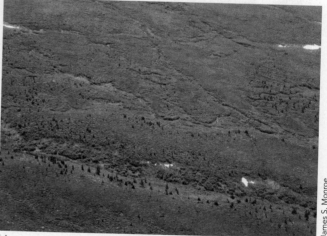

**FIGURE 10.18 Permafrost and Solifluction** (a) Distribution of permafrost areas in the Northern Hemisphere. (b) Solifluction flows in Kluane National Park, Yukon Territory, Canada, show the lobate topography that is characteristic of solifluction conditions.

and sidewalks, or cracked retaining walls or foundations (●FIGURE 10.20). Creep usually involves the whole hillside, and probably occurs, to some extent, on any weathered or soil-covered sloping surface.

**creep** A widespread type of mass wasting in which soil or rock moves slowly downslope.

● **FIGURE 10.19 Permafrost Damage** This house, south of Fairbanks, Alaska, has settled unevenly because the underlying permafrost in fine-grained silts and sands has thawed.

O. J. Ferrains, Jr./USGS

National Geophysical Data Center/NOAA

National Geophysical Data Center/NOAA

David J. Matty

● **FIGURE 10.20 Creep** (a) Some evidence of creep: (A) curved tree trunks, (B) displaced monuments, (C) tilted power poles, (D) displaced and tilted fences, (E) roadways moved out of alignment, (F) hummocky surface. (b) Trees bent by creep, Wyoming. (c) Creep has bent these sandstone and shale beds of the Haymond Formation near Marathon, Texas. (d) Tilted stone wall due to creep in Champion, Michigan.

Creep is difficult not only to recognize but also to control. Although engineers can sometimes slow or stabilize creep, many times the only course of action is to simply avoid the area if at all possible or, if the zone of creep is relatively thin, design structures that can be anchored into the bedrock.

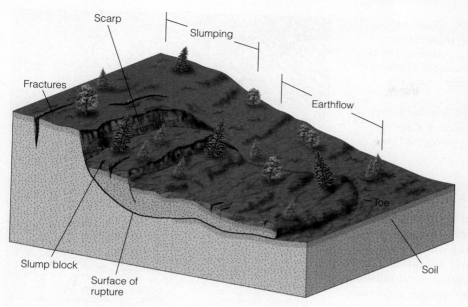

● **FIGURE 10.21 Complex Movement** A complex movement is one in which several types of mass wasting are involved. In this example, slumping occurs at the head, followed by an earthflow.

Scarp

Slumping

Fractures

Earthflow

Slump block

Surface of rupture

Toe

Soil

## Complex Movements

Recall that many mass movements are combinations of different movement types. When one type is dominant, the movement can be classified as one of those described thus far. If several types are more or less equally involved, however, it is called a **complex movement**.

The most common type of complex movement is the *slide-flow*, in which there is sliding at the head and then some type of flowage farther along its course. Most slide-flow landslides involve well-defined slumping at the head, followed by a debris flow or earthflow (●**FIGURE 10.21**). Any combination of different mass movement types is a complex movement.

# RECOGNIZING AND MINIMIZING THE EFFECTS OF MASS WASTING

LO14 Explain how a thorough geologic investigation is the most important factor in eliminating or minimizing the effects of mass wasting

LO15 Explain how geologists identify areas of high potential for slope failure

LO16 Discuss the use of slope-stability maps in deciding the most favorable locations for development and infrastructure siting

LO17 Discuss the various methods geologists and engineers can use to minimize the danger and damage from mass wasting

The most important factor in eliminating or minimizing the damaging effects of mass wasting is a thorough geologic investigation of the region in question. In this way, former landslides and areas susceptible to mass movements can

be identified and perhaps avoided. By assessing the risks of possible mass wasting before construction begins, engineers can take steps to eliminate or minimize the effects of such events.

Identifying areas with a high potential for slope failure is important in any hazard assessment study; these studies include identifying former landslides as well as sites of potential mass movement. Scarps, open fissures, displaced or tilted objects, a hummocky surface, and sudden changes in vegetation are some of the features that indicate former landslides or areas susceptible to slope failure. The effects of weathering, erosion, and vegetation may, however, obscure the evidence of previous mass wasting.

The information derived from a hazard assessment study can be used to produce *slope-stability maps* of an area (●**FIGURE 10.22**). These maps allow planners and developers to make decisions about where to site roads, utility lines, and housing or industrial developments based on the relative stability or instability of a particular location. The maps also indicate the extent of an area's landslide problem and the type of mass movement that may occur.

Although most large mass movements usually cannot be prevented, geologists and engineers can use various methods to minimize the danger and damage resulting from them. Because water plays such an important role in many landslides, one of the most effective and inexpensive ways to reduce the potential for slope failure or to increase existing slope stability is surface and subsurface drainage of a hillside. Drainage serves two purposes: (1) it reduces the weight of the material likely to slide, and (2) increases the shear strength of the slope material by lowering pore pressure.

**complex movement** A combination of different types of mass movements in which no single type is dominant; usually involves sliding and flowing.

● **FIGURE 10.22 Slope-Stability Map** This slope-stability map of part of San Clemente, California, shows areas delineated according to relative stability. Such maps help planners and developers make decisions about where to site roads, utility lines, buildings, and other structures.

**Critical Thinking Question** Locate the line that shows horizontal contact between rocks of different stability. What is the potential for mass wasting along this line, and why?

Surface waters can be drained and diverted by ditches, gutters, or culverts designed to direct water away from slopes. Drainpipes perforated along one surface and driven into a hillside can help remove subsurface water (●**FIGURE 10.23**). Finally, planting vegetation on hillsides helps stabilize slopes by holding the soil together and reducing the amount of water in the soil.

Another way to help stabilize a hillside is to reduce its slope. Recall that overloading and oversteepening by grading are common causes of slope failure. Reducing the angle of a hillside decreases the potential for slope failure. Two methods

● **FIGURE 10.23 Using Drainpipes to Remove Subsurface Water** (a) Driving drainpipes that are perforated on one side into a hillside, with the perforated side up, can remove some subsurface water and help stabilize a hillside. (b) A drainpipe driven into the hillside at Point Fermin, California, helps to reduce the amount of subsurface water in these porous beds.

**Critical Thinking Question** What two features in (b) can you identify that indicate mass wasting at this location is both a current and potential problem?

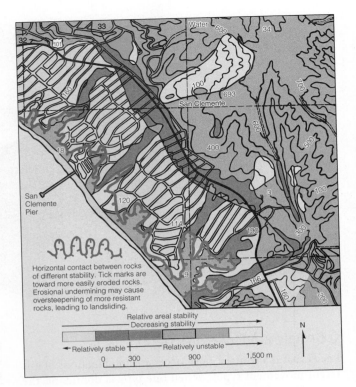

Horizontal contact between rocks of different stability. Tick marks are toward more easily eroded rocks. Erosional undermining may cause oversteepening of more resistant rocks, leading to landsliding.

Relative areal stability
Decreasing stability

←Relatively stable   Relatively unstable→

0    300         900        1,500 m

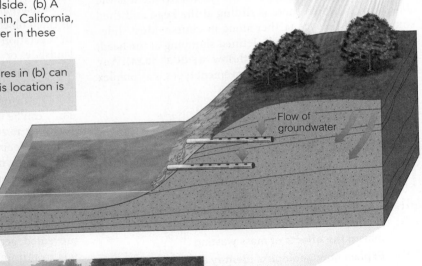

Flow of groundwater

a)

Reed Wicander

b)

are usually employed to reduce a slope's angle. In the *cut-and-fill* method, material is removed from the upper part of the slope and used as fill at the base, thus providing a flat surface for construction and reducing the slope (●**FIGURE 10.24**).

The second method, which is called *benching*, involves cutting a series of benches, or steps, into a hillside (●**FIGURE 10.25**). This process reduces the overall average slope, and the benches serve as collecting sites for small

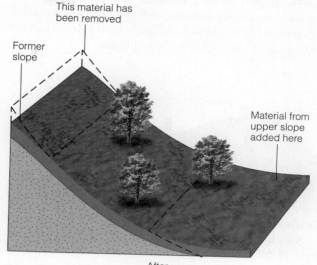

● **FIGURE 10.24 Stabilizing a Hillside by the Cut-and-Fill Method** One common method used to help stabilize a hillside and reduce its slope is the cut-and-fill method. Material from the steeper upper part of the hillside is removed, thereby decreasing the slope angle, and is used to fill in the base. This provides some additional support at the base of the slope.

John D. Cunningham/Visuals Unlimited

● **FIGURE 10.25 Stabilizing a Hillside by Benching** (a) Another common method used to stabilize a hillside and reduce its slope is benching. This process involves making several cuts along a hillside to reduce the overall slope. Therefore, individual slope failures are now limited in size, and the material collects on the benches. (b) Benching is used in many road cuts and can be clearly seen in this photograph.

**Critical Thinking Question** Given the height of the road cut (b), how effective do you think benching will be in helping to stabilize this slope? What other measures can be taken to minimize the damage from potential mass wasting?

landslides or rockfalls that might occur. Benching is most commonly used on steep hillsides in conjunction with a system of surface drains to divert runoff.

In some situations, retaining walls are constructed to provide support for the base of the slope (●FIGURE 10.26). The walls are usually anchored well into bedrock, backfilled with crushed rock, and provided with drain holes to prevent the buildup of water pressure in the hillside.

Recognition, prevention, and control of landslide-prone areas are expensive, but not nearly as expensive as the damage can be when such warning signs are ignored or not recognized. Unfortunately, there are numerous examples of landfill and dam collapses, such as the January, 2019 mining dam collapse in Brazil, that serve as tragic reminders of the price paid in loss of lives and property damage when the warning signs of impending disaster are ignored.

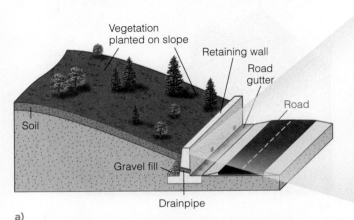

a)

b)

●**FIGURE 10.26 Retaining Walls Help Reduce Landslides** (a) Retaining walls anchored into bedrock, backfilled with gravel, and provided with drainpipes can support a slope's base and reduce landslides. (b) A steel retaining wall built to stabilize the slope and keep falling and sliding rocks off the highway.

# Key Concepts Review

- Mass wasting is the downslope movement of material under the direct influence of gravity. It may result in loss of life, as well as millions to billions of dollars in damage annually.
- Mass wasting occurs when the gravitational force acting parallel to a slope exceeds the slope's shear strength (the resisting forces that help maintain slope stability).
- The major factors causing mass wasting include slope angle, weathering and climate, water content, overloading, and removal of vegetation. It is usually several of these factors in combination that results in slope failure.
- Mass movements are classified on the basis of their rate of movement (rapid versus slow), type of movement (falling, sliding, or flowing), and type of material (rock, soil, or debris).
- Rockfalls are a common mass movement in which rocks free-fall. They are common along steep canyons, cliffs, and road cuts.
- The two types of slides are slumps and rock slides. Slumps, or rotational slides, involve movement along a curved surface and are most common in poorly consolidated or unconsolidated material. Rock slides, also known as block slides, occur when movement takes place along a more or less planar surface, and they usually involve solid pieces of rock.
- Rate of movement (rapid versus slow), type of material (rock, sediment, or soil), and amount of water are the criteria used to recognize the several types of flows.

- Mudflows consist of mostly clay- and silt-size particles and contain up to 30% water. They are most common in semiarid and arid environments and generally follow preexisting channels.
- Debris flows are composed of larger particles and contain less water than mudflows.
- Earthflows move more slowly than either debris flows or mudflows and move downslope as thick, viscous, tongue-shaped masses of wet regolith.
- Quick clays are clays that spontaneously liquefy and flow like water when they are disturbed.
- Solifluction is the slow downslope movement of water-saturated surface material and is most common in areas of permafrost.
- Creep, the slowest type of flow, is the imperceptible downslope movement of soil or rock. It is the most widespread of all types of mass wasting.
- Complex movements are combinations of different types of mass movements in which no single type is dominant. Most complex movements involve sliding and flowing.
- The most important factor in reducing or eliminating the damaging effects of mass wasting is a thorough geologic investigation to outline areas susceptible to mass movements.
- Although mass movement cannot be eliminated, its effects can be minimized by building retaining walls, draining excess water, regrading slopes, and planting vegetation.

# Important Terms

complex movement   217
creep   215
debris flow   211
earthflow   211
mass wasting   202
mudflow   211

permafrost   213
quick clay   212
rapid mass movement   206
rockfall   206
rock slide   209
shear strength   202

slide   207
slow mass movement   206
slump   208
solifluction   213

# Review Questions

1. Mass wasting can occur
   a. on gentle slopes.
   b. on steep slopes.
   c. in flat-lying areas.
   d. all of these.
   e. none of these.

2. Solifluction is
   a. one type of slide.
   b. a type of slow movement of water-saturated surface sediment.
   c. the result of creep.
   d. reduced by overloading a slope.
   e. most common in arid environments.

3. Shear strength is
   a. the force of gravity.
   b. the resisting force that helps maintain a slope's stability.
   c. a state of dynamic equilibrium in a slope.
   d. increased when a slope is subjected to overloading.
   e. unimportant when assessing a slope's potential for mass wasting.

4. Which of the following factors can actually enhance slope stability?
   a. increasing the slope angle
   b. vegetation
   c. overloading
   d. rocks dipping in the same direction as the slope
   e. none of these

5. Former landslides and areas currently susceptible to slope failure can be identified by which of the following features?
   a. tilted objects
   b. open fissures
   c. scarps
   d. hummocky surfaces
   e. all of these

6. If an area has a documented history of mass wasting that has endangered or taken human life, how should people and governments prevent such events from happening again? Are most large mass-wasting events preventable or predictable?

7. What roles do climate and weathering play in mass wasting?

8. Discuss how the different factors that influence mass wasting are interconnected.

9. Why is creep so prevalent? Why does it do so much damage? What are some of the ways that creep might be controlled?

# Creative Thinking Visual Question

What features of slope stabilization do you see in this photograph (●FIGURE 1) of a housing development in Concord, California? You should be able to recognize at least three features.

John Karachewski/Geoscapes Photography

● **FIGURE 1** Concord, California housing development.

David Buch Photography/Alamy Stock Photo

# 11

# RUNNING WATER

View of the Nile River in Egypt.
The Nile River floodplain and its
delta are nearly the only areas in
Egypt that support agriculture.
At 6,650 km long, it is the world's
longest river.

# INTRODUCTION

You have probably experienced the power of running water if you have ever swam or canoed in a fast-flowing stream or river, but the vivid accounts of floods truly impress us with the energy of moving water. For example, at 4:07 p.m. on May 31, 1889, residents of Johnstown, Pennsylvania, heard "a roar like thunder" and within 10 minutes the town was devastated by an 18-m-high wall of water that tore through the town at 60 km/hr, sweeping up debris, houses, and entire families (●FIGURE 11.1). Heavy rainfall and the failure of a dam upstream from the town caused the flood in which at least 2,200 people were killed, making it the deadliest river flood in U.S. history.

In July and August 2010, because of exceptionally heavy monsoon rains, the Indus River in Pakistan flooded about 20% of the entire country, accounting for at least 2,000 deaths. The next year saw extensive flooding in Brisbane, Australia, caused by heavy rain from tropical cyclone Tasha. At least 90 towns in the region and more than 200,000 people were affected, with damage estimated at US$1.8 billion. In the same year (2011), flooding along the Mississippi River was one of the largest and most damaging, rivaling the Great Mississippi River flood of 1927, which is considered the greatest flood in its modern history.

Every year floods cause extensive property damage and fatalities, and yet we derive many benefits from running water, even from some floods. Before the completion of the Aswan High Dam in 1970, Egyptian farmers depended on silt deposited on the Nile River floodplain to replenish their croplands (see the Chapter Opening photo). Indeed, in ancient Egypt, taxes were levied depending on the level of the Nile River.

In addition, running water is one source of freshwater for agriculture, industry, domestic use, and recreation, and about 6–7% of all electricity used in North America is generated by hydroelectric plants. When Europeans first explored North America, they followed the St. Lawrence, Mississippi, Missouri, and Ohio Rivers. Indeed, the Lewis and Clark Expedition (1804–1806), the first to cross the continent, followed the Missouri River from Missouri to its headwaters in Montana, where the explorers crossed the mountains by horse and then followed the Columbia River to the Pacific Ocean. These and similar waterways throughout the world are also major avenues of commerce.

# WATER ON EARTH

**LO1**  Explain how the hydrologic cycle works
**LO2**  Define fluid flow

Most of Earth's 1.34 billion $km^3$ of water is in the oceans (about 97%), and nearly all of the rest (approximately 2%) is frozen in glaciers on land. The remaining 1–2% of Earth's water is in the atmosphere, groundwater, lakes, swamps, and bogs, with only a very small, but important amount, in stream and river channels. Nevertheless, the water in channels is, with few exceptions, the most important geologic agent in modifying the land surface.

Much of our discussion of running water is descriptive, but always be aware that streams and rivers are dynamic systems that must continuously respond to change, be it natural or otherwise. For example, paving in urban areas increases surface runoff to waterways, and other human activities such as building dams and impounding reservoirs, also alter the dynamics of stream and river systems. Natural changes, too, affect the complex interacting parts of these systems.

## The Hydrologic Cycle

The connection between precipitation and clouds is obvious, but where does the moisture for rain and snow come from in the first place? You might immediately suspect that the oceans are the ultimate source of precipitation. In fact, water is continuously recycled from the oceans, through the atmosphere, to the continents, and back to the oceans. This **hydrologic cycle**, as it is called (●FIGURE 11.2), is powered by solar radiation and is possible because water changes easily from liquid to gas (water vapor) under surface conditions. About 85% of all water entering the atmosphere evaporates from the oceans. The remaining 15% comes from water on land, but this water originally came from the oceans as well.

Regardless of its source, water vapor rises into the atmosphere where the complex processes of cloud formation and condensation take place. About 80% of all precipitation falls directly back into the oceans, in which case, the hydrologic cycle is a three-step process of evaporation, condensation, and precipitation. For the 20% of precipitation that falls on land, the hydrologic cycle is more complex, involving evaporation, condensation, movement of water vapor from the oceans to land, precipitation, and runoff. Some precipitation evaporates as it falls and

● **FIGURE 11.1 Aftermath of the Johnstown, Pennsylvania, Flood** On May 31, 1889, an 18-m-high wall of water destroyed Johnstown and killed at least 2,200 people.

National Park Service

**hydrologic cycle**  The continuous recycling of water from the oceans, through the atmosphere, to the continents, and back to the oceans, or from the oceans, through the atmosphere, and back to the oceans.

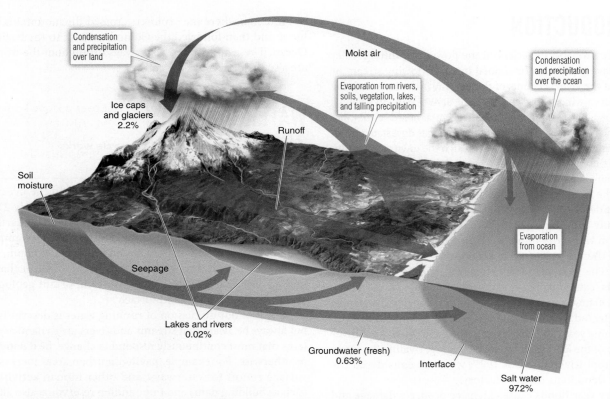

● **FIGURE 11.2 The Hydrologic Cycle** During the hydrologic cycle, water evaporates from the oceans and rises as water vapor to form clouds that release their precipitation over the oceans or land. Some of the precipitation falling on land enters the groundwater system, but much of it returns to the oceans by surface runoff, thus completing the cycle.

reenters the cycle, but about 36,000 km³ of the precipitation falls on land and returns to the oceans by **runoff**, the surface flow in streams and rivers.

Some precipitation is temporarily stored in lakes and swamps, snowfields and glaciers, or seeps below the surface where it enters the groundwater system. Water might remain in these reservoirs for thousands of years, but eventually glaciers melt, lakes and groundwater feed streams and rivers, and this water returns to the oceans. Even the water used by plants evaporates, a process known as *evapotranspiration*, and returns to the atmosphere.

## Fluid Flow

Solids are rigid substances that retain their shapes unless deformed by a force, but fluids—that is, liquids and gases—have no strength, so they flow in response to any force, no matter how slight. Liquid water therefore flows downslope in response to gravity.

Runoff during a rainstorm depends on **infiltration capacity**, the maximum rate at which surface materials absorb water. Several factors control infiltration capacity, including intensity and duration of rainfall. If rain is absorbed as fast as it falls, no surface runoff takes place. Loosely packed dry soil absorbs water faster than tightly packed wet soil, and thus more rain must fall on loose dry soil before runoff begins. Regardless of the initial condition of surface materials, once they are saturated, excess water collects on the surface and, if on a slope, it moves downhill.

# RUNNING WATER

**L03** Discuss the role of running water in modifying Earth's land surface

**L04** Explain the difference between sheet erosion and channel flow

**L05** Define gradient, velocity, and discharge

**L06** Explain how discharge is determined

The term *running water* applies to any surface water that moves from higher to lower areas in response to gravity. We have already noted that running water is very effective in modifying Earth's land surface by erosion and that it is the primary geologic process responsible for sediment transport and deposition in many areas. In fact, it is responsible for everything from the tiniest rills in farmer's fields to such scenic wonders as the Grand Canyon in Arizona (see Figure 1.2), as well as vast deposits such as the Mississippi River delta.

## Sheet Flow and Channel Flow

Even on steep slopes, flow is initially slow and hence causes little or no erosion. As water moves downslope, though, it accelerates and may move by *sheet flow*, a more or less

**runoff** The surface flow in streams and rivers.

**infiltration capacity** The maximum rate at which soil or sediment absorbs water.

continuous film of water flowing over the surface. Sheet flow is not confined to depressions, and it accounts for *sheet erosion*, a particular problem on some agricultural lands.

In *channel flow*, surface runoff is confined to troughlike depressions that vary in size from tiny rills with a trickling stream of water to huge river channels, such as the Mississippi and Amazon rivers. We describe flow in channels with terms such as *rill*, *brook*, *creek*, *stream*, and *river*, most of which are distinguished by size and volume. Here we use the terms *stream* and *river* more or less interchangeably, although the latter usually refers to a larger body of running water.

Streams and rivers receive water from several sources, including sheet flow and rain that falls directly into their channels. Far more important, however, is the water supplied by soil moisture and groundwater, both of which flow downslope and discharge into waterways. In areas where groundwater is plentiful, streams and rivers maintain a fairly stable flow year-round, because their water supply is continuous. In contrast, the amount of water in streams and rivers of arid and semiarid regions fluctuates widely, because they depend more on infrequent rainstorms and surface runoff for their water.

## Gradient, Velocity, and Discharge

Water in any channel flows downhill over a slope known as its **gradient**. Suppose a river has its headwaters (source) 1,000 m above sea level, and it flows 500 km to the sea, so it drops vertically 1,000 m over a horizontal distance of 500 km. Its gradient is found by dividing the vertical drop by the horizontal distance, which in this example is 1,000 m/500 km = 2 m/km on average (●FIGURE 11.3A).

In this example, we calculated the average gradient for a hypothetical river, but gradients vary, not only among

---

**gradient** The slope over which a stream or river flows; expressed in meters per kilometer (m/km) or feet per mile (ft/mi).

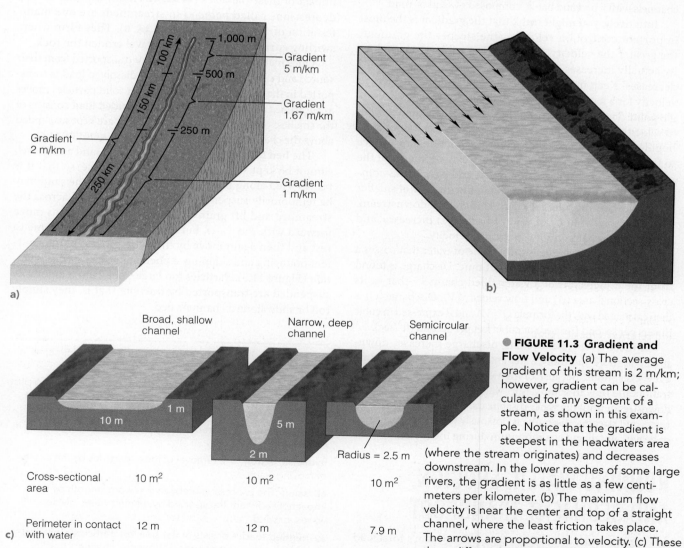

● FIGURE 11.3 Gradient and Flow Velocity (a) The average gradient of this stream is 2 m/km; however, gradient can be calculated for any segment of a stream, as shown in this example. Notice that the gradient is steepest in the headwaters area (where the stream originates) and decreases downstream. In the lower reaches of some large rivers, the gradient is as little as a few centimeters per kilometer. (b) The maximum flow velocity is near the center and top of a straight channel, where the least friction takes place. The arrows are proportional to velocity. (c) These three differently shaped channels have the same cross-sectional area; however, the semicircular one has less water in contact with its perimeter and thus less frictional resistance to flow.

channels but even along the course of a single channel. Rivers and streams are steeper in their upper reaches (near their headwaters) where they may have gradients of several tens of meters per kilometer, but they have gradients of only a few centimeters per kilometer where they discharge into the sea.

The **velocity** of running water is a measure of the downstream distance water travels in a given time. It is usually expressed in meters per second (m/sec) or feet per second (ft/sec), and it varies across a channel's width as well as along its length. Water moves more slowly, and with greater turbulence, near a channel's bed and banks, because friction is greater there than it is some distance from these boundaries (●FIGURE 11.3B). Channel shape and roughness also influence flow velocity. Broad, shallow channels and narrow, deep channels have proportionately more water in contact with their perimeters than do channels with semicircular cross sections (●FIGURE 11.3C). As one would expect, rough channels, such as those strewn with boulders, offer more frictional resistance to flow than do channels with a bed and banks composed of sand or mud.

Intuitively, you might think that the gradient is the most important control on velocity—the steeper the gradient, the greater the velocity. In fact, a channel's average velocity actually increases downstream even though its gradient decreases! Keep in mind that we are talking about average velocity for a long segment of a channel, not velocity at a single point. Two factors account for this downstream increase in velocity. First, the upstream reaches of channels tend to be boulder-strewn, broad, and shallow, so frictional resistance to flow is high there, whereas downstream segments of the same channels are more semicircular and have banks composed of finer materials. And second, the number of smaller tributaries joining a larger channel increases downstream. Thus, the total volume of water (discharge) increases, and increasing discharge results in greater velocity.

Specifically, **discharge** is the volume of water that passes a particular point in a given period of time. Discharge is found from the dimensions of a water-filled channel—that is, its cross-sectional area ($A$) and flow velocity ($V$). Discharge ($Q$) is then calculated with the formula $Q = VA$ and is expressed in cubic meters per second (m³/sec) or cubic feet per second (ft³/sec).

In most rivers and streams, discharge increases downstream as more and more water enters a channel, but there are a few exceptions. Because of high evaporation rates and infiltration, the flow in some desert waterways actually decreases downstream until the water disappears. And even in perennial rivers and streams, discharge is obviously highest during times of heavy rainfall and at a minimum during the dry season.

# RUNNING WATER, EROSION, AND SEDIMENT TRANSPORT

**LO7**  Define each of the three types of a stream's total load

Streams and rivers possess two kinds of energy: potential and kinetic. *Potential energy* is the energy of position, such as the energy of water at high elevation. During stream flow,

potential energy is converted to *kinetic energy*, the energy of motion. Much of this kinetic energy is used up in fluid turbulence, but some is available for erosion and transport. The materials transported by a stream include a *dissolved load*, and a load of solid particles (mud, sand, and gravel).

Because the **dissolved load** of a stream is invisible, it is commonly overlooked, but it is an important part of the total sediment load. Some of it is acquired from a stream's bed and banks, where soluble rocks such as limestone are present, but much of it is carried into waterways by sheet flow and by groundwater. A stream's solid load is made up of particles ranging from clay sized (<1/256 mm) to huge boulders, much of it supplied by mass wasting, but some is eroded directly from a stream's bed and banks. The direct impact of running water, **hydraulic action**, is sufficient to set particles in motion.

Running water carrying sand and gravel erodes by **abrasion**, as exposed rock is worn and scraped by the impact of these particles (●FIGURE 11.4A). Circular to oval depressions called *potholes* in streambeds are one manifestation of abrasion (●FIGURE 11.4A, B). They form where swirling currents with sand and gravel eroded the rock.

Once materials are eroded, they are transported from their source and eventually deposited. The dissolved load is transported in the water itself, but the load of solid particles moves as *suspended load* or *bed load*. The **suspended load** consists of the smallest particles of silt and clay, which are kept suspended above the channel's bed by fluid turbulence (●FIGURE 11.5).

The **bed load** of larger particles, mostly sand and gravel, cannot be kept suspended by fluid turbulence, so that it is transported along the bed. However, some of the sand may be temporarily suspended by currents that swirl across the streambed and lift grains into the water. The grains move forward with the water, but also settle and finally come to rest and then again move by the same process of intermittent bouncing and skipping, a phenomenon known as *saltation* (Figure 11.5). Particles too large to be even temporarily suspended are transported by *traction*; that is, they simply roll or slide along a channel's bed.

---

**velocity**  A measure of distance traveled per unit of time, as in the flow velocity in a stream or river.

**discharge**  The volume of water in a stream or river moving past a specific point in a given interval of time; expressed in cubic meters per second (m³/sec) or cubic feet per second (ft³/sec).

**dissolved load**  The part of a stream's load consisting of ions in solution.

**hydraulic action**  The removal of loose particles by the power of moving water.

**abrasion**  The process whereby rock is worn smooth by the impact of sediment transported by running water, glaciers, waves, or wind.

**suspended load**  Consists of the smallest particles of silt and clay, which are kept suspended above the channel's bed by fluid turbulence.

**bed load**  The part of a stream's sediment load, mostly sand and gravel, transported along its bed.

a)

b)

James S. Monroe

James S. Monroe

● **FIGURE 11.4 Abrasion by Running Water Carrying Sand and Gravel** (a) Potholes in the bed of the McCloud River in California. Notice also that the rock surface has been smoothed and polished by abrasion. (b) View into one of the potholes showing the sand and gravel that swirled around to erode the pothole.

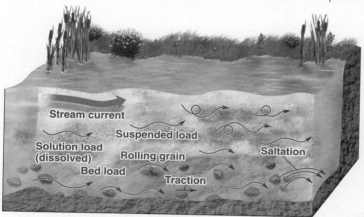

Stream current
Suspended load
Solution load (dissolved)     Rolling grain     Saltation
Bed load     Traction

● **FIGURE 11.5 Sediment Transport by Running Water** Sediment transport as bed load, suspended load, and dissolved load. Flow velocity is highest near the surface, but gravel- and sand-sized particles are too large to be lifted far from the streambed, thus they make up the bed load, whereas silt and clay are in the suspended load.

# DEPOSITION BY RUNNING WATER

**LO8**  Explain the difference between a braided stream and a meandering stream

**LO9**  Explain how point bars and oxbow lakes form

**LO10**  Explain how floodplains form

**LO11**  Define each of the three types of marine deltas

**LO12**  Explain the difference between a delta and an alluvial fan

Rivers and streams constantly erode, transport, and deposit sediment, but they do most of their geologic work when they flood. Consequently, their deposits, collectively called **alluvium**, do not represent the day-to-day activities of running water, but rather the periodic sedimentation that takes place during floods. Recall from Chapter 6 that sediments accumulate in continental, transitional, and marine *depositional environments* (see Figure 6.14). Deposits of rivers and streams are found mostly in the first two of these settings; however, much of the detrital sediment found on continental margins is derived from the land and transported to the oceans by running water.

## The Deposits of Braided and Meandering Streams

A few rivers and streams have single, straight channels for some distance, but most waterways are either braided or meandering. **Braided streams** have an intricate network of dividing and rejoining channels separated from one another by sand and gravel bars (●**FIGURE 11.6**). Braided channels develop when the sediment supply exceeds the transport capacity of running water, resulting in the deposition of sand and gravel bars. Braided streams have broad, shallow channels and are characterized as bed-load transport streams, because they transport and deposit mostly sand and gravel.

In contrast to braided streams, **meandering streams** have a single sinuous channel with broadly looping curves known as *meanders* (●**FIGURE 11.7**). Channels of

**MINDTAP**
From Cengage

Animation: Development of Floodplains, Levees, Terraces, and their Features

**alluvium** A collective term for all detrital sediment transported and deposited by running water.

**braided stream** A stream with multiple dividing and rejoining channels.

**meandering stream** A stream that has a single, sinuous channel with broadly looping curves.

● **FIGURE 11.6 A Braided Stream and Its Deposits** This braided stream is in Denali National Park in Alaska. Its deposits are mostly gravel and sand.

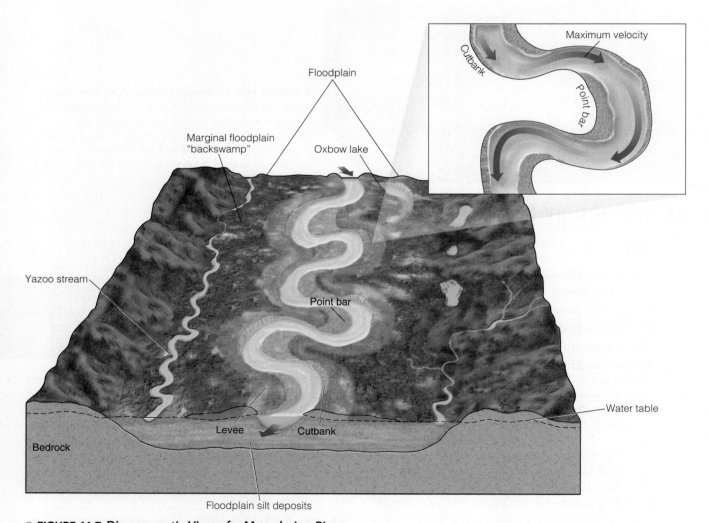

● **FIGURE 11.7 Diagrammatic View of a Meandering River**

**Critical Thinking Question** Given what you know about the dynamics of running water in channels, explain why it is not a good idea to build a house along the cut bank of a river?

● **FIGURE 11.8 Point Bars** Two point bars of sand in Otter Creek in Yellowstone National Park in Wyoming. Point bars form on the gently sloping side of meanders, where flow velocity is lowest. Note that the point bars are inclined into the deeper part of the channel.

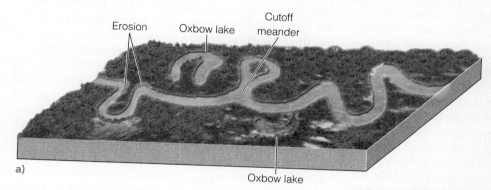

a)

● **FIGURE 11.9 Oxbow Lakes** (a) A meandering stream showing the stages in the evolution of oxbow lakes. A future oxbow will form when the meander on the left side of the illustration is cut off. (b) Oxbow lakes along the Mississippi River in Minnesota.

meandering streams are semicircular in cross section along straight reaches, but markedly asymmetric at meanders, where they vary from shallow to deep across the meander. The deeper side of the channel is known as the *cut bank*, because greater velocity and fluid turbulence erode it. On the opposite side of the channel, though, the water is shallow, flow velocity and fluid turbulence are less, and deposition of a sand body, known as a **point bar**, takes place (●**FIGURE 11.8**).

The broadly looping curves of meandering streams commonly become so sinuous that during a flood, the thin neck of land between adjacent ones is cut off, thereby forming a crescent-shaped **oxbow lake**, which is simply a cutoff meander (**FIGURES 11.7** and ●**11.9**). Oxbow lakes may persist for a long time, but they eventually fill in with organic matter and fine-grained sediments carried by floods.

## Floodplain Deposits

Streams and rivers periodically receive more water than their channels can handle, so they overflow their banks and

---

**point bar** The sediment body deposited on the gently sloping side of a meander loop.

**oxbow lake** A cutoff meander filled with water.

b)

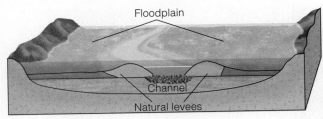

a)

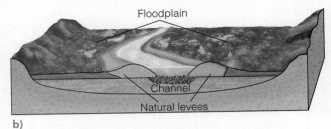

b)

● **FIGURE 11.10 Floodplain Deposits** (a) The origin of vertical floodplain deposits. During floods, streams deposit natural levees, and silt and mud settle from suspension on the floodplain. (b) After flooding.

spread across adjacent low-lying, fairly flat **floodplains** (●**FIGURE 11.10**). Sand and gravel might be deposited on floodplains where a stream or river bursts from it banks, but much more commonly the deposits are silt and clay, or simply mud. During a flood, a stream or river overtops its banks and water pours onto the floodplain, but as it does so, its depth and velocity rapidly decrease. As a result, ridges of sandy alluvium known as **natural levees** are deposited along the channel margins, and mud is carried beyond the natural levees into the floodplain, where it settles from suspension.

Another feature found on floodplains is oxbow lakes; recall that oxbow lakes are cut-off meanders (Figure 11.9). Once isolated from the main channel, oxbow lakes receive water mostly from periodic floods, although groundwater may also contribute. When they fill in with sediments

brought in by floods and accumulating organic matter they are called *meander scars*, some of which are visible in the upper part of the image in Figure 11.9b.

## Deltas

Where a river or stream flows into a lake or the ocean, its flow velocity rapidly diminishes, and any sediment in transport is deposited. Under some circumstances, this deposition creates a **delta**, an alluvial deposit that causes the shoreline to build outward into the lake or sea; this process is called *progradation*. The simplest prograding deltas have a characteristic vertical sequence of *bottomset beds* overlain successively by foreset beds and *topset beds* (●**FIGURE 11.11**). This vertical sequence develops when a river or stream enters another body of water where the finest sediment (silt and clay) is carried some distance out into the lake or sea, where it settles to form bottomset beds. Nearer the shore, foreset beds are deposited as gently inclined layers, and topset beds, consisting of the coarsest sediments, are deposited in a network of *distributary channels* traversing the top of the delta (Figure 11.11).

Small deltas in lakes may have the three-part sequence just described, but deltas deposited along seacoasts are much larger, far more complex, and considerably more important as potential areas of natural resources. Depending on the relative importance of running water, waves, and tides, geologists identify three main types of marine deltas (●**FIGURE 11.12**). *Stream-dominated deltas* have long fingerlike sand bodies, each deposited in a distributary

---

**floodplain** A low-lying, flat area adjacent to a channel that is partly or completely water-covered when a stream or river overflows its banks.

**natural levee** A ridge of sandy alluvium deposited along the margins of a channel during floods.

**delta** An alluvial deposit formed where a stream or river flows into the sea or a lake.

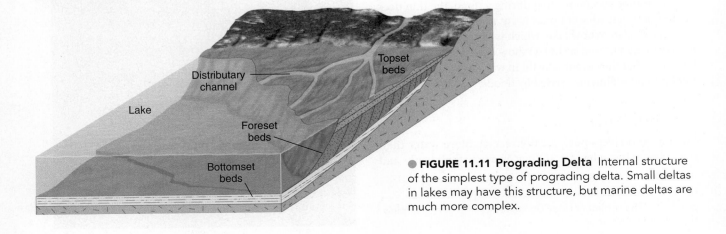

● **FIGURE 11.11 Prograding Delta** Internal structure of the simplest type of prograding delta. Small deltas in lakes may have this structure, but marine deltas are much more complex.

a)

Image courtesy of the Image Science & Analysis Laboratory, NASA Johnson Space Center (http://eol.jsc.nasa.gov)

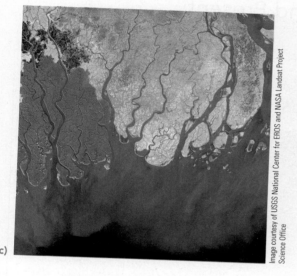

c)

Image courtesy of USGS National Center for EROS and NASA Landsat Project Science Office

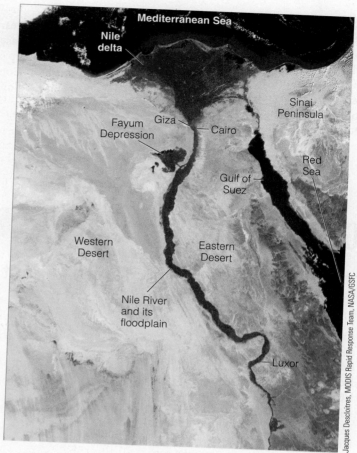

b)

Jacques Descloitres, MODIS Rapid Response Team, NASA/GSFC

● **FIGURE 11.12 Stream-, Wave-, and Tide-Dominated Deltas** (a) The Mississippi River delta on the United States Gulf Coast is stream-dominated. (b) The Nile River delta of Egypt is wave-dominated. (c) The Ganges-Brahmaputra delta of Bangladesh is tide-dominated.

channel that progrades far seaward (see GEO-FOCUS). *Wave-dominated deltas* also have distributary channels, but the seaward margin of the delta consists of islands reworked by waves, and the entire margin of the delta progrades. *Tide-dominated deltas* are continuously modified into tidal sand bodies that parallel the direction of tidal flow.

## Alluvial Fans

Fan-shaped deposits of alluvium on land, known as **alluvial fans,** form best on lowlands with adjacent highlands in arid and semiarid regions where little vegetation exists to stabilize surface materials (●**FIGURE 11.13**). During periodic

---

**alluvial fan** A cone-shaped accumulation of mostly sand and gravel deposited where a stream flows from a mountain valley onto an adjacent lowland.

Dr. Marli Miller/Visuals Unlimited/Getty Images

● **FIGURE 11.13 Alluvial Fans and Their Deposits** This is the Badwater Fan in Death Valley National Park, California. It was deposited where streams and debris flows discharged from the mountain canyon visible in this image. Notice the road for scale.

rainstorms, surface materials are quickly saturated, and surface runoff is funneled into a mountain canyon leading to adjacent lowlands. In the mountain canyon, the runoff is confined so that it cannot spread laterally, but when it discharges onto the lowlands, it quickly spreads out, its velocity diminishes, and deposition ensues. Repeated episodes of sedimentation result in the accumulation of a fan-shaped body of alluvium.

Deposition by running water in the manner just described is responsible for many alluvial fans. In some cases, though, the water flowing through a canyon picks up so much sediment that it becomes a viscous debris flow. Consequently, some alluvial fans consist mostly of debris-flow deposits that show little or no layering.

# GEO-FOCUS

## The Mississippi River Delta—Past and Present

As you now know, deltas form where a stream or river enters a standing body of water and deposits alluvium. However, a developing delta may be dominated by waves, tides, or stream activity (see Figure 11.13). The Mississippi River delta is a classic stream-dominated delta that lies on a shallow shelf where it is mostly protected from waves and tides. On the delta, distributary channels prograde so far seaward that they become inefficient avenues of sediment transport, so that the stream abandons that channel and establishes a new one elsewhere (●FIGURE 1). In fact, as a result of sedimentation, the entire coastline of Louisiana has built seaward as much as 80 km during the last 5,000 years.

In addition to abandoning distributary channels, stream-dominated deltas periodically abandon entire deltaic lobes. The Mississippi River is currently depositing sediment on a single lobe, but the entire delta is made up of several lobes that formed during the past few thousands of years (●FIGURE 2). In point of fact, were it not for the efforts of the Army Corps of Engineers, the currently active Balize lobe may have already been abandoned.

During the devastating flooding that took place on the lower Mississippi River during April and May 2011, the Corp of

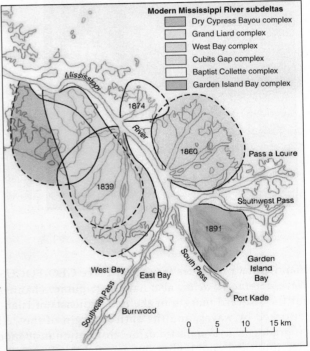

● **FIGURE 1** The present lobe of the Mississippi River delta consists of several subdeltas that formed when distributary channels were abandoned. The area shown here corresponds to the Balize lobe in Figure 2.

Engineers deemed it necessary to divert much of the Mississippi River into the Atchafalaya River at the Morganza Spillway, so as to protect levees and prevent major flooding in Baton Rouge and New Orleans. If the Mississippi were to abandon its present course, it would follow the Atchafalaya River and discharge into the Gulf of Mexico about 100 km west of where it does now. Although such an

event would be economically catastrophic for New Orleans, the river would simply continue to build its delta, but in an area that it occupied long ago (Figure 2).

Stream-dominated deltas grow seaward by progradation, but in the case of the Mississippi River, flood-control projects now trap much of the sediment so that it no longer reaches the delta. As a result, the delta is losing

# CAN FLOODS BE CONTROLLED AND PREDICTED?

**LO13** Discuss some of the practices that can be implemented to help minimize the effects of flooding

**LO14** Discuss the problems associated with predicting floods

When any waterway receives more water than its channel can handle, it floods, occupying part or all of its floodplain. Indeed, floods are so common that unless they cause extensive property damage or fatalities, they rate little more than a passing notice in the news. Dozens of floods occur in the United States every year, but the most extensive river flooding in recent history was The Great Flood of 1993, and the 2011 flooding on the lower Mississippi River (●**FIGURE 11.14**).

Flooding is a fact of life, but there are several practices that can protect people and their property and minimize the impact of floods. Unfortunately, none of these measures is

land along its margin due to wave erosion, and, more importantly, to compaction. Delta deposits, especially mud, compact under their own weight, and thus the surface subsides,* and the sea invades the area where subsidence takes place. Geologists estimate that during the past 4,400 years the region has subsided about 12 cm per century.

Delta subsidence is a major problem along the coastline of the Gulf of Mexico. For example, a large part of New Orleans lies below sea level because of subsidence. Much of the city is located between Lake Pontchartrain and the Mississippi River, so when Hurricane Katrina roared ashore in 2005, nearly 80% of New Orleans was flooded and 1,833 people died. Damages reached $125 billion (tying it with Hurricane Harvey in 2017 as the two most costliest hurricanes in history), mostly in Louisiana and Mississippi.

---

*Pumping oil and natural gas from the delta sediments also contributes to subsidence.

●**FIGURE 2** **The Major Deltaic Lobes of the Mississippi River** Deposition of the present-day Balize lobe began about 5,500 years ago.

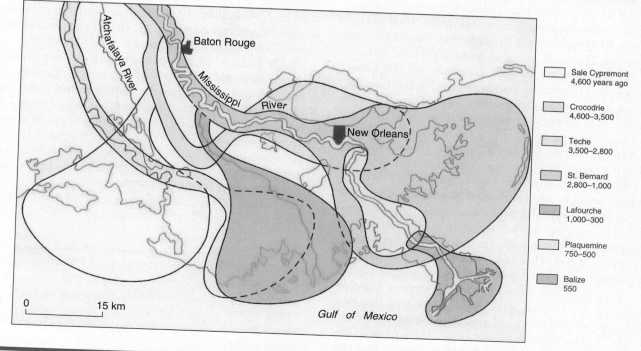

Sale Cypremont
4,600 years ago

Crocodrie
4,600–3,500

Teche
3,500–2,800

St. Bernard
2,800–1,000

Lafourche
1,000–300

Plaquemine
750–500

Balize
550

foolproof and most are expensive. One is to simply move all or parts of small communities to higher ground. Accordingly, following The Great Flood of 1993, parts of some small communities in Missouri and Illinois were moved to higher ground. Martin, Kentucky, for example, has been flooded 37 times since 1862, and parts of the downtown area are being moved.

Another common flood control practice is to construct *dams* that impound excess water during floods (●**FIGURE 11.15A**). Of course dams are expensive, they eventually fill in with sediment unless dredged, and some fail. For example, in 1976, the Teton Dam in Idaho collapsed, killing 11 people, 13,000 cattle, and causing about $2 billion in damage.

Another practice is to construct *levees* that raise the banks of streams and rivers, thereby increasing a channel's capacity (●**FIGURE 11.15B**). Unfortunately, deposition within channels raises streambeds, making the levees less effective unless they are also raised. Levees along the Huang He in China caused the streambed to rise more than 20 m above its surrounding floodplain in 4,000 years. In 1887, when the river breached its levees, more than one million people were killed.

In some areas, state or federal agencies build *floodways*, which are channels used to divert excess water around a community or an area of economic importance (●**FIGURE 11.15C**). Yet another practice is to build *floodwalls* to protect vulnerable areas. These are vertical structures placed along stream or river banks where levees are not practical, such as in cities (●**FIGURE 11.15D**). Reforestation of cleared land also reduces the potential for flooding because vegetated soil absorbs water and decreases runoff.

Flooding streams and rivers accounted for about $8.4 billion in property damage in 2011 in the United States and that figure does not include coastal flooding from hurricanes. And even though more and more flood control projects are completed, the damages from flooding are not decreasing. The combination of fertile soils, level surfaces, and proximity to water for agriculture, industry, and domestic use makes floodplains popular places for development. Unfortunately, urbanization increases surface runoff because soils are compacted or covered by asphalt or concrete, reducing infiltration capacity.

As for predicting floods, the best that can be done is to monitor streams, evaluate their past behavior, and anticipate floods of a given size in a specified period. Most people have heard of 10-year floods, 20-year floods, and so on, but how are such determinations made? The U.S. Geologic Survey, as well as state agencies, record and analyze stream behavior through time in order to anticipate floods of a specified size. Thus, a 20-year flood, for example, is the period during which a flood of a given magnitude can be expected. It does not mean that the river in question will have a flood of that size every 20 years, only that over a long period of time, it will average 20 years. Or we can say that the chances of a 10-year flood taking place in any one year are 1 in 10 (1/10). In fact, it is possible that two

10-year floods could take place in successive years, but then not occur again for several decades.

# DRAINAGE SYSTEMS

**LO15**  Discuss what each of the five drainage patterns reveals about the underlying geology

Thousands of waterways that are parts of larger drainage systems flow directly or indirectly into the oceans. The only exceptions are some rivers and streams that flow into desert basins surrounded by higher areas. But even these are parts of larger systems consisting of a main channel with all its tributaries—that is, streams that contribute water to another stream. The Mississippi River and its tributaries, such as the Ohio, Missouri, Arkansas, and Red rivers and thousands of smaller ones, carry runoff from an area known as a **drainage basin**. A topographically high area called a **divide** separates adjacent drainage basins (●**FIGURE 11.16**). The continental divide along the crest of the Rocky Mountains in North America, for instance, separates drainage in opposite directions; drainage to the west goes to the Pacific, whereas drainage to the east eventually reaches the Gulf of Mexico.

The arrangements of channels within an area are types of **drainage patterns**. The most common is *dendritic drainage*, which consists of a network of channels resembling tree branching (●**FIGURE 11.17A**). It develops on gently sloping surfaces composed of materials that respond more or less homogeneously to erosion, such as areas underlain by nearly horizontal sedimentary rocks.

In dendritic drainage, tributaries join larger channels at various angles, but *rectangular drainage* is characterized by right-angle bends and tributaries joining larger channels at right angles (●**FIGURE 11.17B**). Such regularity in channels is strongly controlled by geologic structures, particularly regional joint systems that intersect at right angles.

*Trellis drainage*, consisting of a network of nearly parallel main streams with tributaries joining them at right angles, is common in some parts of the eastern United States. In Virginia and Pennsylvania, erosion of folded sedimentary rocks developed a landscape of alternating ridges on resistant rocks and valleys underlain by easily eroded rocks. Main waterways follow the valleys, and short tributaries flowing from the nearby ridges join the main channels at nearly right angles (●**FIGURE 11.17C**).

In *radial drainage*, streams flow outward in all directions from a central high point, such as a large volcano

---

**drainage basin**  The surface area drained by a stream or river and its tributaries.

**divide**  A topographically high area that separates adjacent drainage basins.

**drainage pattern**  The regional arrangement of channels in a drainage system.

a)

b)

● **FIGURE 11.14 The Flood of '93** (a) This view shows flooding on July 30, 1993, in the vicinity of the Jefferson City Memorial Airport just north of Jefferson City, Missouri. Iowa and Missouri were particularly hard hit by this flood, but large areas in nine states were also flooded. (b) Flood waters in Portage des Sioux covered the 5.5-m high pedestal of this statue on the bank of the Mississippi River.

a)

b)

c)

d)

● **FIGURE 11.15 Flood Control** Dams and reservoirs, levees, floodways, and floodwalls are some of the structures used to control floods. (a) Oroville Dam in California, at 235 m high, is the highest dam in the United States. It helps control floods, provides water for irrigation, and produces electricity at its power plant. (b) This levee, an artificial embankment along a waterway, helps protect nearby areas from floods. A university campus lies out of view just to the right of the levee. (c) This floodway carries excess water from a river (not visible) around a small community. (d) This floodwall on the bank of the Danube River at Mohács, Hungary, helps protect the city from floods.

**Critical Thinking Question** What are some of the functions of dam and its reservoir?

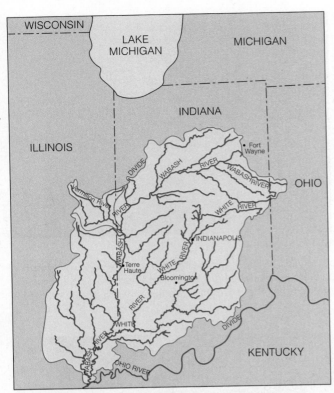

● **FIGURE 11.16 Drainage Basins** A detailed view of the Wabash River's drainage basin, a tributary of the Ohio River. All tributary streams within the drainage basin, such as the Vermilion River, have their own smaller drainage basins. Divides are shown by red lines.

(●**FIGURE 11.17D**). Many of the volcanoes in the Cascade Range of western North America have radial drainage patterns.

In the types of drainage mentioned so far, some kind of pattern is easily recognized. *Deranged drainage*, in contrast, is characterized by irregularity, with streams flowing into and out of swamps and lakes, streams with only a few short tributaries, and vast swampy areas between channels (●**FIGURE 11.17E**). This kind of drainage developed recently and has not yet formed a fully organized drainage system. In parts of Minnesota, Wisconsin, and Michigan, where glaciers obliterated the previous drainage, only 10,000 years have elapsed since the glaciers melted. As a result, drainage systems have not yet fully developed, and large areas remain undrained.

# THE SIGNIFICANCE OF BASE LEVEL

LO16    Define base level

LO17    Define a graded stream

**Base level** is the lowest limit to which a stream or river can erode. With the exception of streams that flow into closed depressions in deserts, all others are restricted ultimately to sea level. That is, they can erode no lower than sea level, because they must have some gradient to maintain flow. So *ultimate*

**base level** The level below which a stream or river cannot erode; sea level is ultimate base level.

● **FIGURE 11.17 Drainage Patterns** (a) Dendritic drainage. (b) Rectangular drainage. (c) Trellis drainage. (d) Radial drainage. (e) Deranged drainage.

Ridges of resistant rock

*base level* is sea level, which is simply the lowest level of erosion for any waterway that flows into the sea (●**FIGURE 11.18**). Ultimate base level applies to an entire stream or river system, but channels may also have a *local* or *temporary base level*. For example, a local base level may be a lake or another stream, or where a stream or river flows across particularly resistant rocks and a waterfall develops (Figure 11.18A).

Ultimate base level is sea level, but suppose that sea level dropped or rose with respect to the land, or presume that the land rose or subsided? In these cases, base level would change and bring about changes in stream and river systems. During the maximum extent of Pleistocene glaciers, sea level

was about 130 m lower than it is now, and streams adjusted by eroding deeper valleys and extending well out onto the continental shelves. Rising sea level at the end of the Ice Age accounted for a rising base level, decreased stream gradients, and deposition within channels.

Geologists and engineers are well aware that building a dam to impound a reservoir creates a local base level. A stream entering a reservoir deposits sediment, so unless dredged, reservoirs eventually fill with sediment. In addition, the water discharged at a dam is largely sediment free, but it still possesses energy to carry a sediment load. As a result, a stream may erode vigorously downstream from a dam to acquire a sediment load.

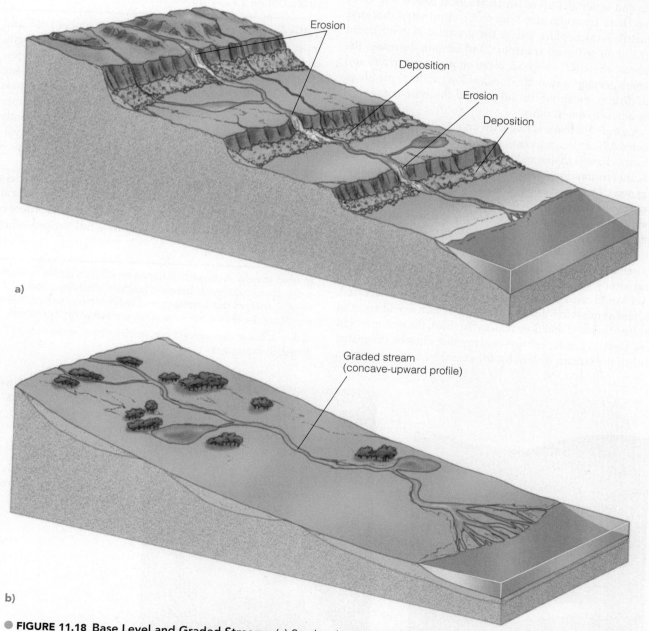

● **FIGURE 11.18 Base Level and Graded Streams** (a) Sea level is ultimate base level, but a resistant rock layer over which a waterfall plunges is a local base level. Also, this stream has several irregularities in its profile, so it is ungraded. (b) Erosion and deposition along its course eliminate irregularities and the stream becomes graded as it develops a smooth, concave profile of equilibrium.

Draining a lake may seem like a small change and well worth the time and expense to expose dry land for agriculture or commercial development. But draining a lake eliminates a local base level, and a stream that originally flowed into the lake responds by rapidly eroding a deeper valley as it adjusts to a new base level.

## What Is a Graded Stream?

The *longitudinal profile* of any waterway shows the elevations of a channel along its length as viewed in cross section (Figure 11.18B). For some rivers and streams, the longitudinal profile is smooth, but others show irregularities such as lakes and waterfalls, all of which are local base levels. Over time, these irregularities tend to be eliminated, because deposition takes place where the gradient is insufficient to maintain sediment transport, and erosion decreases the gradient where it is steep. So, given enough time, rivers and streams develop a smooth, concave longitudinal profile of equilibrium, meaning that all parts of the system dynamically adjust to one another.

A **graded stream** is one with an equilibrium profile in which a delicate balance exists among gradient, discharge, flow velocity, channel shape, and sediment load so that neither significant erosion nor deposition takes place within its channel (Figure 11.18b). Such a delicate balance is rarely attained, so the concept of a graded stream is an ideal. Nevertheless, the graded condition is closely approached in many streams, although only temporarily and not necessarily along their entire lengths.

Even though the concept of a graded stream is an ideal, we can anticipate the response of a graded stream to changes that alter its equilibrium. For instance, a change in base level would cause a stream to adjust as previously discussed. Increased rainfall in a stream's drainage basin would result in greater discharge and flow velocity. In short, the stream would now possess greater energy—energy that must be dissipated within the stream system by, for example, a change from a semicircular to a broad, shallow channel that would dissipate more energy by friction. On the other hand, the stream may respond by eroding a deeper valley, effectively reducing its gradient until it is once again graded.

## THE EVOLUTION OF VALLEYS

LO18  **Define valleys**

LO19  **Discuss the processes involved in a valley's evolution**

LO20  **Explain the formation and evolution of stream terraces**

LO21  **Explain how incised meanders form**

LO22  **Define a superposed stream**

**Valleys** are low areas on land bounded by higher land, and most of them have a river or stream running their length, with tributaries draining the nearby high areas. With few exceptions, valleys form and evolve in response to erosion by running water, although other processes, especially mass wasting, contribute. The shapes and sizes of valleys vary from small, steep-sided *gullies* to those that are broad with gently sloping valley walls (●**FIGURE 11.19**). Steep-walled, deep valleys of vast size are *canyons*, and particularly narrow and deep ones are *gorges*.

A valley might start to erode where runoff has sufficient energy to dislodge surface materials and excavate a small rill. Once formed, a rill collects more runoff and becomes deeper and wider and continues to do so until a full-fledged valley

**graded stream**  A stream that has an equilibrium profile in which a delicate balance exists between gradient, discharge, flow velocity, channel characteristics, and sediment load so that neither significant deposition nor erosion takes place within its channel.

**valley**  A linear depression bounded by higher areas such as ridges or mountains.

●**FIGURE 11.19 Gullies and Valleys** (a) Gullies are small valleys, but they are steep-sided. This gully measures about 15 m across. (b) This valley has walls that descend to a narrow valley bottom.

a)

b)

James S. Monroe

Sue Monroe

develops. *Downcutting* takes place when a river or stream has more energy than it needs to transport sediment, so some of its excess energy is used to deepen its valley. In most cases, the valley walls are simultaneously undercut, a process called *lateral erosion*, creating unstable slopes that may fail by *mass wasting*. Furthermore, erosion by sheetwash and erosion by tributary streams carry materials from the valley walls into the main stream in the valley.

Valleys not only become deeper and wider but they also become longer by *headward erosion*, a phenomenon involving erosion by entering runoff at the upstream end of a valley (●FIGURE 11.20A). Continued headward erosion may result in *stream piracy*, the breaching of a drainage divide and diversion of part of the drainage of another stream (●FIGURE 11.20B).

## Stream Terraces

Rather flat surfaces paralleling a stream or river but at a higher level than the present-day floodplain are **stream terraces**. These surfaces represent an older floodplain that was adjacent to the waterway when it flowed at a higher level but

subsequently eroded down to a lower level (●FIGURE 11.21). Some streams have several steplike surfaces above their

**stream terrace** An erosional remnant of a floodplain that formed when a stream was flowing at a higher level.

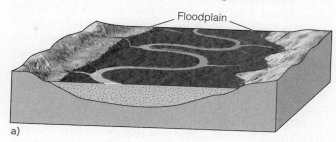
a)

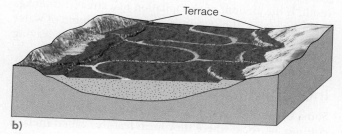
b)

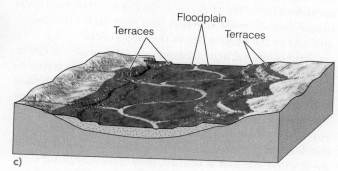
c)

d)

**FIGURE 11.20 Two Stages in the Evolution of a Valley**
(a) This stream widens its valley by lateral erosion and mass wasting while simultaneously extending its valley by headward erosion. (b) As the larger stream continues to erode headward, stream piracy takes place when it captures some of the drainage of the smaller stream. Notice also that the valley is wider than it was in (a).

● **FIGURE 11.21 Origin of Stream Terraces** (a) A stream has a broad floodplain. (b) The stream erodes downward and establishes a new floodplain at a lower level. Remnants of its old, higher floodplain are stream terraces. (c) Another level of stream terraces forms as the stream erodes downward again. (d) Stream terraces along the Madison River in Montana.

present floodplains, indicating the terraces formed more than once.

The formation of stream terraces is preceded by an episode of deposition followed by erosion as the stream or river begins eroding downward. For example, suppose that a stream is graded so that it is adjusted to its present gradient, volume of water, and sediment load. If the land over which the stream flows is uplifted or sea level becomes lower, the stream's gradient is steeper and it has more energy to deepen its valley. When this stream once again reaches a level at which it is graded, downcutting ceases, and it begins to erode laterally, thus establishing a new floodplain at a lower level. Several such episodes account for the multiple stream terraces seen in the valleys of some streams.

Although changes in base level probably account for many stream terraces, a change in climate can have the same result. If the amount of precipitation in a stream's drainage basin increases, the stream has a greater volume of water and the capacity to erode a deeper valley, perhaps leaving remnants of an older floodplain as stream terraces.

## Incised Meanders

Some streams are restricted to deep, meandering canyons cut into bedrock, where they form features called **incised meanders** (●FIGURE 11.22). Streams restricted by rock walls usually cannot erode laterally; thus, they lack a floodplain and occupy the entire width of the canyon floor.

It is not difficult to understand how a stream can cut downward into rock, but how a stream forms a meandering pattern in bedrock is another matter. Because lateral erosion is inhibited once downcutting begins, one must infer that the meandering course was established when the stream flowed across an area covered by alluvium. For example, suppose that a stream near base level has established a meandering pattern. If the land that the stream flows over is uplifted, then erosion begins and the meanders become incised into the underlying bedrock.

## Superposed Streams

Water flows downhill in response to gravity, so the direction of flow in streams and rivers is determined by topography. Yet a number of waterways seem, at first glance, to have defied this fundamental control. For instance, several rivers in the eastern United States flow in valleys that cut directly through ridges that lie in their paths. These are examples of **superposed streams**, all of which once flowed on a surface at a higher level, but as they eroded downward, they eroded into resistant rocks and cut narrow canyons, or what geologists call *water gaps* (●FIGURE 11.23).

---

**incised meander**  A deep, meandering canyon cut into bedrock by a stream or river.

**superposed stream**  A stream that once flowed on a higher surface and eroded downward into resistant rocks while maintaining its course.

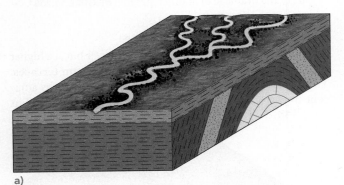

a)

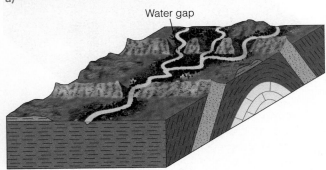

b)

Water gap

c)

●**FIGURE 11.23 Origin of a Superposed Stream** (a) As a stream erodes down and removes the surface layers of rock, it is lowered onto ridges that form when resistant rocks in the underlying structure are exposed. (b) The narrow valleys through the ridges are water gaps. (c) View of a water gap cut by the Jefferson River in Montana.

●**FIGURE 11.22 Incised Meanders** The Colorado River at Dead Horse State Park in Utah is incised to a depth of 600 m.

A water gap has a stream flowing through it, but if the stream is diverted elsewhere, perhaps by stream piracy, the abandoned gap is then called a *wind gap*. The Cumberland Gap in Kentucky is a good example; it was the avenue through which settlers migrated from Virginia to Kentucky from 1790 until well into the 1800s. Furthermore, several water gaps and wind gaps played important strategic roles during the Civil War (1861–1865).

# Key Concepts Review

- Water continuously evaporates from the oceans, rises as water vapor, condenses, and falls as precipitation. About 20% of this precipitation falls on land and eventually returns to the oceans, mostly by surface runoff.
- Runoff takes place by sheet flow, a thin, more or less continuous sheet of water, and by channel flow, which is flow that is confined to long, trough like depressions.
- The vertical drop in a given horizontal distance (gradient) for a channel varies from steep in its upper reaches to more gentle in its lower reaches.
- Flow velocity and discharge are related, so that if one changes, the other changes as well.
- Erosion by running water takes place by hydraulic action, abrasion, and solution.
- The bed load in channels is made up of sand and gravel, whereas suspended load consists of silt- and clay-sized particles. Running water also transports a dissolved load.
- Braided streams have a complex of dividing and rejoining channels, separated from one another by sand and gravel bars. Their deposits are mostly sheets of sand and gravel.
- A single sinuous channel is typical of meandering streams that deposit mostly mud, with subordinate point-bar deposits of sand or, more rarely, gravel.
- Because of the unequal distribution of flow velocity across meanders, point bars are deposited on the gently sloping inner banks, whereas cut bars erode the outer banks.

- Broad, flat floodplains adjacent to channels are the sites of oxbow lakes, which are simply abandoned meanders.
- An alluvial deposit at a river's mouth is a delta. Some deltas conform to the three-part division of bottomset, foreset, and topset beds, but large marine deltas are much more complex and are characterized as stream-, wave-, or tide-dominated.
- Alluvial fans are fan-shaped deposits of sand and gravel on land that form best in semiarid regions. They form mostly by deposition from running water, but debris flows are also important.
- Rivers and streams carry runoff from their drainage basins, which are separated from one another by divides.
- Sea level is ultimate base level, the lowest level to which streams or rivers can erode. Local base levels may be lakes or where streams or rivers flow across resistant rocks.
- Graded streams tend to eliminate irregularities in their channels, so that they develop a smooth, concave profile of equilibrium.
- A combination of processes, including downcutting, lateral erosion, sheetwash, mass wasting, and headward erosion, are responsible for the origin and evolution of valleys.
- Stream terraces and incised meanders usually form when a stream or river that was formerly in equilibrium begins a new episode of downcutting.

# Important Terms

abrasion   226
alluvial fan   231
alluvium   227
base level   236
bed load   226
braided stream   227
delta   230
discharge   226
dissolved load   226
divide   234

drainage basin   234
drainage pattern   234
floodplain   230
graded stream   238
gradient   225
hydraulic action   226
hydrologic cycle   223
incised meander   240
infiltration capacity   224
meandering stream   227

natural levee   230
oxbow lake   229
point bar   229
runoff   224
stream terrace   239
superposed stream   240
suspended load   226
valley   238
velocity   226

# Review Questions

1. Running water carrying sand and gravel effectively erodes by
   a. solution.
   b. sheetwash.
   c. abrasion.
   d. saltation.
   e. alluviation.

2. The deposit that forms on the gently sloping side of a meander where current velocity is low is called a/an
   a. point bar.
   b. natural levee.
   c. valley.
   d. oxbow lake.
   e. alluvial fan.

3. The recycling of water from the oceans to the land and back is known as the
   a. evapo-condensation cycle.
   b. profile of equilibrium.
   c. drainage cycle.
   d. hydrologic cycle.
   e. marine cycle.

4. A stream or river with multiple channels that divide and rejoin is characterized as
   a. incised.
   b. superposed.
   c. braided.
   d. sinuous.
   e. meandering.

5. The suspended load of a stream or river consists of
   a. sand and gravel.
   b. gravel and boulders.
   c. sand and silt.
   d. dissolved materials.
   e. silt and clay.

6. What is a graded stream and how does it develop?

7. The discharge of most streams and rivers increases downstream, but in a few cases, it actually decreases and eventually disappears. Explain why.

8. Calculate the daily discharge of a river 148 m wide and 2.6 m deep, with a flow velocity of 0.7 m/sec.

9. Where and how do point bars form?

# Creative Thinking Visual Question

Explain how this stream in Canada cut this notch through bedrock (●**FIGURE 1**).

● **FIGURE 1 Erosion by a small stream in Canada.**

James S. Monroe

# 12

# GROUNDWATER

Mammoth Cave, Western Australia. Shown are many common cave features, including stalactites, stalagmites, soda straws, columns, and drip curtains.

# INTRODUCTION

Within the limestone region of western Kentucky lies the largest cave system in the world. In 1941, approximately 51,000 acres were set aside and designated as Mammoth Cave National Park. In 1981, it became a World Heritage Site. From ground level, the topography of the area is unimposing, with gently rolling hills. Beneath the surface, however, are more than 540 km of interconnected passageways whose spectacular geologic features have been enjoyed by millions of cave explorers and tourists.

During the War of 1812, approximately 180 metric tons of saltpeter (potassium nitrate—$KNO_3$) used in the manufacture of gunpowder were mined from Mammoth Cave. At the end of the war, the saltpeter market collapsed, and Mammoth Cave was developed as a tourist attraction, easily overshadowing the other caves in the area. During the next 150 years, the discovery of new passageways and links to other caverns helped establish Mammoth Cave as the world's premier cave and the standard against which all others were measured.

The colorful cave deposits are the primary reason that millions of tourists have visited Mammoth Cave over the years. Hanging down from the ceiling and growing up from the floor are spectacular icicle-like structures, as well as columns and curtains in a variety of colors. Moreover, intricate passageways connect rooms of various sizes. The cave is also home to more than 200 species of insects and other animals, including about 45 blind species.

In addition to the beautiful caves, caverns, and cave deposits produced by groundwater movement, groundwater is also an important natural resource. Although groundwater (fresh) constitutes only 0.63% of the world's water (see Figure 11.2), it is nonetheless a significant source of freshwater for agricultural, industrial, and domestic use. More than 65% of the groundwater used in the United States each year goes for irrigation, with industrial use being second, followed by domestic needs. These demands have severely depleted the groundwater supply in many areas and have led to such problems as ground subsidence and saltwater contamination. In other areas, pollution from landfills, toxic waste, and agriculture has rendered the groundwater supply unsafe.

As the world's population and industrial development expand, the demand for water, particularly groundwater, will increase. Not only must new groundwater sources be located, but once found, these sources must be protected from pollution and managed properly to ensure that users do not withdraw more water than can be replenished.

# GROUNDWATER AND THE HYDROLOGIC CYCLE

**L01**  Define groundwater

**L02**  Discuss the role of groundwater in the hydrologic cycle

**Groundwater**—water that fills open spaces in rocks, sediment, and soil beneath Earth's surface—is one reservoir in the hydrologic cycle (see Figure 11.2). Like all other water in the hydrologic cycle, the ultimate source of groundwater is the oceans; however, its more immediate source is the precipitation that infiltrates the ground and seeps down through the voids in soil, sediment, and rocks. Groundwater may also come from water infiltrating from streams, lakes, swamps, artificial recharge ponds, and water-treatment systems.

Regardless of its source, groundwater moving through the tiny openings between soil and sediment particles and the spaces in rocks filters out many impurities, such as disease-causing microorganisms and many pollutants. However, not all soils and rocks are good filters, and sometimes so much undesirable material may be present that it contaminates the groundwater. Groundwater movement and its recovery from wells depend on two critical aspects of the materials that it moves through: *porosity* and *permeability*.

# POROSITY AND PERMEABILITY

**L03**  Define porosity and permeability

**L04**  Discuss the role porosity and permeability play in the movement of groundwater

**L05**  Explain the difference between an aquifer and aquiclude

Porosity and permeability are important physical properties of Earth materials and are largely responsible for the amount, availability, and movement of groundwater. Water soaks into the ground because soil, sediment, and rock have open spaces, or pores. **Porosity** is the percentage of a material's total volume that is pore space. Porosity most often consists of the spaces between particles in soil, sediment, and sedimentary rocks, but other types of porosity include cracks, fractures, faults, and vesicles in volcanic rocks (●**FIGURE 12.1**).

Porosity varies among different rock types and is dependent on the size, shape, and arrangement of the material composing the rock (●**TABLE 12.1**). Most igneous and metamorphic rocks, as well as many limestones and dolostones, have very low porosity, because they consist of tightly interlocking crystals. Their porosity can be increased, however, if they have been fractured or weathered by groundwater. This is particularly true for massive limestone and dolostone, whose fractures can be enlarged by acidic groundwater.

In contrast, detrital sedimentary rocks composed of well-sorted and well-rounded grains can have high porosity, because any two grains touch at only a single point, leaving relatively large open spaces between the grains (●**FIGURE 12.1A**). Poorly sorted sedimentary rocks, on the other hand, typically have low porosity, because smaller grains fill in the spaces between the larger grains, further reducing porosity (●**FIGURE 12.1B**). In addition, the amount of cement between grains can decrease porosity.

---

**groundwater**  Underground water stored in the pore spaces of soil, sediment, and rock.

**porosity**  The percentage of a material's total volume that is pore space.

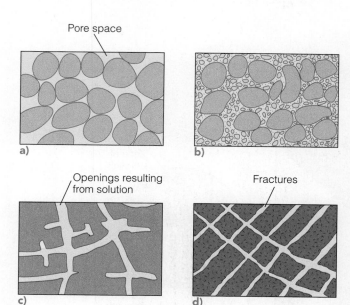

**FIGURE 12.1 Porosity** A rock's porosity depends on the size, shape, and arrangement of the material composing the rock. (a) A well-sorted sedimentary rock has high porosity. (b) A poorly sorted one has lower porosity. (c) In soluble rocks such as limestone, porosity can be increased by solution. (d) Crystalline metamorphic and igneous rocks can be rendered porous by fracturing.

**Critical Thinking Question** How can some Earth materials be porous, yet not permeable? Give an example.

**TABLE 12.1** Porosity Values for Different Materials

| Material | Percentage Porosity |
|---|---|
| UNCONSOLIDATED SEDIMENT | |
| Soil | 55 |
| Gravel | 20–40 |
| Sand | 25–50 |
| Silt | 35–50 |
| Clay | 50–70 |
| ROCKS | |
| Sandstone | 5–30 |
| Shale | 0–10 |
| Solution activity in limestone, dolostone | 10–30 |
| Fractured basalt | 5–40 |
| Fractured granite | 10 |

*Source:* U.S. Geological Survey, Water Supply Paper 2220 (1983) and others.

Porosity determines the amount of groundwater that Earth materials can hold, but it does not guarantee that the water can be easily extracted. So, in addition to being porous, Earth materials must have the capacity to transmit fluids, a property known as **permeability**. Thus, both porosity and permeability play important roles in groundwater movement and recovery.

Permeability is dependent not only on porosity but also on the size of the pores or fractures and their interconnections. For example, deposits of silt or clay are typically more porous than sand or gravel, but they have low permeability, because the pores between the particles are very small and molecular attraction between the particles and water is great, thereby preventing movement of the water. In contrast, the pore spaces between grains in sandstone and conglomerate are much larger, and molecular attraction on the water is therefore low. Chemical and biochemical sedimentary rocks, such as limestone and dolostone, and many igneous and metamorphic rocks that are highly fractured can also be very permeable provided that the fractures are interconnected.

The contrasting porosity and permeability of familiar substances are well demonstrated by sand versus clay. Pour some water on sand and it rapidly sinks in, whereas water poured on clay simply remains on the surface. Furthermore, wet sand dries quickly, but once clay absorbs water, it may take days to dry out because of its low permeability.

A permeable layer transporting groundwater is an **aquifer**, from the Latin *aqua*, "water." The most effective aquifers are deposits of well-sorted and well-rounded sand and gravel. Limestones in which fractures and bedding planes have been enlarged by solution are also good aquifers. Shales and many igneous and metamorphic rocks make poor aquifers because they are typically impermeable unless fractured. Rocks such as these and any other materials that prevent the movement of groundwater are *aquicludes*.

# THE WATER TABLE

**L06** Define the zone of aeration, the zone of saturation, and the water table

**L07** Discuss the relationship between these three entities

Some of the precipitation on land evaporates, and some enters streams and returns to the oceans by surface runoff; the remainder seeps into the ground. As this water moves down from the surface, a small amount adheres to the material it moves through and halts its downward progress. With the exception of this *suspended water*, however, the rest seeps further downward and collects until it fills all of the available pore spaces. Thus, two zones are defined by whether their

**permeability** A material's capacity to transmit fluids.

**aquifer** A permeable layer in which groundwater flows. From the Latin *aqua*, "water."

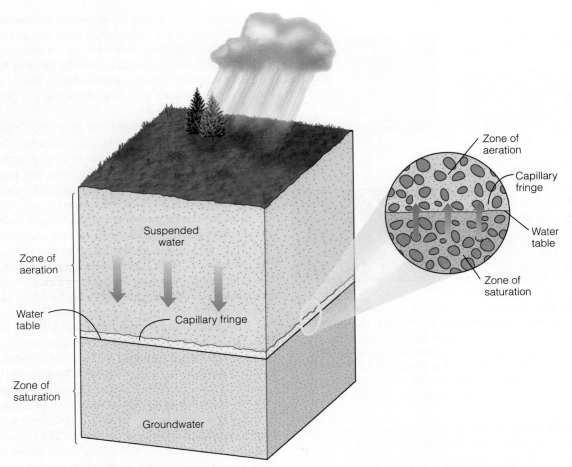

● **FIGURE 12.2  Water Table**  The zone of aeration contains both air and water within its pore spaces, whereas all pore spaces in the zone of saturation are filled with groundwater. The water table is the surface separating the zones of aeration and saturation. Within the capillary fringe, water rises by surface tension from the zone of saturation into the zone of aeration.

pore spaces contain mostly air, the **zone of aeration**, or mostly water, the underlying **zone of saturation**. The surface that separates these two zones is the **water table** (●FIGURE 12.2).

The base of the zone of saturation varies from place to place, but it usually extends to a depth where an impermeable layer is encountered or to a depth where confining pressure closes all open space. Extending irregularly upward a few centimeters to several meters from the zone of saturation is the *capillary fringe*. Water moves upward in this region because of surface tension, much as water moves upward through a paper towel.

In general, the configuration of the water table is a subdued replica of the overlying land surface; that is, it rises beneath hills and has its lowest elevations beneath valleys. Several factors contribute to the overall configuration of a region's water table, including regional differences in the amount of rainfall, permeability, and rate of groundwater movement. During periods of high rainfall, groundwater tends to rise beneath hills because it cannot flow fast enough into adjacent valleys to maintain a level surface.

During droughts, the water table falls and tends to flatten out because it is not being replenished. In arid and semiarid regions, the water table tends to be quite flat regardless of the overlying land surface.

# GROUNDWATER MOVEMENT

**LO8  Explain how groundwater moves**

Gravity provides the energy for the downward movement of groundwater. Water entering the ground moves through the zone of aeration to the zone of saturation (●FIGURE 12.3).

**zone of aeration**  The zone above the water table that contains both air and water within the pore spaces of soil, sediment, or rock.

**zone of saturation**  The area below the water table in which all pore spaces are filled with water.

**water table**  The surface that separates the zone of aeration from the underlying zone of saturation.

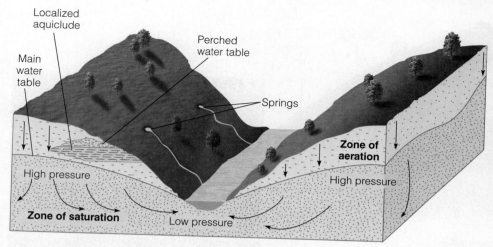

Localized aquiclude

Main water table

Perched water table

Springs

Zone of aeration

High pressure

High pressure

Zone of saturation

Low pressure

● **FIGURE 12.3 Groundwater Movement** Groundwater moves down through the zone of aeration to the zone of saturation. Then some of it moves along the slope of the water table, and the rest moves through the zone of saturation from areas of high pressure toward areas of low pressure. Some water might collect over a local aquiclude, such as a shale layer, thus forming a perched water table.

**Critical Thinking Question** If you were drilling a water well on your property and struck water at a considerably shallower depth than your neighbors, should you consider drilling deeper, or just celebrate your good fortune at not having to pay for a deeper well?

When water reaches the water table, it continues to move through the zone of saturation from areas where the water table is high toward areas where it is lower, such as streams, lakes, or swamps. Only some of the water follows the direct route along the slope of the water table. Most of it takes longer curving paths down and then enters a stream, lake, or swamp from below, because it moves from areas of high pressure toward areas of lower pressure within the saturated zone.

Groundwater velocity varies greatly and depends on many factors. Velocities range from 250 m per day in some extremely permeable material to less than a few centimeters per year in nearly impermeable material. In most ordinary aquifers, the average velocity of groundwater is a few centimeters per day.

# SPRINGS, WATER WELLS, AND ARTESIAN SYSTEMS

**LO9**  Define recharge

**LO10**  Explain how springs form

**LO11**  Explain how a cone of depression forms in a water well

**LO12**  Define an artesian system

**LO13**  Explain how an artesian system works

You can think of the water in the zone of saturation much like a reservoir, whose surface rises or falls depending on additions as opposed to natural and artificial withdrawals. *Recharge*—that is, additions to the zone of saturation—comes from rainfall or melting snow, or water might be added artificially at wastewater-treatment plants or recharge ponds constructed for just this purpose. But if groundwater is discharged naturally, or withdrawn at wells without

sufficient recharge, the water table drops—just as a savings account diminishes if withdrawals exceed deposits. Withdrawals from the groundwater system take place where groundwater flows laterally into streams, lakes, or swamps, where it discharges at the surface as *springs*, and where it is withdrawn from the system at water wells.

## Springs

Places where groundwater flows or seeps out of the ground as **springs** have always fascinated people. The water flows out of the ground for no apparent reason and from no readily identifiable source. So, it is not surprising that springs have long been regarded with superstition and revered for their supposed medicinal value and healing powers. Nevertheless, there is nothing mystical or mysterious about springs.

Although springs can occur under a wide variety of geologic conditions, they all form in basically the same way (●**FIGURE 12.4A**). When percolating water reaches the water table or an impermeable layer, it flows laterally, and if this flow intersects the surface, the water discharges as a spring (●**FIGURE 12.4B**).

Springs can also develop wherever a *perched water table*—a local aquiclude present within a larger aquifer, such as a lens of shale within sandstone—intersects the surface (Figure 12.3). As water migrates through the zone of aeration, it is stopped by the local aquiclude, and a localized zone of saturation "perched" above the main water table forms. Water moving laterally along the perched water table may intersect the surface to produce a spring.

**spring**  A place where groundwater flows or seeps out of the ground.

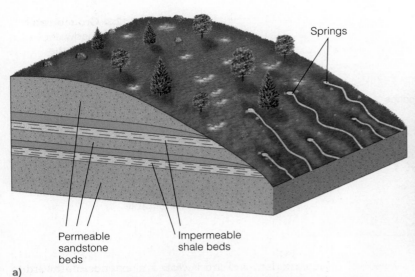

a)

Springs

Permeable
sandstone
beds

Impermeable
shale beds

b)

Reed Wicander

● **FIGURE 12.4  Springs**  Springs form wherever laterally moving groundwater intersects Earth's surface. (a) Most commonly, springs form when percolating water reaches an impermeable layer and migrates laterally until it seeps out at the surface. (b) The Hraunfossar (which translates into lava waterfalls) are a series of waterfalls in Borganfjörður (western Iceland). Water percolating downward through permeable rocks is forced to move laterally when it encounters one of the numerous impermeable lava layers formed from volcanic eruptions beneath the Lanjökull glacier. The laterally flowing groundwater then gushes out along these cliffs. Notice that lack of the vegetation is mostly on the lower lava face.

## Water Wells

**Water wells** are openings made by digging or drilling down into the zone of saturation. Once the zone of saturation has been penetrated, water percolates into the well, filling it to the level of the water table. A few wells are free-flowing; however, for most, the water must be brought to the surface by pumping.

When groundwater is pumped from a well, the water table in the area around the well is lowered, forming a **cone of depression** (●**FIGURE 12.5**). This happens when the rate of water withdrawal from the well exceeds the rate of water inflow to the well, thus lowering the water table around the well. A cone of depression's gradient—that is, whether it is steep or gentle—depends to a great extent on the permeability of the aquifer being pumped. A highly permeable aquifer produces a gentle gradient in the cone of depression, whereas a low-permeability aquifer results in a steep cone of depression, because water cannot easily flow to the well to replace the water being withdrawn.

The formation of a cone of depression does not normally pose a problem for the average domestic well provided that the well is drilled deep enough into the zone of saturation. However, the tremendous amounts of water used by industry and for irrigation may create a large cone of depression that lowers the water table sufficiently to cause shallow wells in the immediate area to go dry (Figure 12.5). This situation is not uncommon and frequently results in lawsuits by the owners of the shallow dry wells.

Lowering of the regional water table because of more groundwater being withdrawn than is being replenished

---

**water well**  A well made by digging or drilling into the zone of saturation.

**cone of depression**  A cone-shaped depression around a well where water is pumped from an aquifer faster than it can be replaced.

---

● **FIGURE 12.5  Cone of Depression**  A cone of depression forms whenever water is withdrawn from a well. If water is withdrawn faster than it can be replenished, the cone of depression will grow in depth and circumference, lowering the water table in the area and causing nearby shallow wells to go dry.

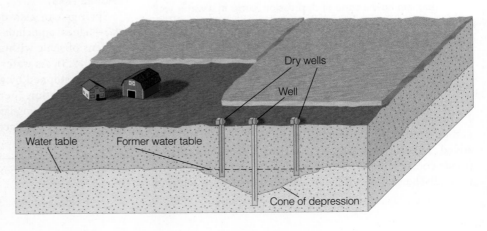

Dry wells

Well

Water table          Former water table

Cone of depression

is becoming a serious problem in many areas, particularly in the southwestern United States, where rapid growth has placed tremendous demands on the groundwater system. Unrestricted withdrawal of groundwater cannot continue indefinitely, and the rising costs and decreasing supply of groundwater should soon limit the growth of some regions in the United States, such as the aforementioned Southwest.

## Artesian Systems

The word *artesian* comes from the French town and province of Artois (called Artesium during Roman times) near Calais, where the first European artesian well was drilled in 1126 and is still flowing today. The term **artesian system** can be applied to any system in which groundwater is confined and builds up high hydrostatic (fluid) pressure (●**FIGURE 12.6**). Water in such a system is able to rise above the level of the aquifer if a well is drilled through the confining layer, thereby reducing the pressure and forcing the water upward. An artesian system can develop when (1) an aquifer is confined above and below by aquicludes; (2) the rock sequence is (usually) tilted to build up hydrostatic pressure; and (3) the aquifer is exposed at the surface, thus enabling it to be recharged.

The elevation of the water table in the recharge area and the distance of the well from the recharge area determine the height to which artesian water rises in a well. The surface defined by the water table in the recharge area, called the *artesian-pressure surface*, is indicated by the sloping dashed line in Figure 12.6. Because friction slightly reduces the pressure of the aquifer water, and consequently the level to which artesian water rises, the artesian-pressure surface therefore slopes.

An artesian well will flow freely at the ground surface only if the wellhead is at an elevation below the artesian-pressure surface. In this situation, the water flows out of the well, because it rises toward the artesian-pressure surface, which is at a higher elevation than the wellhead. In a nonflowing artesian well, the wellhead is above the artesian-pressure surface, and the water will rise in the well only as high as the artesian-pressure surface.

One of the best-known artesian systems in the United States underlies South Dakota and extends southward to central Texas. The majority of the artesian water from this system is used for irrigation. The aquifer of this artesian system, the Dakota Sandstone, is recharged where it is exposed along the margins of the Black Hills of South Dakota. Originally, the hydrostatic pressure in this system was great enough to produce free-flowing wells and to operate

---

**artesian system** A confined groundwater system with high hydrostatic pressure that causes water to rise above the level of the aquifer.

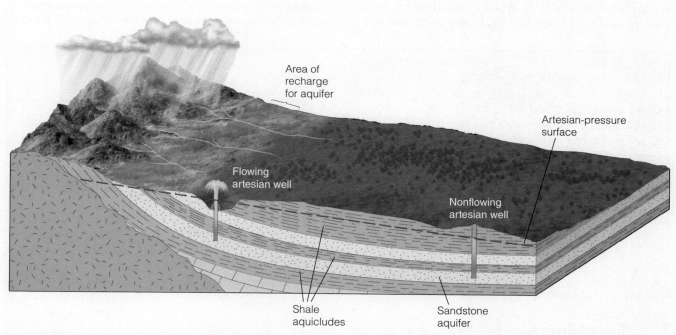

● **FIGURE 12.6 Artesian System** An artesian system must have an aquifer confined above and below by aquicludes; the aquifer must be exposed at the surface; and there must be sufficient precipitation in the recharge area to keep the aquifer filled. The elevation of the water table in the recharge area, which is indicated by a sloping dashed line (the artesian-pressure surface), defines the highest level to which well water can rise.

**Critical Thinking Question**  If the elevation of the wellhead is at or above that of the artesian-pressure surface, why then will the well be nonflowing?

waterwheels. However, because of the extensive use of this groundwater for irrigation over the years, the hydrostatic pressure in many of the wells is so low that they are no longer free-flowing, and the water must be pumped to the surface.

As a final comment on artesian systems, we should mention that it is not unusual for advertisers to tout the quality of artesian water as somehow being superior to other groundwater. Some artesian water might indeed be of excellent quality, but its quality is not dependent on the fact that water rises above the surface of an aquifer. Rather, its quality is a function of dissolved minerals and any introduced substances, so artesian water really is no different from any other groundwater. The myth of its superiority probably arises from the fact that people have always been fascinated by water that flows freely from the ground.

# GROUNDWATER EROSION AND DEPOSITION

LO14   Explain how karst topography forms

LO15   Explain the formation of a cave

LO16   List and define the major depositional features found in a cave

When rainwater begins to seep into the ground, it immediately starts to react with the minerals it contacts, weathering them chemically. In an area underlain by soluble rock, groundwater is the principal agent of erosion and is responsible for the formation of many major features of the landscape.

Limestone, a common sedimentary rock composed primarily of the mineral calcite ($CaCO_3$), underlies large areas of Earth's surface (●FIGURE 12.7). Although limestone is practically insoluble in pure water, it readily dissolves if a small amount of acid is present. Carbonic acid ($H_2CO_3$) is a weak acid that forms when carbon dioxide combines with water ($H_2O + CO_2 \rightarrow H_2CO_3$). Because the atmosphere contains a small amount of carbon dioxide (0.03%) and carbon dioxide is also produced in soil by the decay of organic matter, most groundwater is slightly acidic. When groundwater percolates through the various openings in limestone, the slightly acidic water readily reacts with the calcite to dissolve the rock by forming soluble calcium bicarbonate, which is carried away in solution (see Chapter 6).

## Sinkholes and Karst Topography

In regions underlain by soluble rock, the ground surface may be pitted with numerous depressions that vary in size and shape. These depressions, called **sinkholes** or merely *sinks*, mark areas with underlying soluble rock (●FIGURE 12.8). Most sinkholes form in one of two ways. The first is when soluble rock below the soil is dissolved by seeping water, and openings in the rock are enlarged and filled in by the

---

**sinkhole** A depression in the ground that forms by the solution of the underlying carbonate rocks or by the collapse of a cave roof.

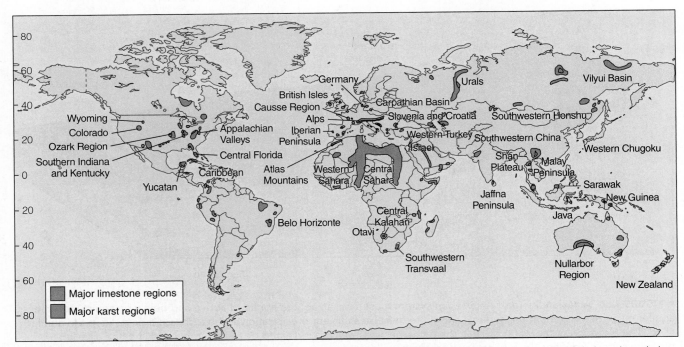

● **FIGURE 12.7 Distribution of the Major Limestone and Karst Areas of the World** Karst topography develops largely by groundwater erosion in areas underlain by soluble rocks.

**Critical Thinking Question** Why is karst topography typically restricted to regions with humid and temperate climates?

● **FIGURE 12.8 Sinkholes** This sinkhole formed on May 8 and 9, 1981, in Winter Park, Florida. It formed in previously dissolved limestone following a drop in the water table. The 100-m-wide, 35-m-deep sinkhole destroyed a house, numerous cars, and a municipal swimming pool.

**Critical Thinking Question** Why are sink holes becoming prevalent in Florida, especially in areas experiencing population growth?

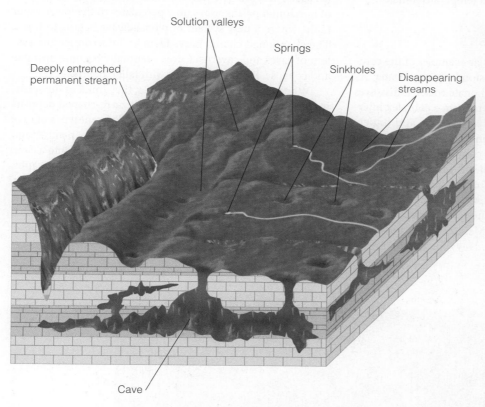

● **FIGURE 12.9 Features of Karst Topography** Erosion of soluble rock by groundwater produces karst topography. Features commonly found include solution valleys, springs, sinkholes, and disappearing streams.

overlying soil. As the groundwater continues to dissolve the rock, the soil is eventually removed, leaving shallow depressions with gently sloping sides. When adjacent sinkholes merge, they form a network of larger, irregular, closed depressions called *solution valleys*.

Sinkholes also form when a cave's roof collapses, usually producing a steep-sided crater. Sinkholes formed in this way are a serious hazard, particularly in populated areas. In regions prone to sinkhole formation, extensive geologic and hydrogeologic investigations must be done to determine the depth and extent of underlying cave systems prior to any site development to ensure that the underlying rocks are thick enough to support planned structures.

**Karst topography,** or simply *karst*, develops largely by groundwater erosion in many areas underlain by soluble rocks (●**FIGURE 12.9**). The name *karst* is derived from the

**karst topography** Landscape consisting of numerous caves, sinkholes, and solution valleys formed by groundwater solution of rocks such as limestone and dolostone.

plateau region of the border area of Slovenia, Croatia, and northeastern Italy, where this type of topography is well developed. In the United States, regions of karst topography include large areas of southwestern Illinois, southern Indiana, Kentucky, Tennessee, southern Missouri, Alabama, and central and northern Florida (Figure 12.7).

Karst topography is characterized by numerous caves, springs, sinkholes, solution valleys, and disappearing streams (Figure 12.9). *Disappearing streams* are so named because they typically flow only a short distance at the surface and then disappear into a sinkhole. The water continues flowing underground through fractures or caves until it surfaces again at a spring or other stream.

Karst topography varies from the spectacular high-relief landscapes of China (●**FIGURE 12.10**) to the subdued and pockmarked landforms of Kentucky. Common to all karst topography, though, is the presence of thick-bedded, readily soluble rock at the surface or just below the soil, and enough water for solution activity to occur. Karst topography is therefore typically restricted to humid and temperate climates.

## Caves and Cave Deposits

Caves are perhaps the most spectacular examples of the combined effects of weathering and erosion by groundwater. As groundwater percolates through carbonate rocks, it dissolves and enlarges fractures and openings to form a complex interconnecting system of crevices, caves, caverns, and underground streams. A **cave** is usually defined as a naturally formed subsurface opening that is generally connected to the surface and is large enough for a person to enter. A *cavern*, on the other hand, is a very large cave or a system of interconnected caves.

More than 17,000 caves are known in the United States. Some of the more famous ones are Mammoth Cave, Kentucky; Carlsbad Caverns, New Mexico; Lewis and Clark Caverns, Montana; Lehman Cave, Nevada; and Meramec

Caverns, Missouri, which Jesse James and his outlaw band often used as a hideout. Canada also has many famous caves, including the 536-m-deep Arctomys Cave in Mount Robson Provincial Park, British Columbia, the deepest known cave in North America. And Mexico is known for its Cueva de los Cristales, or Cave of Crystals, where some of the world's largest crystals of gypsum and selenite have been found.

Caves and caverns form as a result of the dissolution of carbonate rocks by weakly acidic groundwater (●**FIGURE 12.11**). Groundwater percolating through the zone of aeration slowly dissolves the carbonate rock and enlarges its fractures and bedding planes. On reaching the water table, the groundwater migrates toward the region's surface streams. As the groundwater moves through the zone of saturation, it continues to dissolve the rock and gradually forms a system of horizontal passageways through which the dissolved rock is carried to the streams (●**FIGURE 12.11A**).

As the surface streams erode deeper valleys, the water table drops in response to the lower elevation of the streams (●**FIGURE 12.11B**). The water that flowed through the system of horizontal passageways now percolates to the lower water table, where a new system of passageways begins to form. The abandoned channelways form an interconnecting system of caves and caverns. Caves eventually become unstable and collapse, littering the floor with fallen debris.

When most people think of caves, they think of the seemingly endless variety of colorful and bizarre-shaped deposits found in them. Although a great many different types of cave deposits exist, most form in essentially the same manner and are collectively known as *dripstone*. As water seeps into a cave, some of the dissolved carbon dioxide in the water escapes, and a small amount of calcite is precipitated. In this manner, the various dripstone deposits are formed (●**FIGURE 12.11C**).

**cave** A natural subsurface opening generally connected to the surface and large enough for a person to enter.

● **FIGURE 12.10 Karst Landscape southeast of Kunming, China** The Stone Forest, 125 km southeast of Kunming, China, is a high-relief karst landscape formed by the dissolution of carbonate rocks.

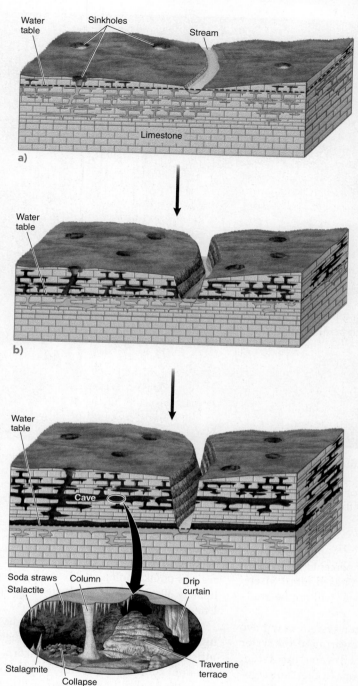

● **FIGURE 12.11 Cave**
**Formation** (a) As groundwater percolates through the zone of aeration and flows through the zone of saturation, it dissolves the carbonate rocks and gradually forms a system of passageways. (b) As the stream erodes more deeply, groundwater moves along the surface of the lower water table, forming a system of horizontal passageways through which dissolved rock is carried to the surface streams, thus enlarging the passageways. (c) As the surface streams erode deeper valleys, the water table drops, and the abandoned channelways form an interconnecting system of caves and caverns.

*Stalactites* are icicle-shaped structures hanging from cave ceilings that form as a result of precipitation from dripping water (●**FIGURE 12.12**). With each drop of water, a thin layer of calcite is deposited over the previous layer, forming a cone-shaped projection that grows down from the ceiling. The water that drips from a cave's ceiling also precipitates a small amount of calcite when it hits the floor. As additional calcite is deposited, an upward-growing projection, called a *stalagmite*, forms (Figure 12.12). If a stalactite and stalagmite meet, they form a *column*. Groundwater seeping from a crack in a cave's ceiling may form a vertical sheet of rock called a *drip curtain*, and water flowing across a cave's floor may produce *travertine terraces* (Figure 12.11c).

# MODIFICATIONS OF THE GROUNDWATER SYSTEM AND ITS EFFECTS

**LO17** Discuss the four major modifications to the groundwater system and the consequences resulting from each modification

Groundwater is a valuable natural resource that is rapidly being exploited with seemingly little regard to the effects of overuse and misuse. Approximately 20% of all water used in

● **FIGURE 12.12  Cave Deposits** Stalactites are the icicle-shaped structures hanging from the cave's ceiling, whereas the upward-pointing structures on the floor are stalagmites. A column (middle of photo) results when stalactites and stalagmites meet. A drip curtain (right side of photo) is a vertical sheet of rock formed by groundwater seeping from a crack in the cave's ceiling. All four structure are present in the Beautiful Stalactites on the Trails inside Postojna Cave Park, Postjna, Slovenia.

the United States is groundwater. This percentage is rapidly increasing, and unless this resource is used more wisely, sufficient amounts of clean groundwater will not be available in the future.

Modifications of the groundwater system may have many consequences, including (1) lowering of the water table, causing wells to dry up; (2) saltwater incursion; (3) subsidence; and (4) contamination. In fact, an ongoing debate pits the benefits of hydraulic fracturing for the extraction of natural gas from organic-rich shales against the possible contamination of the groundwater system in the area where this exploration is occurring (see GEO-FOCUS).

## Lowering the Water Table

Withdrawing groundwater at a significantly greater rate than it is replaced by either natural or artificial recharge can have serious effects. For example, the High Plains aquifer (also referred to as the Ogallala aquifer) is one of the most important aquifers in the United States. It underlies more than 450,000 km² of land, including most of Nebraska, large parts of Colorado and Kansas, portions of South Dakota,

Wyoming, and New Mexico, as well as the panhandle regions of Oklahoma and Texas, and accounts for approximately 30% of the groundwater used for irrigation in the United States (●FIGURE 12.13).

Although the High Plains aquifer has contributed to the high agricultural productivity of the region, providing a significant percentage of the nation's corn, cotton, and wheat, it cannot continue to provide the quantities of water that it has in the past. A study released in 2017 by the United States Geological Survey showed a 9% overall decline in recoverable water in storage from predevelopment (generally before 1950) to 2015.

It must be noted, however, that much of the High Plains aquifer water infiltrated during the wetter glacial climates more than 10,000 years ago. Consequently, most of the water being pumped is fossil water that is not being replenished at anywhere near the same rate as when it formed during the Pleistocene Epoch.

Most users of the aquifer realize that they cannot continue to withdraw the quantities of groundwater that they have in the past, and thus are turning to greater conservation, monitoring of the aquifer, and using new

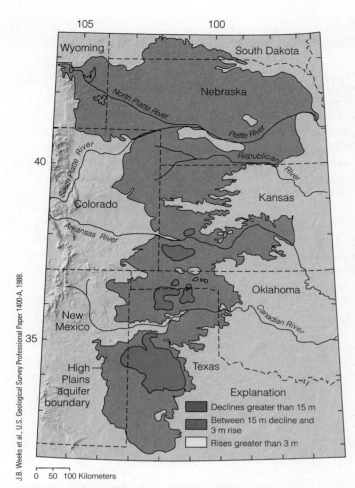

● **FIGURE 12.13 High Plains Aquifer** The geographic extent of the High Plains aquifer (also referred to as the Ogallala aquifer) and changes in water level from predevelopment through 1993. Irrigation from the High Plains aquifer is largely responsible for the region's agricultural productivity.

technologies to try to better balance withdrawal with recharge rates. Nevertheless, the rates of withdrawal of groundwater from some aquifers still exceed their rates of recharge, and population growth in a number of areas is continuing to put significant demands on an already limited water supply.

## Saltwater Incursion

The excessive pumping of groundwater in coastal areas has resulted in *saltwater incursion*, which has become a major problem in many rapidly growing coastal communities where greater demand for groundwater creates an even greater imbalance between withdrawal and recharge. Along coastlines where permeable rocks or sediments are in contact with the ocean, the fresh groundwater, being less dense than seawater, forms a lens-shaped body above the underlying saltwater (●**FIGURE 12.14A**). The weight of the freshwater exerts pressure on the underlying saltwater.

As long as rates of recharge equal rates of withdrawal, the contact between the fresh groundwater and the seawater remains the same.

If excessive pumping occurs, however, a deep cone of depression forms in the fresh groundwater (●**FIGURE 12.14B**). Because some of the pressure from the overlying freshwater has been removed, saltwater forms a *cone of ascension* as it rises to fill the pore space that formerly contained freshwater. When this occurs, wells become contaminated with saltwater and remain contaminated until recharge by freshwater restores the former level of the fresh groundwater water table.

To counteract the effects of saltwater incursion, recharge wells are often drilled to pump water back into the groundwater system (●**FIGURE 12.14C**). Recharge ponds that allow large quantities of fresh surface water to infiltrate the groundwater supply may also be constructed.

## Subsidence

As excessive amounts of groundwater are withdrawn from poorly consolidated sediments and sedimentary rocks, the water pressure between grains is reduced, and the weight of the overlying materials causes the grains to pack more closely together, resulting in *subsidence* of the ground. As greater amounts of groundwater are pumped to meet the burgeoning needs of agriculture, industry, and population growth, subsidence is becoming more prevalent, and an issue that is receiving increasingly more attention.

The San Joaquin Valley of California is a major agricultural region that relies largely on groundwater for irrigation. Since the 1920s, excessive groundwater withdrawals in parts of the valley for irrigation have caused subsidence of nearly 9 m (●**FIGURE 12.15**). Groundwater withdrawals continue today (2017), with some areas subsiding up to 0.6 m per year, causing damage to public and private wells, as well as jeopardizing the region's infrastructure.

Looking elsewhere in the world, the tilt of the Leaning Tower of Pisa in Italy is partly a result of groundwater withdrawal (●**FIGURE 12.16**). The tower started tilting soon after construction began in 1173 because of differential compaction of the foundation. During the 1960s, the city of Pisa withdrew ever-greater amounts of groundwater, causing the ground to subside further; as a result, the tilt of the tower increased until it was in danger of falling over. Strict control of groundwater withdrawal, stabilization of the foundation, and recent renovations have reduced the amount of tilting to about 1 mm per year, thus ensuring that the tower should stand for several more centuries.

A spectacular example of continuing subsidence is taking place in Mexico City, which is built on a former lake bed of fine-grained deposits. As groundwater is removed for the increasing needs of the city's 21.2 million people, these lake deposits are compacted, resulting in subsidence. During the past 60 years, the water table has been lowered almost 10 m and is continuing to sink at an average annual rate of 1 m (●**FIGURE 12.17**). The fact that 72% of the city's water comes

# GEO-FOCUS

## Hydraulic Fracturing: Pros and Cons

The debate over hydraulic fracturing, popularly known as "fracking," is certainly a contentious one with strong feelings on both sides. Those in favor of this new technology point to its benefits in reducing energy dependence, lowering energy costs, and creating new jobs in the energy sector. Those opposed to it cite environmental concerns, such as increased pollution to the environment, particularly to groundwater supplies, increased health risks, as well as a possible increase in global warming due to the leaking of methane, a potent greenhouse gas.

What is hydraulic fracturing? Hydraulic fracturing has been around for many years and involves forcing fluids under high pressure into organic-rich impermeable rocks like shale, or into already oil- or gas-producing rock formations to create fractures so that the oil or gas can more freely flow into the well. Drilling a conventional well and fracturing the producing zone only allows tapping the oil or gas from the small fractured area around the well. The advent of horizontal drilling, combined with hydraulic fracturing, soon made tapping thin layers of oil- and particularly gas-rich shales profitable. This combination of horizontal drilling and hydraulic fracturing has allowed the energy industry to open up vast deposits of untapped shale gas and oil in the United States that were previously unprofitable, and fuel the "shale revolution." This technology has allowed the United States to become a major exporter of natural gas, and the largest producer of crude oil in 2018.

How does horizontal drilling and hydraulic fracturing work? Most of the gas-rich shale presently being explored and produced, such as the Bakken Shale in the Williston Basin, the Barnett Shale in Texas, and the Marcellus Shale in the Appalachian Basin of the United States, are typically thin, less than 100 m thick, and not worth the cost of drilling a conventional vertical well. These shales, however, extend over vast areas, and by drilling vertically to them, and then drilling horizontally through them and fracturing the shale formation, large quantities of gas can be released (●**FIGURE 1**).

The output of these gas- and oil-producing shales is enormous. In the United States alone, dry natural gas production reached almost 27 trillion $ft^3$ (736 billion $m^3$) per day in 2017 according to the United States Energy Administration's *Annual Energy Outlook 2018*. An examination of currently targeted and prospective shale plays shows huge areas of shale gas and oil reserves in North America (●**FIGURE 2**). There are equally large areas of potential and producing areas elsewhere in the world. Exploiting these reserves depends, in part, on price, infrastructure, and government regulations. There is no debate that there are vast oil and gas shales that can be produced by current technology, but at what potential cost to the environment?

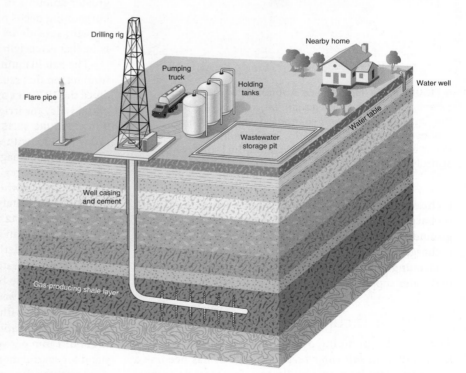

● **FIGURE 1** A diagrammatic representation of how horizontal drilling and hydraulic fracturing are used to extract gas, from a gas-rich shale layer.

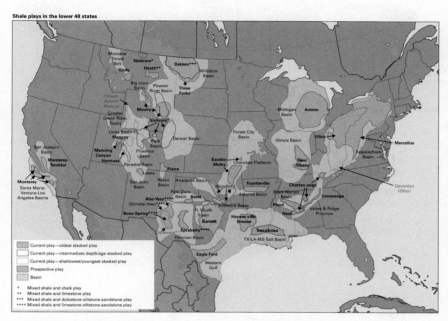

Shale plays in the lower 48 states

Current play—oldest stacked play
Current play—intermediate depth/age stacked play
Current play—shallowest/youngest stacked play
Prospective play
Basin

\*     Mixed shale and chalk play
\*\*    Mixed shale and limestone play
\*\*\*   Mixed shale and dolostone-siltstone-sandstone play
\*\*\*\* Mixed shale and limestone-siltstone-sandstone play

● **FIGURE 2** Map showing the current gas shale plays and prospective gas shale plays in North America as of April 2015, according to the United States Energy Information Administration. Also shown are the various sedimentary basins in which these shales are found.

Probably the biggest concern is polluting the groundwater system. Because the well is drilled through the ground-water aquifer, the well must be cased with cement and steel to guard against leakage from the pipe into the groundwater. Although this is standard procedure, accidents do happen, which over time can result in pollution of the groundwater system by drilling and hydraulic fluids (Figure 1).

A similar worry is that in those wells that have wastewater storage, the wastewater might leak into the ground, contaminating the soil, streams, and groundwater, as well as emitting noxious fumes into the surrounding area.

The problem of methane leakage and pollution has also been raised.

Methane is the primary gas in natural gas, and there are concerns about its leakage into the groundwater system, where it can build up to high levels. Likewise, methane is a potent greenhouse gas, and when released into the atmosphere, it acts as a powerful greenhouse gas that can trap 20 to 25 times as much heat as carbon dioxide (see Figure 1.6 on the greenhouse effect and global warming). Certainly continuing research is needed to determine the various sources of methane emissions from natural gas production, to evaluate their impact on climate and atmospheric chemistry.

Lastly, the question has been raised about whether hydraulic fracturing contributes to earthquakes in the

region being drilled. As was discussed in the section on earthquake control in Chapter 8, it has been noted that the number of earthquakes in Texas, Oklahoma, and the Dakotas has increased since fracking in these areas began. The rise in small earthquakes in areas of increased drilling activity appears to be linked to wastewater injection wells. Although the earthquakes are small, usually less than magnitude 3, there have been a number of higher magnitude earthquakes recorded, including a 5.6-magnitude earthquake in Oklahoma in September 2016. While evidence continues to mount of a link between fracking and earthquakes, studies are ongoing regarding as to whether the increase in earthquake activity directly correlates to the fracked wells, or whether it is linked to wastewater injection wells situated nearby that are used to dispose of drilling wastewater.

What now needs to be asked is whether hydraulic fracturing is worth it. Like many technologies, it comes with both risks and benefits. The risks are damage to the environment in the form of pollution to the atmosphere, groundwater supplies, as well as the release of greenhouse gases, and health concerns resulting from the chemicals used in the fracking fluids, which are not always disclosed. The benefits are that oil and gas shales can go a long way toward continued energy independence, deliver a relatively inexpensive source of energy for the foreseeable future, and provide a boost to national economies. Whether the "shale revolution" is indeed the answer to our energy needs, and can be exploited with minimal damage to the environment, is yet to be determined.

● **FIGURE 12.14 Saltwater Incursion**
(a) Because freshwater is not as dense as saltwater, it forms a lens-shaped body above the underlying saltwater. (b) If excessive pumping occurs, a cone of depression develops in the fresh groundwater, and a cone of ascension forms in the underlying salty groundwater, which may result in saltwater contamination of the well. (c) Pumping water back into the groundwater system through recharge wells can help lower the interface between the fresh groundwater and the salty groundwater and reduce saltwater incursion.

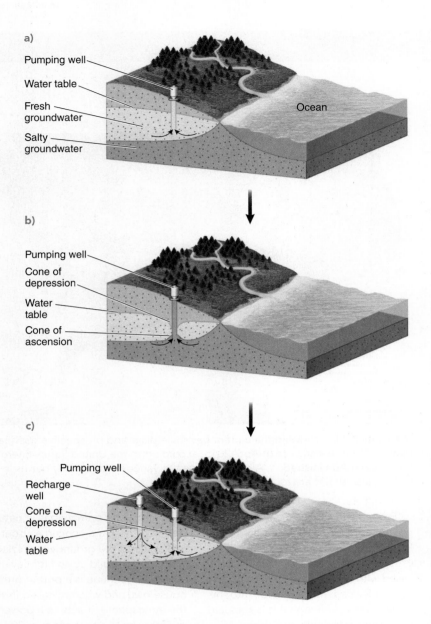

from the aquifer beneath the metropolitan area ensures that problems of subsidence will continue.

The extraction of oil can also cause subsidence. For example, Long Beach, California, has subsided 9 m as a result of many decades of oil and gas extraction from the giant Wilmington Oil Field. In addition to the pumping of groundwater, damages due to subsidence in the region reached billions of dollars. As a result of drilling restrictions and water injection into the oil reservoir, subsidence has essentially been halted.

## Groundwater Contamination

A major problem facing our society is the safe disposal of the numerous pollutant by-products of an industrialized economy. We are becoming increasingly aware that streams, lakes, and oceans are not unlimited reservoirs for waste and that we must find new, safe ways to dispose of pollutants.

The most common sources of groundwater contamination are sewage, landfills, toxic waste disposal sites, and agriculture. Once pollutants get into the groundwater system, they spread wherever groundwater travels, which can make their containment difficult. Furthermore, because groundwater moves so slowly, it takes a long time to cleanse a groundwater reservoir once it has become contaminated.

In many areas, septic tanks are the most common way of disposing of sewage. A septic tank slowly releases sewage into the ground, where it is decomposed by oxidation and microorganisms and filtered by the sediment as it percolates through the zone of aeration. In most situations, by the time the water from the sewage reaches the zone of saturation, it has been cleansed of impurities and is safe to use (●**FIGURE 12.18A**). If the water table is close to the surface, or if the rocks are very permeable, water entering the zone of saturation may still be contaminated and unfit to use.

● **FIGURE 12.15 Subsidence in the San Joaquin Valley, California** The dates on this power pole dramatically illustrate the amount of subsidence in the San Joaquin Valley, California. Because of groundwater withdrawals and subsequent sediment compaction, the ground subsided nearly 9 m between 1925 and 1977.

● **FIGURE 12.16 The Leaning Tower of Pisa, Italy** The tilting is partly the result of subsidence due to the removal of groundwater. Strict control of groundwater withdrawal, recent stabilization of the foundation, and renovation of the structure itself have ensured that the Leaning Tower will continue leaning for many more centuries.

Landfills are also potential sources of groundwater contamination (●**FIGURE 12.18B**). Not only does liquid waste seep into the ground but rainwater also carries dissolved chemicals and other pollutants down into the groundwater reservoir. Unless the landfill is carefully designed and lined with an impermeable layer such as clay, many toxic compounds such as paints, solvents, cleansers, pesticides, and battery acid will find their way into the groundwater system.

Toxic waste sites where dangerous chemicals are either buried or pumped underground are an increasing source of groundwater contamination. The United States alone must dispose of several thousand metric tons of hazardous chemical waste per year. Unfortunately, much of this waste has been, and still is being, improperly dumped and is contaminating the surface water, soil, and groundwater.

Examples of indiscriminate dumping of dangerous and toxic chemicals can be found in every state. Perhaps the most infamous and notorious environmental tragedy is

Love Canal, located near Niagara Falls, New York. During the 1940s, the Hooker Chemical Company dumped approximately 19,000 tons of chemical waste into Love Canal. In 1953, Hooker covered one of the dump sites with dirt and sold it for $1.00 to the Niagara Falls Board of Education, which built an elementary school and playground on the site. Heavy rains and snow during the winter of 1976–1977 raised the water table and turned the area into a muddy swamp in the spring of 1977. Mixed with the mud were thousands of toxic, noxious chemicals that formed puddles in the playground, oozed into people's basement, and covered gardens and lawns. Trees, lawns, and gardens began to die, and many of the residents of the area suffered from serious illnesses. The cost of cleaning up the Love Canal site took 21 years and close to $400 million. Since then, a number of houses have been refurbished north of the canal (renamed Black Creek Village), and sold to new owners. In addition, about 150 acres east of the canal have been sold for light industrial usage.

● **FIGURE 12.17 Subsidence in Mexico City** Excessive withdrawal of groundwater from beneath Mexico City has resulted in subsidence and uneven settling of buildings, such as Our Lady of Guadalupe, in Mexico City, Mexico.

● **FIGURE 12.18 Groundwater Contamination** (a) A septic system slowly releases sewage into the zone of aeration. Oxidation, bacterial degradation, and filtering usually remove impurities before they reach the water table. However, if the rocks are very permeable or the water table is too close to the septic system, contamination of the groundwater can result. (b) Unless there is an impermeable barrier between a landfill and the water table, pollutants can be carried into the zone of saturation and contaminate the groundwater supply: (1) Infiltrating water leaches contaminants from the landfill; (2) the polluted water enters the water table and moves away from the landfill; (3) wells may tap the polluted water and thus contaminate drinking water supplies; and (4) the polluted water may emerge into streams and other water bodies downslope from the landfill.

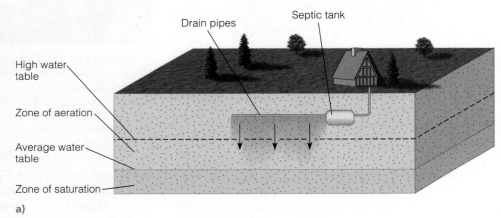

a)

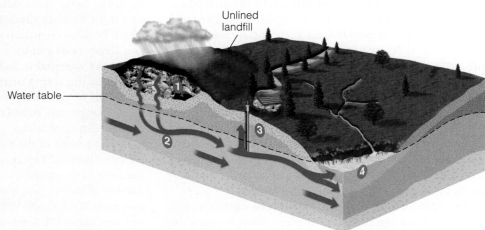

b)

# HYDROTHERMAL ACTIVITY

L018  Define the term hydrothermal

L019  Define a hot spring

L020  Explain how geysers form

L021  Define geothermal activity

L022  Discuss the potential of geothermal energy as an alternative to coal, oil, and natural gas in the production of energy

**Hydrothermal** is a term referring to hot water. Some geologists restrict the meaning to include only water heated by magma, but here we use it to refer to any hot subsurface water and the surface activity resulting from its discharge. One manifestation of hydrothermal activity in areas of active or recently active volcanism is the discharge of gases, such as steam, at vents known as *fumeroles* (●**FIGURE 12.19**). Of more immediate concern here is the groundwater that rises to the surface as *hot springs* or *geysers*. It may be heated by its proximity to magma or by Earth's geothermal gradient because it circulates deeply.

## Hot Springs

A **hot spring** (also called a *thermal spring* or *warm spring*) is any spring in which the water temperature is higher than 37°C, the temperature of the human body (●**FIGURE 12.20**). Some hot springs are much hotter, with temperatures up to the boiling point in many instances. Of the more than 1,100 known hot springs in the United States, many of them are in the far West, with the others in the Black Hills of South Dakota, Georgia, the Ouachita region of Arkansas, and the Appalachian region.

Hot springs are also common in other parts of the world. One of the most famous is in Bath, England, where shortly after the Roman conquest of Britain in A.D. 43, numerous bathhouses and a temple were built around the hot springs (●**FIGURE 12.21**).

The heat for most hot springs comes from magma or cooling igneous rocks. The geologically recent igneous activity in the western United States accounts for the large number of hot springs in that region. The water in some hot springs, however, circulates deep into Earth, where it is warmed by the normal increase in temperature, the geothermal gradient. For example, the spring water of Warm Springs, Georgia, is heated in this manner. This hot spring was a health and bathing resort long before the Civil War (1861–1865); later, with the establishment of the Georgia Warm Springs Foundation, it was used to help treat polio victims.

● **FIGURE 12.20  The Morning Glory Hot Spring, Yellowstone National Park, Wyoming** One of the most colorful hot springs in Yellowstone National Park, Wyoming, the Morning Glory hot spring, is fringed with multicolored mats of heat-loving cyanobacteria and algal mats. Each color represents a certain temperature range that allows for specific bacterial species to thrive in this extreme environment.

**hydrothermal** A term referring to hot water, as in hot springs and geysers.

**hot spring** A spring in which the water temperature is warmer than the temperature of the human body (37°C).

● **FIGURE 12.19  Fumeroles** Gases emitted from a fumerole in Hveravellir, Iceland.

Rachelle Burnside/Shutterstock.com

● **FIGURE 12.21 Bath, England** One of the many bathhouses in Bath, England, that were built around hot springs shortly after the Roman conquest in A.D. 43.

## Geysers

Hot springs that intermittently eject hot water and steam with tremendous force are known as **geysers**. The word comes from the Icelandic *geysir*, "to gush" or "to rush forth." One of the most famous geysers in the world is Old Faithful in Yellowstone National Park, Wyoming (●FIGURE 12.22). With a thunderous roar, it erupts a column of hot water and steam every 30 to 90 minutes. Other well-known geyser areas are found in Iceland and New Zealand.

Geysers are the surface expression of an extensive underground system of interconnected fractures within hot igneous rocks (●FIGURE 12.23). Groundwater percolating down into the network of fractures is heated as it comes into contact with the hot rocks. Because the water near the bottom of the fracture system is under higher pressure than the water near the top, it must be heated to a higher temperature before it will boil. Thus, when the deeper water is heated to near the boiling point, a slight rise in temperature or a drop in pressure, such as from escaping gas, will instantly change it to steam. The expanding steam quickly pushes the water above it out of the ground and into the air, producing

James S. Monroe

● **FIGURE 12.22 Old Faithful Geyser** Old Faithful Geyser in Yellowstone National Park, Wyoming, is one of the world's most famous geysers, erupting faithfully every 30 to 90 minutes and spewing water 32 to 56 m high.

a geyser eruption. After the eruption, relatively cool groundwater starts to seep back into the fracture system, where it heats to near its boiling temperature and the eruption cycle begins again. Such a process explains how geysers erupt with some regularity.

Hot spring and geyser water typically contains large quantities of dissolved minerals, because most minerals dissolve more rapidly in warm water than in cold water. Due of this high mineral content, some believe that the waters of many hot springs have medicinal properties. Numerous spas and bathhouses have been built at hot springs throughout the world to take advantage of these supposed healing properties.

When the highly mineralized water of hot springs or geysers cools at the surface, some of the material in solution is precipitated, forming various types of deposits. The amount and type of precipitated minerals depend on the solubility and composition of the material that the groundwater flows through. If the groundwater contains dissolved calcium carbonate ($CaCO_3$), then *travertine* or *calcareous tufa* (both of which are varieties of limestone) are precipitated. Spectacular examples of hot spring travertine deposits

---

**geyser** A hot spring that periodically ejects hot water and steam.

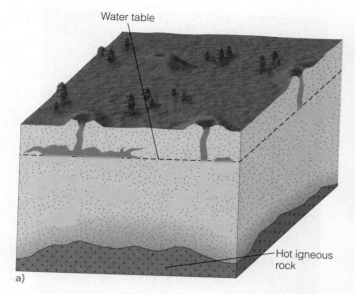

Water table

Hot igneous rock

a)

a)

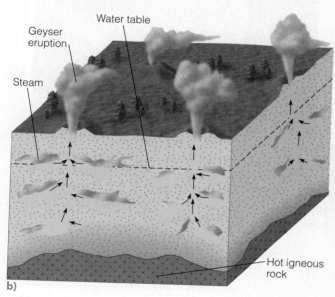

Geyser eruption

Water table

Steam

Hot igneous rock

b)

● **FIGURE 12.23  Anatomy of a Geyser** (a) The eruption of a geyser starts when groundwater percolates down into a network of interconnected openings and is heated by the hot igneous rocks. The water near the bottom of the fracture system is under higher pressure than the water near the top and consequently must be heated to a higher temperature before it will boil. (b) Any rise in the temperature of the water above its boiling point or a drop in pressure will cause the water to change to steam, which quickly pushes the water above it up and out of the ground, producing a geyser eruption.

are found at Pamukhale in Turkey and at Mammoth Hot Springs in Yellowstone National Park (●**FIGURE 12.24A**). Groundwater containing dissolved silica will, upon reaching the surface, precipitate a soft, white, hydrated mineral called *siliceous sinter* or *geyserite*, which can accumulate around a geyser's opening (●**FIGURE 12.24B**).

b)

● **FIGURE 12.24  Hot Spring Deposits in Yellowstone National Park, Wyoming** (a) Minerva Terrace, formed when calcium carbonate-rich hot spring water cooled, precipitating travertine. (b) Liberty Cap is a geyserite mound formed by numerous geyser eruptions of silicon-dioxide-rich hot spring water.

**GEOTHERMAL ENERGY** Geothermal energy is any energy produced from Earth's internal heat. In fact, the term *geothermal* comes from *geo*, "Earth," and *thermal*, "heat." Several forms of internal heat are known, such as hot dry rocks and magma, but so far, only hot water and steam are used in the production of geothermal energy.

**geothermal energy** Energy that comes from steam and hot water trapped within Earth's crust.

● **FIGURE 12.25 The Geysers, Sonoma County, California** Steam rising from one of the geothermal power plants at The Geysers in Sonoma County, California. Steam from wells drilled into the geothermal region, about 116 km north of San Francisco, is piped directly to electricity-generating turbines to produce electricity that is distributed throughout the area.

**Critical Thinking Question** Although geothermal energy is a relatively nonpolluting form of energy used as a source of heat and to generate electricity, why is it typically a more expensive form of energy? (Hint: Which contains more dissolved minerals, hot or cold water?)

© John Karachewski/Geoscapes Photography

As oil reserves decline, geothermal energy is becoming an attractive alternative. In those areas where it is plentiful, geothermal energy can supply most, if not all, of the energy needs, sometimes, but not always, at a fraction of the cost of other types of energy. For example, in 2017, geothermal plants in the United States produced about 0.4% of utility-scale electricity generation. The top ten countries in 2017 in terms of geothermal installed power generation capacity were: United States, Philippines, Indonesia, Turkey, New Zealand, Mexico, Italy, Iceland, Kenya, and Japan.

In the United States, the first commercial geothermal electricity-generating plant, The Geysers, was built in 1960 in the Mayacamas Mountains, approximately 116 km north of San Francisco, California (●**FIGURE 12.25**). This complex of 22 geothermal power plants, drawing steam from more than 350 wells, is the largest geothermal field in the world. Here, wells are drilled into the numerous near-vertical fractures underlying the region, and as pressure on the rising groundwater decreases, the water changes to steam, which is piped directly to electricity-generating turbines and generators. The Geysers currently have the capacity to power 900,000 homes in the area.

# Key Concepts Review

- Groundwater is part of the hydrologic cycle and an important natural resource. It consists of all subsurface water trapped in the pores and other open spaces in rocks, sediment, and soil.
- Porosity is the percentage of a material's total volume, that is, pore space. Permeability is the capacity to transmit fluids and is dependent not only on porosity but also on the size of the pores or fractures and their interconnections.
- The water table is the surface separating the zone of aeration (in which the pores are filled with air and water) from the underlying zone of saturation (in which the pores are filled with water). The water table is a subdued replica of the overlying land surface in most places.
- Groundwater moves slowly downward under the influence of gravity through the zone of aeration to the zone of saturation. Some of it then moves along the surface of the water table, and the rest moves from areas of high pressure to areas of low pressure.

- Groundwater velocity varies greatly and depends on a number of factors. Generally, the average velocity of groundwater is a few centimeters per day.
- Springs are found wherever the water table intersects the ground surface. Some springs are the result of a perched water table—that is, a localized aquiclude within an aquifer and above the regional water table.
- Water wells are openings made by digging or drilling down into the zone of saturation. When water is pumped from a well, the water table in the area around the well is lowered, forming a cone of depression.
- In an artesian system, confined groundwater builds up high hydrostatic pressure. For an artesian system to develop, an aquifer must be confined above and below by aquicludes; the aquifer is usually tilted so that it can build up hydrostatic pressure; and the aquifer must be exposed at the surface so that it can be recharged.
- Karst topography develops by groundwater erosion in many areas underlain by soluble rocks. It is

characterized by sinkholes, caves, solution valleys, and disappearing streams.
- Caves form when groundwater in the zone of saturation weathers and erodes soluble rock such as limestone. Common cave deposits include stalactites, stalagmites, columns, drip curtains, and travertine terraces.
- Modification of the groundwater system can cause serious problems such as lowering of the water table, saltwater incursion, subsidence, and contamination.

- Hydrothermal refers to hot water, typically heated by magma. Manifestations of hydrothermal activity include fumaroles, hot springs, and geysers.
- Geothermal energy is energy produced from Earth's internal heat and comes from the steam and hot water trapped within Earth's crust. It is a relatively nonpolluting form of energy that is used as a source of heat and to generate electricity.

# Important Terms

aquifer   245

artesian system   249

cave   252

cone of depression   248

geothermal energy   263

geyser   262

groundwater   244

hot spring   261

hydrothermal   261

karst topography   251

permeability   245

porosity   244

sinkhole   250

spring   247

water table   246

water well   248

zone of aeration   246

zone of saturation   246

# Review Questions

1. Which of the following is not an example of groundwater erosion?
   a. karst topography
   b. stalactites
   c. sinkholes
   d. caves
   e. disappearing streams

2. The water table is a surface separating the
   a. zone of porosity from the underlying zone of permeability.
   b. capillary fringe from the underlying zone of aeration.
   c. zone of aeration from the underlying zone of saturation.
   d. capillary fringe from the underlying zone of saturation.
   e. zone of saturation from the underlying zone of aeration.

3. The porosity of Earth materials is defined as
   a. their ability to transmit fluids.
   b. the depth of the zone of saturation.
   c. the percentage of void space.
   d. their solubility in the presence of weak acids.
   e. the temperature of groundwater.

4. A cone of depression forms when
   a. a stream flows into a sinkhole.
   b. water in the zone of aeration is replaced by water from the zone of saturation.
   c. a spring forms where a perched water table intersects the surface.
   d. water is withdrawn from a well faster than it can be replaced.
   e. the ceiling of a cave collapses, forming a steep-sided crater.

5. Which of the following conditions must exist for an artesian system to form?
   a. Groundwater must circulate near magma.
   b. The water table must be at or very near the surface.
   c. The rocks below the surface must be especially resistant to solution.
   d. Water must rise very high in the capillary fringe.
   e. An aquifer must be confined above and below by aquicludes.

6. Explain how groundwater weathers and erodes Earth materials.

7. Why should we be concerned about how fast the groundwater supply is being depleted in some areas?

8. What is the difference between an artesian system and a water well? Is there any difference between the quality of the water obtained from an artesian system and water from a water well?

9. One concern geologists have about burying nuclear waste in present-day arid regions such as Nevada is that the climate may change during the next several thousand years and become more humid, thus allowing more water to percolate through the zone of aeration. Why is this a concern? What would the average rate of groundwater movement have to be during the next 5,000 years to reach canisters containing radioactive waste buried at a depth of 400 m?

# Creative Thinking Visual Question

Withdrawal of large quantities of groundwater in the Las Vegas, Nevada, area has resulted in differential subsidence and damage to roads and buildings. What evidence of subsidence is visible in this photograph (●**FIGURE 1**)?

● **FIGURE 1** Differential subsidence in Las Vegas, Nevada.

Marii Bryant Mille

13

The Bering Glacier in Alaska is the longest glacier in North America. At its greatest extent in 1996, it measured 190 km long, but since 2005, its terminus has retreated approximately 4.8 km, and it has thinned by more than 60 m.

# GLACIERS AND GLACIATION

# INTRODUCTION

At present, glaciers cover about 10% of Earth's land surface, but during the Ice Age, or what geologists call the Pleistocene Epoch, they were much more extensive. During that time, glaciers covered vast areas, especially on the Northern Hemisphere continents, and small glaciers in mountain valleys were much more numerous and larger than they are now (see the chapter opening photo). During this comparatively brief interval of geologic time (2.6 million to 10,000 years ago) glaciers waxed and waned several times, and as they did so, they deeply eroded some areas and deposited huge quantities of sediments elsewhere.

Since the Ice Age, Earth has gone through several climatic changes. About 6,000 years ago, during the Holocene Maximum, average temperatures were slightly higher than they are now, and some of today's arid regions were much more humid. Today the only arable land in Egypt is along the Nile River (see Chapter Opening photo in Chapter 11), but during the Holocene Maximum much of North Africa was covered by grasslands, swamps, and lakes.

Following the Holocene Maximum, a time of cooler temperatures prevailed, but from about A.D. 1000 to 1300, Europe went through the Medieval Warm Period, when wine grapes grew 480 km farther north than they do now. Then a cooling trend began about A.D. 1300 that led to the **Little Ice Age**, which lasted from 1500 to the middle or late 1800s. During the Little Ice Age, glaciers expanded to their greatest historic extent; summers were cooler and wetter; winters were colder; and sea ice persisted longer around Greenland, Iceland, and the Canadian Arctic islands. The cooler, wetter summers were a problem because the growing seasons were much shorter, resulting in several famines.

During the coldest part of the Little Ice Age (1680–1730), the growing season in England was five weeks shorter than during the 1900s. Occasionally, Eskimos following the southern edge of the sea ice paddled their kayaks as far south as Scotland, and the canals in Holland froze over in some winters. In 1608, the first recorded Frost Fair was held in London, England, on the Thames River, and the last one in 1814. In 1816, known as the "year without a summer," unusually cold temperature persisted into June and July in New England and northern Europe. The eruption of Tambora in 1815 contributed to the cold spring and summer of 1816 (see Chapter 5).

Many of you have probably heard of the Ice Age and have some idea of what a glacier is, but it is doubtful that you know about the dynamics of glaciers, how they form, and what causes ice ages. All *glaciers* are moving bodies of ice on land, and as moving solids, they have a tremendous capacity to erode, transport, and deposit rock and sediment. Indeed, glaciers are responsible for many easily recognizable landforms and account for some of the most spectacular scenery in our National Parks and elsewhere in the world. And even though they are more restricted than during the Ice Age, glaciers remain an important geologic agent for modifying Earth's surface, especially in high mountains and at high latitudes.

Why study glaciers? Of course, glaciers are part of the hydrologic cycle, illustrating again the complex interactions between Earth's systems. In addition, a large part of the water in some countries, such as Nepal, Tibet, and Pakistan, comes from melting glaciers. Even in the United States and Canada, some areas in the West depend partly on water stored in glaciers. And, finally, glaciers are very sensitive to climatic changes, so scientists are interested in the fluctuations of glaciers as possible indications of global warming.

# THE KINDS OF GLACIERS

**L01** Define a valley glacier, continental glacier, and an ice cap

A **glacier** is a moving body of ice on land that flows downslope or outward from areas of accumulation. Our definition of a glacier excludes frozen seawater, as in the North Polar region, and sea ice that forms yearly adjacent to Greenland and Iceland. Drifting icebergs are not glaciers either, although they may have come from glaciers that flowed into lakes or the sea. The critical points in the definition are *moving* and *on land*. Accordingly, permanent snowfields in high mountains, though on land, are not glaciers because they do not move. All glaciers share several characteristics, but they differ enough in size and location for scientists to define two specific types, *valley glaciers* and *continental glaciers*, and several subvarieties of them.

## Valley Glaciers

**Valley glaciers** are confined to mountain valleys, where they flow from higher to lower elevations (●**FIGURE 13.1**), whereas continental glaciers cover vast areas, are not confined by the underlying topography, and flow outward in all directions from areas of snow and ice accumulation. We use the term *valley glacier*, but some geologists prefer the synonym *alpine glacier* or *mountain glacier*. Valley glaciers commonly have tributaries, just as streams and rivers do, thereby forming a network of glaciers in an interconnected system of mountain valleys.

Valley glaciers are common in many of the mountain regions of the world. However, Australia is the only continent with no glaciers of any kind. A valley glacier's shape is controlled by the shape of the valley it occupies, so they tend to be long and narrow tongues of moving ice. Some valley glaciers flow into the ocean and are called *tidewater glaciers*;

---

**Little Ice Age** An interval from about 1500 to the mid- to late-1800s during which glaciers expanded to their greatest historic extent.

**glacier** A mass of ice on land that moves by plastic flow and basal slip.

**valley glacier** A glacier confined to a mountain valley or an interconnected system of mountain valleys.

● **FIGURE 13.1 Valley Glaciers in Alaska** Smaller glaciers join to form the Sustina Glacier in Alaska.

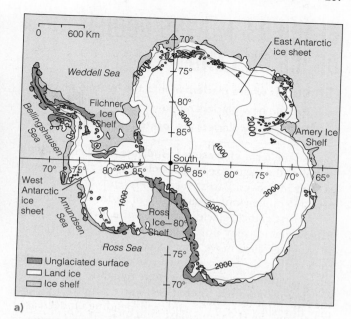

a)

they differ from other valley glaciers only in that their terminus is in the sea rather than on land.

Valley glaciers are rather small compared with the much more extensive continental glaciers, but even so, they may be several kilometers across, tens of kilometers long, and hundreds of meters thick. Erosion and deposition by valley glaciers account for much of the spectacular scenery in several U.S. and Canadian national parks, most notably Yosemite in California, Glacier in Montana, and Banff-Jasper in Canada.

## Continental Glaciers

**Continental glaciers**, also known as *ice sheets*, are vast, covering at least 50,000 km$^2$, and they are unconfined by topography. That is, their shape and movement are not controlled by the underlying landscape, as is the case with valley glaciers, which are long, narrow tongues of ice with the existing slope determining their direction of flow. In contrast, continental glaciers flow outward in all directions from a central area or areas of accumulation in response to variations in ice thickness.

In Earth's two areas of continental glaciation, Greenland and Antarctica, the ice is more than 3,000 m thick and covers all but the highest mountains (●**FIGURE 13.2A**). The continental glacier in Greenland covers about 1,800,000 km$^2$, and in Antarctica the East and West Antarctic glaciers merge to form a continuous ice sheet blanketing more than 12,650,000 km$^2$. In short, about 90% of all glacial ice on Earth is in Antarctica, with most of the rest in Greenland. The glaciers in Antarctica flow into the sea, where the buoyant effect of water causes the ice to float in vast *ice shelves*; the Ross Ice Shelf alone covers more than 547,000 km$^2$ (Figure 13.2a).

An **ice cap**, a dome-shaped mass of glacial ice, is similar to, but smaller than a continental glacier, covering less than 50,000 km$^2$ (●**FIGURE 13.2B**). Some ice caps form where valley glaciers grow and overtop the divides and passes between adjacent valleys, then coalesce to form a continuous ice cover. They also form on fairly flat terrain, such as in Iceland and some of the islands in the Canadian Arctic.

b)

● **FIGURE 13.2 Continental Glaciers and Ice Caps** (a) The West and East Antarctic ice sheets merge to form a nearly continuous ice cover that averages 2,160 m thick. The blue lines are lines of equal thickness. (b) View of the Penny Ice Cap on Baffin Island, Canada. It covers about 6,000 km$^2$.

**continental glacier** A glacier that covers a vast area (at least 50,000 km$^2$) and is not confined by topography; also called an *ice sheet*.

**ice cap** A dome-shaped mass of glacial ice that covers less than 50,000 km$^2$.

# GLACIERS: MOVING BODIES OF ICE ON LAND

LO2 Discuss where glaciers fit into the hydrologic cycle
LO3 Explain how glaciers move
LO4 Discuss why glaciers are located where they are

We use the term **glaciation** to indicate all glacial activity, including the origin, expansion, and retreat of glaciers, as well as their impact on Earth's surface. Presently, glaciers cover nearly 15 million km², or about 10% of Earth's land surface.

At first glance, glaciers appear static. Even briefly visiting a glacier may not dispel this impression because, although glaciers move, they usually do so slowly. Nevertheless, they do move, and just like other geologic agents such as running water, glaciers are dynamic systems that continuously adjust to changes. For example, a glacier may flow slower or more rapidly depending on decreased or increased amounts of snow or the absence or presence of water at its base. Glaciers may also expand or contract depending on climatic changes.

## Glaciers: Part of the Hydrologic Cycle

Glaciers make up one reservoir in the hydrologic cycle where water is stored for long periods, but even this water eventually returns to its original source, the oceans (see Figure 11.2). Many glaciers at high latitudes, as in Alaska, northern Canada, and Scandinavia, flow directly into the oceans (tidewater glaciers), where they melt, or icebergs break off (a process known as *calving*) and drift out to sea, where they eventually melt. At low latitudes or areas remote from the oceans, glaciers flow to lower elevations, where they melt and the water enters the groundwater system (another reservoir in the hydrologic cycle), or it returns to the seas by surface runoff.

In addition to melting, glaciers lose water by *sublimation*, when ice changes to water vapor without an intermediate liquid phase. The water vapor so derived, enters the atmosphere, where it may condense and fall as rain or snow, but in the long run, this water also returns to the oceans.

## How Do Glaciers Originate and Move?

Ice is a crystalline solid with characteristic physical properties and a specific chemical composition, and thus is a mineral. Accordingly, *glacial ice* is a type of metamorphic rock, but one that is easily deformed. Glaciers form in any area where more snow falls than melts during the warmer seasons, and a net accumulation takes place. Freshly fallen snow has about 80% air-filled pore space and 20% solids, but it compacts as it accumulates, partially thaws, and refreezes, converting to a granular type of snow known as **firn**. As more snow accumulates, the firn is buried and further compacted and

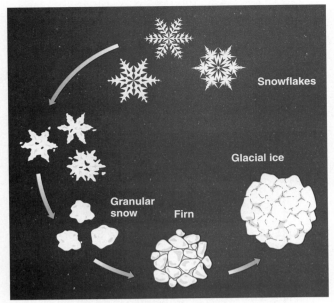

a)

b)

● **FIGURE 13.3 Glacial Ice** (a) The conversion of freshly fallen snow to firn and then to glacial ice. (b) This iceberg in Portage Lake in Alaska shows the blue color of glacial ice. The longer wavelengths of white light are absorbed by the ice, but blue (short wavelength) is transmitted into the ice and scattered, accounting for the blue color.

recrystallized until it is transformed into **glacial ice**, consisting of about 90% solids and 10% air (●**FIGURE 13.3**).

Now you know how glacial ice forms, but we still have not addressed how glaciers move. At this time, it is useful to recall some terms from Chapter 9. Remember that *stress* is force per unit area, and *strain* (or *deformation*) is a change in the shape or volume of solids. When accumulating snow

---

**glaciation** Refers to all aspects of glaciers, including their origin, expansion, and retreat, and their impact on Earth's surface.

**firn** Granular snow formed by partial melting and refreezing of snow; transitional material between snow and glacial ice.

**glacial ice** Water in the solid state within a glacier; forms as snow partially melts and refreezes and compacts so that it is transformed first to firn and then to glacial ice.

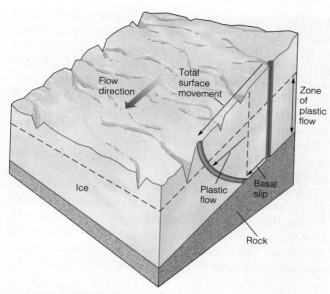

**● FIGURE 13.4 Part of a Glacier Showing Movement by a Combination of Plastic Flow and Basal Slip** Plastic flow involves internal deformation within the ice, whereas basal slip is sliding over the underlying surface. If a glacier is solidly frozen to its bed, it moves only by plastic flow. Notice that the top of the glacier moves farther in a given amount of time than the bottom does.

and ice reach a critical thickness of about 40 m, the stress on the ice at depth is great enough to induce **plastic flow**, a type of permanent deformation involving no fracturing. Glaciers move mostly by plastic flow, but they may also slide over their underlying surface by **basal slip** (●FIGURE 13.4). Liquid water facilitates basal slip because it reduces friction between a glacier and the surface over which it moves.

The total movement of a glacier in a given time is a result of plastic flow and basal slip, although the former occurs continuously, whereas the latter varies, depending on the season, latitude, and elevation. Indeed, if a glacier is solidly frozen to the surface below, as in the case of many polar environments, it moves only by plastic flow. Furthermore, basal slip is far more important in valley glaciers, as they flow from higher to lower elevations whereas continental glaciers need no slope for flow.

Although glaciers move by plastic flow, the upper 40 m or so of ice behaves like a brittle solid and fractures if subjected to stress. Large crevasses commonly develop in glaciers where they flow over an increase in slope of the underlying surface, or where they flow around a corner (●FIGURE 13.5). In either case, the ice is stretched (subjected to tension) and crevasses open, which extend down to the zone of plastic flow. In some cases, a glacier descends over such a steep precipice that crevasses break up the ice into a jumble of blocks and spires, and an icefall develops.

## Distribution of Glaciers

As you might suspect, the amount of snowfall and temperature are important factors in determining where glaciers form. Parts of northern Canada are cold enough to support glaciers but receive too little snowfall, whereas some mountain areas in California receive huge amounts of snow but are too warm for glaciers. Of course, temperature varies with

---

**plastic flow** The flow that takes place in response to pressure and causes deformation with no fracturing.

**basal slip** Movement involving a glacier sliding over its underlying surface.

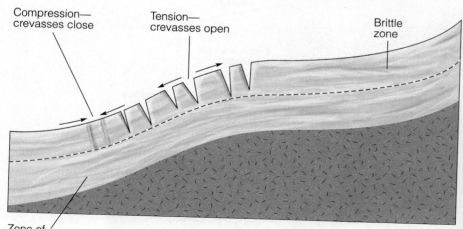

**● FIGURE 13.5 Crevasses** Crevasses are common in the upper parts of glaciers when the ice is subjected to tension. (a) Crevasses open where the brittle part of a glacier is stretched as it moves over a steeper slope in its valley. (b) These crevasses are on the Byron Glacier in Kenai Fjords National Park and Preserve, near Seward, Alaska. Note the two hikers for scale.

elevation and latitude, so we would expect to find glaciers in high mountains and at high latitudes if these areas receive enough snow.

Many small glaciers are present in the Sierra Nevada of California, but only at elevations exceeding 3,900 m. In fact, the high mountains in California, Oregon, and Washington all have glaciers because they are high and receive so much snow.

Glaciers are also found in the mountains along the Pacific Coast of Canada, which also receive considerable snowfall, and, of course, they are farther north. Some of the higher peaks in the Rocky Mountains in both the United States and Canada also support glaciers. At even higher latitudes, as in Alaska, northern Canada, and Scandinavia, glaciers exist at sea level.

# THE GLACIAL BUDGET

**L05**  Define the term glacial budget

**L06**  Explain the difference between the zone of accumulation and the zone of wastage

**L07**  Explain how the zones of accumulation and wastage affect a glacial budget

**L08**  Discuss the parameters that determine how fast a glacier moves

Just as a savings account grows and shrinks as funds are deposited and withdrawn, a glacier expands and contracts in response to accumulation and wastage. We describe a glacier's behavior in terms of a **glacial budget**, which is essentially a balance sheet of accumulation and wastage. For instance, the upper part of a valley glacier is a **zone of accumulation**, where additions exceed losses and the surface is perennially snow covered. In contrast, the lower part of the same glacier is a **zone of wastage**, where losses from melting, sublimation, and calving of icebergs exceed the rate of accumulation (●**FIGURE 13.6**).

At the end of winter, a valley glacier's surface is completely covered with the accumulated seasonal snowfall. During the spring and summer, the snow begins to melt, first at lower elevations and then progressively higher up the glacier. The elevation to which snow recedes during a wastage season is the *firn limit* (Figure 13.6). You can easily

**glacial budget**  The balance between expansion and contraction of a glacier in response to accumulation versus wastage.

**zone of accumulation**  The part of a glacier where additions exceed losses and the glacier's surface is perennially covered with snow. Also refers to horizon B in soil where soluble material leached from horizon A accumulates as irregular masses.

**zone of wastage**  The part of a glacier where losses from melting, sublimation, and calving of icebergs exceed the rate of accumulation.

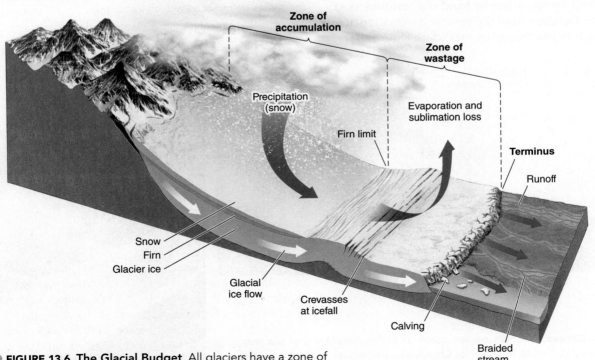

● **FIGURE 13.6  The Glacial Budget** All glaciers have a zone of accumulation, where additions exceed losses and the glacier's surface is perennially covered by snow. They also have a zone of wastage, where losses from melting, calving of icebergs, evaporation, and sublimation exceed gains. If a glacier's budget is balanced, its terminus remains in the same position. However, should the budget be positive, its terminus advances, and if the budget is negative, the terminus retreats.

identify the zones of accumulation and wastage by noting the location of the firn limit.

The firn limit on a glacier may change yearly, but if it does not change or shows only minor fluctuations, the glacier has a balanced budget. That is, additions in the zone of accumulation are exactly balanced by losses in the zone of wastage, and the distal end, or terminus, of the glacier remains stationary (Figure 13.6). If the firn limit moves up the glacier, indicating a negative budget, the glacier's terminus retreats. If the firn limit moves down the glacier, however, the glacier has a positive budget, additions exceed losses, and its terminus advances.

Even though a glacier may have a negative budget and a retreating terminus, the glacial ice continues to move toward the terminus by plastic flow and basal slip. If a negative budget persists long enough, though, the glacier continues to recede, and it thins until it is no longer thick enough to maintain flow. It then ceases moving and becomes a *stagnant glacier*; if wastage continues, the glacier eventually disappears.

We used a valley glacier as an example, but the same budget considerations control the flow of ice caps and continental glaciers as well. The entire Antarctic ice sheet is in the zone of accumulation, but it flows into the ocean, where wastage occurs.

## How Fast Do Glaciers Move?

In general, valley glaciers move more rapidly than continental glaciers, but the rates for both vary from centimeters to tens of meters per day. Valley glaciers moving down steep slopes flow more rapidly than glaciers of comparable size on gentle slopes, assuming that all other variables are the same. The main glacier in a valley glacier system contains a greater volume of ice, and thus has a greater discharge and flow velocity than its tributaries (Figure 13.1). Temperature exerts a seasonal control on valley glaciers because, although plastic flow remains rather constant year-round, basal slip is more important during warmer months when meltwater is abundant.

Flow rates also vary within the ice itself. For example, flow velocity increases downslope in the zone of accumulation until the firn limit is reached; from that point, the velocity becomes progressively slower toward the glacier's terminus. Valley glaciers are similar to streams in that the valley walls and floor cause frictional resistance to flow, so the ice in contact with these boundaries moves more slowly than the ice some distance away (●**FIGURE 13.7A**).

Notice in Figure 13.7a that flow velocity in the interior of a glacier increases upward until the top few tens of meters of ice are reached, but little or no additional increase occurs after that point. This upper ice layer constitutes the rigid part of the glacier that is moving as a result of basal slip and plastic flow below.

Continental glaciers ordinarily flow at a rate of centimeters to meters per day. One reason continental glaciers move comparatively slowly is that they exist at higher latitudes and are frozen to the underlying surface much of the time, which

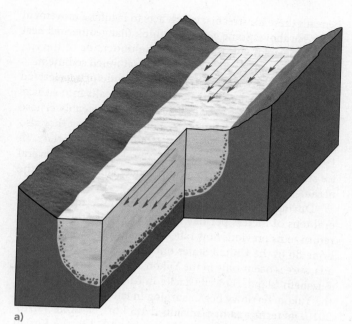
a)

b)

● **FIGURE 13.7 Flow Velocity in Valley Glaciers** (a) Flow velocity in a valley glacier varies horizontally and vertically. Velocity is greatest at the top center because friction with the walls and floor of the trough slows the flow adjacent to these boundaries. The lengths of the arrows are proportional to velocity. (b) Terminus of the Lowell Glacier in Kluane National Park, Yukon Territory, Canada. The glacier was surging when this image was taken on July 2, 2010. Its surge began probably in October 2009 when its terminus was more than 1.5 km farther up the valley.

limits the amount of basal slip. Nevertheless, some parts of continental glaciers manage to achieve extremely high flow rates. Near the margins of the Greenland ice sheet, the ice is forced between mountains in what are called *outlet glaciers*. In some of these outlets, flow velocities exceed 100 m per day.

In parts of the continental glacier covering West Antarctica, scientists have identified ice streams in which flow rates are considerably higher than in adjacent glacial ice. Drilling has revealed a 5-m-thick layer of water-saturated sediment beneath these ice streams, which acts to facilitate movement of the ice above. Some geologists think that geothermal heat from subglacial volcanism melts the underside of the ice, thus accounting for the layer of water-saturated sediment.

A **glacial surge** is a short-lived episode of accelerated flow during which the glacier's surface breaks into a maze of crevasses and its terminus advances noticeably. These brief episodes are best known in valley glaciers, but they also take place in ice caps and even in continental glaciers. In 1995, a huge ice shelf in Antarctica broke apart and several ice streams from the Antarctic ice sheet surged toward the ocean.

During a surge, a glacier's terminus may advance several tens of meters per day for weeks or months and then return to its previous flow rate. Not many glaciers surge, and none do in the United States outside Alaska. Even in Canada, surges occur only in the Yukon Territory and the Queen Elizabeth Islands. Lowell Glacier in Kluane National Park in the Yukon Territory began surging in late 2009, and by May 2010, its terminus had advanced 1.5 km (●**FIGURE 13.7B**). The fastest surge ever recorded was in 1953 in the Kutiah Glacier in Pakistan; the glacier's terminus advanced 12 km in three months, or 130 m per day on average.

# EROSION AND SEDIMENT TRANSPORT BY GLACIERS

**LO9**   Discuss how glaciers erode bedrock

**LO10**   Discuss the features produced by valley glaciers

**LO11**   Explain why areas eroded by continental glaciers are different from those eroded by valley glaciers

As moving solids, glaciers erode, transport, and eventually deposit huge quantities of sediment and soil. Indeed, they have the capacity to transport boulders the size of a house, as well as clay-sized particles. Important processes of erosion include bulldozing, plucking, and abrasion.

Although *bulldozing* is not a formal geologic term, it is fairly self-explanatory; glaciers shove or push unconsolidated materials in their paths. *Plucking*, also called *quarrying*, results when glacial ice freezes in the cracks and crevices of a bedrock projection and eventually pulls it loose.

Bedrock over which sediment-laden glacial ice moves is effectively eroded by **abrasion** and develops a **glacial polish**, a smooth surface that glistens in reflected light (●**FIGURE 13.8A**). Abrasion also yields **glacial striations**, rather straight scratches rarely more than a few millimeters deep on rock surfaces. Abrasion thoroughly pulverizes rocks, yielding an aggregate of clay- and silt-sized particles that have the consistency of flour—hence, the name *rock flour* (●**FIGURE 13.8B**). Rock flour is so common in streams discharging from glaciers that the water has a milky appearance.

a)

b)

● **FIGURE 13.8 Glacial Striations, Polish, and Rock Flour** (a) Abrasion produced glacial polish and striations—the straight scratches—on this basalt at Devils Postpile National Monument in California. (b) The water in this stream in Switzerland is milky because it is discolored by rock flour, small particles produced by glacial abrasion.

Critical Thinking Question Why is the basalt in image (a) broken into five- and six-sided polygons?

Continental glaciers derive sediment from mountains projecting through them, and windblown dust settles on their surfaces, but most of their sediment comes from the

**glacial surge**   A time of greatly accelerated flow in a glacier. Commonly results in displacement of the glacier's terminus by several kilometers.

**abrasion**   The process whereby rock is worn smooth by the impact of sediment transported by running water, glaciers, waves, or wind.

**glacial polish**   A smooth, glistening rock surface formed by the movement of sediment-laden ice over bedrock.

**glacial striation**   A straight scratch rarely more than a few millimeters deep on a rock caused by the movement of sediment-laden glacial ice.

● **FIGURE 13.9 Sediment Transport by Valley Glaciers**
Debris on the surface of the Mendenhall Glacier in Alaska.
The largest boulder is about 2 m across. Notice the icefall
in the background. The person left of center provides scale.

surface they move over. As a result, most sediment is transported in the lower part of the ice sheet. In contrast, valley
glaciers carry sediment in all parts of the ice, but it is concentrated at the base and along the margins (●**FIGURE 13.9**).
Some of the marginal sediment is derived by abrasion and
plucking, but much of it is supplied by mass wasting, as when
soil, sediment, or rock falls or slides onto the glacier's surface.

## Erosion by Valley Glaciers

When mountains are eroded by valley glaciers, they take
on a unique appearance of angular ridges and peaks in the
midst of broad, smooth valleys with near-vertical walls. The
erosional landforms produced by valley glaciers are easily
recognized and enable us to appreciate the tremendous ero-
sive power of moving ice.

**U-SHAPED GLACIAL TROUGHS** A **U-shaped glacial
trough** is one of the most distinctive features of valley glaciation. Mountain valleys eroded by running water are typically V-shaped in cross section; that is, they have valley walls
that descend to a narrow valley bottom (●**FIGURE 13.10A**).
In contrast, valleys scoured by glaciers (●**FIGURE 13.10B**) are
deepened, widened, and straightened so that they have very
steep or vertical walls, but broad, rather flat valley floors;
thus, they exhibit a U-shaped profile (●**FIGURE 13.11**).

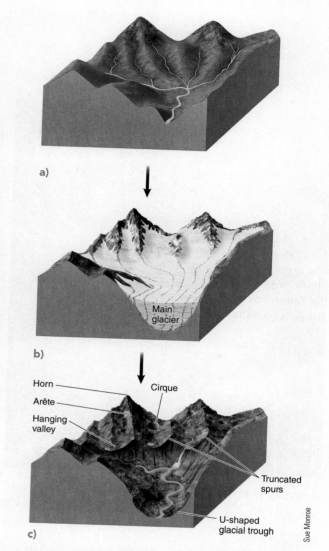

● **FIGURE 13.10 Erosional Landforms Produced by Valley
Glaciers** (a) A mountain area before glaciation. (b) The
same area during the maximum extent of valley glaciers.
(c) After glaciation.

During the Pleistocene, when glaciers were more extensive, sea level was as much as 130 m lower than at present.
Thus, glaciers flowing into the sea eroded their valleys below
present sea level. When the glaciers melted at the end of
the Pleistocene, sea level rose and the ocean filled the lower
ends of the glacial troughs, so that now they are long, steep-
walled embayments called **fiords**.

Fiords are restricted to high latitudes where glaciers exist
at low elevations, such as Alaska, western Canada, Scandinavia, Greenland, southern New Zealand, and southern Chile.
Lower sea level during the Pleistocene was not entirely

**U-shaped glacial trough** A valley with steep or vertical walls
and a broad, rather flat floor formed by the movement of a
glacier through a stream valley.

**fiord** An arm of the sea extending into a glacial trough eroded
below sea level.

Sue Monroe

● **FIGURE 13.11 U-Shaped Glacial Trough** This U-shaped glacial trough is in the Cascade Range in Washington State.

**Critical Thinking Question** How does a U-shaped glacial trough differ from a mountain valley eroded by running water?

responsible for the formation of all fiords. Unlike running water, glaciers can erode a considerable distance below sea level. In fact, a glacier 500 m thick can stay in contact with the seafloor and effectively erode it to a depth of about 450 m before the buoyant effects of water cause the glacial ice to float. The depth of some fiords is impressive; some in Norway and southern Chile are as much as 1,300 m deep!

**HANGING VALLEYS** Some of the world's highest and most spectacular waterfalls are found in recently glaciated areas. Bridalveil Falls in Yosemite National Park, California, plunge 188 m from a **hanging valley**, which is a tributary valley whose floor is at a higher level than that of the main valley (●**FIGURE 13.12**). Where the two valleys meet, the mouth of the hanging valley is perched far above the main valley's floor due to the main valley deepening more rapidly than its tributary valley (Figure 13.10c). Accordingly, streams flowing through hanging valleys plunge over vertical or steep precipices.

James F. Petersen

● **FIGURE 13.12 Hanging Valley** Bridalveil Falls plunge 188 m from a hanging valley in Yosemite National Park, California. The hanging valley was occupied by a glacier that was a tributary to a much larger glacier in Yosemite Valley in the foreground, which is a U-shaped glacial trough.

Although not all hanging valleys form by glacial erosion, many do. As Figure 13.10 shows, the large glacier in the main valley vigorously erodes, whereas the smaller glaciers in tributary valleys are less capable of erosion. When the glaciers disappear, the smaller tributary valleys remain as hanging valleys.

**CIRQUES, ARÊTES, AND HORNS** Perhaps the most spectacular erosional landforms in areas of valley glaciation are at the upper ends of glacial troughs and along the divides that separate adjacent glacial troughs. Valley glaciers form and move out from steep-walled, bowl-shaped depressions called **cirques** at the upper end of their troughs (●**FIGURES 13.10C** and **13.13A**). Cirques are typically steepwalled on three sides, but one side is open and leads into the glacial trough. Some cirques slope continuously into

**hanging valley** A tributary glacial valley whose floor is at a higher level than that of the main glacial valley.

**cirque** A steep-walled, bowl-shaped depression on a mountainside at the upper end of a glacial valley.

James S. Monroe

a)

● **FIGURE 13.13 Cirques, Arêtes, and Horns** (a) A cirque in the St. Elias Mountains of the Yukon Territory of Canada. (b) This view in Switzerland shows two small valley glaciers, the headwall of a cirque, and an arête. (c) The Matterhorn in Switzerland is a well-known horn.

James S. Monroe

b)

c)

Swiss National Tourist Office

the glacial trough, but many have a lip, or threshold, at their lower end.

The details of cirque origin are not fully understood, but they probably form by erosion of a preexisting depression on a mountainside. As snow and ice accumulate in the depression, frost wedging and plucking, combined with glacial erosion, enlarge and transform the head of a steep mountain valley into a typical amphitheater-shaped cirque. Small lakes of meltwater, called *tarns*, often form on the floors of cirques.

**Arêtes**—narrow, serrated ridges—form in two ways. In many cases, cirques form on opposite sides of a ridge, and headward erosion reduces the ridge until only a thin partition of rock remains (Figures 13.10c and ●**13.13B**). The same effect occurs when erosion in two parallel glacial troughs reduces the intervening ridge to a thin spine of rock.

The most majestic of all mountain peaks are **horns**, steep-walled, pyramidal peaks formed by headward erosion

**arête** A narrow, serrated ridge between two glacial valleys or adjacent cirques.

**horn** A steep-walled, pyramid-shaped peak formed by the headward erosion of at least three cirques.

of cirques. For a horn to form, a mountain peak must have at least three cirques on its flanks, all of which erode headward (Figure 13.10). Excellent examples of horns are Mount Assiniboine in the Canadian Rockies, the Grand Tetons in Wyoming, and the most famous of all, the Matterhorn in Switzerland (●**FIGURE 13.13C**).

## Continental Glaciers and Erosional Landforms

Areas eroded by continental glaciers tend to be smooth and rounded, because these glaciers bevel and abrade high areas that project into the ice. Rather than yielding the sharp, angular landforms typical of valley glaciation, they produce a landscape of subdued topography interrupted by rounded hills, because they bury landscapes entirely during their development.

In a large part of Canada, particularly the vast Canadian shield region, continental glaciers have stripped off the soil and unconsolidated surface sediment, revealing extensive exposures of striated and polished bedrock. These areas also have deranged drainage (see Figure 11.17e), numerous lakes and swamps, low relief, extensive bedrock exposures, and little or no soil, and are referred to as **ice-scoured plains**. Similar, though smaller bedrock exposures are also widespread in the northern United States from Maine through Minnesota.

# DEPOSITS OF GLACIERS

LO12  Define the term glacial drift

LO13  Define the two types of glacial drift

LO14  Define the different types of moraines—end, ground, recessional, terminal, lateral, and medial form

LO15  Define a drumlin

LO16  Discuss the two hypotheses for the formation of drumlins

LO17  Define the four major glacial landforms that are composed of stratified drift and formed by running water

LO18  Explain the significance of varves and dropstones

## Glacial Drift

Given that glaciers erode and transport gravel, sand, and mud, it follows that they must also deposit sediment. All glacial deposits go by the general term **glacial drift**, but geologists recognize two types of drift—*till* and *stratified drift*. **Till** is made up of any sediment deposited directly by glacial ice, as, for example, at the terminus of a glacier.

Till is not sorted by particle size and it shows no stratification (●**FIGURE 13.14A**). The till of both valley and continental glaciers is similar, but that of continental glaciers is much more extensive and usually has been transported much further. As opposed to till, **stratified drift**, as the name implies, is layered or stratified, and invariably it shows some degree of sorting by particle size. As a matter of fact, most stratified drift is layers of gravel and sand or mixtures thereof deposited in braided streams that discharge from melting glaciers.

The appearance of till and stratified drift may not be as inspiring as some landforms resulting from glacial erosion, but they are important groundwater reservoirs, and they are exploited for their sand and gravel. In fact, sand and gravel, mostly for construction, are the most valuable mineral commodities in many areas. It is true that all glaciers deposit stratified drift, but, as you would expect, those of continental glaciers are far more extensive. In contrast, the deposits of a valley glacier tend to be restricted to the lower parts of the valley occupied by the glacier.

One conspicuous aspect of glacial drift is rock fragments of various sizes that were obviously not derived from the underlying bedrock. These fragments, called **glacial erratics**, came from some distant source and were then transported and deposited in their present location (●**FIGURE 13.14B**). Some glacial erratics are gigantic, for example, the 15,000-metric-ton Big Rock, or Okotoks erratic in Alberta, Canada, is the largest one known!

## Landforms Composed of Till

Remember, till is deposited directly by glacial ice, and it shows no sorting or stratification. Till is found in several types of deposits collectively called *moraines* and elongated hills known as *drumlins*.

**END MORAINES** If a glacier has a balanced budget, its terminus may become stabilized in one position for some period of time, perhaps a few years or even decades. When an ice front is stationary, flow within the glacier continues, and any sediment transported within or upon the ice is dumped

---

**ice-scoured plain** A low relief bedrock surface with glacial striations and polish eroded by a glacier.

**glacial drift** A collective term for all sediment deposited directly by glacial ice (till) and by meltwater streams (outwash).

**till** All sediment deposited directly by glacial ice.

**stratified drift** Glacial deposits that show both stratification and sorting; deposited by streams that discharge from glaciers.

**glacial erratic** A rock fragment carried some distance from its source by a glacier and usually deposited on bedrock of a different composition.

**● FIGURE 13.14 Glacial Drift and Glacial Erratics** (a) This glacial drift from the Matanuska Glacier, near Palmer, Alaska, is till, because it is unsorted and shows no stratification. (b) Glacial erratic that was deposited by the Exit Glacier in Kenai Fjords National Park near Seward, Alaska.

as a pile of rubble at the glacier's terminus (**●FIGURE 13.15**). These deposits are **end moraines**, which continue to grow as long as the ice front remains stationary. End moraines of valley glaciers are crescent-shaped ridges of till spanning the

**● FIGURE 13.15 End Moraine** End moraines are accumulations of unsorted, nonstratified till deposited at the end of a glacier. End moraine of the Salmon Glacier in the Coast Mountains of British Columbia, Canada.

valley occupied by the glacier. Those of continental glaciers similarly parallel the ice front but are much more extensive.

Following a period of stabilization, a glacier may advance or retreat depending on changes in its budget. If it advances, the ice front overrides and modifies its former moraine. If it has a negative budget, though, the ice front retreats toward the zone of accumulation. As the ice front recedes, till is deposited as it is liberated from the melting ice and forms a layer of **ground moraine**. Ground moraine has an irregular, rolling topography, whereas an end moraine consists of long, ridgelike accumulations of sediment. After a glacier has retreated for some time, its terminus may once again stabilize, and it deposits another end moraine. Because the ice front has receded, such moraines are called **recessional moraines**. The outermost end moraines, marking the greatest extent of the glaciers, go by the special name **terminal moraine**.

**LATERAL AND MEDIAL MORAINES** Valley glaciers transport considerable sediment along their margins, much of it abraded and plucked from the valley walls, but a significant amount falls or slides onto the glacier's surface by mass wasting processes. In any case, this sediment is transported and deposited as long ridges of till called **lateral moraines** along the margin of the glacier (**●FIGURE 13.16**).

**end moraine** A pile or ridge of rubble deposited at the terminus of a glacier.

**ground moraine** The layer of sediment released from melting ice as a glacier's terminus retreats.

**recessional moraine** An end moraine that forms when a glacier's terminus retreats, then stabilizes, and a ridge or mound of till is deposited.

**terminal moraine** An end moraine consisting of a ridge or mound of rubble marking the farthest extent of a glacier.

**lateral moraine** A ridge of sediment deposited along the margin of a valley glacier.

Sue Monroe

Lateral
moraine

Medial
moraine

● **FIGURE 13.16 Types of Moraines** A moraine is a mound or ridge of unstratified till. Moraines are called end, lateral, and medial depending on their position. End moraines are deposited at a glacier's terminus (Figure 13.15), but a lateral moraine is found along the margin of a glacier, and a medial moraine, which forms where two lateral moraines merge, is on the more central part of a glacier.

**Critical Thinking Question** In addition to moraines, what other features of glacial erosion can be identified in this figure?

Where two lateral moraines merge, as when a tributary glacier flows into a larger glacier, a **medial moraine** forms (Figure 13.16). A large glacier will often have several dark stripes of sediment on its surface, each of which is a medial moraine. One can determine how many tributaries a valley glacier has by the number of its medial moraines.

**DRUMLINS** In many areas where continental glaciers deposited till, the till has been reshaped into elongated hills known as **drumlins** (●**FIGURE 13.17**). Some drumlins are 50 m high and 1 km long, but most are much smaller. From the side, a drumlin looks like an inverted spoon, with the steep end on the side from which the glacial ice advanced and the gently sloping end pointing in the direction of ice movement. Drumlins are rarely found as single, isolated hills; instead, they occur in *drumlin fields* that contain hundreds or thousands of drumlins. Drumlin fields are found in several states and Ontario, Canada, but perhaps the finest example is near Palmyra, New York.

According to one hypothesis, drumlins form when till beneath a glacier is reshaped into streamlined hills as the ice moves over it by plastic flow. Another hypothesis holds that huge floods of glacial meltwater modify till into drumlins.

## Landforms Composed of Stratified Drift

Stratified drift exhibits sorting and layering, both indications that it was deposited by running water. In fact, it is deposited by streams discharging from valley and continental glaciers.

**OUTWASH PLAINS AND VALLEY TRAINS** Glaciers discharge meltwater most of the time, except perhaps during the coldest months. This meltwater forms a series of braided streams that radiate out from the front of continental glaciers over a wide region. So much sediment is supplied to these streams that much of it is deposited within their channels as sand and gravel bars. This vast blanket of sediment so formed is an **outwash plain** (●**FIGURE 13.17A**). Valley glaciers also discharge large amounts of meltwater and have braided streams extending from them. However, these streams are confined to the lower parts of glacial troughs, and their long, narrow deposits of stratified drift are known as **valley trains** (●**FIGURE 13.18**).

Outwash plains, valley trains, and some moraines commonly contain numerous circular to oval depressions, many of which contain small lakes. These depressions, or *kettles*, form when a retreating glacier leaves a block of ice that is subsequently partly or wholly buried (Figures 13.17 and 13.18). When the ice block eventually melts, it leaves a depression, and if the depression extends below the water table, it becomes the site of a small lake. Some outwash plains have so many kettles that they are called *pitted outwash plains*.

**KAMES AND ESKERS** Kames are conical hills of stratified drift up to 50 m high (●**FIGURE 13.17B**). Many form when a stream deposits sediment in a depression on a glacier's surface; as the ice melts, the deposit is lowered to the land surface. Kames also form in cavities within or beneath stagnant ice.

Long sinuous ridges of stratified drift, many of which meander and have tributaries, are **eskers** (Figure 13.17b). Most eskers have sharp crests and sides that slope at about 30 degrees. Some are as high as 100 m and can be traced for more than 500 km. The sorting and stratification of the sediments in eskers clearly indicate deposition by running water. The features of ancient eskers and observations of present-day glaciers show that they form in tunnels beneath stagnant ice. Excellent examples of eskers can be seen at Kettle Moraine State Park in Wisconsin, and in several other states. But the most extensive eskers in the world are in northern Canada.

## Deposits in Glacial Lakes

Some lakes in areas of glaciation formed as a result of glaciers scouring out depressions; others occurred where a

**medial moraine** A moraine carried on the central surface of a glacier; formed where two lateral moraines merge.

**drumlin** An elongated hill of till formed by the movement of a continental glacier or by floods.

**outwash plain** The sediment deposited by meltwater discharging from a continental glacier's terminus.

**valley train** A long, narrow deposit of stratified drift confined within a glacial valley.

**kame** A conical hill of stratified drift originally deposited in a depression on a glacier's surface.

**esker** A long, sinuous ridge of stratified drift deposited by running water in a tunnel beneath stagnant ice.

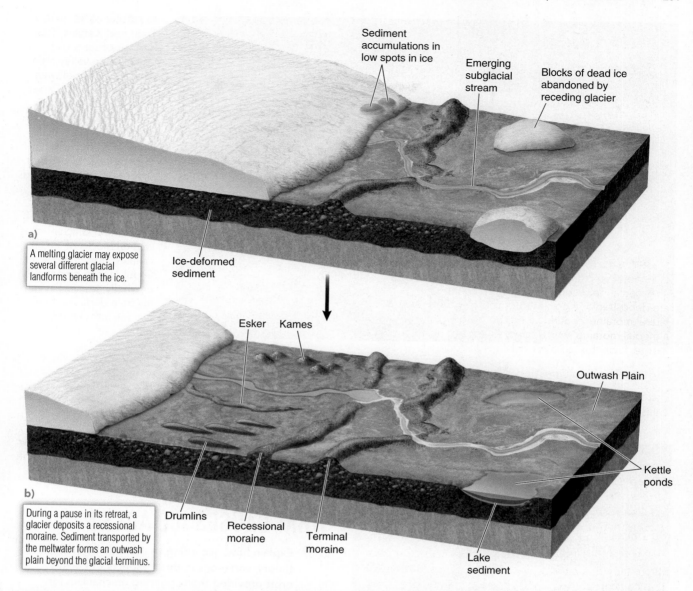

a)

A melting glacier may expose several different glacial landforms beneath the ice.

Sediment accumulations in low spots in ice

Emerging subglacial stream

Blocks of dead ice abandoned by receding glacier

Ice-deformed sediment

b)

During a pause in its retreat, a glacier deposits a recessional moraine. Sediment transported by the meltwater forms an outwash plain beyond the glacial terminus.

Esker    Kames

Outwash Plain

Kettle ponds

Drumlins

Recessional moraine

Terminal moraine

Lake sediment

● **FIGURE 13.17 Development of Features Produced by Past Continental Glaciers** (a) This retreating glacier once covered a larger area as indicated by the terminal moraine. (b) Moraines, eskers, drumlins, kettles, and outwash are all features found in areas once overlain by continental glaciers.

**Critical Thinking Question** Was running water important in the origin of any of the features in this illustration?

stream's drainage was blocked; and others are the result of water accumulating behind moraines or in kettles. Regardless of how they formed, glacial lakes, like all lakes, are areas of deposition. Sediment may be carried into them and deposited as small deltas, but of special interest are the fine-grained deposits.

Mud deposits in glacial lakes are commonly finely laminated (having layers less than 1 cm thick) and consist of alternating light and dark layers known as **varves** (●**FIGURE 13.19**), which represents an annual episode of deposition. The light layer formed during the spring and summer and consists of silt and clay, whereas the dark layer

formed during the winter when the smallest particles of clay and organic matter settled from suspension as the lake froze over.

Another distinctive feature of glacial lakes with varves is **dropstones** (Figure 13.19). These are pieces of gravel, some

**varve** Alternating finely laminated deposits of light and dark layers of sediment that represent an annual episode of deposition.

**dropstone** Pieces of gravel, some of boulder size, found in otherwise very fine-grained deposits.

James S. Monroe

● **FIGURE 13.18 Valley Train and Kettles** This deposit of sand and gravel is the valley train deposited by meltwaters from the Salmon Glacier in the Coast Mountains of British Columbia, Canada. Notice the circular, water-filled depressions; these are kettle lakes, or simply kettles if not filled with water. Also, notice in Figure 13.15—which shows the terminus of the Salmon Glacier—that several blocks of ice are isolated from the glacier. Should these be partially or wholly covered in valley train deposits, they too will form kettles.

**Critical Thinking Question** Why do you think the water in the stream has a milky appearance?

Natural Resources Canada 2011, Courtesy of the Geological Survey of Canada

● **FIGURE 13.19 Varves and a Dropstone in Glacial Deposits** These varves have a dropstone that was probably liberated from floating ice.

of boulder size, in otherwise very fine-grained deposits. The presence of varves indicates that currents and turbulence in these lakes were minimal; otherwise, clay and organic matter would not have settled from suspension. How then can we account for dropstones in a low-energy environment?

Most of them were probably carried into the lakes by icebergs that eventually melted and released sediment contained in the ice.

# WHAT CAUSES ICE AGES?

**LO19** Explain how, according to the Milankovitch theory, variations in three aspects of Earth's orbit provided the triggering mechanism for the glacial–interglacial episodes during the Pleistocene Epoch

**LO20** Discuss the various hypotheses put forth to explain short-term climatic changes

We discussed the conditions necessary for a glacier to form earlier in this chapter in which more snow falls than melts during the warm season, thus accounting for a net accumulation of snow and ice over the years. But this does not address the broader question of what causes ice ages—that is, times of much more extensive glaciation. Actually, we need to address not only what causes ice ages but also why there have been so few episodes of widespread glaciation in all of Earth history.

Only during the Late Proterozoic Eon, the Late Ordovician, Carboniferous, and Permian periods, and the Pleistocene Epoch has Earth had glaciers on a grand scale. Additionally, widespread glaciation occurred several times during the Pleistocene, with each glacial episode separated by a long *interglacial stage*, during which glaciers were restricted in their distribution.

For more than a century, scientists have attempted to develop a comprehensive theory explaining all aspects of ice ages, but they have not yet been completely successful. One reason for their lack of success is that the climatic changes responsible for glaciation, the cyclic occurrence of glacial–interglacial episodes, and short-term events such as the Little Ice Age operate on vastly different time scales.

Only a few periods of glaciation are recognized in the geologic record, each separated from the others by long intervals of mild climate. Such long-term climatic changes probably result from slow geographic changes related to plate tectonic activity. Moving plates carry continents to high latitudes where glaciers exist, provided they receive enough precipitation as snow. Plate collisions, the subsequent uplift of vast areas far above sea level, and the varying atmospheric and oceanic circulation patterns caused by the changing shapes and positions of plates also contribute to long-term climate change.

## The Milankovitch Theory

During the 1920s, the Serbian astronomer Milutin Milankovitch proposed that minor irregularities in Earth's rotation and orbit are sufficient to alter the amount of solar radiation received at any given latitude and hence bring about climate changes. Now called the **Milankovitch theory**, this reasoning was initially ignored but has received renewed interest since the 1970s and is now widely accepted.

Milankovitch attributed the onset of the glacial episodes to variations in three aspects of Earth's orbit. The first is *orbital eccentricity*, which is the degree to which Earth's orbit around the Sun changes over time (●**FIGURE 13.20A**). When the orbit is nearly circular, both the Northern and Southern Hemispheres have similar contrasts between the seasons. However, if the orbit is more elliptical, hot summers and cold winters will occur in one hemisphere, whereas warm summers and cool winters will take place in the other hemisphere. Calculations indicate a roughly 100,000-year cycle between times of maximum eccentricity, which corresponds closely to the 20 warm–cold climatic cycles that took place during the Pleistocene.

Milankovitch also pointed out that the angle between Earth's axis and a line perpendicular to the plane of Earth's orbit shifts about 1.5 degrees from its current value of 23.5 degrees during a 41,000-year cycle (●**FIGURE 13.20B**). Although changes in *axial tilt* have little effect on equatorial latitudes, they strongly affect the amount of solar radiation received at high latitudes and the duration of the dark period at and near Earth's poles. Coupled with the third aspect of Earth's orbit, precession of the equinoxes (Figure 13.10c), high latitudes might receive as much as 15% less solar radiation, certainly enough to affect glacial growth and melting.

*Precession of the equinoxes*, the last aspect of Earth's orbit that Milankovitch cited, refers to a change in the time of the

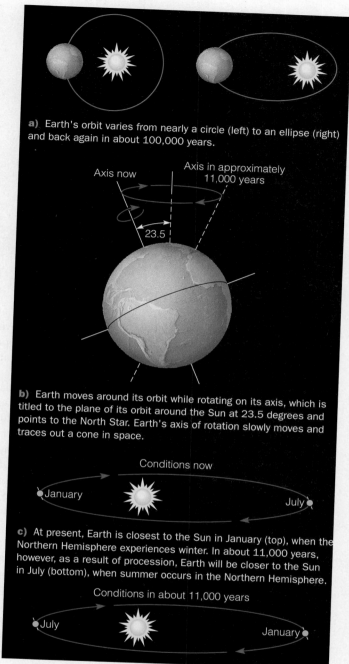

a) Earth's orbit varies from nearly a circle (left) to an ellipse (right) and back again in about 100,000 years.

b) Earth moves around its orbit while rotating on its axis, which is tilted to the plane of its orbit around the Sun at 23.5 degrees and points to the North Star. Earth's axis of rotation slowly moves and traces out a cone in space.

Conditions now

c) At present, Earth is closest to the Sun in January (top), when the Northern Hemisphere experiences winter. In about 11,000 years, however, as a result of procession, Earth will be closer to the Sun in July (bottom), when summer occurs in the Northern Hemisphere.

Conditions in about 11,000 years

● **FIGURE 13.20 Milankovitch Theory** According to the Milankovitch theory, minor irregularities in Earth's rotation and orbit may affect climatic changes.

equinoxes (●**FIGURE 13.20C**). Presently, the equinoxes take place on, or about, March 21 and September 21, when the Sun is directly over the equator. But as Earth rotates on its axis, it also wobbles as its axial tilt varies 1.5 degrees from

**Milankovitch theory** An explanation for the cyclic variations in climate and the onset of ice ages as a result of irregularities in Earth's rotation and orbit.

its current value, thus changing the time of the equinoxes. Taken alone, the time of the equinoxes has little climatic effect, but changes in Earth's axial tilt also change the times of *aphelion* and *perihelion*, which are, respectively, when Earth is farthest from and closest to the Sun during its orbit (Figure 13.20c). Earth is now at perihelion, closest to the Sun, during Northern Hemisphere winters, but in about 11,000 years, perihelion will be in July. Accordingly, Earth will be at aphelion, farthest from the Sun, in January and have colder winters.

Continuous variations in Earth's orbit and axial tilt cause the amount of solar heat received at any latitude to vary slightly through time. The total heat received by the planet changes little, but according to Milankovitch, and now many scientists agree, these changes cause complex climatic variations and provided the triggering mechanism for the glacial–interglacial episodes of the Pleistocene.

## Short-Term Climatic Events

Climatic events with durations of several centuries, such as the Little Ice Age, are too short to be accounted for by plate tectonics or Milankovitch cycles (see GEO-FOCUS).

# GEO-FOCUS

## Glaciers and Global Warming

As you are well aware, global warming is a phenomenon of a warming trend in Earth's lower atmosphere during the past 100 years or so. Many scientists are convinced that warming is caused by the concentration of greenhouse gases in the atmosphere, especially carbon dioxide, methane, and nitrous oxide. They further think that the source of the increased quantities of these gases is the burning of fossil fuels, especially coal and petroleum. Of course, there are dissenters, most of whom acknowledge that Earth's surface temperatures are warmer than in the past, but attribute the increase to normal climatic variations.

Some ask, "So what if global temperatures increase by a few degrees?" However, temperature increase has far-reaching effects. Remember that during the Little Ice Age, temperatures were only a little cooler than they are now, and yet the cooler weather, especially during the summers, had tragic consequences (see Introduction).

Whatever the cause of global warming, no one doubts that glaciers are good indicators of short-term climatic changes. Two factors account for the health of a glacier—the amount of snowfall and temperature. If there is considerable snowfall and cold temperatures, glaciers thicken and

advance (they have a positive budget), whereas if the opposite conditions prevail, they lose mass and their termini retreat.

So, what do glaciers tell us about climate? We already noted that snowfall and temperature are controls on a glacier's budget (Figure 13.6). It is true that not many of the 160,000 or so glaciers on Earth have been studied, but those that have been studied show an alarming trend: many are thinning, their termini are retreating, and in many cases, they have nearly or completely disappeared. This is especially true of the ice sheets covering Greenland and Antarctica, and the sea ice extending from the coasts of the Arctic, Antarctica, Greenland, and northern Canada. Examples of glacial retreat, thinning ice sheets, and the loss of sea ice are numerous, and the several selected here, serve to illustrate the extent of this phenomenon.

Although Nisqually Glacier on Mount Rainier in Washington State has shown a reversal in its overall trend of retreat, it is not nearly as large as it was just a few decades ago. Furthermore, most of the other glaciers in Washington State show an overall retreat, even following years with exceptionally heavy snowfall. Glacier Peak in Washington, a volcano that last erupted in

1880, has more than a dozen glaciers, all of which are retreating.

Of the 150 or so glaciers that were in Glacier National Park in Montana about a century ago, perhaps 25 remain, and many of those are such shrunken remnants, that it is difficult to determine if they are active glaciers or not (●**FIGURE 1**). Indeed, the consensus is that they will all be gone in a decade or two. The warming in Glacier National Park is slight, but there are fewer days per year of below-freezing temperatures, and there are now more warmer days.

In Africa, the area covered by glaciers on Mount Kilimanjaro has decreased by 85%, and a similar trend is seen on Mount Kenya. In the Himalayas of Asia, it is true that some glaciers are advancing or holding their own, but mostly in the Karakoram Range, whereas those in other parts of this mountain system are retreating.

In the Arctic region, the extent of Arctic ice has been decreasing due to warming, such that the extent of its ice cover has declined almost 2.6 million km$^2$ between 1979 and 2016. In fact, the Arctic is warming faster than elsewhere on Earth, further exacerbating the shrinking of ice cover in this area. Some scientists are predicting that at its current rate,

Several hypotheses have been proposed to account for these short-term events, including variations in solar energy and volcanism.

Variations in solar energy could result from changes within the Sun itself or from anything that would reduce the amount of energy Earth receives from the Sun. The latter could result from the solar system passing through clouds of interstellar dust and gas or from substances in the atmosphere reflecting solar radiation back into space. Records kept during the past century indicate that during this time the amount of solar radiation has varied only slightly. Although variations in solar energy may influence short-term climatic events, such a correlation has not yet been demonstrated.

During large volcanic eruptions, tremendous amounts of ash and gases are spewed into the atmosphere, where they reflect incoming solar radiation and thus reduce atmospheric temperatures. Small droplets of sulfur gases remain in the atmosphere for years and can have a significant effect on climate. Several large-scale volcanic events have occurred in the past, such as the 1815 eruption of Tambora, and are now known to have had climatic effects. However, no relationship between periods of volcanic activity and periods of glaciation has yet been established.

summer sea ice in the Arctic could disappear this century.

In Antarctica, the ice shelfs around its coastlines are crumbling and breaking apart due to warming waters, with West Antarctica experiencing the highest rate of shrinking. Of the 674 glaciers in West Antarctica, 90% are retreating and breaking off pieces that float away as icebergs

From 2000 to the present, Earth has experienced many of the hottest years on record. The heat waves in 2016 were the highest in recorded history. This record beat the previous 2015 heat waves as the hottest on record, which had broken the 2014 record. One consequence of higher temperatures is, of course, retreating glaciers as we have discussed. But remember that glaciers are one reservoir in the hydrologic cycle, so that as they waste away, their water returns to the oceans and sea level rises. We know that during the Pleistocene Epoch (Ice Age), sea level was as much as 130 m lower than it is today, and we know that it has risen on average nearly 2 mm per year during the past 100 years; the current rate is estimated at about 3 mm per year. This increase may not sound like much, but even a slight rise in sea level will eventually be a real problem for many of Earth's coastal regions.

Climate change is a complex problem in which there is no simple answer. The melting and retreat of glaciers is just one example, albeit one that can be quantitatively measured. This brief discussion is in no means a comprehensive analysis of the problem, but merely an introduction and starting point for further study.

● **FIGURE 1** Views of Shepherd Glacier in Glacier National Park, Montana, taken in 1913 (top) and 2005 (bottom). By 2010, Shepherd Glacier was too small to qualify as a glacier, as was another glacier in the park.

Glacier National Park Archives

# Key Concepts Review

- Glaciers currently cover about 10% of the land surface.
- The Little Ice Age was a period of time between 1500 and the middle to late 1800s during which glaciers expanded to their greatest historical extent.
- A glacier forms when winter snowfall exceeds summer melt and accumulates year after year. Snow is compacted and converted to glacial ice, and when the ice is about 40 m thick, pressure causes it to flow.
- Glaciers move by plastic flow and basal slip.
- Valley glaciers are confined to mountain valleys and flow from higher to lower elevations, whereas continental glaciers cover vast areas and flow outward in all directions from a zone of accumulation.
- The behavior of a glacier depends on its budget, which is the relationship between accumulation and wastage. If a glacier has a balanced budget, its terminus remains stationary; a positive or negative budget results in the advance or retreat of the terminus, respectively.
- Glaciers move at varying rates depending on slope, discharge, and season. Valley glaciers tend to flow more rapidly than continental glaciers.
- Glaciers effectively erode and transport because they are solids in motion. They are particularly effective at eroding soil and unconsolidated sediment, and they can transport any size sediment supplied to them.
- Continental glaciers transport most of their sediment in the lower part of the ice, whereas valley glaciers may carry sediment in all parts of the ice.
- Erosion of mountains by valley glaciers yields several sharp, angular landforms, including cirques, arêtes,

and horns. U-shaped glacial troughs, fiords, and hanging valleys are also products of valley glaciation.
- Continental glaciers abrade and bevel high areas, producing a smooth, rounded landscape known as an ice-scoured plain.
- Depositional landforms include moraines, which are ridgelike accumulations of till. The several types of moraines are terminal, recessional, lateral, and medial.
- Drumlins are composed of till that was apparently reshaped into streamlined hills by continental glaciers or floods.
- Stratified drift in outwash plains and valley trains consists of sand and gravel deposited by meltwater streams issuing from glaciers. Ridges known as eskers, and conical hills called kames are also composed of stratified drift.
- Major glacial intervals separated by tens or hundreds of millions of years probably occur as a result of the changing positions of tectonic plates, which in turn cause changes in oceanic and atmospheric circulation patterns.
- Currently, the Milankovitch theory is widely accepted as the explanation for glacial–interglacial intervals during the Pleistocene Epoch.
- The reasons for short-term climatic changes, such as the Little Ice Age, are not understood. Two proposed causes are changes in the amount of solar energy received by Earth and volcanism.

# Important Terms

abrasion   274
arête   277
basal slip   271
cirque   276
continental glacier   269
dropstone   281
drumlin   280
end moraine   279
esker   280
fiord   275
firn   270
glacial budget   272
glacial drift   278
glacial erratic   278

glacial ice   270
glacial polish   274
glacial striation   274
glacial surge   274
glaciation   270
glacier   268
ground moraine   279
hanging valley   276
horn   277
ice cap   269
ice-scoured plain   278
kame   280
lateral moraine   279
Little Ice Age   268

medial moraine   280
Milankovitch theory   283
outwash plain   280
plastic flow   271
recessional moraine   279
stratified drift   278
terminal moraine   279
till   278
U-shaped glacial trough   275
valley glacier   268
valley train   280
varve   281
zone of accumulation   272
zone of wastage   272

# Review Questions

1. A medial moraine forms where
   a. snow is converted to glacial ice.
   b. erosion by a valley glacier forms a cirque.
   c. two lateral moraines merge.
   d. isolated blocks of ice are buried in outwash.
   e. a continental glacier abrades its underlying surface.

2. The line on a glacier that separates the zone of accumulation from the zone of wastage is the
   a. drift area.
   b. firm limit.
   c. glacial termination.
   d. glacial ice limit.
   e. Milankovitch limit.

3. Glaciers move mostly by
   a. plastic flow.
   b. abrasion.
   c. plucking.
   d. basal surge.
   e. elastic rebound.

4. A distinctive erosional feature of mountains that have been eroded by valley glaciers is a/an
   a. valley train.
   b. kame.

   c. ice-scoured plain.
   d. cirque.
   e. dropstone.

5. When freshly fallen snow compacts and partly melts and refreezes, it forms granular ice known as
   a. till.
   b. outwash.
   c. glacial frost.
   d. firn.
   e. glacial drift.

6. How does the Milankovitch theory account for the onset of glacial episodes?

7. What kinds of evidence would indicate that an ice-free mountain area was once glaciated?

8. A valley glacier has a cross-sectional area of 400,000 $m^2$ and a flow velocity of 2 m per day. How long will it take for 1 $km^3$ of ice to move past a given point?

9. Explain in terms of the glacial budget how a once-active glacier becomes stagnant.

# Creative Thinking Visual Question

This image (●FIGURE 1) shows the Chugach Mountains in Alaska. Identify the land forms that resulted from valley glaciation.

● FIGURE 1 Glaciers in the Chugach Mountains of Alaska.

Bram Reusen/Shutterstock.com

# 14

# THE WORK OF WIND AND DESERTS

The Mesquite Flat Sand Dunes in Death Valley, California, are a mix of dominantly transverse, crescent-type, and star dunes.

# INTRODUCTION

During the past several decades, deserts have been advancing across millions of acres of productive land, destroying rangeland, croplands, and even villages. Such expansion has exacted a terrible toll in human suffering. Because of the relentless advance of deserts, and the resulting degradation of crop and grazing lands, hundreds of thousands of people have died of starvation or been forced to migrate as "environmental refugees." It is estimated that by 2020, 60 million people from sub-Saharan Africa will most likely have to migrate, many to North Africa or Europe, because of the relentless advance of deserts. This expansion of deserts into formerly productive lands is called **desertification** and is a major problem in many countries. Those areas currently most affected and likely to be affected by desertification are sub-Saharan Africa, southern Iraq, Afghanistan, and southern Asia.

Most regions undergoing desertification lie along the margins of existing deserts, where a delicately balanced ecosystem serves as a buffer between the desert on one side and a more humid environment on the other. These regions have limited potential to adjust to increasing environmental pressures from natural causes as well as human activity. Ordinarily, desert regions expand and contract gradually in response to natural processes such as climatic changes, but much of the recent desertification has been greatly accelerated by human activities.

In many areas, the natural vegetation has been cleared as crop cultivation has expanded into increasingly drier desert fringes to support growing populations. Because grasses are the dominant natural vegetation in most desert fringe areas, raising livestock is a common economic activity. However, increasing numbers of livestock in many areas have greatly exceeded the land's ability to support them. Consequently, the vegetation cover that protects the soil has diminished, causing the soil to crumble and be stripped away by wind and water, which results in increased desertification and the many problems associated with it.

There are many important reasons to study deserts and the processes that are responsible for their formation. First, deserts cover large regions of Earth's surface. More than 40 percent of Australia is desert, and the Sahara occupies a vast part of northern Africa. Many of these places already have problems associated with increasing population and the strains it places on the environment, particularly the need for greater amounts of groundwater (see Chapter 12).

Furthermore, with the current debate about global warming, it is important to understand how desert processes operate and how global climate changes affect the various Earth systems and subsystems. Learning about the underlying causes of climate change by examining ancient desert regions may provide insight into the possible duration and severity of future climatic changes.

As an example, more than 6,000 years ago, the Sahara was a fertile savannah supporting a diverse fauna and flora, including humans. Then the climate changed, and the area became a desert. How did this happen? Will this region change back again in the future? These are some of the questions geoscientists hope to answer by studying deserts. By understanding the underlying causes of desertification, it might be possible to implement steps to reduce the destruction done by desertification, particularly in terms of human suffering.

# SEDIMENT TRANSPORT BY WIND

**L01**  Explain how bed load and suspended load sediments are moved by wind

Wind is a turbulent fluid and therefore transports sediment in much the same way as running water. Although wind typically flows at a greater velocity than water, it has a lower density and thus can carry only clay- and silt-size particles as *suspended load*. Sand and larger particles are moved along the ground as *bed load*.

## Bed Load

Sediments that are too large or heavy to be carried in suspension by water or wind are moved as **bed load** either by *saltation* or by rolling and sliding. Saltation on land occurs when wind starts sand grains rolling and lifts and carries some grains short distances before they fall back to the surface. As the descending sand grains hit the surface, they strike other grains, causing them to bounce along (●**FIGURE 14.1**). Wind-tunnel experiments show that once sand grains begin moving, they

---

**desertification**  Sediments, mostly sand and gravel, that are moved by wind along the ground surface by saltation, rolling, or sliding.

**bed load**  The part of a stream's sediment load, mostly sand and gravel, transported along its bed.

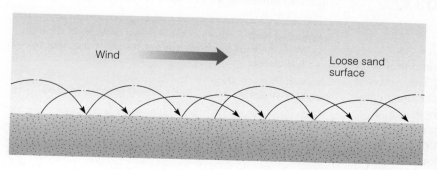

Wind          Loose sand surface

● **FIGURE 14.1 Saltation**  Most sand is moved near the ground surface by saltation. Sand grains are picked up by the wind and carried a short distance before falling back to the ground, where they usually hit other grains, causing them to bounce and move in the direction of the wind.

continue to move, even if the wind drops below the speed necessary to start them moving. This happens because once saltation starts, it sets off a chain reaction of collisions between sand grains that keeps the grains in constant motion.

Saltating sand usually moves near the surface, and even when winds are strong, grains are rarely lifted higher than about a meter. If the winds are very strong, these wind-whipped grains can cause extensive abrasion. It is not uncommon during a severe sandstorm to have a car's paint removed by sandblasting in a short time, and its windshield become completely frosted and translucent from pitting.

## Suspended Load

Silt- and clay-sized particles constitute most of a wind's **suspended load**. Even though these particles are much smaller and lighter than sand-sized particles, wind usually starts the latter moving first. The reason for this phenomenon is that a very thin layer of motionless air lies next to the ground, where the small silt and clay particles remain undisturbed. The larger sand grains, however, stick up into the turbulent air zone, where they can be moved. Unless the stationary air layer is disrupted, the silt and clay particles remain on the ground, providing a smooth surface.

This phenomenon can be observed on a dirt road on a windy day. Unless a vehicle travels over the road, little dust is raised even though it is windy. When a vehicle moves over the road, however, it breaks the calm boundary layer of air and disturbs the smooth layer of dust, which is picked up by the wind and forms a dust cloud in the vehicle's wake.

In a similar manner, when a sediment layer is disturbed, silt- and clay-sized particles are easily picked up and carried in suspension by the wind, creating clouds of dust or even dust storms. Once these fine particles are lifted into the atmosphere, they may be carried thousands of kilometers from their source.

# WIND EROSION

**LO2**  **Explain the difference between abrasion and deflation and the resulting desert features produced by each process**

Although wind action produces many distinctive erosional features, and is an extremely efficient sorting agent, running water is responsible for most erosional landforms in arid regions despite the fact that stream channels are typically dry. Wind erodes material in two ways: *abrasion* and *deflation*.

## Abrasion

**Abrasion** involves the impact of saltating sand grains on an object and is analogous to sandblasting. The effects of abrasion are usually minor, because sand, the most common agent of abrasion, is rarely carried more than a meter above the surface. Rather than creating major erosional features, wind abrasion typically modifies existing features by etching, pitting, smoothing, or polishing. Nonetheless, wind

● **FIGURE 14.2**  **Wind Abrasion**  Wind abrasion has formed these bizarre-shaped structures by eroding the lower part of the exposed limestone in the Libyan Desert, Egypt.

**Critical Thinking Question**  Why is only the lower part of these outcrops eroded by wind abrasion?

abrasion can produce many strange-looking and bizarre-shaped features (●**FIGURE 14.2**).

**Ventifacts** are a common product of wind abrasion; these are stones whose surfaces have been polished, pitted, grooved, or faceted by the wind (●**FIGURE 14.3**). If the wind blows from different directions, or if the stone is moved, the ventifact will have multiple facets. Ventifacts are most common in deserts, yet they can form wherever stones are exposed to saltating sand grains—for example, on beaches in humid regions and some outwash plains, such as in New England.

## Deflation

Another important mechanism of wind erosion is **deflation**, which is the removal of loose surface sediment by wind. Among the characteristic features of deflation in many arid and semiarid regions are *deflation hollows*, or *blowouts* (●**FIGURE 14.4**). These shallow depressions of variable dimensions result from differential erosion of surface materials. Ranging in size from several kilometers in diameter and tens of meters deep to small depressions only a few meters wide and less than a meter deep, deflation hollows are common in the southern Great Plains region of the United States.

In many dry regions, the removal of sand-sized and smaller particles by wind leaves a surface of pebbles, cobbles, and boulders. As the wind removes the fine-grained

**suspended load**  Consists of the smallest particles of silt and clay, which are kept suspended above the ground surface by wind turbulence.

**abrasion**  The process whereby rock is worn smooth by the impact of sediment transported by running water, glaciers, waves, or wind.

**ventifact**  A stone with a surface polished, pitted, grooved, or faceted by wind abrasion.

**deflation**  The removal of sediment and soil by wind.

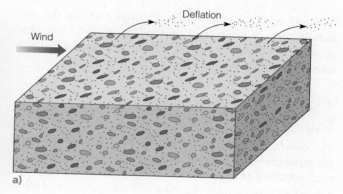

● **FIGURE 14.3 Ventifacts** (a) A ventifact forms when windborne particles (1) abrade the surface of a rock, (2) forming a flat surface. If the rock is moved, (3) additional flat surfaces are formed. (b) Numerous ventifacts are visible in this photo, which also shows desert pavement in Death Valley, California. Desert pavement prevents further erosion and transport of a desert's surface materials by forming a protective layer of close-fitting, larger rocks.

● **FIGURE 14.4 Deflation Hollow** A deflation hollow, shown here as the low area between two sand dunes in Death Valley, California, results when loose surface sediment is differentially removed by wind.

material from the surface, the effects of gravity and occasional heavy rain, and even the swelling of clay minerals, rearrange the remaining coarse particles into a mosaic of close-fitting rocks called **desert pavement** (**FIGURES 14.3B** and ●**14.5**). Once desert pavement forms, it protects the underlying material from further deflation.

# WIND DEPOSITS

**L03** Explain how dunes form and move
**L04** Name the four major types of dunes and explain how each of them forms
**L05** Define loess
**L06** List the three main sources from which loess is derived

Although wind is of minor importance as an erosional agent, it is responsible for impressive deposits, which are primarily of two types. The first, *dunes*, occur in several distinctive

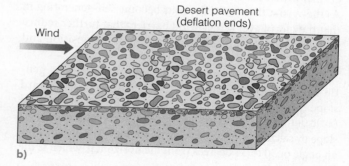

● **FIGURE 14.5 Desert Pavement** (a) Fine-grained material is removed by wind. (b) This leaves a concentration of larger particles that form desert pavement.

**Critical Thinking Question** Why is desert pavement important in desert environments?

**desert pavement** A surface mosaic of close-fitting pebbles, cobbles, and boulders found in many dry regions; results from wind erosion of sand and smaller particles.

● **FIGURE 14.6  Sand Dunes**  Large sand dunes in Death Valley, California. The prevailing wind direction is from left to right, as indicated by the sand dunes in which the gentle windward side is on the left and the steeper leeward slope is on the right.

types, all of which consist of sand-sized particles that are usually deposited near their source. The second, *loess*, consists of layers of windblown silt and clay deposited over large areas downwind and commonly far from their source.

## The Formation and Migration of Dunes

The most characteristic features in sand-covered regions are **dunes**, which are mounds or ridges of wind-deposited sand (●**FIGURE 14.6**). Dunes form when wind flows over and around an obstruction, resulting in the deposition of sand grains, which accumulate and build up a deposit of sand. As they grow, these sand deposits become self-generating in that they form ever-larger wind barriers that further reduce the wind's velocity, resulting in more sand deposition and growth of the dune.

Most dunes have an asymmetrical profile, with a gentle windward slope and a steeper downwind or leeward slope that is inclined in the direction of the prevailing wind (●**FIGURE 14.7A**). Sand grains move up the gentle windward slope by saltation and accumulate on the leeward side, forming an angle of 30 to 34 degrees from the horizontal, which is the

angle of repose of dry sand (see Figure 10.1b). When this angle is exceeded by accumulating sand, the slope collapses, and the sand slides down the leeward slope, coming to rest at its base. As sand moves from a dune's windward side and periodically slides down its leeward slope, the dune slowly migrates in the direction of the prevailing wind (●**FIGURE 14.7B**). When preserved in the geologic record, dunes help geologists determine the prevailing direction of ancient winds (●**FIGURE 14.8**).

## Dune Types

Geologists recognize four major dune types (barchan, longitudinal, transverse, and parabolic), although intermediate forms also exist. The size, shape, and arrangement of dunes result from the interaction of such factors as sand supply, the direction and velocity of the prevailing wind, and the amount of vegetation. Although dunes are usually found in deserts, they can also develop wherever sand is abundant, such as along the upper parts of many beaches.

● **FIGURE 14.8  Cross-Bedding**  Ancient cross-bedding in sandstone beds in Zion National Park, Utah, helps geologists determine the prevailing direction of the wind that formed these ancient sand dunes.

**Critical Thinking Question**  Did the prevailing direction of wind change during the time that the sand forming these sandstone beds was deposited? Explain.

**dunes**  A mound or ridge of wind-deposited sand.

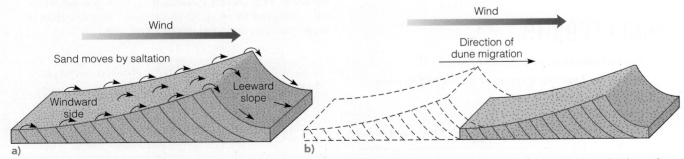

● **FIGURE 14.7  Dune Migration**  (a) Profile of a sand dune. (b) Dunes migrate when sand moves up the windward side and slides down the leeward slope. Such movement of the sand grains produces a series of cross-beds that slope in the direction of wind movement.

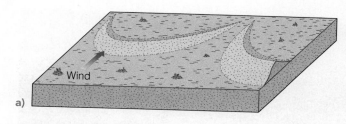

**Barchan dunes** are crescent-shaped dunes whose tips point downwind (●FIGURE 14.9). They form in areas that have a generally flat, dry surface with little vegetation, a limited supply of sand, and a nearly constant wind direction. Most barchans are small, with the largest reaching about 30 m high. Barchans are the most mobile of the major dune types, moving at rates that can exceed 10 m per year.

**Longitudinal dunes** (also called *seif dunes*) are long, parallel ridges of sand aligned generally parallel to the direction of the prevailing winds; they form where the sand supply is somewhat limited (●FIGURE 14.10). Longitudinal dunes result when winds converge from slightly different directions to produce the prevailing wind. They range in height from about 3 m to more than 100 m, and some stretch for more than 100 km. Longitudinal dunes are especially well developed in central Australia, where they cover nearly one-fourth of the continent. They also cover extensive areas in Saudi Arabia, Egypt, and Iran.

**Transverse dunes** form long ridges perpendicular to the prevailing wind direction in areas that have abundant sand and little or no vegetation (●FIGURE 14.11). When viewed from the air, transverse dunes have a wavelike appearance and are therefore sometimes called *sand seas*. The crests of transverse dunes can be as high as 200 m, and the dunes can be as wide as 3 km.

**Parabolic dunes** are most common in coastal areas with abundant sand, strong onshore winds, and a partial cover of vegetation (●FIGURE 14.12). Although parabolic dunes have a crescent shape like barchan dunes, their tips point upwind.

---

**barchan dune** A crescent-shaped sand dune with its tips pointing downwind.

**longitudinal dune** A long ridge of sand generally parallel to the direction of the prevailing wind.

**transverse dune** A ridge of sand with its long axis perpendicular to the wind direction.

**parabolic dune** A crescent-shaped dune with its tips pointing upwind.

●**FIGURE 14.9 Barchan Dunes** (a) Barchan dunes form in areas that have a limited amount of sand, a nearly constant wind direction, and a generally flat, dry surface with little vegetation. The tips of barchan dunes point downward. (b) A ground-level view of several barchan dunes.

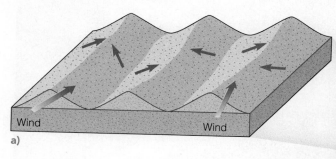

●**FIGURE 14.10 Longitudinal Dunes** (a) Longitudinal dunes form long, parallel ridges of sand aligned roughly parallel to the prevailing wind direction. They typically form where sand supplies are limited. (b) Longitudinal dunes, 15 m high, in the Gibson Desert, west central Australia. The bright blue areas between the dunes are shallow pools of rainwater, and the darkest patches are areas where the Aborigines (Australian Indigenous peoples) have set fires to encourage the growth of spring grasses.

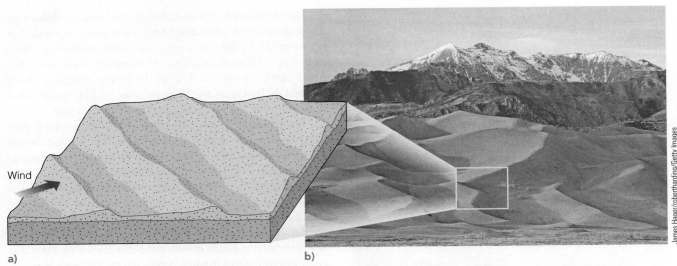

● **FIGURE 14.11 Transverse Dunes** (a) Transverse dunes form long ridges of sand that are perpendicular to the prevailing wind direction in areas of little or no vegetation and abundant sand. (b) Transverse dunes, Great Sand Dunes National Monument, Colorado.

● **FIGURE 14.12 Parabolic Dunes** (a) Parabolic dunes typically form in coastal areas that have a partial cover of vegetation, a strong onshore wind, and abundant sand. (b) A parabolic dune developed along the Lake Michigan shoreline west of St. Ignace, Michigan.

**Critical Thinking Question** Why do the tips of a parabolic dune point upwind, and those of a barchan dune point downwind?

Parabolic dunes form when the vegetation cover is broken and deflation produces a deflation hollow, or blowout. As the wind transports the sand out of the depression, it builds up on the convex downwind dune crest. The central part of the dune is excavated by the wind while vegetation holds the ends and sides fairly well in place.

## Loess

Wind-blown silt and clay deposits composed of angular quartz grains, feldspars, micas, and calcite are known as **loess**. The distribution of loess shows that it is derived from three main sources: deserts, Pleistocene glacial outwash deposits, and the floodplains of rivers in semiarid regions. Loess must be stabilized by moisture and vegetation in order to accumulate. Consequently, loess is not found in deserts, even though deserts provide much of its material. Because of its unconsolidated nature, loess is easily eroded, and as a result, loess areas are characterized by steep cliffs and rapid lateral and headward stream erosion (●**FIGURE 14.13**).

Presently, loess deposits cover approximately 10% of Earth's land surface and 30% of the United States. The most extensive and thickest loess deposits are found in northeast China, where accumulations greater than 30 m thick are common. Loess-derived soils are some of the world's most fertile. It is therefore not surprising that the world's major grain-producing regions correspond to large loess deposits such as the North European Plain, Ukraine, and the Great Plains of North America.

# AIR-PRESSURE BELTS AND GLOBAL WIND PATTERNS

**L07** Explain how the combination of latitudinal pressure differences and the Coriolis effect produce the worldwide pattern of east-west-oriented wind belts

To understand the work of wind and the distribution of deserts, we need to consider the global pattern of air-pressure belts and winds, which are responsible for Earth's atmospheric circulation patterns. Air pressure is the density of air exerted on its surroundings (i.e., its weight). When air is heated, it expands and rises, reducing its mass for a given volume and causing a decrease in air pressure. Conversely, when air is cooled, it contracts and air pressure increases. Therefore, those areas of Earth's surface that receive the most solar radiation, such as the equatorial regions, have low air pressure, whereas the colder areas, such as the polar regions, have high air pressure.

Air flows from high-pressure zones to low-pressure zones. If Earth did not rotate, winds would move in a straight line from one zone to another. Because Earth rotates, however, winds are deflected to the right of their direction of motion (clockwise) in the Northern Hemisphere and to the left of their direction of motion (counterclockwise) in the Southern Hemisphere. This deflection of air between latitudinal zones resulting from Earth's rotation is known as the **Coriolis effect**. The combination of latitudinal pressure differences and the Coriolis effect produces a worldwide pattern of east-west–oriented wind belts (●**FIGURE 14.14**).

Earth's equatorial zone receives the most solar energy, which heats the surface air and causes it to rise. As the air rises, it cools and releases moisture that falls as rain in the equatorial region (Figure 14.14). The rising air is now much drier as it moves northward and southward toward each pole. By the time it reaches 20 to 30 degrees north and south latitudes, the air has become cooler and denser and begins to descend. Compression of the atmosphere warms the descending air mass and produces a warm, dry, high-pressure area, the perfect conditions for the formation of the low-latitude deserts of the Northern and Southern Hemispheres (●**FIGURE 14.15**).

Lowell Georgia/Corbis Documentary/Getty Images

●**FIGURE 14.13 Terraced Wheat Fields in the Loess Soil at Tangwa Village, China** Because of the unconsolidated nature of loess, many farmers live in hillside caves they carved from the loess.

**Critical Thinking Question** Although it is easy to carve hillside caves in loess because it is unconsolidated material, what are the hazards of living in such a structure?

**loess** A wind-blown deposit of silt and clay.

**Coriolis effect** The apparent deflection of a moving object from its anticipated course because of Earth's rotation. Winds and oceanic currents are deflected clockwise in the Northern Hemisphere and counterclockwise in the Southern Hemisphere.

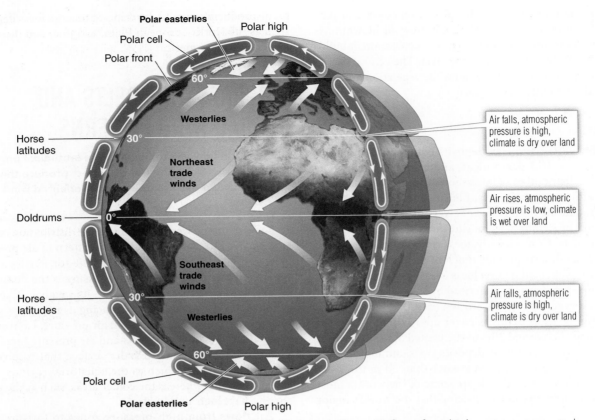

● **FIGURE 14.14 The General Circulation Pattern of Earth's Atmosphere** Air flows from high-pressure zones to low-pressure zones, and the resulting winds are deflected to the right of their direction of movement (clockwise) in the Northern Hemisphere, and to the left of their direction of movement (counterclockwise) in the Southern Hemisphere. This deflection of air between latitudinal zones resulting from Earth's rotation is known as the Coriolis effect.

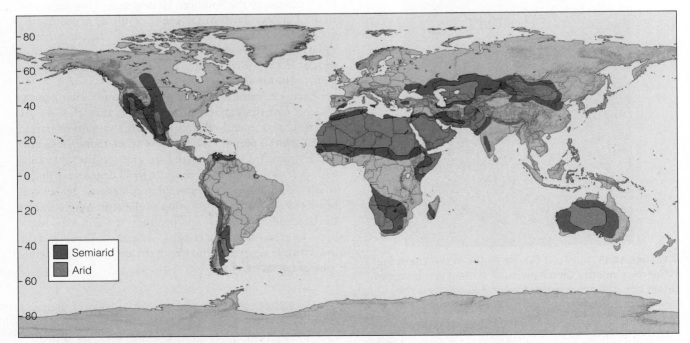

● **FIGURE 14.15 The Distribution of Earth's Arid and Semiarid Regions** Semiarid regions receive more precipitation than arid regions, yet they are still moderately dry. Arid regions, generally described as deserts, are dry and receive less than 25 cm of rain per year. The majority of the world's deserts are located in the dry climates of the low and middle latitudes.

# THE DISTRIBUTION OF DESERTS

**LO8**   Define a desert

**LO9**   Describe the difference between arid and semiarid regions

**LO10**  Locate on a map the world's arid and semiarid regions

Dry climates occur in the low and middle latitudes, where the potential loss of water by evaporation may exceed the yearly precipitation (Figure 14.15). Dry climates cover 30% of Earth's land surface and are subdivided into semiarid and arid regions. *Semiarid regions* receive more precipitation than arid regions, yet are moderately dry. Their soils are usually well developed and fertile and support a natural grass cover. *Arid regions*, generally described as **deserts**, are dry; they receive less than 25 cm of rain per year, have high evaporation rates, typically have poorly developed soils, and are mostly or completely devoid of vegetation.

The majority of the world's deserts are in the low and middle latitude, dry-climate zone between 20 and 30 degrees north and south latitudes (Figure 14.15). The remaining deserts of the world are found in the middle or high latitudes, mostly within the continental interiors in the middle latitudes of the Northern Hemisphere, or in the dry polar regions of the high latitudes. Many of these middle latitude areas are dry because of their remoteness from moist maritime air and the presence of mountain ranges that produce a **rain-shadow desert** (●**FIGURE 14.16**). When moist marine air moves inland and meets a mountain range, it is forced upward. As it rises,

it cools, forming clouds and producing precipitation that falls on the windward side of the mountains. The air that descends on the leeward side of the mountain range is much warmer and drier, producing a rain-shadow desert.

# CHARACTERISTICS OF DESERTS

**LO11**  Describe the temperature ranges, amount of precipitation, and features of vegetation found in deserts

**LO12**  Discuss which of the two types of weathering is dominant in deserts and how this relates to the type of soil formed

**LO13**  Describe the type of mass wasting that takes place in deserts

**LO14**  Describe the type of drainage found in deserts

**LO15**  Explain how wind is an effective geologic agent in deserts

To people who live in humid regions, deserts may seem stark and inhospitable. Instead of a landscape of rolling hills and gentle slopes with an almost continuous cover of vegetation,

**desert**  Any area that receives less than 25 cm of rain per year and that has a high evaporation rate.

**rain-shadow desert**  A desert found on the lee side of a mountain range because precipitation falls mostly on the windward side of the range.

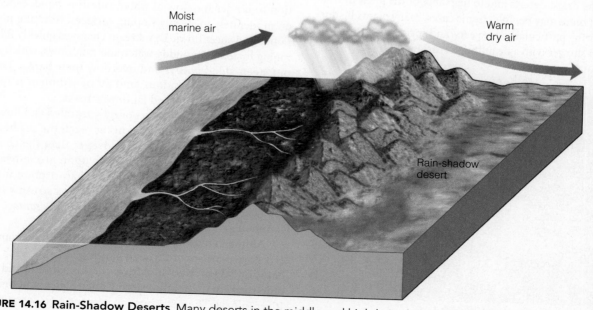

Moist marine air

Warm dry air

Rain-shadow desert

●**FIGURE 14.16 Rain-Shadow Deserts** Many deserts in the middle and high-latitudes are rain-shadow deserts, so named because they form on the leeward side of mountain ranges. When moist marine air moving inland meets a mountain range, it is forced upward where it cools and forms clouds that produce rain. This rain falls on the windward side of the mountains. The air descending on the leeward side is much warmer and drier, producing a rain-shadow desert.

deserts are dry, have little vegetation, and consist of nearly continuous rock exposures, desert pavement, or sand dunes. Yet, despite the great contrast between deserts and more humid areas, the same geologic processes are at work, only operating under different climatic conditions.

## Temperature, Precipitation, and Vegetation

The heat and dryness of deserts are well known. Many of the deserts of the low latitudes have average summer temperatures that range between 32°C and 38°C. It is not uncommon for some low-elevation inland deserts to record daytime highs of 46°C to 50°C for weeks at a time. During the winter months, when the Sun's angle is lower and there are fewer daylight hours, daytime temperatures average between 10°C and 18°C.

Although deserts are defined as regions that receive, on average, less than 25 cm of rain per year, the amount of rain that falls each year is unpredictable and unreliable. It is not uncommon for an area to receive more than an entire year's average rainfall in one cloudburst and then to receive little rain for several years. Thus, yearly rainfall averages can be misleading.

Deserts display a wide variety of vegetation (●FIGURE 14.17). Although the driest deserts, or those with large areas of shifting sand, are almost devoid of vegetation, most deserts support at least a sparse plant cover that, when closely examined, reveals a surprising diversity of plants that have evolved the ability to live in the near absence of water.

Desert plants are widely spaced, typically small, and grow slowly. Their stems and leaves are usually hard and waxy to minimize water loss by evaporation and to protect the plant from sand erosion. Most plants have a widespread shallow root system to absorb the dew that forms each morning in all but the driest deserts and to help anchor the plant in what little soil there may be. In extreme cases, many plants lie dormant during particularly dry years, and spring to life after the first rain shower with a beautiful profusion of flowers.

● **FIGURE 14.17 Desert Vegetation** Desert vegetation is typically sparse, widely spaced, and characterized by slow growth rates. The vegetation shown here is in Death Valley, California.

## Weathering and Soils

Mechanical weathering is dominant in desert regions. Daily temperature fluctuations and frost wedging are the primary forms of mechanical weathering (see Chapter 6). The breakdown of rocks by roots and from salt crystal growth is of minor importance. Some chemical weathering does occur, but its rate is greatly reduced by aridity and the scarcity of organic acids produced by the sparse vegetation. Most chemical weathering takes place during the winter months, when there is more precipitation, particularly in the mid-latitude deserts.

Desert soils, if developed, are usually thin and patchy because the limited rainfall and the resultant scarcity of vegetation reduce the efficiency of chemical weathering and hence soil formation. Furthermore, the sparseness of the vegetative cover enhances wind and water erosion of what little soil actually forms.

## Mass Wasting, Streams, and Groundwater

When traveling through a desert, most people are impressed by such wind-formed features as moving sand, sand dunes, and sand and dust storms. They may also notice the dry washes and dry streambeds. Because of the lack of running water, most people would conclude that wind is the most important erosional agent in deserts. They would be wrong! Running water, even though it occurs infrequently, causes most of the erosion in deserts. The dry conditions and sparse vegetation characteristic of deserts enhance water erosion.

Most of a desert's average annual rainfall of 25 cm or less comes in brief, heavy, localized cloudbursts. During these times, considerable erosion takes place, because the ground cannot absorb all of the rainwater. With so little vegetation to hinder the flow of water, runoff is rapid, especially on moderately to steeply sloping surfaces, resulting in flash floods and sheet flows. Dry stream channels quickly fill with raging torrents of muddy water and mudflows, which carve out steep-sided gullies and overflow their banks. During these times, a tremendous amount of sediment is rapidly transported and deposited far downstream.

Most desert streams are poorly integrated and flow only intermittently. Many of them never reach the sea because, as the water table is usually far deeper than the channels of most streams, they cannot draw upon groundwater to replace water lost to evaporation and absorption into the ground. This type of drainage, in which a stream's load is deposited within the desert, is called *internal drainage* and is common in most arid regions.

Although most deserts have internal drainage, some deserts have permanent through-flowing streams, such as the Nile and Niger rivers in Africa, the Rio Grande and Colorado rivers in the southwestern United States, and the Indus River in Asia. These streams can flow through desert regions because (1) their headwaters are well outside the desert and (2) water is plentiful enough to offset losses resulting from evaporation and infiltration.

## Wind

Although running water does most of the erosional work in deserts, wind can also be an effective geologic agent capable of producing a variety of distinctive erosional (Figure 14.2) and depositional features (Figures 14.9 through 14.13). Wind is effective in transporting and depositing unconsolidated sand-, silt-, and dust-sized particles. Contrary to popular belief, most deserts are not sand-covered wastelands, but rather vast areas of rock exposures and desert pavement (Figure 14.3). Sand-covered regions, or sandy deserts, constitute less than 25% of the world's deserts. The sand in these areas has accumulated primarily by the action of wind.

Wind is not only an effective erosional and depositional agent in deserts but it is also becoming an important resource in generating electricity in many parts of the world (see GEO-FOCUS). The same wind that erodes, transports, and deposits materials is increasingly being harnessed to produce electricity for an ever more energy-hungry world.

# DESERT LANDFORMS

**LO16** Define playa, alluvial fan, pediment, mesa, and butte

**LO17** Explain how each of the aforementioned features in LO16 forms

Because of differences in temperature, precipitation, and wind, as well as the underlying rocks and recent tectonic events, landforms in arid regions vary considerably. Running water, although infrequent in deserts, is responsible for producing and modifying many distinctive landforms found there.

After an infrequent and particularly intense rainstorm, excess water not absorbed by the ground may accumulate in low areas and form *playa lakes* (●FIGURE 14.18A). These lakes are temporary, lasting from a few hours to several months.

Most of them are shallow and have rapidly shifting boundaries as water flows in or leaves by evaporation and seepage into the ground, and the water in a playa lake is often very saline.

When a playa lake evaporates, the dry lake bed is called a **playa** or *salt pan* and is characterized by mud cracks and precipitated salt crystals (●FIGURE 14.18B). Salts in some playas are thick enough to be mined commercially. For example, borates have been mined in Death Valley, California, for more than 100 years.

Another common feature of deserts, particularly in the Basin and Range Province of the western United States, are **alluvial fans**. Alluvial fans form when sediment-laden streams flowing out from the generally straight, steep mountain fronts deposit their load on the relatively flat desert floor. Once beyond the mountain front, where no valley walls confine streams, the sediment spreads out laterally, forming a gently sloping and poorly sorted fan-shaped sedimentary deposit (●FIGURE 14.19). Although alluvial fans are similar in origin and shape to deltas (see Chapter 11), they are formed entirely on land.

Most mountains in desert regions, including those of the Basin and Range Province, rise abruptly from gently sloping surfaces called **pediments**. Pediments are erosional bedrock surfaces of low relief that slope gently away from mountain bases, and are typically covered by a thin layer of debris, or alluvial fans (●FIGURE 14.20).

Rising conspicuously above the flat plains of many deserts are isolated, steep-sided erosional remnants called *inselbergs*, a German word meaning "island mountain."

---

**playa** A dry lake bed found in deserts.

**alluvial fan** A cone-shaped accumulation of mostly sand and gravel deposited where a stream flows from a mountain valley onto an adjacent lowland.

**pediment** An erosion surface of low relief gently sloping away from the base of a mountain range.

a)

b)

**●FIGURE 14.18 Playas and Playa Lakes** (a) A playa lake formed after a rainstorm near Badwater, Death Valley National Park, California. Playa lakes are ephemeral features, lasting from a few hours to several months. (b) Salt deposits and salt ridges cover the floor of this playa in the Mojave Desert, California. Salt crystals and mud cracks are characteristic features of playas.

# GEO-FOCUS

## Windmills and Wind Power

Whoosh, whoosh, whoosh. Ah, the gentle sound of a windmill's blades turning in the wind. The image most people associate with windmills is one of a pastoral landscape in Holland dominated by a classic Dutch windmill crafted of wood (●**FIGURE 1**), or perhaps Don Quixote tilting at windmills in the famous novel *Don Quixote de la Mancha* by Miguel Cervantes. Today, instead of that whoosh, whoosh, whoosh sound, modern electricity-generating windmills in a wind farm sound more like woomph, woomph, woomph, although even that sound is getting quieter with better technology (●**FIGURE 2**).

As early as 5000 B.C., people began to harness the power of wind to propel boats along the Nile River. The Chinese used windmills to pump water for irrigating crops as long ago as 2000 B.C. Wind power was used in the Middle Ages in Europe, particularly in Holland, where windmills have played an important role in society.

Windmills were first used to grind corn, which is where the term *windmill* originally came from. Later, windmills were used to drain lakes and marshes from low-lying areas and to saw timber. Settlers in the United States in the late 19th and early 20th centuries used this technology to pump water and generate electricity in the Great Plains.

With the application of steam power and industrialization in Europe, and later in the United States, the use of windmills rapidly declined. However, industrialization led to the development of larger and more efficient windmills exclusively designed to generate electricity. Denmark began using such windmills as early as 1890, and other countries soon followed suit. Interest in electricity-generating windmills has always mirrored the price of fossil fuels. When the price of petroleum, natural gas, and coal is low, it is cheaper to use these fuels to generate electricity. When the price

of fossil fuels goes up, interest in wind power also increases.

Today, the use of wind power to generate electricity is increasing, in part, because of the realization that burning fossil fuels contributes to climate change. Furthermore, as fossil fuel resources decline, renewable resources such as wind- and solar-generated power can play an increasingly significant role in reducing greenhouse gases. As wind turbine technology has increased the efficiency of wind-generated electricity, the cost of producing electricity has decreased greatly, such that wind farms are no longer a novelty. However, it should be pointed out that without federal subsidies to promote the use of wind energy, wind-generated electric power is still too expensive in most areas to compete with traditional fossil fuel–burning plants, particularly those using natural gas.

How do windmills produce electricity? Simply stated, wind turbines (the term commonly used to describe electricity-producing windmills) convert the kinetic energy (the energy of an object due to motion) of the wind into mechanical power, in this case, the generation of electricity. This electricity is then sent to the local power grid where it is distributed throughout the area.

To be effective, numerous wind turbines are clustered together in wind farms that are located in areas with relatively strong, steady winds. The number of turbines on wind farms can range from several to thousands, as in Texas (Figure 2).

Offshore wind farms are becoming more common as technological innovations allow offshore platforms to operate in deeper waters farther from shore. In the northeastern and Great

George Pachantouris/Moment/Getty Images

●**FIGURE 1** Five traditional Dutch windmills line up along a canal at Kinderdiik, the Netherlands.

David Sucsy/Photographer's Choice/Getty Images

● **FIGURE 2** A wind farm in Texas. In 2016, Texas had the most operating wind-generation capacity in the United States.

Lakes regions where dense populations make land wind farms impractical, the use of offshore wind resources are increasing. One advantage to offshore wind power is that winds are typically stronger during the day, thus allowing for a more stable and efficient production of electricity when consumer and industrial demands are highest. Furthermore, offshore windfarms are generally far enough from land that they cannot be seen, which has been a major complaint by many residents living along the coast. In addition, offshore windfarms are usually far enough from the shore so as not to interfere with fishing, navigation, and recreational boating areas.

In 2016, the 30-megawatt Block Island wind farm off the coast of Rhode Island began operation as the first commercial offshore wind farm in the United States, generating enough electricity to power about 17,000 homes. With the success of the Block Island wind farm, other states along the east coast are moving ahead with projects to develop additional offshore windfarms.

Elsewhere, 3.7% of the world's electricity was supplied by wind power in 2015. The United Kingdom and Germany, which account for more than 66% of offshore wind power installed, were first and second, respectively, in the world in terms of the largest offshore wind farms. Furthermore, 49% of the electricity used in South Australia in 2016 was generated by wind power, and 38% of Denmark's electricity consumption in 2016 was supplied by wind power. In contrast, 8% of electricity-generating capacity in the United States in 2016 was supplied by wind generators, although that percentage is growing.

What are the advantages and disadvantages of wind power? First of all, the wind is a free, renewable energy source, so it cannot be used up. It is also a clean source of energy that does not pollute the water or atmosphere, or contribute to greenhouse gases. Thus, it reduces the consumption of fossil fuels. The land on which windmills are sited can still be used for farming and ranching, thereby increasing the productivity of the land and providing an additional source of income to the landowner who leases the land to utilities. Furthermore, wind farms can benefit the local economy of rural and remote areas by supplying wind energy for local consumption.

There are some disadvantages to wind-generated electricity. The major disadvantage is that wind does not always blow with sufficient strength to be totally reliable, thereby necessitating backup generation. Furthermore, good wind sites are frequently located in remote areas, far from the areas where large quantities of electricity are needed, or in coastal areas, where land is expensive and local residents do not want large wind turbines as neighbors. The initial start-up cost of a wind farm is usually higher than the cost of building a conventional power plant. However, as the cost of wind power has decreased because of better technology, wind-generated electricity is beginning to compete favorably with traditional power plants in many areas.

The "not-in-my-backyard" opposition to wind farms can make siting a wind farm difficult. The major objection to wind farms is the noise generated by the turbines, although as the windmills are built taller and the turbines are more efficient at noise reduction, that is not the major concern it once was.

As the price of fossil fuel continues to rise, the use of a centuries-old staple, the windmill, albeit modernized, will continue to gain in popularity.

● **FIGURE 14.19 Alluvial Fan** A ground view of an alluvial fan, Death Valley, California. Alluvial fans form when sediment-laden streams flowing out from a mountain deposit their load on the desert floor, forming a gently sloping, fan-shaped, sedimentary deposit.

**Critical Thinking Question** Why are alluvial fans that form on land the same general shape as deltas that form in water?

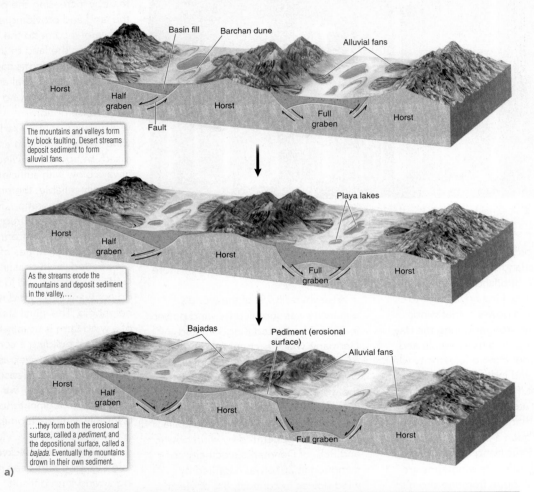

Basin fill    Barchan dune    Alluvial fans

Horst    Half graben    Horst    Full graben    Horst

Fault

The mountains and valleys form by block faulting. Desert streams deposit sediment to form alluvial fans.

Playa lakes

Horst    Half graben    Horst    Full graben    Horst

As the streams erode the mountains and deposit sediment in the valley,...

Bajadas    Pediment (erosional surface)    Alluvial fans

Horst    Half graben    Horst    Full graben    Horst

...they form both the erosional surface, called a *pediment*, and the depositional surface, called a *bajada*. Eventually the mountains drown in their own sediment.

a)

b)

● **FIGURE 14.20 Pediment** (a) Pediments are erosional bedrock surfaces formed by erosion along a mountain front. (b) A pediment north of Mesquite, Nevada.

Inselbergs have survived for a longer period of time than other mountains because of their greater resistance to weathering. Uluru (formerly known as Ayers Rock, Australia) is an excellent example of an inselberg (●FIGURE 14.21).

Other easily recognized erosional remnants common to arid and semiarid regions are mesas and buttes (●FIGURE 14.22). A **mesa** is a broad, flat-topped erosional remnant bounded on all sides by steep slopes. Continued weathering and stream erosion form isolated pillar-like structures known as **buttes**. Buttes and mesas consist of relatively easily weathered sedimentary rocks capped by nearly horizontal, resistant rocks such as sandstone, limestone, or basalt. They form when the resistant rock layer is breached, which allows rapid erosion of the less resistant underlying sediment.

● **FIGURE 14.22 Buttes** Left Mitten Butte and Right Mitten Butte in Monument Valley on the border of Arizona and Utah.

● **FIGURE 14.21 Uluru at Sunset** Contrary to popular belief, Uluru, Australia, is not a giant boulder. Rather, it is the exposed portion of the nearly vertically tilted Uluru Arkose, which can clearly be seen in this image. Differential weathering of the sedimentary layers has produced the distinct parallel ridges and other features characteristic of Uluru.

**mesa** A broad, flat-topped erosional remnant bounded on all sides by steep slopes.

**butte** An isolated, steep-sided, pillar-like hill formed when resistant cap rock is breached, allowing erosion of less resistant underlying rocks.

# Key Concepts Review

- Desertification is the expansion of deserts into formerly productive lands. It destroys croplands and rangelands, causing massive starvation and forcing hundreds of thousands of people to migrate from their homelands.
- Wind transports sediment in suspension or as bed load. Suspended load is the material that is carried in suspension by water or wind. Silt- and clay-sized particles constitute most of a wind's suspended load. Bed load is the material that is too large or heavy to be carried in suspension and is thus moved along the surface by saltation, and by rolling or sliding.
- Wind erodes material by either abrasion or deflation. Abrasion is the impact of saltating sand grains on an object. Ventifacts are common products of wind abrasion.
- Deflation is the removal of loose surface material by wind. Deflation hollows resulting from differential erosion of surface material are common features of many deserts, as is desert pavement, which effectively protects the underlying surface from additional deflation.
- Dunes are mounds or ridges of wind-deposited sand that form when wind flows over and around an obstruction, resulting in the deposition of sand grains, which accumulate and build up a deposit of sand.

- Barchan, longitudinal, transverse, and parabolic are the four major dune types. The amount of sand available, the prevailing wind direction and velocity, and the amount of vegetation determine which type of dune will form.
- Loess consists of wind-blown deposits of silt and clay that is derived from deserts, Pleistocene glacial outwash deposits, or river floodplains in semiarid regions. It covers approximately 10% of Earth's land surface and weathers to a rich, productive soil.
- The winds of the major air-pressure belts, oriented east–west, result from the rising and cooling of air. The winds are deflected clockwise in the Northern Hemisphere and counterclockwise in the Southern Hemisphere by the Coriolis effect to produce Earth's global wind patterns.
- Dry climates, located in the low and middle latitudes where the potential loss of water by evaporation exceeds the yearly precipitation, cover 30% of Earth's land surface and are subdivided into semiarid and arid regions.
- Semiarid regions receive more precipitation than arid regions, yet are moderately dry. Arid regions, generally described as deserts, are dry and receive less than 25 cm of rain per year.

- The majority of the world's deserts are in the dry climates of the low and middle latitudes. The remaining dry climates of the world are found in the middle and high latitudes, mostly within continental interiors in the Northern Hemisphere.
- Deserts are characterized by high temperatures, little precipitation, and sparse plant cover. Rainfall is unpredictable and, when it does occur, tends to be intense and of short duration.
- Mechanical weathering is the dominant form of weathering in deserts and, coupled with slow rates of chemical weathering, results in poorly developed soils.
- Running water is the major agent of erosion in deserts.

- Wind is also an erosional agent in deserts and is very effective in transporting and depositing unconsolidated, fine-grained sediments.
- Desert landforms include playas, which are dry lakebeds, but when temporarily filled with water, they form playa lakes. Alluvial fans are fan-shaped sedimentary deposits. Pediments are erosional bedrock surfaces of low relief that slope gently away from mountain bases, and which are typically covered by a thin layer of debris, or alluvial fans.
- Mesas are flat-topped erosional remnants bounded on all sides by steep slopes, whereas buttes are isolated, steep-sided, pinnaclelike hills.

# Important Terms

abrasion   290
alluvial fan   299
barchan dune   293
bed load   289
butte   303
Coriolis effect   295
deflation   290

desert   297
desertification   289
desert pavement   291
dune   292
loess   295
longitudinal dunes   293
mesa   303

parabolic dune   293
pediment   299
playa   299
rain-shadow desert   297
suspended load   290
transverse dune   293
ventifact   290

# Review Questions

1. What type of dune is crescent-shaped, and its tips point downwind?
   a. parabolic
   b. star
   c. longitudinal
   d. barchan
   e. transverse

2. The majority of the world's deserts are located between what latitudes?
   a. 10 and 20 degrees
   b. 20 and 30 degrees
   c. 30 and 40 degrees
   d. 40 and 60 degrees
   e. 60 and 80 degrees

3. Which of the following is a feature produced by wind deposition?
   a. ventifact
   b. loess
   c. butte
   d. mesa
   e. inselberg

4. The Coriolis effect causes wind to be deflected
   a. to the right in the Northern Hemisphere and to the left in the Southern Hemisphere.
   b. to the left in the Northern Hemisphere and to the right in the Southern Hemisphere.
   c. only to the left in both hemispheres.
   d. only to the right in both hemispheres.
   e. not at all.

5. The major agent of erosion in deserts today is
   a. glaciers.
   b. wind.
   c. abrasion.
   d. running water.
   e. none of the previous answers.

6. Desertification is caused by climate change and exacerbated by human activities. Although we cannot effectively affect short-term climate change on a human time scale, what steps could be taken to reduce the desertification resulting primarily from human activities?

7. As more people move into arid and semiarid areas, an increasing strain is placed on the environment of these areas. What are some of the issues that government entities must face in dealing with these population increases? What are some of the problems that must be dealt with from a geologic perspective?

8. Using what you now know about deserts, their location, how they form, and the various landforms found in them, what evidence in the rock record might allow you to determine where deserts existed in the past?

9. As more images of Mars surface are sent back to Earth from the various landers, a number of features suggest that running water, as well as wind deposition and erosion, have played a major role in shaping the Martian landscape. What type of evidence would you look for in these images that show wind has been an important agent in the formation of the Martian surface?

# Creative Thinking Visual Question

What desert landforms or features of a desert can you identify in this image (●**FIGURE 1**) of the Mesquite Flat sand dunes in Death Valley, California? What types of sand dunes are present?

●**FIGURE 1** Mesquite Flat sand dunes, Death Valley, California.

Jef Wodniack/Shutterstock.com

# 15

# OCEANS, SHORELINES, AND SHORELINE PROCESSES

White chalk cliffs near Etretat Normandy, France. An excellent example of a sea arch is present on the left side of the photo, and a portion of a sea stack can be seen along the left margin.

# INTRODUCTION

In Chapter 2 we discussed several aspects of the seafloor such as oceanic ridges, deep-sea hydrothermal vents, seamounts, and guyots. We also discussed the nature of the continental margins, which include the continental shelf, slope, and rise. Our interest here is in the oceanic waters themselves, and their impact on shorelines.

You already know that the hydrosphere consists of all water on Earth, most of which (97.2%) is in the oceans. This vast interconnected body of saltwater covers 71% of Earth's surface, but parts of it are distinct enough for us to recognize the Pacific, Atlantic, Indian, and Arctic oceans (●FIGURE 15.1). Seas, on the other hand, are marginal parts of oceans, as in the Sea of Japan, the Mediterranean Sea, and the Black Sea. The oceans and seas are underlain by oceanic crust, but the same is not true of the Dead Sea, the Caspian Sea, or the Salton Sea; these are actually large saline lakes on the continents.

What lies beneath the ocean's surface is largely hidden from view, which accounts for why so many sensational stories have persisted for centuries. In about 350 B.C., the Greek Philosopher Plato claimed that a continent called *Atlantis* existed in the Atlantic Ocean west of what we now call the Strait of Gibraltar (●FIGURE 15.2). Plato claimed that following the conquest of *Atlantis* by Athens, this vast continent supposedly

sank, and now only "mud shallows" mark its former location. No geologic evidence indicates that *Atlantis* ever existed, so why has this story persisted for so long? One reason is that stories of lost civilizations are popular, and another is that until fairly recently little was known beneath the ocean's surface.

In this enormous body of water we call oceans, wave energy is transferred through the water to shorelines, where it has a tremendous impact (see the Chapter Opening photo). Accordingly, understanding shoreline processes is important to oceanographers, geologists, and coastal engineers, as well as elected officials and city planners of coastal communities. Indeed, tourism is an important part of the economies of many coastal communities, such as Myrtle Beach, South Carolina; Fort Lauderdale, Florida; and Padre Island, Texas, to name a few.

Another important aspect of shorelines is rising sea level, because buildings or even entire communities that were once inland are now in peril, must be protected or moved, or have already been destroyed. Furthermore, hurricanes expend much of their energy on shorelines, resulting in extensive coastal flooding, numerous fatalities, and widespread property damage. Two of the most destructive and deadly recent hurricanes are Hurricane Katrina, which struck the Gulf Coast in August, 2015, and Hurricane Harvey, in the Houston, Texas area in August, 2017.

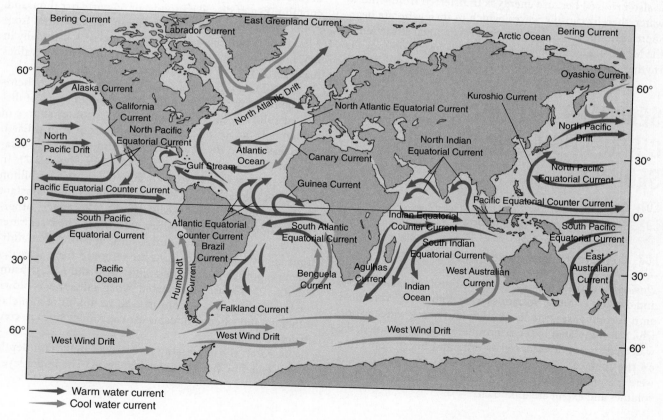

● **FIGURE 15.1 The Oceans** Map showing the Atlantic, Pacific, Indian, and Arctic oceans, and the ocean currents.

**Critical Thinking Question** How do oceans modify Earth's climate?

● **FIGURE 15.2 Atlantis** According to Plato, *Atlantis* was a continent west of the Pillars of Hercules, now called the Strait of Gibraltar. In this map from Anthanasium Kircher's *Mundus Subterraneous* (1664), north is toward the bottom of the map. The Strait of Gibraltar is the narrow area between Hispania (Spain) and Africa.

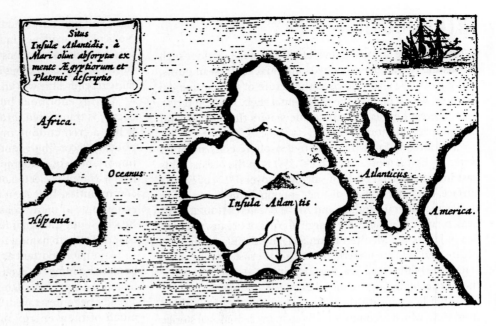

The study of oceans and shorelines provides another excellent example of systems interactions—in this case, between the hydrosphere and solid Earth. The atmosphere is also involved because energy is transferred from wind to water, thereby causing waves, which in turn generate near-shore currents. And, of course, the gravitational attraction of the Moon and Sun on the ocean waters is responsible for the rhythmic rise and fall of tides.

# SEAWATER, OCEANIC CIRCULATION, AND SEAFLOOR SEDIMENTS

**LO1**  Define salinity

**LO2**  Explain why gyres are important as a world temperature control

**LO3**  Compare the three basic types of reefs

During its earliest history, Earth was probably hot, airless, and dry, but erupting volcanoes were ubiquitous. Volcanoes emit several gases, the most abundant being water vapor, which accumulated in the atmosphere. As Earth cooled, the water vapor condensed, fell as rain, and began collecting on the surface. Geologic evidence indicates that oceans were present by at least 3.5 billion years ago, although their volumes and extent are unknown.

## Seawater—Its Composition

Of the more than 70 chemical elements in solution in seawater, the most common ones are chloride ions and

sodium ions, which together make up 85.6% of all dissolved substances and give seawater its most distinctive feature—its saltiness, or **salinity**, which is a measure of the total quantity of dissolved solids. On average, 1 kg (1,000 g) of seawater contains 35 g of dissolved solids, or 35 parts per thousand, symbolized as 35‰. In the open ocean, salinity varies from 32‰ to 37‰, although in some marginal seas, especially in dry, hot areas nearly isolated from the open ocean, values may exceed 40‰.

Runoff from the continents, the source of most chemical elements in seawater, adds about 4 billion tons of dissolved solids to the oceans each year. Another source of elements is *outgassing*, in which gases from within Earth are released into the oceans and the atmosphere by volcanoes at deep-sea hydrothermal vents (see Figure 2.7). In any case, the oceans have been salty for at least 1.5 billion years, but the salinity of seawater remains rather constant because continuous recycling of ions takes place; otherwise seawater would get saltier with time. Thus, seawater is in a state of dynamic equilibrium, meaning that additions are offset by losses. Ions are removed from seawater when evaporites such as rock salt (NaCl) and rock gypsum ($CaSO_4 \cdot 2H_2O$) are precipitated, when salt spray is blown onshore, when magnesium is used in dolomite and clay minerals, and when organisms use calcium or silica to construct their shells.

Based on the decreasing intensity of light with depth, scientists define two layers in the oceans: The upper layer,

---

**salinity** A measure of the dissolved solids in seawater, commonly expressed in parts per thousand.

called the **photic zone**, is usually 100 m or less deep and receives enough light for organisms to photosynthesize. Below is the **aphotic zone**, where too little light is available for photosynthesis, and most organisms depend directly or indirectly on organic substances that "rain" down from the photic zone.

## Oceanic Circulation

As wind blows over a water surface, some of its energy is transferred to the water, which generates surface currents and waves. Figure 15.1 shows the global surface-water current patterns averaged over a long time. Notice in Figure 15.1 that surface currents in the Northern Hemisphere are deflected to the right (clockwise) of their direction of motion and to the left (counterclockwise) of their direction of motion in the Southern Hemisphere. This deflection is the Coriolis effect that results from Earth's rotation (see Chapter 14). The combination of wind and the Coriolis effect produces large-scale water circulation systems known as **gyres** between the 60 degree parallels in the Atlantic, Pacific, and Indian oceans (Figure 15.1). One of the best-known ocean currents is the *Gulf Stream*, which is actually part of the much larger North Atlantic gyre.

Gyres are important as a world temperature control, because seawater near the equator absorbs huge amounts of heat and transports it to high latitudes in warm currents. Cold currents originate at high latitudes and flow toward the equator. Indeed, the reason that some rather northerly countries, such as Scotland, have mild climates is related to these currents. On the other hand, the current along the northern and central coasts of California is cold, which keeps that area much cooler than the inland parts of the state.

In addition to surface currents, horizontal circulation of water also takes place in the deep ocean basins because of differences in temperature and density of adjacent water masses; a water mass of greater density (colder or saltier) will displace and flow beneath a water mass of lesser density. Deep-ocean circulation affects about 90% of all ocean water, but because studying it is expensive and time-consuming, scientists know less about it than about other circulation patterns.

Vertical circulation occurs in the oceans when **upwelling** slowly transfers cold water from depth to the surface, and when **downwelling** carries warm water from the surface to depth. Upwelling is by far the most important process, because as it transfers water upward, it also carries nutrients, particularly nitrates and phosphate, into the photic zone. Here, high concentrations of plankton are sustained that in turn support other organisms. In fact, less than 1% of the ocean surface is in regions of upwelling, yet they support more than 50% by weight of all fishes. In addition, most of Earth's sedimentary rocks containing phosphate were deposited along continental margins where upwelling takes place.

## Seafloor Sediments

Much of the sediment eroded from continents is deposited on the continental margins, but some, mostly silt-and clay-sized particles, is carried into the deep ocean basins where it is deposited. Most of this sediment is *pelagic*, meaning that it settled from suspension far from land. **Pelagic clay** is brown or red, and composed mostly of clay-sized particles, whereas **ooze** is made up of the tiny shells of marine organisms. If dominated by calcium carbonate ($CaCO_3$) skeletons, it is *calcarious ooze*, or if mostly siliceous ($SiO_2$) skeletons, it is *siliceous ooze*.

The term **reef** has many meanings, but here we are concerned with reefs defined as mound-like, wave-resistant structures composed of the shells of marine organisms (●**FIGURE 15.3**). Although commonly called *coral reefs*, they actually have a solid framework of skeletons of corals and clams and encrusting organisms such as sponges and algae. Most reefs are found in tropical seas, where the water temperature does not fall below about 20°C, and rarely at depths of more than 50 m because many corals rely on symbiotic algae that must have enough sunlight for photosynthesis.

Most reefs are one of three basic types: fringing, barrier, and atoll (Figure 15.3). *Fringing reefs* are up to 1 km wide, solidly attached to a landmass, have a rough, tablelike surface, and on their seaward side, slope steeply down to the seafloor. *Barrier reefs* are similar, except that a lagoon separates them from the mainland. The 2,000-km-long Great Barrier Reef of Australia is a good example. Circular to oval reefs surrounding a lagoon are *atolls*. They form around volcanic islands that subside below sea level as the plate they rest on moves into progressively deeper water. As subsidence occurs, the reef-building organisms grow upward so that the living part of the reef remains in shallow water.

---

**photic zone**  The sunlit layer in the oceans where plants photosynthesize.

**aphotic zone**  The depth in the ocean below which sunlight does not penetrate.

**gyre**  A system of ocean currents rotating clockwise in the Northern Hemisphere, and counterclockwise in the Southern Hemisphere.

**upwelling**  The slow circulation of ocean water from depth to the surface.

**downwelling**  The slow transfer of ocean surface water to depth.

**pelagic clay**  Brown or red deep-sea sediment composed of clay-sized particles.

**ooze**  Deep-sea sediment composed mostly of shells of marine animals and plants.

**reef**  A moundlike, wave-resistant structure composed of the skeletons of organisms.

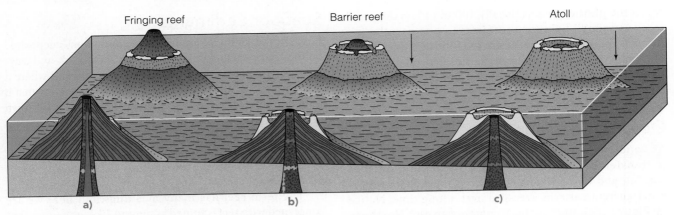

Fringing reef    Barrier reef    Atoll

a)    b)    c)

d)

Courtesy of Carl Roessler

**FIGURE 15.3 The Evolution of Coral Reefs** The origin of a fringing reef, barrier reef, and atoll as the plate upon which the fringing reef is carried into deeper water. (a) A fringing reef forms around an island in the tropics. (b) The island sinks as the oceanic plate on which it rides moves away from a spreading center. In this case, the island does not sink at a rate faster than coral organisms can build upward. (c) The island eventually disappears beneath the surface, but the coral remains at the surface as an atoll. (d) An underwater view of a coral reef in the Red Sea.

# SHORELINES AND SHORELINE PROCESSES

**LO4** Explain how tides form

**LO5** Define the following terms as they relate to waves: crest, trough, wavelength, wave height, celerity, and wave base

**LO6** Explain how waves are generated

**LO7** Explain how longshore currents form

**LO8** Discuss why a swimmer should swim parallel to the shoreline to escape a rip tide

A **shoreline** is the area of land in contact with the ocean or a lake. But we can expand this definition by noting that ocean shorelines include the land between low tide and the highest level on land affected by storm waves. How does a shoreline differ from a coast? Actually, the terms are commonly used interchangeably, but *coast* is more inclusive, and it includes the shoreline as well as an area of indefinite width seaward and landward of the shoreline (●FIGURE 15.4). Our main concern here is with tides, nearshore currents, and waves, the processes that modify seashores.

In the marine realm, several biological, chemical, and physical processes are operating continuously. For example, organisms change the local chemistry of seawater and contribute their skeletons to nearshore sediments. However, the processes most important for modifying shorelines are purely physical ones, especially tides, waves, and nearshore currents.

## Tides

The surface of the oceans rise and fall twice daily in response to the gravitational attraction of the Moon and Sun. These regular fluctuations in the ocean's surface, or **tides**, result in most seashores having two daily high tides and two low tides as sea level rises and falls from a few centimeters to more than 15 m (●FIGURE 15.5). A complete tidal cycle includes a *flood tide* that progressively covers more and more of a nearshore area until high tide is reached, followed by an *ebb tide*, during which the nearshore area is once again exposed (Figure 15.5). These regular fluctuations in sea level constitute one largely untapped source of energy, as do waves, ocean currents, and temperature differences in seawater (see GEO-FOCUS).

---

**shoreline** The area between mean low tide and the highest level on land affected by storm waves.

**tide** The regular fluctuation of the sea's surface in response to the gravitational attraction of the Moon and Sun.

● **FIGURE 15.4 Shorelines and Coasts** The shoreline of this part of the U.S. Pacific Coast consists of the area from about where the waves break to the base of the sea cliffs. However, the coast extends farther seaward and also includes the sea cliffs and an area some distance inland.

a) Low tide.

b) High tide.

● **FIGURE 15.5 Low and High Tides** Low tide (a) and high tide (b) in Turnagain Arm, part of Cook Inlet in Alaska. The tidal range here is about 10 m. Turnagain Arm is a huge fiord now being filled with sediment carried in by rivers. Notice the mudflats in (a).

# GEO-FOCUS

## Energy from the Oceans

If we could harness the energy of waves, ocean currents, and temperature differences in oceanic waters, and tides, an almost limitless, largely nonpolluting energy supply would be ensured. Unfortunately, ocean energy is diffuse, meaning that the energy for a given volume of water is small and thus difficult to concentrate and use.

Of the several sources of ocean energy briefly discussed below, only tides show much promise for the near future.

Ocean thermal energy conversion (OTEC) exploits the temperature difference between surface waters and those at depth to run turbines and generate electricity. The amount of energy from this source is enormous, but a number of practical problems still need to be solved. The most difficult is that large quantities of water must be circulated through a power plant, which requires large surface areas devoted to this purpose. Unfortunately, despite several decades of research, only a few facilities have been tested in Hawaii and Japan.

Ocean currents also possess energy that might be tapped to generate electricity. Regrettably, these currents flow at only a few kilometers per hour at most, whereas hydroelectric power plants on land rely on water moving rapidly from higher to lower elevations. Furthermore, ocean currents cannot be dammed, their energy is diffuse, and any power plant would have to contend with unpredictable changes in flow direction.

Harnessing wave energy to generate electricity is not a new idea, and, in fact, is now used on a limited scale. Any facility using wave energy obviously must be designed to withstand the impact of waves, especially storm waves; be resistant to saltwater corrosion; and be situated where wave energy is vigorous enough for the task. Although no large-scale wave-energy plants are presently operating, Japanese researchers at Okinawa Institute of Science and Technology have shown that electricity can be generated using specially designed small turbines placed where waves break along a coastline. Commercial development of this technology is currently being studied.

The world's first commercial wave power plant, abbreviated Limpet for Land Installed Marine Powered Energy Transformer, went operational in 2000 on the Scottish island of Islay. This is a shoreline device that uses the energy of oscillating waves to generate electricity and has been supplying electricity to the electrical grid in the United Kingdom since its commission (●FIGURE 1).

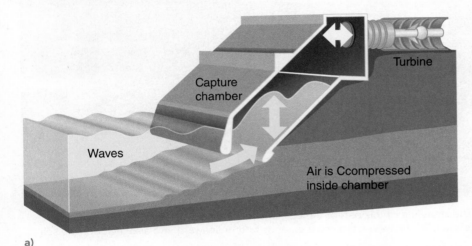

a)

b)

SCIENCE PHOTO LIBRARY/Science Source

● **FIGURE 1 Diagram showing the Limpet (Land Installed Marine Powered Energy Transformer).** (a) Diagram showing a Land Installed Marine Powered Energy Transformer (Limpet). The Limpet, set into a rock face along a shoreline, has a chamber with a turbine that operates when waves alternately compress and decompress air. (b) Waves striking the Limpet 500, the world's first commercial-scale wave powered station. Located on the coast of Islay, a Scottish Hebridean Island, the Limpet generates 500 kilowatts of electricity, enough to power 300 homes.

The success of this unattended power plant demonstrates the potential of shoreline energy as another source of renewable and nonpolluting power. Areas that have the potential for harnessing wave energy include the west coasts of the United States and Europe, as well as the coasts of Japan and New Zealand.

In addition to wave energy, tidal power has potential, but putting it into practice is not easy. First, a dam with sluice gates to regulate water flow must be built across the entrance to a bay or estuary. When the water level has risen sufficiently high during flood tide, the sluice gates are closed. Water held on the landward side of the dam is then released and electricity is generated just as it is at a hydroelectric dam. Furthermore, a tidal power plant can operate during both flood and ebb tides (●FIGURE 2).

The first tidal power-generating facility was constructed in 1966 at the La Ranee River estuary in France. It has a peak output of 240 megawatts, and supplies 0.12% of France's electrical demands. In North America, a much smaller tidal power plant, capable of generating 2 megawatts of electricity, enough to power 500 homes, has been operating in the Bay of Fundy, Nova Scotia, since 2016, where the tidal range, the greatest in the world, exceeds 16 m.

Although tidal power shows some promise, most analysts think there are only 100 to 150 sites worldwide that have sufficiently high tidal ranges and the appropriate coastal configuration to exploit this energy resource. This coupled with the fact that construction costs are high and tidal energy systems can have disastrous effects on the ecology (biosphere) of estuaries makes it unlikely that tidal energy will ever contribute more than a small percentage of all energy production.

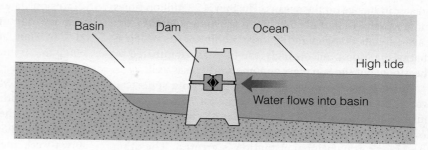

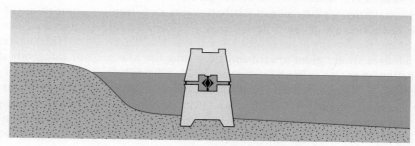

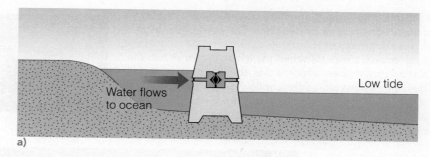

a)

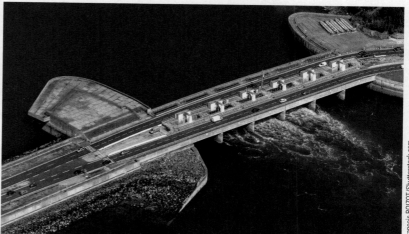

b)

● **FIGURE 2 Rising and falling tides produce electricity by spinning turbines connected to generators, just as at hydroelectric plants.** (a) Cross-sectional diagram showing how water flows into and out of a basin, during high and low tides, thus producing electricity, just as it would at a hydroelectric dam. (b) Aerial view of the Rance tidal power plant in the La Ranee River estuary, Brittany, France.

Both the Moon and the Sun have sufficient gravitational attraction to exert tide-generating forces strong enough to deform the solid body of Earth, but they have a much greater influence on the oceans. The Sun is 27 million times more massive than the Moon, but it is 390 times as far from Earth, and its tide-generating force is only 46% as strong as that of the Moon. Accordingly, the tides are dominated by the Moon, but the Sun plays an important role as well.

If we consider only the Moon acting on a spherical, water-covered Earth, its tide-generating forces produce two bulges on the ocean surface (●FIGURE 15.6). One bulge points toward the Moon, because it is on the side of Earth where the Moon's gravitational attraction is greatest. The other bulge is on the opposite side of Earth; it points away from the Moon, because of centrifugal force due to Earth's rotation, and the Moon's gravitational attraction is less. These two bulges always point toward and away from the Moon (●FIGURE 15.6A), so as Earth rotates and the Moon's position changes, an observer at a particular shoreline location experiences the rhythmic rise and fall of tides twice daily, but the heights of two successive high tides may vary depending on the Moon's inclination with respect to the equator.

The Moon revolves around Earth every 28 days, so its position with respect to any latitude changes slightly each day. That is, as the Moon moves in its orbit and Earth rotates on its axis, it takes the Moon 50 minutes longer each day to return to the same position it was in the previous day. Thus, an observer would experience a high tide at 1:00 p.m. on one day, for example, and at 1:50 p.m. on the following day.

When the Moon and Sun are aligned every two weeks, their forces added together generate *spring tides*, which are about 20% higher than average tides (●FIGURE 15.6B). When the Moon and Sun are at right angles to each another, also at two-week intervals, the Sun's tide-generating force cancels some of the Moon's, and *neap tides*, which are about 20% lower than average, occur (●FIGURE 15.6C).

Tidal ranges are also affected by shoreline configuration. Broad, gently sloping continental shelves as in the Gulf of Mexico have low tidal ranges, whereas steep, irregular shorelines experience much greater rise and fall of tides. Tidal ranges are greatest in some narrow, funnel-shaped bays and inlets. The Bay of Fundy in Nova Scotia has a tidal range of 16.5 m, and ranges greater than 10 m occur in several other areas.

● **FIGURE 15.6 Tidal Bulges** The gravitational attraction of the Moon and Sun causes tides. The sizes of the tidal bulges are greatly exaggerated. (a) Tidal bulges if only the Moon caused them. (b) When the Moon is new or full, the solar and lunar tides reinforce one another, causing spring tides, the highest high tides and lowest low tides. (c) During the Moon's first and third quarters, the Moon, Sun, and Earth form right angles, causing neap tides, the lowest high tides and highest low tides.

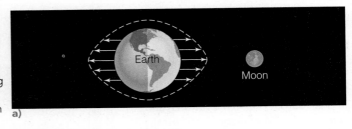
a)

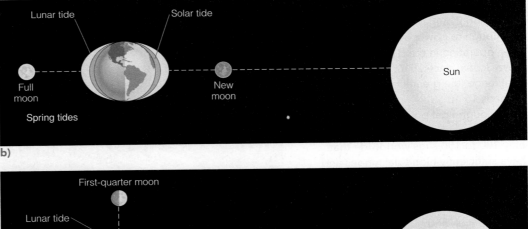
b)

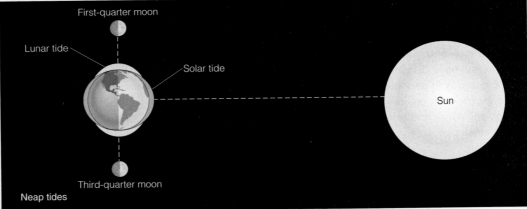
c)

Tides have an important impact on shorelines, because the area of wave attack constantly shifts onshore and offshore as the tides rise and fall. Tidal currents themselves, however, have little modifying effect on shorelines, except in narrow passages where tidal current velocity is great enough to erode and transport sediment.

## Waves

You can see **waves**, or oscillations of a water surface, on all bodies of water, but they are best developed in the oceans. In fact, waves are directly or indirectly responsible for most erosion, sediment transport, and deposition along shorelines. Wave terminology is illustrated with a typical series of waves in ●**FIGURE 15.7A**. A *crest*, as you would expect, is the highest part of a wave, whereas the low area between crests is a *trough*. The distance from crest to crest (or trough to trough) is the *wavelength*, and the vertical distance from trough to crest is *wave height*. You can calculate the speed at which a wave advances, called *celerity* (C), by the formula $C = L/T$, where $L$ is wavelength and $T$ is wave period—that is, the time it takes for two successive wave crests, or troughs, to pass a given point.

The speed of wave advance (C) is actually a measure of the velocity of the wave form rather than the speed of the molecules of water in a wave. When waves move across a water surface, the water moves in circular orbits but shows little or no net forward movement (Figure 15.7a). Only the wave form moves forward, and as it does, it transfers energy in the direction of wave movement.

The diameters of the orbits that water follows in waves diminish rapidly with depth, and at a depth of about one-half wavelength ($L/2$), called **wave base**, they are essentially zero. Thus, at a depth exceeding wave base, the water and seafloor or lake floor are unaffected by surface waves (Figure 15.7a).

**WAVE GENERATION** Most geologic work on shorelines is accomplished by wind-generated waves, especially storm waves. When wind blows over water—that is, one fluid (air) moves over another fluid (water)—friction between the two transfers energy to the water, causing the water surface to oscillate.

In areas where waves are generated, such as beneath a storm center at sea, sharp-crested, irregular waves called *seas* develop. Seas have various heights and lengths, and one wave cannot be easily distinguished from another. But as seas move out from their area of generation, they are sorted into broad *swells* with rounded, long crests, and all are about the same size (Figure 15.7a).

The harder and longer the wind blows, the larger are the waves, but these are not the only factors that control wave size. High-velocity wind blowing over a small pond will never generate large waves regardless of how long it blows. In fact, waves on ponds and most lakes appear only while the wind is blowing. In contrast, the surface of the ocean is always in motion, and waves with heights of 34 m have been recorded during storms in the open sea.

The reason for the disparity between wave sizes on ponds and lakes and on the oceans is the **fetch**, which is the distance the wind blows over a continuous water surface. So, on ponds and lakes, fetch corresponds to their length or width, depending on wind direction. To produce waves of greater length and height, more energy must be transferred from wind to water; hence large waves form beneath large storms at sea.

**SHALLOW-WATER WAVES AND BREAKERS** Swells moving out from an area of wave generation lose little energy as they travel long distances across the ocean. In these deep-water swells, the water surface oscillates and water moves in circular orbits, but little net displacement of water takes place in the direction of wave travel (Figure 15.7a). Of course, wind blows some water from wave crests, thus forming whitecaps with foamy white crests, and surface currents transport water for great distances, but deepwater waves accomplish little actual water movement. When these waves enter progressively shallower water, however, the wave shape changes, and water is displaced in the direction of wave advance.

Broad, undulating deepwater waves are transformed into sharp-crested waves as they enter shallow water. This transformation begins at a water depth corresponding to wave base—that is, one-half wavelength (Figure 15.7a). At this point, the waves "feel" the seafloor, and the orbital motion of water within the waves is disrupted. As waves continue moving shoreward, the speed of wave advance and wavelength decrease, but wave height increases. Thus, as they enter shallow water, waves become oversteepened as the wave crest advances faster than the wave form, and eventually the crest plunges forward as a **breaker** (●**FIGURE 15.7B**). Breaking waves might be several times higher than their deepwater counterparts, and when they break, they expend their kinetic energy on the shoreline.

The waves just described are the classic *plunging breakers* (Figure 15.7b) that crash onto shorelines with steep offshore slopes, such as those on the north shore of Oahu in the Hawaiian Islands. In contrast, shorelines where the offshore slope is more gentle usually have *spilling breakers*, where the waves build up slowly and the wave's crest spills down the wave front (●**FIGURE 15.7C**).

## Nearshore Currents

The area extending seaward from the upper limit of the shoreline to just beyond the area of breaking waves is conveniently designated as the *nearshore zone* (Figure 15.7a). Within the nearshore zone are the breaker zone and a surf zone, where water from breaking waves rushes forward and then flows seaward as backwash. The nearshore zone's width varies, depending on

---

**wave** An undulation on the surface of a body of water, resulting in the water surface rising and falling.

**wave base** The depth corresponding to about one-half wavelength, below which water in unaffected by surface waves.

**fetch** The distance the wind blows over a continuous water surface.

**breaker** A wave that steepens as it enters shallow water until its crest plunges forward.

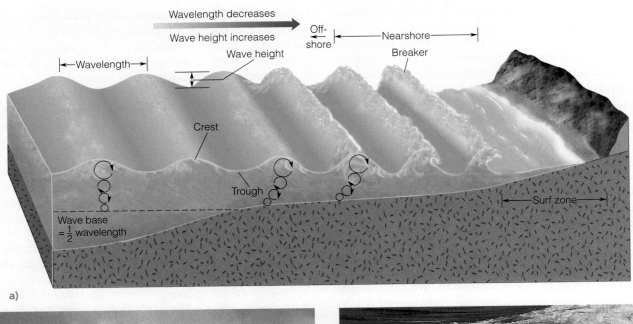

● **FIGURE 15.7 Waves and Wave Terminology** (a) Waves and the terminology applied to them. Note that swells are disrupted when they encounter water shallower than wave base, and they eventually form breakers. (b) A plunging breaker on the north shore of Oahu, Hawaii. (c) A spilling breaker.

the length of approaching waves, because long waves break at a greater depth, and thus farther offshore, than do short waves. Incoming waves are responsible for two types of currents in the nearshore zone: *longshore currents* and *rip currents*.

**WAVE REFRACTION AND LONGSHORE CUR-RENTS** Deepwater waves have long, continuous crests, but rarely are their crests parallel with the shoreline (●**FIGURE 15.8**). In other words, they seldom approach a shoreline head-on, but rather at some angle. Thus, one part of a wave enters shallow water, where it encounters wave base and begins breaking before other parts of the same wave. As a wave begins to break, its velocity diminishes, but the part of the wave still in deep water races ahead until it too encounters wave base. The net effect of this oblique approach is that waves bend so that they more nearly parallel the shoreline, a phenomenon known as **wave refraction** (Figure 15.8).

Even though waves are refracted, they still usually strike the shoreline at some angle, causing the water between the breaker zone and the beach to flow parallel to the shoreline. These

**longshore currents**, as they are called, are long and narrow and flow in the same general direction as the approaching waves (Figure 15.8). These currents are particularly important in transporting and depositing sediment in the nearshore zone.

**RIP CURRENTS** Waves carry water into the nearshore zone, so there must be a mechanism for mass transfer of water back out to sea. One way in which water moves seaward from the nearshore zone is in **rip currents**, narrow surface currents that flow out to sea through the breaker zone (●**FIGURE 15.9**). Surfers commonly take advantage of rip currents for an easy ride out beyond the breaker zone, but

**wave refraction**   The bending of waves so that they move nearly parallel to the shoreline.

**longshore current**   A current resulting from wave refraction found between the breaker zone and a beach that flows parallel to the shoreline.

**rip current**   A narrow surface current that flows out to sea through the breaker zone.

● **FIGURE 15.8 Wave Refraction** These waves are refracted (wave crests are indicated by dashed lines) as they enter shallow water and more nearly parallel the shoreline. The waves generate a longshore current that flows in the direction of wave approach, from upper left to lower right (arrow) in this example.

Sue Monroe

a)

b)

Photo by Steve Elgar, plane piloted by Kimball Millikan

● **FIGURE 15.9 Rip Currents** (a) Rip currents are fed on each side by currents moving parallel to the shoreline. (b) Suspended sediment, indicated by discolored water, is being carried seaward in these rip currents.

these currents pose a danger to inexperienced swimmers. Some rip currents flow at several kilometers per hour, so if a swimmer is caught in one, it is useless to try to swim directly back to shore. Instead, because rip currents are narrow and usually nearly perpendicular to the shore, one can swim parallel to the shoreline for a short distance and then turn shoreward with little difficulty.

Rip currents are circulating cells fed by longshore currents that increase in velocity from midway between each rip current (Figure 15.9). When waves approach a shoreline, the amount of water builds up until the excess moves out to sea through the breaker zone.

Rip currents commonly develop where wave heights are lower than in adjacent areas, and differences in wave height are controlled by variations in water depth. For instance, if waves move over a depression, the height of the waves over the depression tends to be less than in adjacent areas, forming the ideal environment for rip currents.

# SHORELINE EROSION AND DEPOSITION

**LO9**  Explain how a wave-cut platform develops and what its relationship is to a marine terrace

**LO10**  Explain how sea caves, sea arches, and sea stacks form

**LO11**  Define a beach and list its component parts

**LO12**  Discuss the relationship between a spit, baymouth bar, and a tombolo

**LO13**  Discuss the two models for the formation of barrier islands

**LO14**  Explain what processes affect a nearshore sediment budget

Erosion and deposition by waves and nearshore currents account for many interesting seashore features, such as sea stacks, arches, beaches, and spits, all of which are easy to recognize.

## Erosion and Wave-Cut Platforms

On many seashores, erosion creates steep or vertical slopes known as *sea cliffs*, which during storms are pounded by waves (hydraulic action), worn by the impact of sand and gravel (abrasion) (●**FIGURE 15.10**), and eroded by dissolution involving the solvent action of seawater. Tremendous energy from waves is concentrated on the lower parts of sea cliffs and is most effective on those composed of sediment or highly fractured rocks. In any case, the net effect is erosion of the sea cliff and the landward migration of the cliff face.

As sea cliffs are undercut by hydraulic action and abrasion at their bases, their upper parts are left unsupported and susceptible to mass wasting processes. Thus, sea cliffs retreat landward little by little, and as they do so, they leave

a)

b)

●**FIGURE 15.10 Wave Erosion by Abrasion and Hydraulic Action** (a) The rocks in the lower part of this image on a small island in the Irish Sea have been smoothed by abrasion, but the rocks higher up are out of the reach of waves. (b) Hydraulic action and abrasion have undercut these sea cliffs near Bodega Bay, California. Erosion was particularly intense during storms of February 1998. Attempts to stabilize the shoreline have failed and the homes are now gone, and others are now threatened.

a beveled surface called a **wave-cut platform** that slopes gently seaward (●**FIGURE 15.11A**). Broad wave-cut platforms are common in many areas, but invariably the water over them is shallow, because the abrasive planing action of waves is effective to a depth of only about 10 m. The sediment eroded from sea cliffs is transported seaward until it reaches deeper water at the edge of the wave-cut platform. There it is deposited and forms a *wave-built platform*, which is a seaward extension of the wave-cut platform (Figure 15.11a). If a wave cut platform is uplifted above sea level, it is called a **marine terrace** (●**FIGURE 15.11B**).

## Sea Caves, Arches, and Stacks

Sea cliffs do not retreat uniformly, because some shoreline materials are more resistant to erosion than others. **Headlands** are seaward-projecting parts of the shoreline that are eroded on both sides by wave refraction (●**FIGURE 15.12A**). *Sea caves* form on opposite sides of a headland, and if these caves join, they form a *sea arch* (●**FIGURE 15.12B**). Continued erosion causes the span of an arch to collapse, creating isolated *sea stacks* on wave-cut platforms (●**FIGURE 15.12C**).

In the long run, erosion tends to straighten an initially irregular shoreline. Wave refraction causes more wave energy to be expended on headlands and less on embayments. Thus, headlands erode, and some of the sediment formed by erosion is deposited in the embayments.

## Deposition

We mentioned in a previous section that longshore currents are effective at transporting sediment. In fact, we can think of the area from the breaker zone to the upper limit of the surf zone as a "river" that flows along the shoreline. Unlike rivers on land, though, this shoreline river's direction of flow

changes if waves approach from a different direction. Nevertheless, the analogy is apt, and just like rivers on land, a longshore current's capacity for transport varies with flow velocity and water depth.

Wave refraction and the resulting longshore currents are the primary agents of sediment transport and deposition on shorelines, but tides also play a role, because, as they rise and fall, the position of wave attack shifts onshore and offshore. Rip currents play no role in shoreline deposition, but they do transport fine-grained sediment (silt and clay) offshore through the breaker zone.

## Beaches

By definition, a **beach** is a deposit of unconsolidated sediment extending landward from low tide to a change in topography, such as a line of sand dunes, a sea cliff, or the point where permanent vegetation begins. Typically, a beach has several component parts (●**FIGURE 15.13A**), including a *backshore* that is usually dry, being covered by water only during storms or exceptionally high tides. The backshore consists of one or more *berms*, platforms composed of sediment deposited by waves; the berms are either nearly horizontal or slope gently landward. The sloping area below a berm exposed to wave swash is the *beach face*. The beach face is part of the *foreshore*, an area covered by water during high tide but exposed during low tide.

---

**wave-cut platform** A beveled surface that slopes gently seaward; formed by the erosion and retreat of a sea cliff.

**marine terrace** A wave-cut platform now above sea level.

**headland** Part of a shoreline, commonly bounded by cliffs, that extends out into the sea or a lake.

**beach** Any deposit of sediment extending landward from low tide to a change in topography or where permanent vegetation begins.

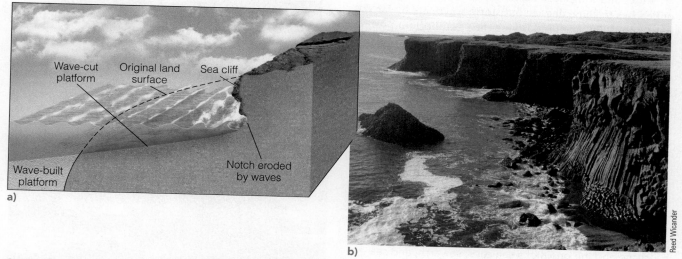

a)

b)

Reed Wicander

●**FIGURE 15.11 Origin of a Wave-Cut Platform** (a) Wave erosion causes a sea cliff to migrate landward, leaving a gently sloping surface, called a wave-cut platform. A wave-built platform originates by deposition at the seaward margin of the wave-cut platform. (b) These marine terraces and the large sea stack are near the village of Arnarstapi on the Snæfellsnes peninsula in West Iceland. Notice also the columnar basalts on the right side of this image (see Figure 5.7).

Depending on shoreline materials and wave intensity, beaches may be discontinuous, existing as only *pocket beaches* in protected areas such as embayments, or they may be continuous for long distances (●**FIGURES 15.13B** and **15.13C**). Some of the sediment on beaches is derived from weathering and wave erosion of the shoreline, but most of it is transported to the coast by streams and redistributed by longshore currents. As we previously noted, waves usually strike beaches at some angle, causing the sand grains to move up the beach face at a similar angle; as the sand grains are carried seaward in the backwash, however, they move perpendicular to the long axis of the beach. Thus, individual sand grains move in a zigzag pattern in the direction of longshore currents, by what is called *longshore drift*. This movement is not restricted to the beach; it extends seaward to the outer edge of the breaker zone (●**FIGURE 15.14A**).

In an attempt to widen a beach or prevent erosion, shoreline residents often build *groins*, structures that project seaward at right angles from the shoreline. A groin interrupts the flow of longshore currents, causing sand deposition on the upcurrent side and widening of the beach at that location. However, erosion inevitably occurs on the downcurrent side of a groin (●**FIGURE 15.14B**).

## Seasonal Changes in Beaches

The loose grains on beaches are constantly moved by waves, but the overall configuration of a beach remains unchanged as long as equilibrium conditions persist. We can think of the beach profile consisting of a berm or berms and a beach face, as a profile of equilibrium; that is, all parts of the beach are adjusted to the prevailing conditions of wave intensity, nearshore currents, and materials composing the beach (Figure 15.13a).

Tides and longshore currents affect the configuration of beaches to some degree, but storm waves are by far the most important agent modifying their equilibrium profile. In many areas, beach profiles change with the seasons, so we recognize *summer beaches* and *winter beaches*, each of which is adjusted to the conditions prevailing at those times. Summer beaches are sand-covered and have a wide berm, a gently sloping beach face, and a smooth offshore profile. Winter beaches, in contrast, tend to be coarser grained and steeper; they have a small berm or none at all, and their offshore profiles reveal sandbars paralleling the shoreline.

Seasonal changes in beach profiles are related to changing wave intensity. During the winter, energetic storm waves erode the sand from beaches and transport

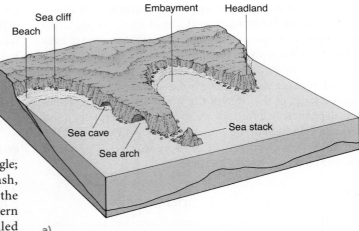

a)

b)

c)

●**FIGURE 15.12 Erosion of a Headland** (a) Erosion of a headland and the origin of a sea cave, sea arch, and sea stacks. (b) The Tasman Arch at Tasman National Park in the southeast corner of Tasmania, Australia, is a spectacular example of a sea arch. (c) Sea stacks at Ruby Beach in Olympic National Park, Washington.

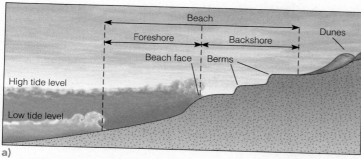

a)

Michael Slear

**FIGURE 15.13 Beaches** (a) Diagram of a beach showing its component parts. (b) The Grand Strand of South Carolina, shown here at Myrtle Beach, is 100 km of nearly continuous beach. (c) A pocket beach near Seal Rock, Oregon.

b)

c)

James S. Monroe

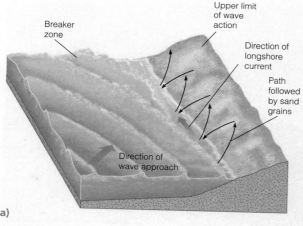

a)

b)

U.S. Geological Survey Circular 1075

**FIGURE 15.14 Longshore Currents and Longshore Drift** (a) Longshore currents transport sediment along the shoreline between the breaker zone and the upper limit of wave action. (b) These groins along the shoreline at Norfolk, Virginia, trap sand to maintain the beach in this area.

**Critical Thinking Question** What is the direction of the longshore currents in (b)?

it offshore, where it is stored in sandbars. The same sand that was eroded from a beach during the winter returns the next summer when it is driven onshore by more gentle swells. The volume of sand in the system remains more or less constant; it simply moves farther offshore or onshore depending on wave energy.

## Spits, Baymouth Bars, and Tombolos

Beaches are the most familiar depositional features of coasts, but spits, baymouth bars, and tombolos are also common.

In fact, these features are simply continuations of a beach. A **spit**, for instance, is a fingerlike projection of a beach into a body of water such as a bay, and a **baymouth bar** is a spit that has grown until it completely closes off a bay from the open sea (**FIGURES 15.15, 15.16A** and **15.16B**). Both are

---

**spit** A fingerlike projection of a beach into a body of water such as a bay.

**baymouth bar** A spit that has grown until it closes off a bay from the open sea or lake.

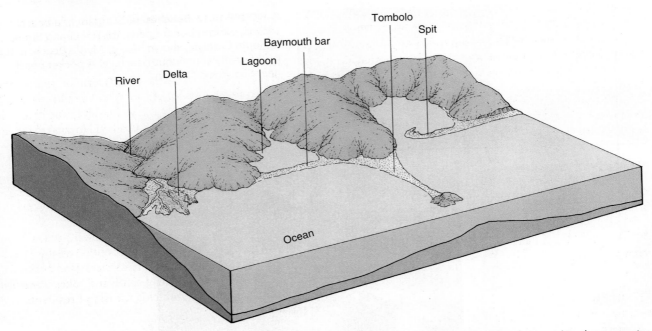

● **FIGURE 15.15 Spits, Baymouth Bars, and Tombolos** Spits form where longshore currents deposit sand in deeper water, as at the entrance to a bay. A spit that is curved at its free end goes by the name of recurved spit as shown here. A baymouth bar is a spit that has grown until it closes off the mouth of a bay. Wave refraction around an island causes converging currents to deposit a sand bar nearly perpendicular to the shoreline, forming a tombolo.

● **FIGURE 15.16 Spits, Baymouth Bars, and Tombolos** (a) A spit at the mouth of the Russian River near Jenner, California. (b) Rodeo Beach north of San Francisco, California, is a baymouth bar. (c) A tombolo along the Oregon Coast.

composed of sand, more rarely gravel, that was transported and deposited by longshore currents at the point where they weakened as they entered the deeper water of a bay's opening. Some spits are modified by waves so that their free ends are curved; they go by the name *hook* or *recurved spit* (Figure 15.15).

A **tombolo** is a type of spit that extends out from the shoreline to an island (●**FIGURE 15.16C**). A tombolo forms on the shoreward side of an island as wave refraction around the island creates converging currents that turn seaward and deposit a sandbar.

Spits and baymouth bars constitute a continuing problem where bays must be kept open for pleasure boating, commercial shipping, or both. Obviously, a bay closed off by a sandbar is of little use for either endeavor, so a bay must be regularly dredged or protected from deposition by longshore currents.

## Barrier Islands

Long, narrow islands of sand lying a short distance offshore from the mainland are **barrier islands** (●**FIGURE 15.17**). On their seaward sides, they are smoothed by waves, but their landward margins are irregular because storm waves carry sediment over the island and deposit it in a lagoon, where it is little modified by further wave activity. The component parts of a barrier island include a beach, windblown sand dunes, and a marshy area on their landward sides.

Everyone agrees that barrier islands form on gently sloping continental shelves where abundant sand is available and where both wave energy and the tidal range are low. For these reasons, many barrier islands are located along the United States' Atlantic and Gulf Coasts. But even though it is well known where barrier islands form, the details of their origin are still unresolved. According to one model, they formed as spits that became detached from land, whereas another model holds that they formed as beach ridges that subsequently subsided.

Most barrier islands are migrating landward as a result of erosion on their seaward sides and deposition on their landward sides. This is a natural part of barrier island evolution, and it takes place rather slowly, but it still takes place fast enough to cause many problems for island residents and communities.

## The Nearshore Sediment Budget

We can think of the gains and losses of sediment in the nearshore zone in terms of a **nearshore sediment budget** (●**FIGURE 15.18**). If a nearshore system has a balanced budget, sediment is supplied as fast as it is removed, and the volume of sediment remains more or less constant, although sand may shift offshore and onshore with the changing seasons. A *positive budget* means that gains exceed losses, whereas a *negative budget* means that losses exceed gains. If a negative budget prevails long enough, a nearshore system is depleted, and beaches may disappear.

Although there are a few exceptions, most sediment on beaches is transported to the shoreline by streams and then redistributed along the shoreline by longshore currents. Thus, longshore currents play a role in the nearshore sediment budget, because they continuously move sediment into and away from beach systems.

The primary ways that a nearshore system loses sediment are offshore transport, wind, and deposition into submarine canyons. Offshore transport mostly involves fine-grained sediment carried seaward, where it eventually settles in deeper water. Wind is an important process, because it removes sand from beaches and blows it inland, where it piles up as sand dunes.

If the heads of submarine canyons are near the shore, huge quantities of sand are funneled into them and deposited

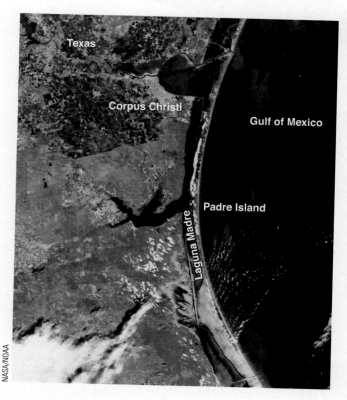

Texas

Corpus Christi

Gulf of Mexico

Laguna Madre

Padre Island

●**FIGURE 15.17 Barrier Islands** View from space of the barrier islands along the Gulf Coast of Texas. Notice that a lagoon up to 20 km wide separates the long, narrow barrier islands from the mainland.

**tombolo** A type of spit that extends out from the shoreline and connects the mainland with an island.

**barrier island** A long, narrow island of sand parallel to a shoreline but separated from the mainland by a lagoon.

**nearshore sediment budget** The balance between additions and losses of sediment in the nearshore zone.

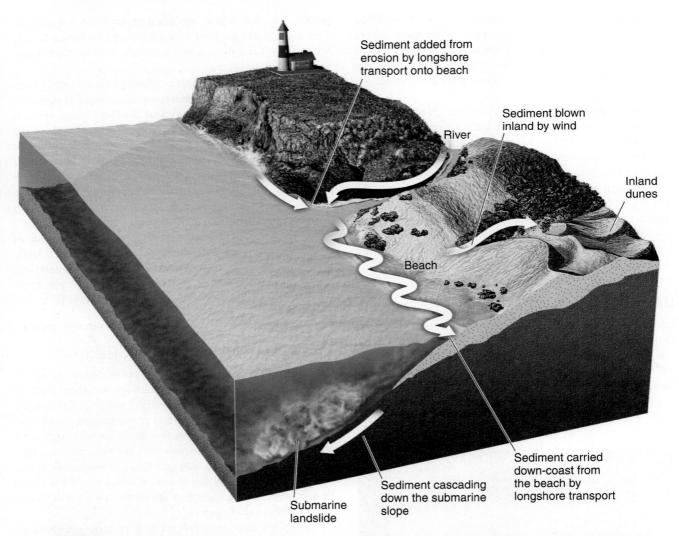

Sediment added from
erosion by longshore
transport onto beach

Sediment blown
inland by wind

River

Inland
dunes

Beach

Sediment carried
down-coast from
the beach by
longshore transport

Sediment cascading
down the submarine
slope

Submarine
landslide

● **FIGURE 15.18 The Nearshore Sediment Budget** The long-term sediment budget can be assessed by considering inputs versus outputs. If inputs and outputs are equal, the system is in a steady state, or equilibrium. If outputs exceed inputs, however, the beach has a negative budget and erosion occurs. Accretion takes place when the beach has a positive budget, with inputs exceeding outputs.

Critical Thinking Question What will happen to the beaches in this area if a dam is built across the river?

in deeper water. La Jolla and Scripps submarine canyons off the coast of southern California funnel an estimated 2 million m³ of sand each year. In most areas, however, submarine canyons are too far offshore to interrupt the flow of sand in the nearshore zone.

It should be apparent from this discussion that if a nearshore system is in equilibrium, its incoming supply of sediment exactly offsets its losses. Such a delicate balance tends to continue unless the system is somehow disrupted. One change that affects this balance is the construction of dams across the streams that supply sand. Once dams have been built, all sediment from the upper reaches of the drainage systems is trapped in reservoirs and thus cannot reach the shoreline.

# TYPES OF COASTS

LO15 **Explain how submergent and emergent coasts form**

Coasts are difficult to classify because of variations in the factors that control their development and variations in their composition and configuration. Rather than attempt to categorize all coasts, we shall simply note that two types of coasts have already been discussed: those dominated by deposition and those dominated by erosion.

In addition, we will examine coasts in terms of their changing relationship to sea level. But note that although some coasts, such as those in southern California, are described as emergent (uplifted), these same coasts may be erosional as well. In other words, coasts commonly possess features that allow them to be classified in more than one way.

## Depositional and Erosional Coasts

Depositional coasts, such as much of the Atlantic and Gulf coasts, are characterized by an abundance of detrital sediment and such depositional landforms as wide sandy beaches, deltas, and barrier islands (Figure 15.13b). In contrast, erosional coasts are steep and irregular and typically lack well-developed beaches, except in protected areas. They are further characterized by sea cliffs, wave-cut platforms, and sea stacks. Many of the coasts along the West Coast of North America fall into this category.

## Submergent and Emergent Coasts

If sea level rises with respect to the land or the land subsides, coastal regions are flooded and said to be **submergent coasts**

or *drowned coasts* (●**FIGURE 15.19A**). Much of the East Coast of North America from Maine southward through South Carolina was flooded during the rise in sea level following the Pleistocene Epoch, so it is extremely irregular. Recall that during the expansion of glaciers during the Pleistocene, sea level was as much as 130 m lower than at present, and that streams eroded their valleys more deeply and extended across continental shelves. When sea level rose, the lower ends of these valleys were drowned, forming *estuaries* such as Delaware and Chesapeake bays (Figure 15.19a). Estuaries are simply the seaward ends of river valleys where seawater and freshwater mix. The divide between adjacent drainage systems on submergent coasts project seaward as broad headlands or a line of islands.

**Emergent coasts** are found where the land has risen with respect to sea level (●**FIGURE 15.19B**). Emergence takes

---

**submergent coast** A coast along which sea level rises with respect to the land or the land subsides.

**emergent coast** A coast where the land has risen with respect to sea level.

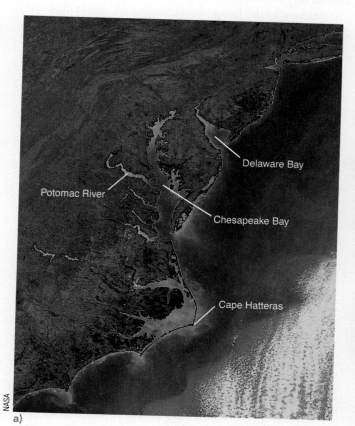

a)

b)

●**FIGURE 15.19 Submergent and Emergent Coasts** (a) Submergent coasts tend to be extremely irregular, with estuaries such as Chesapeake and Delaware bays. They formed when the East Coast of the United States was flooded as sea level rose following the Pleistocene Epoch. (b) Emergent coasts tend to be steep and straighter than submergent coasts. Notice the several sea stacks and the sea arch. Also, a marine terrace is visible in the distance.

**Critical Thinking Question** Do you think a wave-cut platform is forming in (b)? If so, where?

place when water is withdrawn from the oceans, as occurred during the Pleistocene expansion of glaciers. Presently, coasts are emerging as a result of isostasy or tectonism. In northeastern Canada and the Scandinavian countries, for instance, the coasts are irregular, because isostatic rebound is elevating formerly glaciated terrain from beneath the sea.

# THE PERILS OF LIVING ALONG A SHORELINE

**LO16**   Discuss what steps can be taken, if any, to minimize the damage resulting from rising sea level and shoreline erosion

No matter where we live, there are risks. Great Plains residents worry about violent storms, whereas those living near active faults or active volcanoes must be aware of the potential danger of earthquakes, and people living in forests or brush land are cognizant of wildfires, such as the devastating fires in California in 2017 and 2018. Likewise, living along a shoreline poses certain risks, the most obvious being storm waves and coastal flooding. Unfortunately, we have vivid reminders of this risk from large storms such as Hurricane Katrina in August 2005, Hurricane Sandy in October 2012, and Hurricane Harvey in August 2017. Strong winds during hurricanes also pose a risk, but much of the damage and most fatalities results from coastal flooding.

## Storm Waves and Coastal Flooding

Coastal flooding occurs when large waves are driven onshore by hurricanes and by heavy rainfall, as much as 60 cm in as little as 24 hours. In addition, as a hurricane moves over the ocean, low atmospheric pressure beneath the eye of the storm, causes the ocean's surface to bulge upward as much as 0.5 m. When the eye of the storm reaches the shoreline, the bulge, coupled with wind-driven waves, piles up in a **storm surge** that may rise several meters above normal high tide and flood areas far inland.

When Hurricane Katrina roared ashore on August 29, 2005, high winds, a huge storm surge, and coastal flooding destroyed nearly everything in an area of 230,000 km². Gulfport and Biloxi in Mississippi were mostly leveled (●FIGURE 15.20); however, the public's attention was mainly focused on New Orleans, Louisiana.

As Hurricane Katrina moved across Louisiana, the levees surrounding New Orleans initially held, but on the next day, some of the floodwalls were breached, and about 80% of the city was flooded. Because New Orleans is mostly below sea level, the floodwaters could not drain out naturally. In fact, the city had 22 pumping stations to remove water that builds up from normal rainstorms, but as the city flooded, the pumps were overwhelmed, and when the electricity

●**FIGURE 15.20 Hurricane Katrina, 2005** Destruction caused by the storm surge from Hurricane Katrina. This image shows a man in Biloxi, Mississippi, trying to find his house.

failed, they were useless. All in all, Hurricane Katrina caused an estimated $108 billion in damages, and approximations of the number of people who died range between 986 and 1,838, mostly in Louisiana.

Hurricane Sandy swept through the Caribbean and up the eastern coast of the United States in October 2012, where it caused $359 billion in damages in the Caribbean and $62 billion in the United States, with a total death toll of 285 people, which includes at least 125 in the United States. In New York and New Jersey, storm surges reached 4 m above average low tide. New York City was most impacted due to damage to its subways and roadway tunnels. In all, more than 7.5 million people were without electricity.

Lastly, Hurricane Harvey, which affected Florida and Texas, in October 2017, is frequently cited as tied with Hurricane Katrina as the costliest hurricane on record (when adjusted for inflation), primarily due to catastrophic flooding in Houston, Texas (●FIGURE 15.21). Fortunately, the death toll, at least 88 people known dead, was significantly lower than hurricanes Katrina and Sandy. Like Hurricane Katrina, the major cause of damage in Houston was the result primarily of catastrophic rainfall. Because the infrastructure was unable to handle the amounts of rain produced in such a short time, devastating flooding resulted. In fact, some parts of Houston recorded up to 127 cm of rainfall, and the total amount of rain dumped on Texas from Hurricane Harvey was 102 trillion liters of water! Furthermore, Hurricane Harvey is now the wettest Atlantic hurricane ever measured.

---

**storm surge**   The surge of water onto a shoreline as a result of a bulge in the ocean's surface beneath the eye of a hurricane and wind-driven waves.

● **FIGURE 15.21 Hurricane Harvey, 2017** Aerial view of the flooding in Houston, Texas, August 29, 2017 resulting from Hurricane Harvey.

## Coastal Management as Sea Level Rises

From 1870 to 2013, sea level has risen almost 240 mm based on coastal tide gauges, and between 1993 and 2018, sea level has risen 3.2 mm per year as measured by satellites. By all accounts, sea level will continue to rise. Perhaps such a "slow" rate of sea-level change seems insignificant. However, in gently sloping coastal areas, such as in the eastern United States from New Jersey southward, even a slight rise in sea level will eventually have widespread effects.

The cause of sea-level rise depends on two factors, both of which are related to global warming. The first is the added volume of water from melting ice sheets and glaciers, and the second is the expansion of seawater as it warms.

An additional factor that can affect sea level along a coastline is the rate of uplift, or subsidence. In some areas, uplift is occurring fast enough that sea level is actually falling with respect to the land. In other areas, sea level is rising while the coastal region is simultaneously subsiding, resulting in a net change in sea level.

How do these factors affect coastlines? Many of the nearly 300 barrier islands along the East and Gulf Coasts of the United States are migrating landward as sea level rises (●**FIGURE 15.22**). Landward migration of barrier islands would pose few problems were it not for the numerous communities, resorts, and vacation homes located on them. Furthermore, rising sea level threatens many beaches upon which communities depend for revenue. Other problems associated with rising sea level include increased coastal flooding during storms and saltwater incursions that may threaten groundwater supplies.

As dire as these problems are to coastal areas, rising sea level poses even a greater threat to entire nations. For example, many low-lying island chains in the central Pacific Ocean, such as Kiribati, are only a few meters above high tide. At the current rate of sea level rise, these islands will become inhabitable in as little as a single generation.

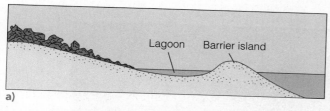

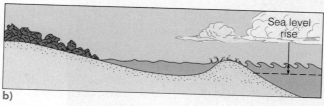

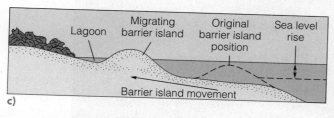

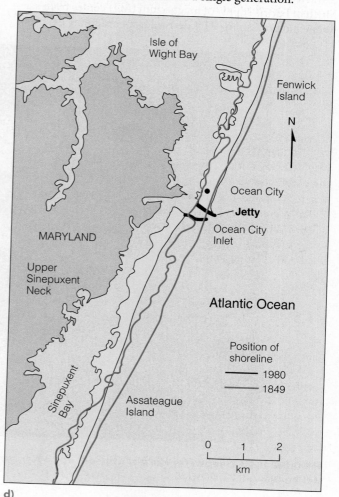

● **FIGURE 15.22 Barrier Island Migration** (a) A barrier island. (b) A barrier island migrates landward as sea level rises and storm waves carry sand from its seaward side into its lagoon. (c) Over time, the entire island shifts toward the land. (d) Jetties were constructed during the 1930s to protect the inlet at Ocean City, Maryland, but they disrupted the net southerly longshore drift and Assateague Island, starved of sediment, has migrated 500 m landward. Beginning in the fall of 2002, beach sand was artificially replenished in an effort to stabilize the island.

Because sea level will continue to rise in the immediate future, thus affecting coastal areas, engineers, scientists, planners, and political leaders must examine what they can do to prevent, or minimize, the effects of shoreline erosion. Presently, only a few viable options exist. One is to put strict controls on coastal development. North Carolina, for example, permits large structures to be built no closer to the shoreline than 60 times the annual erosion rate. Although a growing awareness of shoreline processes has resulted in similar legislation elsewhere, some states have virtually no restrictions on coastal development, and as a result of continued overdevelopment, we will see a continued increase in property damage and human suffering.

# RESOURCES FROM THE OCEANS

LO17    List some of the economically valuable resources found in oceans and the sea floor

LO18    Define the Exclusive Economic Zone

Seawater contains many elements in solution, the most common being sodium (Na) and chlorine (Cl), which are extracted for salt by evaporation or by mining salt deposits. Much of Earth's supply of magnesium (Mg) comes from seawater, and potassium (K) is derived from evaporate deposits.

Recall from Chapter 6 that most limestone (used in cement) and gypsum (used for wallboard) come mainly from deposits that formed from ocean waters.

Seafloor deposits are becoming important, and as a result, most nations bordering the oceans claim those resources within their adjacent waters. The United States, by presidential proclamation issued in 1983, claims sovereign rights over an area designated the **Exclusive Economic Zone (EEZ)**, which extends seaward 371 km. Other ocean-bordering nations make similar claims.

Resources within the United States EEZ include sand and gravel for construction, and a large percentage of all domestic oil production comes from wells on the continental shelf, especially in the Gulf of Mexico. Although this is an important source of oil and natural gas, offshore drilling is not without risk. The *Deepwater Horizon* drill rig that exploded and sank in the Gulf of Mexico, about 66 km off the Louisiana coast, was a semi-submersible offshore rig that was equipped to operate in water 2,100 m deep. Unfortunately, on April 20, 2010, as it was drilling in 1,524 m of water, it exploded, killing 11 workers and beginning the largest oil spill in the history of the petroleum industry (●**FIGURE 15.23**).

---

**Exclusive Economic Zone (EEZ)** An area extending 370 km seaward from the coast of the United States and its possessions in which the United States claims rights to all resources.

U.S. Coast Guard photo

● **FIGURE 15.23** *Deepwater Horizon* **Oil Spill** The *Deepwater Horizon* drilling platform shortly before it sank in the Gulf of Mexico on April 22, 2010.

Other resources within the EEZ include spherical objects called manganese nodules, which are composed of manganese and iron oxides, as well as copper, nickel, and cobalt. Because the United States imports most of the manganese and cobalt it needs, these nodules have attracted some attention as a potential resource. Another important resource found in shallow marine deposits is phosphate-rich sedimentary rock known as *phosphorite*; it is used for chemical fertilizers and animal feed supplements, as well as matches, metallurgy, preserved foods, and ceramics.

Another potential resource within the EEZ is methane hydrate, a chief constituent of natural gas, consisting of single methane molecules bound up in networks formed by frozen water. It is found beneath the ocean floor and in Arctic permafrost. Methane hydrate is stable at water depths of more than 500 m and near-freezing temperatures. Although estimates vary considerably, the energy content in methane hydrate is immense, potentially exceeding that of all coal, oil, and natural gas combined. Other than small-scale field experiments, it is not known if methane hydrate can be effectively recovered in commercial-size quantities. Another factor that must be evaluated in terms of potential effects on climate is that the volume of methane hydrate is about 3,000 times as much as in the atmosphere, and methane is 10 times more effective than carbon dioxide as a greenhouse gas.

# Key Concepts Review

- The upper 100 m or so of the oceans is the photic zone where sunlight is sufficient for photosynthesizing organisms. The aphotic zone lies below.
- The Coriolis effect is the result of Earth's rotation. Winds and oceanic currents are deflected clockwise in the Northern Hemisphere, and counterclockwise in the Southern Hemisphere.
- Oceanic circulation is mostly horizontal in surface currents and deep-sea currents, but vertical circulation also takes place.
- The sediments on the seafloor are mostly pelagic clay and ooze, which consists of the skeletons of tiny organisms.
- Mound-like, wave-resistant structures consisting of animal skeletons are reefs. Most reefs are fringing reefs, barrier reefs, or atolls.
- Tides are caused by the combined effects of the Moon and Sun on the oceans.
- As wind-generated waves enter shallow water, they become oversteepened and plunge forward as breakers or spill onto the shoreline, thus expending their kinetic energy.

- Longshore currents resulting from waves approaching a shoreline at an angle erode, transport, and deposit sediment.
- Rip currents carry excess water from the nearshore zone seaward through the breaker zone.
- Erosional coasts have sea cliffs, wave-cut platforms, and sea stacks, whereas depositional coasts have long sandy beaches, deltas, and barrier islands.
- Beaches are continuously modified by waves and nearshore currents, and their profiles usually show seasonal changes.
- Spits, baymouth bars, and tombolos all form and grow as a result of longshore transport and deposition.
- The sediment budget of a nearshore system remains rather constant unless the system is disrupted, as when dams are built across streams supplying sand to the system.
- Submergent and emergent coasts are defined on the basis of their relationship to changes in sea level.
- Coastal flooding during storms by waves and storm surges is an ongoing problem in many areas.
- The United States claims rights to all resources within 370 km of its shorelines, or what is called its Exclusive Economic Zone (EEZ).

# Important Terms

aphotic zone   309

barrier island   323

baymouth bar   321

beach   319

breaker   315

downwelling   309

emergent coast   325

Exclusive Economic Zone (EEZ)   328

fetch   315

gyre   309

headland   319

longshore current   316

marine terrace   319

nearshore sediment budget   323

ooze   309

pelagic clay   309

photic zone   309

reef   309

rip current   316

salinity   308

shoreline   310

spit   321

storm surge   326

submergent coast   325

tide   310

tombolo   323

upwelling   309

wave   315

wave base   315

wave-cut platform   319

wave refraction   316

# Review Questions

1. The depth in the oceans below about 100 m where there is too little sunlight for photosynthesis is called the
   a. bathosphere.
   b. aphotic zone.
   c. abyssal zone.
   d. tidal zone.
   e. photic zone.

2. Because most waves approach a shoreline at an angle, they generate
   a. an undertow.
   b. gyres.
   c. longshore currents.
   d. waves.
   e. a Coriolis effect.

3. Waves approaching a shoreline have a wavelength of 15 m, so the depth of the wave base is about
   a. 30 m.
   b. 5 m.
   c. 45 m.
   d. 7.5 m.
   e. 150 m.

4. Erosion by waves is caused by the impact of water on shorelines and the wearing action of water carrying sand and gravel. These processes are
   a. hydraulic action/abrasion.
   b. chemical weathering/dissolution.
   c. fetch/surge.
   d. downwelling/gyre.
   e. longshore drift/storm surges.

5. The distance that wind blows over a continuous water surface is called the
   a. oceanic budget.
   b. celerity.
   c. fetch.
   d. wavelength.
   e. surge.

6. While driving along North America's West Coast, you notice a broad surface above sea level that slopes gently toward the ocean and has several masses of rock rising above it. How would you explain the origin of this land-form to your children?

7. What is upwelling and why is it important?

8. What is wave base and how does it affect waves as they enter shallow water?

9. As a member of a planning commission for your coastal community where shoreline erosion is a continuing problem, what recommendations would you make to remedy or at least mitigate the problem?

# Creative Thinking Visual Question

In this image (●FIGURE 1), identify a headland and a pocket beach. Explain where you would expect a sea cave, sea arch, and sea stack to form.

●**FIGURE 1** Aerial view of the Pacific coast of the United States.

James S. Monroe

dibrova/Shutterstock.com

The Grand Canyon, Arizona. Major John Wesley Powell led two expeditions down the Colorado River and through the canyon in 1869 and 1871. He was struck by the seemingly limitless time represented by the rocks exposed in the canyon walls and by the recognition that these rock layers, like the pages in a book, contain the geologic history of this region.

# 16

# GEOLOGIC TIME
## Concepts and Principles

# INTRODUCTION

In 1869, Major John Wesley Powell, a Civil War veteran who lost his right arm in the battle of Shiloh, led a group of hardy explorers down the uncharted Colorado River through the Grand Canyon. With no maps or other information, Powell and his group ran the many rapids of the Colorado River in fragile wooden boats, hastily recording what they saw. Powell wrote in his diary that "all about me are interesting geologic records. The book is open and I read as I run."

Probably no one has contributed as much to the understanding of the Grand Canyon as Major Powell. In recognition of his contributions, the Powell Memorial was erected on the South Rim of the Grand Canyon in 1969 to commemorate the 100th anniversary of this history-making first expedition.

Most tourists today, like Powell and his fellow explorers in 1869, are astonished by the seemingly limitless time represented by the rocks exposed in the walls of the Grand Canyon. For most visitors, viewing a 1.5-km-deep cut into Earth's crust is the only encounter they'll ever have with the magnitude of geologic time. When standing on the rim and looking down into the Grand Canyon, we are really looking far back in time, more than 1 billion years—all the way back to the early history of our planet.

Vast periods of time set geology apart from most of the other sciences, and an appreciation of the immensity of geologic time is fundamental to understanding the physical and biological history of our planet. In fact, understanding and accepting the magnitude of geologic time is one of the major contributions geology has made to the sciences.

Besides gaining an appreciation of the immensity of geologic time, why is the study of geologic time important? One reason is that Earth has undergone periods of warmer and colder conditions in the past, and an understanding of what caused these climate oscillations might be helpful in the debate concerning global climate change. Second, one of the most valuable lessons you will learn in this chapter is how to reason and apply some of geology's fundamental principles to past geologic events. The logic used in applying these principles to interpret the geologic history of an area involves basic reasoning skills that can be transferred and used in almost any profession or discipline.

# HOW IS GEOLOGIC TIME MEASURED?

**L01**    Explain each of the two methods used to measure geologic time

In some respects, time is defined by the methods used to measure it. Geologists use two different frames of reference when discussing geologic time: relative dating and numerical dating. **Relative dating** is placing geologic events in a sequential order as determined from their position in the geologic record. Relative dating will not tell us how long ago a particular event took place, only that one event preceded another.

The principles used to determine relative dating were formulated hundreds of years ago, and since then, they have been used to construct the *relative geologic time scale* (•FIGURE 16.1). Furthermore, these principles are still widely used by geologists today, and especially, in reconstructing the geologic history of the terrestrial planets and their moons.

**Numerical dating** (also called *absolute dating*) provides dates for rock units or events expressed in years before the present. *Radiometric dating* is the most common method used to obtain numerical dates. In this method, dates are calculated from the natural decay rates of various radioactive elements present in trace amounts in some rocks. It was not until the discovery of radioactivity near the end of the 19th century that numerical ages could be accurately applied to the relative geologic time scale. Today, the geologic time scale is really a dual scale: a relative scale based on rock sequences with radiometric dates expressed as years before the present (Figure 16.1).

Advances and refinements in numerical dating techniques have changed the way we view Earth in terms of when events occurred in the past and the rates of geologic change through time. The ability to accurately determine past climatic changes and their causes has important implications in regards to climate change and its effect on Earth's biota. The contribution to climate change by human activity and its consequences have led some geologists to propose a new geologic epoch (see GEO-FOCUS).

# EARLY CONCEPTS OF GEOLOGIC TIME AND EARTH'S AGE

**L02**    Discuss some of the early methods for determining the age of Earth

The concept of geologic time and its measurement have changed throughout human history. Some early Christian scholars and clerics tried to establish the date of creation by analyzing historical records and the genealogies found in Scripture. Based on their analyses, they generally believed that Earth and all its features were no more than about 6,000 years old. Thus, the idea of a very young Earth provided the

---

**relative dating** The process of determining the age of an event as compared to other events; involves placing geologic events in their correct chronologic order, but does not involve consideration of when the events occurred in number of years ago.

**numerical dating** Uses various radioactive decay dating techniques to assign ages to rocks in years before the present.

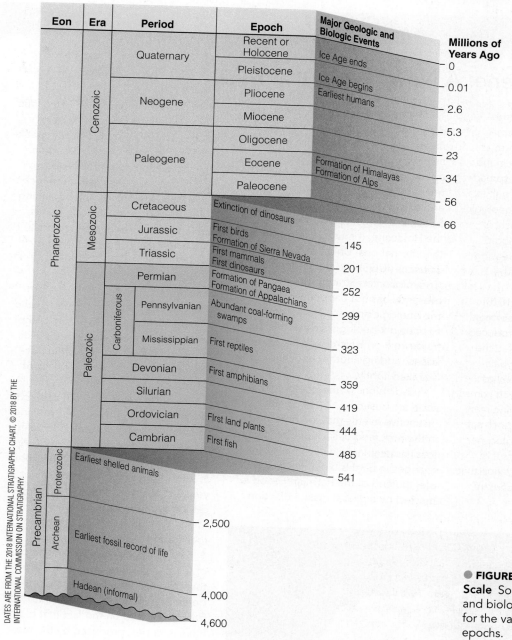

| Eon | Era | Period | | Epoch | Major Geologic and Biologic Events | Millions of Years Ago |
|---|---|---|---|---|---|---|
| Phanerozoic | Cenozoic | Quaternary | | Recent or Holocene | Ice Age ends | 0 |
| | | | | Pleistocene | | 0.01 |
| | | Neogene | | Pliocene | Ice Age begins / Earliest humans | 2.6 |
| | | | | Miocene | | 5.3 |
| | | Paleogene | | Oligocene | | 23 |
| | | | | Eocene | Formation of Himalayas / Formation of Alps | 34 |
| | | | | Paleocene | | 56 |
| | Mesozoic | Cretaceous | | | Extinction of dinosaurs | 66 |
| | | Jurassic | | | First birds / Formation of Sierra Nevada | 145 |
| | | Triassic | | | First mammals / First dinosaurs | 201 |
| | Paleozoic | Permian | | | Formation of Pangaea / Formation of Appalachians | 252 |
| | | Carboniferous | Pennsylvanian | | Abundant coal-forming swamps | 299 |
| | | | Mississippian | | First reptiles | 323 |
| | | Devonian | | | First amphibians | 359 |
| | | Silurian | | | | 419 |
| | | Ordovician | | | First land plants | 444 |
| | | Cambrian | | | First fish | 485 |
| Precambrian | Proterozoic | | | | Earliest shelled animals | 541 |
| | Archean | | | | Earliest fossil record of life | 2,500 |
| | | | | | Hadean (informal) | 4,000 |
| | | | | | | 4,600 |

DATES ARE FROM THE 2018 INTERNATIONAL STRATIGRAPHIC CHART, © 2018 BY THE INTERNATIONAL COMMISSION ON STRATIGRAPHY.

● **FIGURE 16.1 The Geologic Time Scale** Some of the major geologic and biologic events are indicated for the various eras, periods, and epochs.

basis for most Western chronologies of Earth history prior to the 18th century.

During the 18th and 19th centuries, several attempts were made to determine Earth's age on the basis of scientific evidence rather than revelation. The French zoologist Georges Louis de Buffon (1707–1788) assumed that Earth gradually cooled to its present condition from a molten beginning. To simulate this history, he melted iron balls of various diameters and allowed them to cool to the surrounding temperature. By extrapolating their cooling rate to a ball the size of Earth, he determined that Earth was at least 75,000 years old. Although this age was much older than that derived from Scripture, it was still vastly younger than we now know our planet to be.

Other scholars were equally ingenious in attempting to calculate Earth's age. For example, if deposition rates could be determined for various sediments, geologists reasoned that they could calculate how long it would take to deposit any rock layer. They could then extrapolate how old Earth was from the total thickness of sedimentary rock in its crust. Rates of deposition vary, however, even for the same type of rock. Furthermore, it is impossible to estimate how much of a rock has been removed by erosion, or how much a rock sequence has been reduced by compaction. As a result of these variables, estimates of Earth's age ranged from younger than 1 million years to older than 2 billion years.

Besides trying to determine Earth's age, the naturalists of the 18th and 19th centuries were also formulating some

# GEO-FOCUS

## The Anthropocene: A New Geologic Epoch?

Since being coined by Nobel Prize–winning Dutch atmospheric chemist Paul Crutzen, in 2000, the term "anthropocene" has caught on in the scientific community as well as spreading to the much wider general public. The idea of formalizing a new geologic epoch, the Anthropocene Epoch, is being widely discussed in geologic circles and is based on the increasing impact humans are having on the planet.

Interestingly, as far back as 1873, Antonio Stoppani, an Italian geologist, proposed that humans had introduced a new geologic era that he called the "anthropozoic," and over the years, other terms followed. Why then has Crutzen's "anthropocene," which comes from the Latin *antropos*, meaning man or human, and the geologic epoch suffix *cene*, as in Paleocene or Holocene, gained so much favor? Perhaps it is because of an increasing realization that humans are altering the environment, and at an alarmingly rapid rate.

Just because a term is widely used in the scientific literature does not mean that it will be formally accepted as the name for a new geologic epoch. For this to happen, many criteria and formal steps must be taken and met. For the Anthropocene to become a formal Epoch, following the Holocene, it must be adopted by the International Union of Geological Sciences (IUGS) and the International Commission on Stratigraphy, which presently has not happened. Because the geologic time scale is fundamental to the work geologists do and is the framework on which Earth history is based, adding another epoch is not to be taken lightly.

Subdivisions of the geologic time scale are based on the recognition of distinctive events that are preserved in the rock record, such as changes in fossil assemblages, which represent changes in Earth's biota. For example, the end of the Permian Period is marked by a global mass extinction event, as is the end of the Cretaceous Period. Even the boundaries of epochs, which are, geologically speaking, short in duration, are based on distinctive and recognizable geologic events that are preserved in the rock record. So, what marks the beginning of the Anthropocene Epoch, that is, at what point does human impact to the environment show up in the rock record?

There has been much debate as to when human activities began affecting Earth's natural systems, and thus marking a definitive start of the Anthropocene. For many geologists the beginning of the Industrial Revolution, with its increasing levels of carbon dioxide and methane, resulting from the exploitation of coal, oil, and natural gas to fuel a burgeoning population and industrialized society marks the start of the Anthropocene Epoch. However, because the rock record of the present-day does not yet exist, what evidence of our impact can be found in present-day sediments that will make up this future rock record?

The Anthropocene Working Group of the IUGS recommended in 2016 that the year 1950 formally mark the beginning of the Anthropocene. This date was based on the fact that plutonium isotopes, produced by the testing of nuclear weapons (●**FIGURE 1**), and their associated fallout from the atmosphere would be concentrated enough to be detected in the sediment and consequently in future rock strata.

Although much work remains to be done before formalizing the Anthropocene Epoch, for now and the foreseeable future, humans and their global impact will remain firmly part of the Holocene Epoch.

●**FIGURE 1** The mushroom cloud resulting from the first test of a hydrogen bomb in 1952 on Enewetak, an atoll in the Pacific Ocean.

of the fundamental geologic principles that are used in deciphering Earth history. From the evidence preserved in the geologic record, it was clear to them that Earth is very old and that geologic processes have operated over long periods of time.

# JAMES HUTTON AND THE RECOGNITION OF GEOLOGIC TIME

**LO3**  Explain what James Hutton's contribution is to geology

Many consider the Scottish geologist James Hutton (1726–1797) to be the father of modern geology (●FIGURE 16.2). His detailed studies and observations of rock exposures and present-day geologic processes were instrumental in establishing the principle of *uniformitarianism* (see Chapter 1), the concept that the same processes seen today have operated throughout geologic time. Because Hutton relied on known processes to account for Earth history, he concluded that Earth must be very old and wrote that "we find no vestige of a beginning, and no prospect of an end."

Pictorial Press Ltd/Alamy Stock Photo

●**FIGURE 16.2 James Hutton** James Hutton, a Scottish geologist, originated the principle of uniformitarianism and, through his writings, profoundly influenced the course of geologic thinking.

In 1830, Charles Lyell published a landmark book, *Principles of Geology*, in which he championed Hutton's concept of uniformitarianism. Instead of relying on catastrophic events to explain various Earth features, Lyell recognized that imperceptible changes brought about by present-day processes could, over long periods of time, have tremendous cumulative effects. Through his writings, Lyell firmly established uniformitarianism as the guiding principle of geology. In addition, the recognition of vastly long periods of time was also necessary for, and instrumental in, the acceptance of Darwin's 1859 theory of organic evolution.

# RELATIVE DATING METHODS

**LO4**  Explain each of the five fundamental principles of relative dating

**LO5**  Define an unconformity

**LO6**  Define each of the three unconformities

**LO7**  Explain how the principles of relative dating can be applied in reconstructing the geologic history of an area

Before the development of radiometric dating techniques, geologists had to depend solely on relative dating methods. Recall that relative dating places events in sequential order but does not tell us how long ago an event took place. Although the principles of relative dating may now seem self-evident, their discovery was an important scientific achievement because they provided geologists with a means to interpret geologic history and develop a *relative geologic time scale*.

## Fundamental Principles of Relative Dating

The 17th century was an important time in the development of geology as a science because of the widely circulated writings of the Danish anatomist Nicolas Steno (1638–1686). Steno observed that when streams flood, they spread out across their floodplains and deposit layers of sediment that bury organisms dwelling on the floodplain. Subsequent floods produce new layers of sediments that are deposited or superposed over previous deposits. When lithified, these layers of sediment become sedimentary rock.

Thus, in an undisturbed succession of sedimentary rock layers, the oldest layer is at the bottom and the youngest layer is at the top. This **principle of superposition** is the basis for relative-age determinations of strata and their contained fossils (●FIGURE 16.3A).

**principle of superposition**  A principle holding that in a vertical sequence of undeformed sedimentary rocks, the relative ages of the rocks can be determined by their position in the sequence—oldest at the bottom followed by successively younger layers.

a)

● **FIGURE 16.3 The Principles of Original Horizontality, Super-position, and Lateral Continuity** (a) The sedimentary rocks of Bryce Canyon National Park, Utah, illustrate three of the six fundamental principles of relative dating. These rocks were originally deposited horizontally in a variety of continental environments (principle of original horizontality). The oldest rocks are at the bottom of this highly dissected landscape, and the youngest rocks are at the top, forming the rims (principle of superposition). The exposed rock layers extend laterally in all directions for some distance (principle of lateral continuity). (b) These shales and limestones of the Postolonnec Formation, at Postolonnec Beach, Crozon Peninsula, France, were originally deposited horizontally but have been significantly tilted since their formation.

b)

Steno also observed that, because sedimentary particles settle from water under the influence of gravity, sediment is deposited in essentially horizontal layers, thus illustrating the **principle of original horizontality** (Figure 16.3a). Therefore, a sequence of sedimentary rock layers that is steeply inclined from the horizontal must have been tilted after deposition and lithification (●**FIGURE 16.3B**).

Steno's third principle, the **principle of lateral continuity**, states that sediment extends laterally in all directions until it thins and pinches out, or terminates, against the edge of the depositional basin (Figure 16.3a).

James Hutton is credited with formulating the **principle of cross-cutting relationships**. Based on his detailed studies and observations of rock exposures in Scotland, Hutton recognized that an igneous intrusion or fault must be younger than the rocks it intrudes or displaces (●**FIGURE 16.4**).

---

**principle of original horizontality** According to this principle, sediments are deposited in horizontal or nearly horizontal layers.

**principle of lateral continuity** A principle holding that rock layers extend outward in all directions until they terminate.

**principle of cross-cutting relationships** A principle holding that an igneous intrusion or fault is younger than the rocks it intrudes or cuts across.

● **FIGURE 16.4 The Principle of Cross-Cutting Relationships** A small fault (arrows show direction of movement) cuts across, and thus displaces, tilted sedimentary beds along the Templin Highway in Castaic, California. The fault is therefore younger than the youngest beds that are displaced.

Although this principle illustrates that an intrusive igneous structure is younger than the rocks it intrudes, the association of sedimentary and igneous rocks may cause problems in relative dating. For example, buried lava flows and sills look very similar in a sequence of strata (●FIGURE 16.5). A buried lava flow, however, is older than the rocks above it (principle of superposition), whereas a sill, resulting from later igneous intrusion, is younger than all of the beds below it and younger than the immediately overlying bed as well.

To resolve such relative-age problems as these, geologists observe whether the sedimentary rocks in contact with

Strands and shreds of underlying sedimentary material may be in the base of the flow

Baking zone at bottom of flow

Lava flow

Clasts of lava may be present in the overlying, younger layer

Rubble zones may be present at the top and bottom of the flow

a)

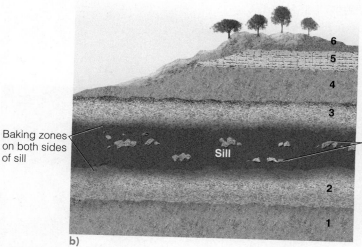

Baking zones on both sides of sill

Sill

Inclusions of rock from both layers above 3 and below 2 may exist in the sill

b)

Lava flow

c)

● FIGURE 16.5 Differentiating Between a Buried Lava Flow and a Sill (a) A buried lava flow has baked underlying bed 2 when it flowed over it. Clasts of the lava were deposited along with other sediments during deposition of bed 3. The lava flow is younger than bed 2 and older than beds 3, 4, and 5. (b) The rock units above and below the sill have been baked, indicating that the sill is younger than beds 2 and 3, but its age relative to beds 4-6 cannot be determined. (c) This buried lava flow in Yellowstone National Park, Wyoming, displays columnar jointing. A baked zone is present below, but not above the columnar joints, indicating it is a buried lava flow and not a sill.

Reed Wicander

**Critical Thinking Question** In (b), it is stated that the sill is younger than beds 2 and 3, but its age relative to beds 4–6 cannot be determined. Why is this?

the igneous rocks show signs of baking or alteration by heat (contact metamorphism). A sedimentary rock that shows such effects must be older than the igneous rock with which it is in contact. In Figure 16.5, for example, a sill produces a zone of baking immediately above and below it because it intruded into previously existing sedimentary rocks. A lava flow, in contrast, bakes only those rocks below it.

The **principle of inclusions** is yet another way to determine relative ages. This principle holds that inclusions, or fragments of one rock contained within a layer of another, are older than the rock layer itself. The batholith shown in ●**FIGURE 16.6A** contains sandstone inclusions, and the sandstone unit shows the effects of baking. Accordingly, we conclude that the sandstone is older than the batholith. In ●**FIGURE 16.6B**, however, the sandstone contains granite rock pieces, indicating that the batholith was the source rock for the inclusions and is therefore older than the sandstone.

Fossils have been known for centuries, yet their utility in relative dating and geologic mapping was not fully appreciated until the early 19th century. William Smith (1769–1839), an English civil engineer involved in surveying and building canals in southern England, independently recognized the principle of superposition by reasoning that the fossils at the bottom of a sequence of strata are older than those at the top of the sequence. This recognition served as the basis for the **principle of fossil succession** or the *principle of faunal and floral succession*, as it is sometimes called (●**FIGURE 16.7**).

According to this principle, fossil assemblages succeed one another through time in a regular and predictable order.

---

**principle of inclusions**  A principle holding that inclusions or fragments in a rock unit are older than the rock unit itself; for example, granite inclusions in sandstone are older than the sandstone.

**principle of fossil succession**  A principle holding that fossils, and especially groups or assemblages of fossils, succeed one another through time in a regular and predictable order.

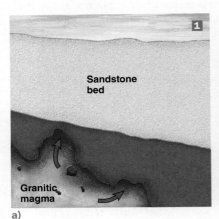
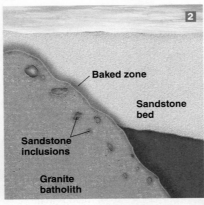

a)

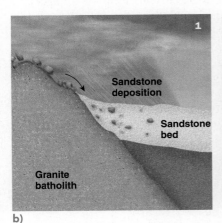
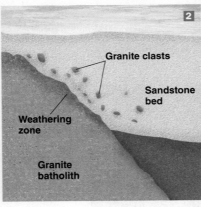

b)

c)

●**FIGURE 16.6 The Principle of Inclusions** (a) The sandstone is older than the granite batholith because there are inclusions of sandstone inside the granite. The sandstone also shows evidence of having been baked along its contact with the granite batholith when the granitic magma intruded the overlying sedimentary beds. (b) The sandstone is younger than the granite batholith because it contains pieces (clasts) of granite. The granite is also weathered along the contact with the sandstone indicating that it was the source of the granite clasts, and must therefore be older than the sandstone. (c) Outcrop in northern Wisconsin showing basalt inclusions (dark gray) in granite (white). Accordingly, the basalt inclusions are older than the granite.

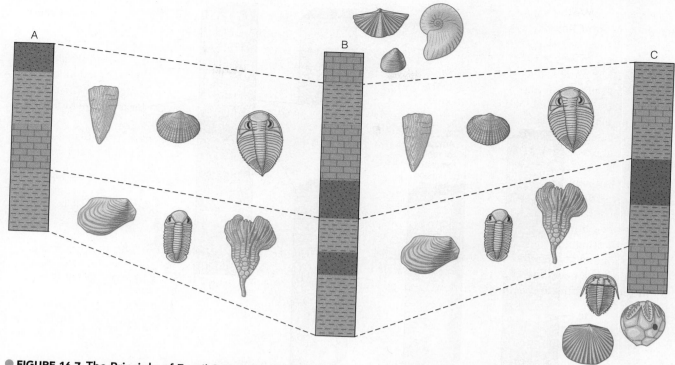

● **FIGURE 16.7 The Principle of Fossil Succession** This generalized diagram shows how geologists use the principle of fossil succession to determine the relative ages of rocks in widely separated areas. The rocks in the three sections encompassed by the dashed lines contain similar fossils and are therefore the same age. Note that the youngest rocks in this region are in section B, whereas the oldest rocks are in section C.

The validity and successful use of this principle depends on three points: (1) life has varied through time, (2) fossil assemblages are recognizably different from one another, and (3) the relative ages of the fossil assemblages can be determined. Observation of fossils in older versus younger strata clearly demonstrate that life-forms have changed through time. Because this is true, fossil assemblages (point 2) are recognizably different. Furthermore, superposition can be used to demonstrate the relative ages of the fossil assemblages.

## Unconformities

Our discussion so far has been concerned with *conformable strata*—sequences in which no depositional breaks of any consequence occur; that is, sedimentation was more or less continuous. A bedding plane between strata may represent a depositional break of anywhere from minutes to tens or hundreds of years, but it is inconsequential in the context of geologic time. However, in sequences of strata, surfaces known as **unconformities** may be present, which represent times of nondeposition, erosion, or both. Unconformities thus encompass long periods of geologic time, perhaps millions or tens of millions of years. Accordingly, the geologic record is incomplete wherever an unconformity is present, just as a book with missing pages is incomplete, and the

interval of geologic time not represented by strata is called a *hiatus* (●**FIGURE 16.8**).

Geologists recognize three types of unconformities. First, a **disconformity** is a surface of erosion, or nondeposition, separating younger from older rocks, both of which are parallel with one another (●**FIGURE 16.9**). Unless the erosional surface separating the older from the younger parallel beds is well defined, or distinct, the disconformity frequently resembles an ordinary bedding plane. Hence, many disconformities are difficult to recognize and must be identified on the basis of fossil assemblages.

Second, an **angular unconformity** is an erosional surface on tilted or folded strata over which younger strata were deposited (●**FIGURE 16.10A**). The strata below the unconformable surface generally dip more steeply than those above, producing an angular relationship.

The most famous angular unconformity in the world is probably the one at Siccar Point, Scotland (●**FIGURE 16.10B**).

---

**unconformity** A break in the geologic record represented by an erosional surface separating younger rocks from older rocks.

**disconformity** An unconformity above and below which the rock layers are parallel.

**angular unconformity** An unconformity below which older rocks dip at a different angle (usually steeper) than overlying strata.

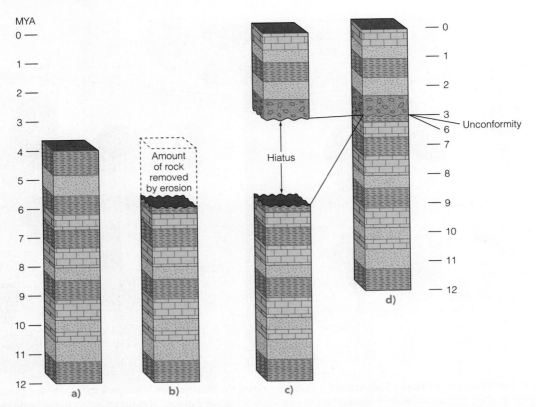

**● FIGURE 16.8 The Development of a Hiatus and an Unconformity** (a) Deposition began 12 million years ago (mya) and continued more or less uninterrupted until 4 mya. (b) Between 3 and 4 mya, an episode of erosion occurred. During that time some of the strata deposited earlier was eroded. (c) A hiatus of 3 million years thus exists between the older strata and the strata that formed during a renewed episode of deposition that began 3 mya. (d) The actual stratigraphic record as seen in an outcrop today. The unconformity is the surface separating the strata and represents a major break in our record of geologic time.

It was here that James Hutton realized that severe upheavals had tilted the lower rocks and formed mountains that were then worn away and covered by younger, flat-lying rocks. The erosional surface between the older, tilted rocks and the younger, flat-lying strata meant that a significant gap existed in the geologic record. Although Hutton did not use the term *unconformity*, he was the first to understand and explain the significance of such discontinuities in the geologic record.

A **nonconformity** is the third type of unconformity. Here, an erosional surface cut into metamorphic or igneous rocks is covered by sedimentary rocks (**●FIGURE 16.11**). This type of unconformity closely resembles an intrusive igneous contact with sedimentary rocks. The principle of inclusions (Figure 16.6) is helpful in determining whether the relationship between the underlying igneous rocks and the overlying sedimentary rocks is the result of an intrusion or erosion. In the case of an intrusion, the igneous rocks are younger, whereas in the case of erosion, the sedimentary rocks are younger. Being able to distinguish between a nonconformity and an intrusive contact is important because they represent completely different sequences of events.

## Applying the Principles of Relative Dating

We can decipher the geologic history of the area represented by the block diagram in **●FIGURE 16.12** by applying the various relative dating principles just discussed. The methods and logic used in this example are the same as those applied by 19th-century geologists in constructing the geologic time scale.

According to the principles of superposition and original horizontality, beds A–G were deposited horizontally; then either they were tilted, faulted (H), and eroded, or after deposition they were faulted (H), tilted, and then eroded (**●FIGURE 16.13A–C**). Because the fault cuts beds A–G, it must be younger than the beds according to the principle of cross-cutting relationships.

---

**nonconformity** An unconformity in which stratified sedimentary rocks overlie an erosion surface cut into igneous or metamorphic rocks.

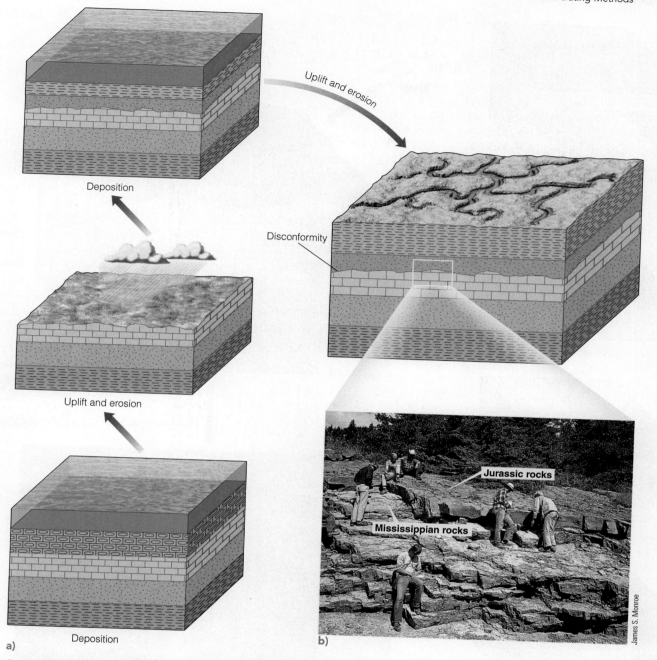


Deposition

Uplift and erosion

Uplift and erosion

Deposition

Disconformity

Jurassic rocks

Mississippian rocks

James S. Monroe

a)

b)

● **FIGURE 16.9 Formation of a Disconformity** (a) Formation of a disconformity. (b) Disconformity between Mississippian and Jurassic strata in Montana. The geologist at the upper left is sitting on Jurassic strata, and his right foot is resting on Mississippian rocks. This disconformity represents approximately 165 million years.

Beds J–L were then deposited horizontally over this erosional surface, producing an angular unconformity (I) (●**FIGURE 16.13D**). Following deposition of these three beds, the entire sequence was intruded by a dike (M), which, according to the principle of cross-cutting relationships, must be younger than all of the rocks that it intrudes (●**FIGURE 16.13E**).

The entire area was then uplifted and eroded; next, beds P and Q were deposited, producing a disconformity (N) between beds L and P and a nonconformity (O) between the igneous intrusion M and the sedimentary bed P (●**FIGURE 16.13F, G**). We know that the relationship between igneous intrusion M and the overlying sedimentary bed P is a nonconformity because of the inclusions of M in P (principle of inclusions).

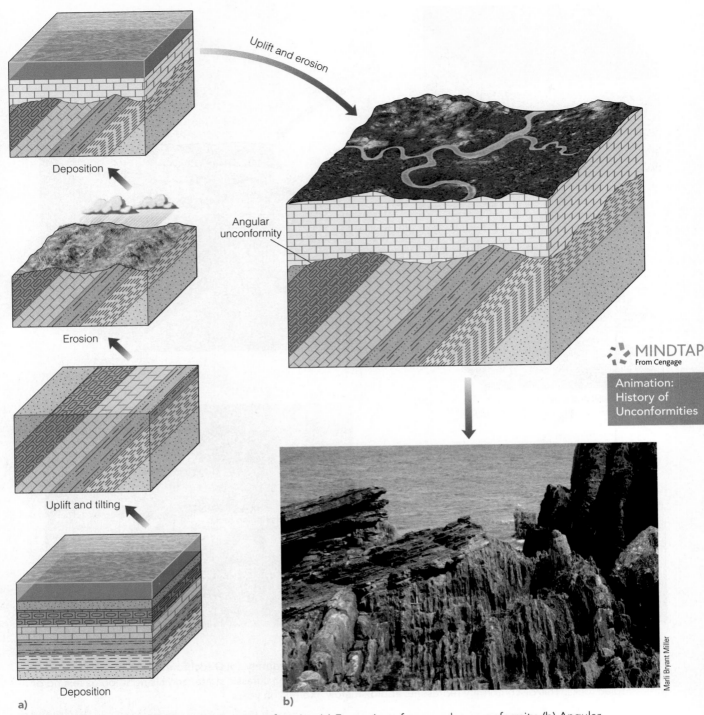

Deposition

Uplift and erosion

Erosion

Angular unconformity

Uplift and tilting

Deposition

a)

b)

Marli Bryant Miller

● **FIGURE 16.10 Formation of an Angular Unconformity** (a) Formation of an angular unconformity. (b) Angular unconformity at Siccar Point, Scotland. James Hutton first realized the significance of unconformities at this site in 1788.

**Critical Thinking Question** From what you can see in the photo in (b), which is the same scene Hutton viewed in 1788, what evidence is there that this outcrop represents an ancient mountain range? Were the forces compressional or tensional?

MINDTAP
From Cengage

Animation: History of Unconformities

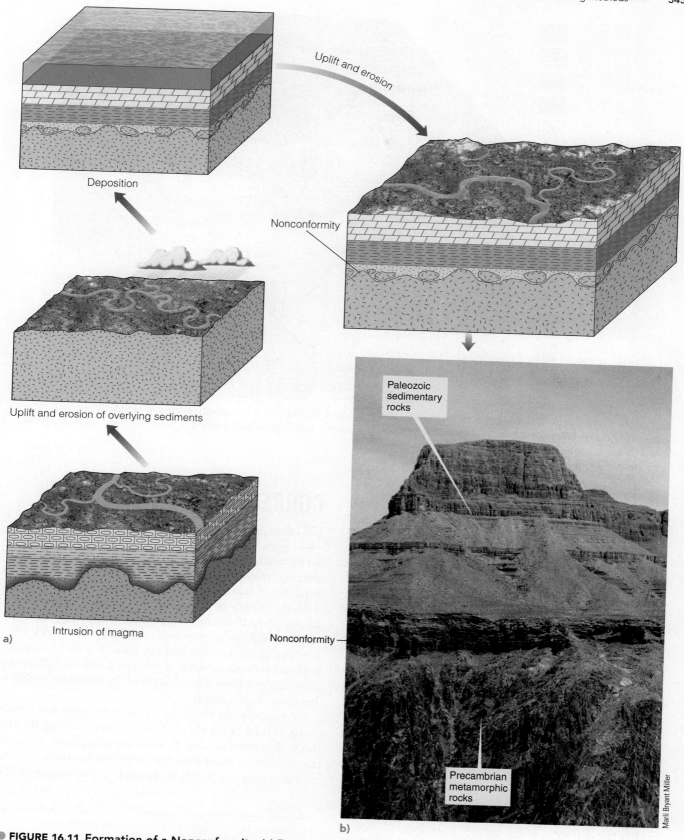

Deposition

Uplift and erosion

Nonconformity

Uplift and erosion of overlying sediments

Intrusion of magma

a)

Paleozoic
sedimentary
rocks

Nonconformity

Precambrian
metamorphic
rocks

Marli Bryant Miller

b)

● **FIGURE 16.11 Formation of a Nonconformity** (a) Formation of a nonconformity. (b) Nonconformity between Precambrian metamorphic rocks and the overlying Paleozoic sedimentary rocks in the Grand Canyon, Arizona.

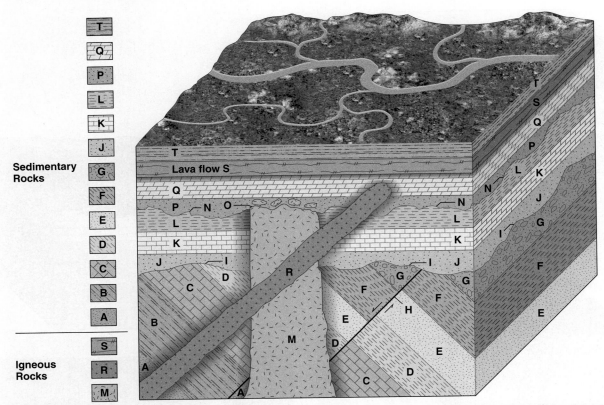

**Sedimentary Rocks**

**Igneous Rocks**

● **FIGURE 16.12 Block Diagram of a Hypothetical Area** A block diagram of a hypothetical area in which the various relative dating principles can be applied to determine its geologic history. See Figure 16.13 to learn how the geologic history was determined using relative dating principles.

At this point, there are several possibilities for reconstructing the geologic history of this area. According to the principle of cross-cutting relationships, dike R must be younger than bed Q because it intrudes into it. It could have intruded anytime *after* bed Q was deposited; however, we cannot determine whether R was formed right after Q, right after S, or after T was formed. For purposes of this history, we will say that it intruded after the deposition of bed Q (●**FIGURE 16.13G, H**).

Following the intrusion of dike R, lava S flowed over bed Q, followed by the deposition of bed T (●**FIGURE 16.13I, J**). Although the lava flow (S) is not a sedimentary unit, the principle of superposition still applies because it flowed onto the surface, just as sediments are deposited on Earth's surface.

We have established a relative chronology for the rocks and events of this area by using the principles of relative dating. Remember, however, that we have no way of knowing how many years ago these events occurred unless we can obtain radiometric dates for the igneous rocks. With these dates, we can establish the range of years between which the different sedimentary units were deposited and also determine how much time is represented by the unconformities.

# CORRELATING ROCK UNITS

**L08   Explain how rock units can be correlated**

To decipher Earth history, geologists must demonstrate the time equivalency of rock units in different areas. This process is known as **correlation**. If surface exposures are adequate, units may simply be traced laterally (principle of lateral continuity), even if occasional gaps exist (●**FIGURE 16.14**). Other criteria used to correlate units are similarity of rock type, position in a sequence, and key beds. *Key beds* are units, such as coal beds or volcanic ash layers, that are sufficiently distinctive to allow identification of the same unit in different areas (Figure 16.14).

Generally, no single location in a region has a geologic record of all events that occurred during its history; therefore, geologists must correlate from one area to another to determine the complete geologic history of the region. An excellent example is the history of the

---

**correlation** Demonstration of the physical continuity of rock units or biostratigraphic units, or demonstration of time equivalence as in time-stratigraphic correlation.

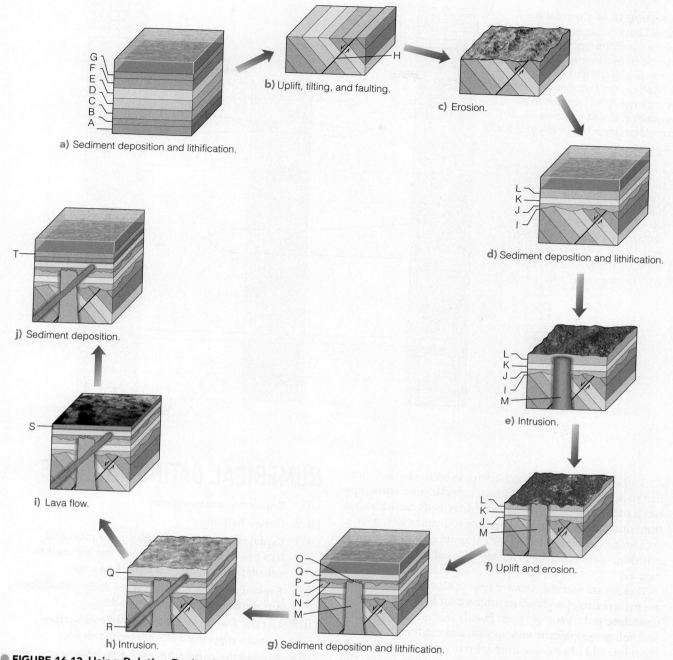

**FIGURE 16.13 Using Relative Dating Principles to Interpret the Geologic History of a Hypothetical Area** (a) Beds A-G are deposited and lithified. (b) The preceding beds are tilted and faulted. (c) Erosion. (d) Beds J-L are deposited and lithified, producing an angular unconformity (I). (e) The entire sequence is intruded by a dike (M). (f) The entire sequence is uplifted and eroded. (g) Beds P and Q are deposited and lithified, producing a disconformity (N) and a nonconformity (O). (h) Dike R intrudes. (i) Lava (S) flows over bed Q, baking it. (j) Bed T is deposited.

Colorado Plateau (●**FIGURE 16.15**). This region provides a record of events occurring over approximately 2 billion years, but because of the forces of erosion, the entire record is not preserved at any single location. Within the walls of the Grand Canyon are rocks of the Precambrian Eon and Paleozoic Era, whereas Paleozoic and Mesozoic

Era rocks are found in Zion National Park, and Mesozoic and Cenozoic Era rocks are exposed in Bryce Canyon National Park (Figure 16.15). By correlating the uppermost rocks at one location with the lowermost equivalent rocks of another area, geologists can decipher the history of the entire region.

● **FIGURE 16.14 Correlating Rock Units** In areas of adequate exposure, rock units can be traced laterally, even if occasional gaps exist, and correlated on the basis of similarity in rock type and position in a sequence. Rocks can also be correlated by a key bed—in this case, volcanic ash.

**Critical Thinking Question** Can you give at least one explanation for why the lower tongue of sandstone present in the left and center columns does not reach all the way to the column on the right?

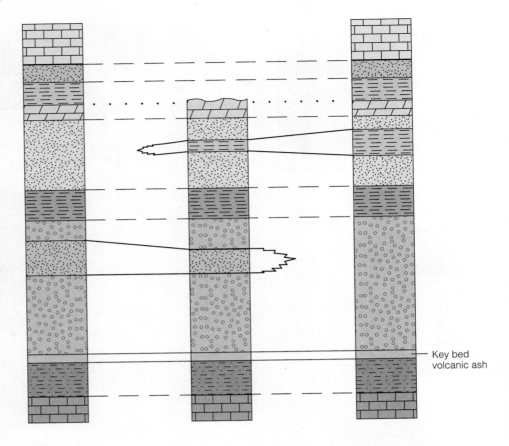

Key bed
volcanic ash

Although geologists match up rocks on the basis of similar rock type and superposition, correlation of this type can be done only in a limited area where beds can be traced from one site to another. To correlate rock units over a large area, or to correlate age-equivalent units of different composition, fossils and the principle of fossil succession must be used.

Fossils are useful as relative time indicators because they are the remains of organisms that lived for a certain length of time during the geologic past. Fossils that are easily identified, are geographically widespread, and existed for a rather short interval of geologic time are particularly useful. Such fossils are **guide fossils**, or *index fossils* (●**FIGURE 16.16**). The trilobite *Paradoxides* and the brachiopod *Atrypa* meet these criteria and are therefore good guide fossils. In contrast, the brachiopod *Lingula* is easily identified and widespread, but its long geologic range of Ordovician to Recent makes it of little use in correlation.

Because most fossils have fairly long geologic ranges, geologists construct *concurrent range zones* to determine the age of the sedimentary rocks containing the fossils. Concurrent range zones are established by plotting the overlapping geologic ranges of two or more fossils that have different geologic ranges (●**FIGURE 16.17**). The first and last occurrences of fossils are used to determine zone boundaries. Correlating concurrent range zones is probably the most accurate method of determining time equivalence.

# NUMERICAL DATING METHODS

**LO9**  Define radioactive decay

**LO10**  Define half-life

**LO11**  Explain how the parent–daughter ratio and half-life of the parent element can be used to calculate the numerical age of a sample

**LO12**  Explain why the most accurate radiometric dates are obtained from igneous rocks

**LO13**  List the five principal long-lived radioactive isotope pairs used in radiometric dating

**LO14**  Explain the carbon-14 dating technique and why it is only effective for specimens younger than about 70,000 years

Although most of the isotopes of the 92 naturally occurring elements (although some sources say there are 94 or even 98) are stable, some are radioactive and spontaneously decay to other, more stable isotopes of elements, releasing energy in the process. The discovery in 1903 by Pierre and Marie Curie that radioactive decay produces heat meant that geologists finally had a mechanism for explaining Earth's

**guide fossil**  Any easily identified fossil with an extensive geographic distribution and short geologic range useful for determining the relative ages of rocks in different areas.

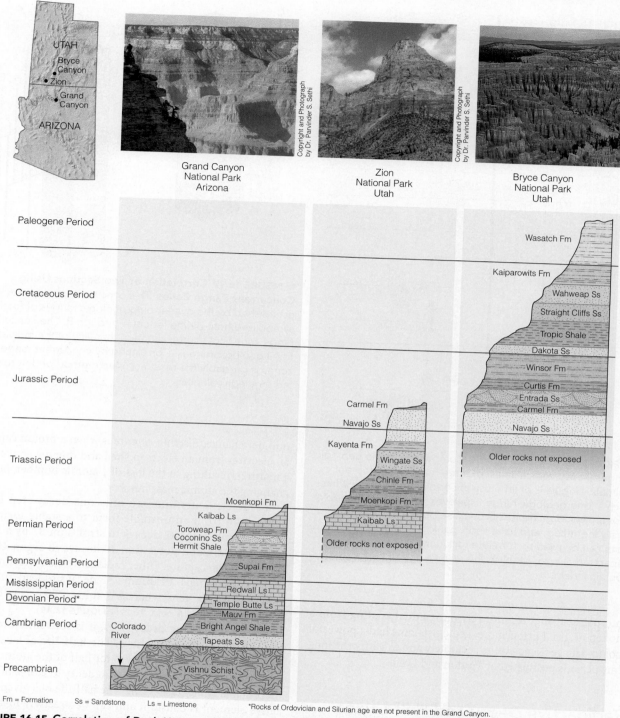

**FIGURE 16.15 Correlation of Rock Units within the Colorado Plateau** At each location only a portion of the geologic record of the Colorado Plateau is exposed. By correlating the youngest rocks at one exposure with the oldest rocks at another exposure, geologists can determine the entire history of the region. For example, the rocks forming the rim of the Grand Canyon, Arizona, are the Kaibab Limestone and Moenkopi Formation—the youngest rocks exposed in the Grand Canyon. The Kaibab Limestone and Moenkopi Formation are the oldest rocks exposed in Zion National Park, Utah, and the youngest rocks are the Navajo Sandstone and Carmel Formation. The Navajo Sandstone and Carmel Formation are the oldest rocks exposed in Bryce Canyon National Park, Utah. By correlating the Kaibab Limestone and Moenkopi Formation between the Grand Canyon and Zion National Park, geologists have extended the geologic history from the Precambrian to the Jurassic. And by correlating the Navajo Sandstone and Carmel Formation between Zion and Bryce Canyon National Parks, geologists can extend the geologic history through the Paleogene Period. Thus, by correlating the rock exposures between these areas and applying the principle of superposition, geologists can reconstruct the geologic history of the region.

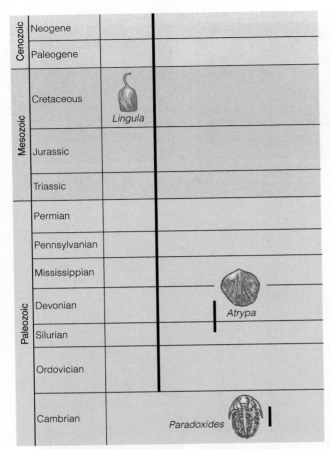

● **FIGURE 16.16 Guide Fossils** Comparison of the geologic ranges (heavy vertical lines) of three marine invertebrate animals. *Lingula* is of little use in correlation because it has such a long geologic range. However, *Atrypa* and *Paradoxides* are good guide fossils because both are widespread, easily identified, and have short geologic ranges. Thus, both can be used to correlate rock units that are widely separated and to establish the relative age of a rock that contains them.

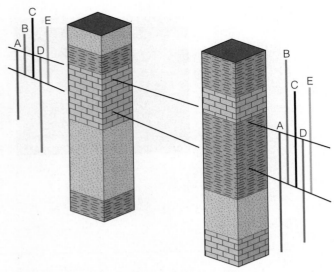

● **FIGURE 16.17 Correlation of Two Sections Using Concurrent Range Zones** This concurrent range zone was established by the overlapping geologic ranges of fossils symbolized here by the letters A through E. The concurrent range zone is of shorter duration than any of the individual fossil geologic ranges. Correlating by concurrent range zones is probably the most accurate method of determining time equivalence.

internal heat that did not rely on residual cooling from a molten origin. Furthermore, geologists now had a powerful tool to date geologic events accurately and to verify the long time periods postulated by Hutton and Lyell.

## Radioactive Decay and Half-Lives

**Radioactive decay** is the process whereby an unstable atomic nucleus is spontaneously transformed into an atomic nucleus of a different element. Three types of radioactive decay are recognized, all of which result in a change of atomic structure (●**FIGURE 16.18**).

In *alpha decay*, 2 protons and 2 neutrons are emitted from the nucleus, resulting in the loss of 2 atomic numbers and 4 atomic mass numbers. In *beta decay*, a fast-moving electron is emitted from a neutron in the nucleus, changing that neutron to a proton and consequently increasing the atomic number by 1, with no resultant atomic mass

number change. *Electron capture* is when a proton captures an electron from an electron shell and thereby converts to a neutron, resulting in the loss of 1 atomic number, but not changing the atomic mass number.

Some elements undergo only one decay step in the conversion from an unstable form to a stable form. For example, rubidium 87 decays to strontium 87 by a single beta emission, and potassium 40 decays to argon 40 by a single electron capture. Other radioactive elements undergo several decay steps. Uranium 235 decays to lead 207 by 7 alpha and 6 beta steps, whereas uranium 238 decays to lead 206 by 8 alpha and 6 beta steps (●**FIGURE 16.19**).

When discussing decay rates, it is convenient to refer to them in terms of half-lives. The **half-life** of a radioactive element is the time it takes for half of the atoms of the original unstable *parent element* to decay to atoms of a new, more stable *daughter element*. The half-life of a given radioactive element is constant and can be precisely measured. Half-lives of various radioactive elements range from less than a billionth of a second to 49 billion years.

**radioactive decay** The spontaneous change of an atom to an atom of a different element by emission of a particle from its nucleus (alpha and beta decay) or by electron capture.

**half-life** The time necessary for half of the original number of radioactive atoms of an element to decay to a stable daughter product; for example, the half-life for potassium 40 is 1.3 billion years.

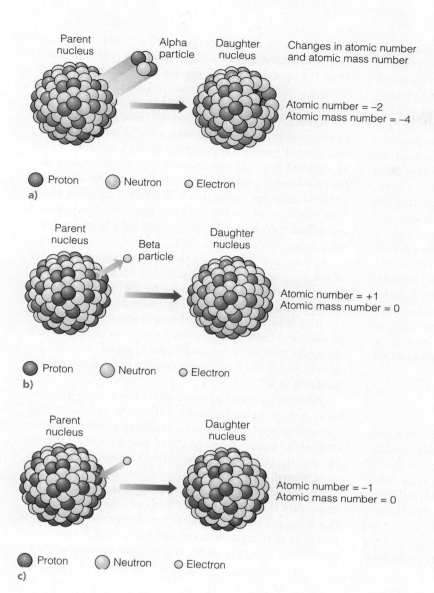

a)

Parent nucleus    Alpha particle    Daughter nucleus    Changes in atomic number and atomic mass number

Atomic number = −2
Atomic mass number = −4

● Proton    ● Neutron    ○ Electron

b)

Parent nucleus    Beta particle    Daughter nucleus

Atomic number = +1
Atomic mass number = 0

● Proton    ● Neutron    ○ Electron

c)

Parent nucleus    Daughter nucleus

Atomic number = −1
Atomic mass number = 0

● Proton    ● Neutron    ○ Electron

● **FIGURE 16.18 Three Types of Radioactive Decay** (a) Alpha decay, in which an unstable parent nucleus emits 2 protons and 2 neutrons. (b) Beta decay, in which an electron is emitted from the nucleus. (c) Electron capture, in which a proton captures an electron and is thereby converted to a neutron.

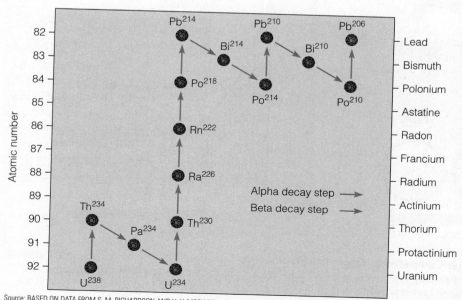

● **FIGURE 16.19 Radioactive Decay Series for Uranium 238 to Lead 206** Radioactive uranium 238 decays to its stable daughter product, lead 206, by 8 alpha and 6 beta decay steps. A number of different isotopes are produced as intermediate steps in the decay series.

Source: BASED ON DATA FROM S. M. RICHARDSON AND H. Y. MCSWEEN, JR., GEOCHEMISTRY—PATHWAYS AND PROCESSES, PRENTICE-HALL.

Radioactive decay occurs at a geometric rate rather than a linear rate (●**FIGURE 16.20A**). Therefore, a graph of the decay rate produces a curve rather than a straight line (●**FIGURE 16.20B**). For example, an element with *1,000,000* parent atoms will have *500,000* parent atoms and *500,000* daughter atoms after one half-life. After two half-lives, it will have *250,000* parent atoms (one-half of the previous parent atoms, which is equivalent to one-fourth of the original parent atoms) and *750,000* daughter atoms. After three half-lives, it will have *125,000* parent atoms (one-half of the previous parent atoms, or one-eighth of the original parent atoms) and *875,000* daughter atoms, and so on, until the number of parent atoms remaining is so few that they cannot be accurately measured by present-day instruments.

By measuring the parent–daughter ratio and knowing the half-life of the parent (which has been determined in the laboratory), geologists can calculate the age of a sample that contains the radioactive element. The parent–daughter ratio is usually determined by a *mass spectrometer*, an instrument that measures the proportions of atoms of different masses.

## Sources of Uncertainty

The most accurate radiometric dates are obtained from igneous rocks. As magma cools and begins to crystallize, radioactive parent atoms are separated from previously formed daughter atoms. Because they are the right size, some radioactive parent atoms are incorporated into the crystal structure of certain minerals. The stable daughter atoms, though, are a different size from the radioactive parent atoms and consequently cannot fit into the crystal structure of the same mineral as the parent atoms. Therefore, a mineral crystallizing in cooling magma will contain radioactive parent atoms but no stable daughter atoms (●**FIGURE 16.21**). Thus, the time that is being measured is the time of crystallization of the mineral that contains the radioactive atoms and *not* the time of formation of the radioactive atoms.

To obtain accurate radiometric dates, geologists must be sure that they are dealing with a *closed system*, meaning that neither parent nor daughter atoms have been added or removed from the system since crystallization, and that the ratio between them results only from radioactive decay. Otherwise, an inaccurate date will result. If daughter atoms have leaked out of the mineral being analyzed, the calculated age will be too young; if parent atoms have been removed, the calculated age will be too old.

Leakage may take place if the rock is heated or subjected to intense pressure as can sometimes occur during metamorphism. If this happens, some of the parent or daughter atoms may be driven from the mineral being analyzed, resulting in an inaccurate age determination. If the daughter product was completely removed, then one would be measuring the time since metamorphism (a useful measurement itself) and not the time since crystallization of the mineral (●**FIGURE 16.22**).

Because heat and pressure affect the parent–daughter ratio, metamorphic rocks are difficult to date accurately. Remember that although the resulting parent–daughter ratio of the sample being analyzed may have been affected by heat, the decay rate of the parent element remains constant, regardless of any physical or chemical changes.

To obtain an accurate radiometric date, geologists must make sure that the sample is fresh and unweathered and that it has not been subjected to high temperature or intense pressures after crystallization. Furthermore, it is sometimes possible to cross-check the radiometric date obtained by

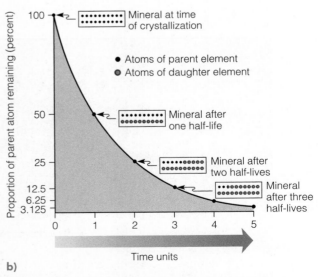

a)

b)

● **FIGURE 16.20 Uniform, Linear Change Compared to Geometric Radioactive Decay** (a) Uniform, linear change is characteristic of many familiar processes. In this example, water is being added to a glass at a constant rate. (b) A geometric radioactive decay curve, in which each time unit represents one half-life, and each half-life is the time it takes for half of the parent element to decay to the daughter element.

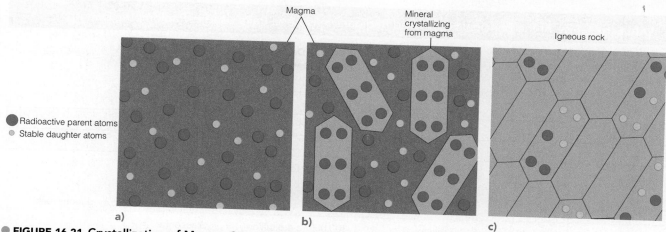

Magma

Mineral crystallizing from magma

Igneous rock

● Radioactive parent atoms
○ Stable daughter atoms

a)    b)    c)

● **FIGURE 16.21 Crystallization of Magma Containing Radioactive Parent and Stable Daughter Atoms** (a) Magma contains both radioactive parent atoms and stable daughter atoms. The radioactive parent atoms are larger than the stable daughter atoms. (b) As magma cools and begins to crystallize, some of the radioactive atoms are incorporated into certain minerals because they are the right size and can fit into the crystal structure. In this example, only the larger radioactive parent atoms fit into the crystal structure. Therefore, at the time of crystallization, minerals in which the radioactive parent atoms can fit into the crystal structure will contain 100% radioactive parent atoms and 0% stable daughter atoms. (c) After one half-life, 50% of the radioactive parent atoms will have decayed to stable daughter atoms such that those minerals that had radioactive parent atoms in their crystal structure will now have 50% radioactive parent atoms and 50% stable daughter atoms.

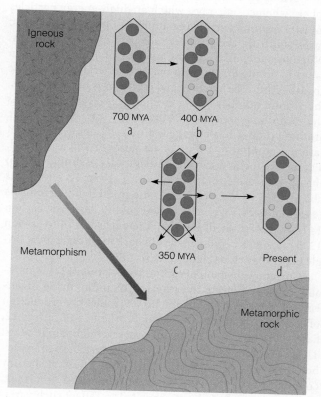

Igneous rock

700 MYA    400 MYA
a    b

Metamorphism

350 MYA    Present
c    d

Metamorphic rock

● **FIGURE 16.22 Effects of Metamorphism on Radiometric Dating** The effect of metamorphism in driving out daughter atoms from a mineral that crystallized 700 million years ago (mya). The mineral is shown immediately after crystallization (a), then at 400 million years (b), when some of the parent atoms had decayed to daughter atoms. Metamorphism at 350 mya (c) drives the daughter atoms out of the mineral into the surrounding rock. If the rock has remained a closed chemical system throughout its history, dating the mineral today (d) yields the time of metamorphism, whereas dating the whole rock provides the time of its crystallization, 700 million years ago.

systems, the ages obtained from each parent–daughter ratio should agree closely. If they do, they are said to be *concordant*, thus reflecting the time of crystallization of the magma. If the ages do not closely agree, then they are said to be *discordant*, and other samples must be taken and ratios measured to which, if either, date is correct.

## Long-Lived Radioactive Isotope Pairs

● **TABLE 16.1** shows the five common, long-lived parent–daughter isotope pairs used in radiometric dating. Long-lived pairs have half-lives of millions or billions of years. All of these were present when Earth formed and are still present in measurable quantities. Other shorter-lived radioactive isotope pairs have decayed to the point that only small quantities near the limit of detection remain.

measuring the parent–daughter ratio of two different radioactive elements in the same material.

For example, naturally occurring uranium consists of both uranium 235 and uranium 238 isotopes. Through various decay steps, uranium 235 decays to lead 207, whereas uranium 238 decays to lead 206 (Figure 16.19). If the minerals that contain both uranium isotopes have remained closed

**TABLE 16.1**  Five of the Principal Long-Lived Radioactive Isotope Pairs Used in Radiometric Dating

| Isotopes | | Half-Life of Parent (years) | Effective Dating Range (years) | Minerals and Rocks That Can Be Dated |
|---|---|---|---|---|
| Parent | Daughter | | | |
| Uranium 238 | Lead 206 | 4.5 billion | 10 million to 4.6 billion | Zircon<br>Uraninite |
| Uranium 235 | Lead 207 | 704 million | | |
| Thorium 232 | Lead 208 | 14 billion | | |
| Rubidium 87 | Strontium 87 | 48.8 billion | 10 million to 4.6 billion | Muscovite<br>Biotite<br>Potassium feldspar<br>Whole metamorphic or igneous rock |
| Potassium 40 | Argon 40 | 1.3 billion | 100,000 to 4.6 billion | Glauconite<br>Hornblende<br>Muscovite<br>Whole volcanic rock<br>Biotite |

The most commonly used isotope pairs are the uranium–lead and thorium–lead series, which are used principally to date ancient igneous intrusives, lunar samples, and some meteorites. The rubidium–strontium pair is also used for very old samples and has been effective in dating the oldest rocks on Earth, as well as meteorites.

The potassium–argon method is typically used for dating fine-grained volcanic rocks from which individual crystals cannot be separated; hence, the whole rock is analyzed. Because argon is a gas, great care must be taken to ensure that the sample has not been subjected to heat, which would allow argon to escape; such a sample would yield an age that is too young. Other long-lived radioactive isotope pairs exist, but they are rather rare and are used only in special situations.

## Carbon-14 Dating Method

Carbon is an important element in nature and is one of the basic elements found in all forms of life. It has three isotopes; two of these, carbon 12 and 13, are stable, whereas carbon 14 is radioactive (see Figure 3.3). Carbon 14 has a half-life of 5,730 years plus or minus 30 years. The **carbon-14 dating technique** is based on the ratio of carbon 14 to carbon 12 and is used to date formerly living material.

The short half-life of carbon 14 makes this dating method practical up to about 50,000 years, although new techniques have extended the range to 60,000 years, and in some cases to about 70,000 years. Consequently, the carbon-14 dating method is especially useful in archaeology and has greatly helped unravel the events of the latter portion of the Pleistocene Epoch.

Carbon 14 is constantly formed in the upper atmosphere when cosmic rays, which are high-energy particles (mostly protons), strike the atoms of upper-atmospheric gases, splitting their nuclei into protons and neutrons. When a neutron strikes the nucleus of a nitrogen atom (atomic number 7, atomic mass number 14), it may be absorbed into the nucleus and a proton is emitted. Thus, the atomic number of the atom decreases by 1, whereas the atomic mass number stays the same. Because the atomic number has changed, a new element, carbon 14 (atomic number 6, atomic mass number 14), is formed. The newly formed carbon 14 is rapidly assimilated into the carbon cycle and, along with carbon 12 and 13, is absorbed in a nearly constant ratio by all living organisms (●**FIGURE 16.23**). When an organism dies, however, carbon 14 is not replenished, and the ratio of carbon 14 to carbon 12 decreases as carbon 14 decays back to nitrogen by a single beta decay step (Figure 16.23).

Currently, the ratio of carbon 14 to carbon 12 is remarkably constant both in the atmosphere and in living organisms. There is good evidence, however, that the production of carbon 14, and thus the ratio of carbon 14 to carbon 12, has varied somewhat during the past several thousand years. This was determined by comparing ages established by carbon-14 dating of wood samples with ages established by counting annual tree rings in the same samples. As a result, carbon-14 ages have been corrected to reflect such variations in the past.

---

**carbon-14 dating technique**  A numerical dating technique relying on the ratio of carbon 12 to carbon 14 in an organic substance; useful back to about 70,000 years ago.

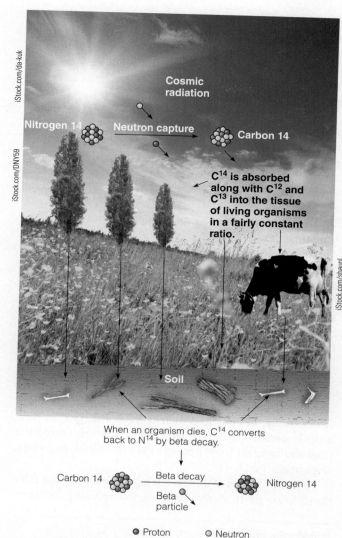

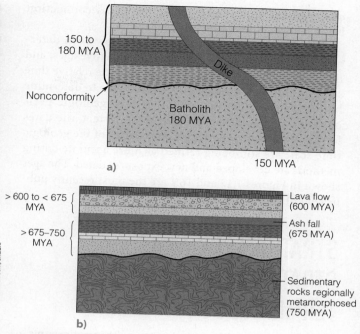

When an organism dies, C$^{14}$ converts back to N$^{14}$ by beta decay.

Carbon 14 → Beta decay → Nitrogen 14

Beta decay
Beta particle

○ Proton    ○ Neutron

● **FIGURE 16.23 Carbon-14 Dating Method** The carbon cycle involves the formation of carbon 14 in the upper atmosphere, its dispersal and incorporation into the tissues of all living organisms, and its decay back to nitrogen 14 by beta decay.

**Critical Thinking Question** Why do you think the ratio of carbon 14 to carbon 12 is constant in living organisms?

● **FIGURE 16.24 Determining Numerical Dates for Sedimentary Rocks** The numerical ages of sedimentary rocks can be determined by dating associated igneous rocks. In (a) and (b), sedimentary rocks are bracketed by rock bodies for which numerical ages have been determined.

# DEVELOPMENT OF THE GEOLOGIC TIME SCALE

**LO15**    Describe the development of the geologic time scale

The geologic time scale is a hierarchical scale in which the 4.6-billion-year history of Earth is divided into time units of varying duration (Figure 16.1). It did not result from the work of any one individual, but rather evolved, primarily during the 19th century, through the efforts of many people.

By applying relative dating methods to rock outcrops, geologists in England and Western Europe defined the major geologic time units without the benefit of radiometric dating techniques. Using the principles of superposition and fossil succession, they correlated various rock exposures and pieced together a composite geologic section. This composite section is, in effect, a relative time scale because the rocks are arranged in their correct sequential order.

By the beginning of the 20th century, geologists had developed a relative geologic time scale but did not yet have any numerical dates for the various time–unit boundaries. Following the discovery of radioactivity near the end of the 19th century, radiometric dates were added to the relative geologic time scale (Figure 16.1).

Because sedimentary rocks, with rare exceptions, cannot be radiometrically dated, geologists have had to rely on interbedded volcanic rocks and igneous intrusions to apply numerical dates to the boundaries of the various subdivisions of the geologic time scale (●**FIGURE 16.24**).

An ashfall or lava flow provides an excellent marker bed that is a time-equivalent surface, supplying a minimum age for the sedimentary rocks below and a maximum age for the rocks above. Ashfalls are particularly useful because they may fall over both marine and nonmarine sedimentary environments and therefore can provide a connection between these different environments.

Thousands of numerical ages have now been determined for sedimentary rocks of known relative ages, and these numerical dates have been added to the relative time scale. In this way, geologists have been able to determine both the numerical ages of the various geologic periods and their durations (Figure 16.1). In fact, the dates for the era, period, and epoch boundaries of the geologic time scale are still being refined as more accurate dating methods are developed and new exposures dated. The ages shown in Figures 1.17 and 16.1 are the most recently published (2018).

# STRATIGRAPHY AND STRATIGRAPHIC TERMINOLOGY

**LO16   Explain the difference between lithostratigraphic units, time-stratigraphic units, and time units**

The recognition of a relative geologic time scale brought some order to *stratigraphy* (the study of the composition, origin, areal distribution, and age relationships of layered rocks); however, problems remained because many sedimentary rock units are time transgressive. This means that they were deposited during one geologic period in a particular area and during another period elsewhere. Therefore, modern stratigraphic terminology includes two fundamentally different kinds of units to deal with both rocks and time: those defined by their content and those related to geologic time (●**TABLE 16.2**).

Units defined by their content include lithostratigraphic and biostratigraphic units. **Lithostratigraphic units** (lith- and litho- are prefixes meaning "stone" or "stonelike") are defined by the physical attributes of the rocks, such as rock type (e.g., sandstone or limestone), with no consideration of time of origin. The basic lithostratigraphic unit is the *formation*, which is a mapable body of rock with distinctive upper and lower boundaries (●**FIGURE 16.25**).

Formations may consist of a single rock type, for example, the Redwall Limestone (Figure 16.15), or a variety of related rock types such as the Morrison Formation. Formations are commonly subdivided into smaller units known as *members* and *beds*, and they may be parts of larger units known as *groups* and *supergroups* (Table 16.2).

A body of strata recognized only on the basis of its fossil content is a **biostratigraphic unit**, the boundaries of which do not necessarily correspond to those of lithostratigraphic boundaries (●**FIGURE 16.26**). The fundamental biostratigraphic unit is the *biozone*. Several types of biozones are recognized, one of which, the *concurrent range zone*, was discussed in the section on correlation (Figure 16.17).

The category of units expressing or related to geologic time includes (1) time-stratigraphic units (also known as *chronostratigraphic units*) and (2) time units (Table 16.2). **Time-stratigraphic units** consist of rocks deposited during a particular interval of geologic time. The *system*, the basic time-stratigraphic unit, is based on a *stratotype*, which consists of rocks in an area where the system was first described. Systems are recognized beyond their stratotype area by their fossil content.

**Time units** are simply designations for certain parts of geologic time. The basic time unit is the *period*; however, two or more periods may be designated as an *era*, and two or more eras constitute an *eon*. Periods also consist of shorter designations such as *epoch* and *age*. The time units known as period, epoch, and age correspond to the time-stratigraphic units known as system, series, and stage, respectively (Table 16.2).

These two types of units referring to time and their relationship are particularly confusing to beginning students.

---

**lithostratigraphic unit**  A body of rock, such as a formation, defined solely by its physical attributes.

**biostratigraphic unit**  An association of sedimentary rocks defined by its fossil content.

**time-stratigraphic unit**  A body of strata that was deposited during a specific interval of geologic time; for example, the Devonian System (a time-stratigraphic unit) was deposited during the Devonian Period (a time unit).

**time unit**  Any of the units such as eon, era, period, epoch, and age, used to refer to specific intervals of geologic time.

---

**TABLE 16.2** Classification of Stratigraphic Units

| Units Defined By Content | | Units Expressing or Related To Geologic Time | |
| --- | --- | --- | --- |
| Lithostratigraphic Units | Biostratigraphic Units | Time-Stratigraphic Units | Time Units |
| Supergroup | Biozones | Eonothem —————————— Eon | |
| Group | | Erathem —————————— Era | |
| Formation | | System —————————— Period | |
| Member | | Series —————————— Epoch | |
| Bed | | Stage —————————— Age | |

Quaternary Peak in Palo Duro Canyon State Park, Texas, is composed of the

**Dockman Group**

Trujillo Sandstone

Tecovas Formation

Quartermaster Formation

● **FIGURE 16.25 Lithostratigraphic Units** Triassic Peak in Palo Duro Canyon State Park, Texas, is composed of the Quartermaster Formation (Permian), followed by the Tecovas Formation and Trujillo Sandstone, both of Triassic age. The Dockman Group includes the Tecovas Formation and Trujillo Sandstone.

Remember though that time-stratigraphic units are material bodies of rock that occupy a position in a sequence of strata (e.g., the Devonian System), whereas time units refer only to time (the Devonian Period). Thus, a system is a body of rock with lower, medial, and upper parts. In contrast, a time unit, such as the Devonian Period, is the time during which rocks of the Devonian System were deposited, and we divide time units into early, middle, and late subdivisions.

## GEOLOGIC TIME AND CLIMATE CHANGE

LO17  Discuss why it is important to understand the causes and durations of past climate changes in terms of how they might apply to present-day climatic changes

Given the debate concerning global warming and its possible implications, it is extremely important to be able to reconstruct past climatic regimes as accurately as possible. To model how Earth's climate system has responded to changes in the past and to use that information for simulations of future climate scenarios, geologists must have a geologic calendar that is as precise and accurate as possible.

New dating techniques with greater precision are providing geologists with more accurate dates for when and how long ago past climate changes occurred. The ability to accurately determine when past climate changes occurred helps geologists correlate these changes with regional and global geologic events to see whether there are any possible connections.

In addition to short-term climatic changes, it is also important to step back a bit and look at climate change from a geologic perspective. We know that Earth has experienced

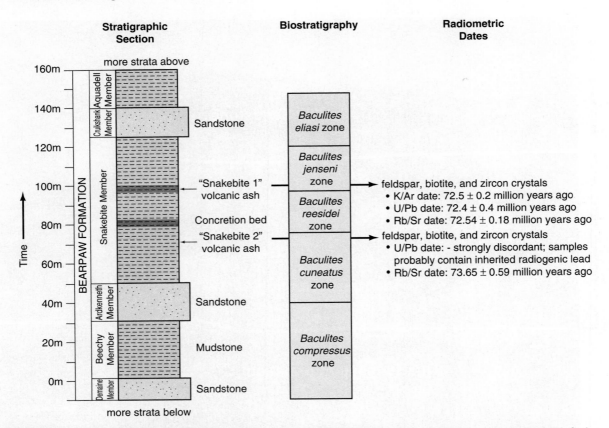

**Stratigraphic Section**

**Biostratigraphy**

**Radiometric Dates**

● **FIGURE 16.26 Rocks and Fossils of the Bearpaw Formation in Saskatchewan, Canada** The column on the left shows formation and members that are lithostratigraphic units. Notice that the biozone boundaries do not correspond with lithostratigraphic boundaries. The numerical ages for the two volcanic ash layers indicate that the *Baculites reesidei* zone is about 72 to 73 million years old.

**Critical Thinking Question**  Why don't the biozone boundaries correspond with the lithostratigraphic boundaries?

periods of glaciation in the past—for instance, during the Proterozoic, at the end of the Ordovician Period, and most recently during the Pleistocene Epoch (see Figure 2 in Creative Thinking Visual Question, Chapter 1). Earth has also undergone large-scale periods of soaring global temperature, such as during what is known as the *Paleocene–Eocene Thermal Maximum*, beginning around 56 million years ago and lasting more the 150,000 years. During this time, Earth experienced drought, a rise in sea level, and changes to its biota (●**FIGURE 16.27**).

Even though we cannot physically travel back in time, geologists can reconstruct what the climate was like in the past. The distribution of plants and animals is controlled, in part, by climate. Plants are particularly sensitive to climate change, and many can live only in particular environments. The fossils of plants and animals can tell us something about the environment and climate at the time these organisms were living.

Furthermore, climate-sensitive sedimentary rocks can be used to interpret past climatic conditions. Desert dunes are typically well sorted and exhibit large-scale cross-bedding (see Figure 14.8). Coals form in fresh-water swamps where climatic conditions promote abundant plant growth (see Figure 17.13). Evaporites such as rock salt result when evaporation exceeds precipitation, such as in desert regions or along hot, dry shorelines. Tillites (glacial sediments) result from glacial activity and indicate cold, wet environments.

By combining all relevant geologic and paleontologic information (●**FIGURE 16.28**), geologists can reconstruct what the climate was like in the past, and how it has changed over time in a particular area. Furthermore, geologists hope that by analysis of this data, they can, at some time in the near future, predict and even possibly modify regional climate changes, such as by reducing the burning of fossil fuels which increase the greenhouse effect, and thus contribute to global warming.

Sean Mcnaughton/National Geographic Image Collection

SEAN MCNAUGHTON, NGM STAFF
SOURCES: CLINTON CROWLEY; C. R. SCOTESE, PALEOMAP PROJECT

● **FIGURE 16.27 Earth at the End of the Paleocene Epoch** At the end of the Paleocene Epoch, Earth was hot and ice-free. During the Paleocene–Eocene Thermal Maximum (PETM), the average temperature was 25°C, compared with 20°C before (late Paleocene) and after (early Eocene) the PETM, and 7.2°C today. Furthermore, sea level was an estimated 67 m higher than presently.

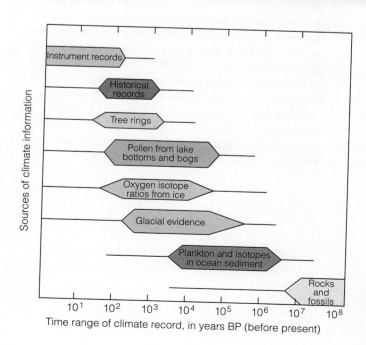

Sources of climate information

- Instrument records
- Historical records
- Tree rings
- Pollen from lake bottoms and bogs
- Oxygen isotope ratios from ice
- Glacial evidence
- Plankton and isotopes in ocean sediment
- Rocks and fossils

$10^1$  $10^2$  $10^3$  $10^4$  $10^5$  $10^6$  $10^7$  $10^8$

Time range of climate record, in years BP (before present)

● **FIGURE 16.28 Sources of Climate Data** Some of the methods used by scientists to determine historical and ancient climates. Each method has its own useful time range and accuracy within that time range.

# Key Concepts Review

- Time is defined by the methods used to measure it. Relative dating places geologic events in sequential order as determined from their position in the geologic record. Numerical dating provides specific dates for geologic rock units or events that are expressed in years before the present.
- During the 18th and 19th centuries, attempts were made to determine Earth's age based on scientific evidence rather than revelation. Although some attempts were ingenious, they yielded a variety of ages that now are known to be much too young.
- James Hutton, considered by many to be the founder of modern geology, thought that present-day processes operating over long periods of time could explain all of the geologic features of Earth. His observations were instrumental in establishing the principle of uniformitarianism and the fact that Earth was much older than earlier scientists thought.
- Uniformitarianism, as articulated by Charles Lyell, soon became the guiding principle of geology. It holds that the laws of nature have been constant through time and that the same processes operating today have also operated in the past, although not necessarily at the same rates.
- Besides uniformitarianism, the principles of superposition, original horizontality, lateral continuity, cross-cutting relationships, inclusions, and fossil succession are basic for determining relative geologic ages and for interpreting Earth history.
- An unconformity is a surface of erosion, nondeposition, or both, separating younger strata from older strata. These surfaces encompass long periods of geologic time for which there is no geologic record at that location.
- Three types of unconformities are recognized. A disconformity separates younger from older sedimentary strata that are parallel to each other. An angular unconformity is an erosional surface on tilted or folded rocks, over which younger sedimentary rocks were deposited. A nonconformity is an erosional surface cut into igneous or metamorphic rocks and overlain by younger sedimentary rocks.
- Correlation is the demonstration of time equivalency of rock units in different areas. Similarity of rock type, position within a rock sequence, key beds, and fossil assemblages can all be used to correlate rock units.
- Radioactivity was discovered during the late 19th century, and soon thereafter, radiometric dating techniques enabled geologists to determine numerical ages for rock units and geologic events.
- Numerical (sometimes referred to as absolute) dates for rocks are usually obtained by determining how many half-lives of a radioactive parent element have elapsed since the sample originally crystallized. A half-life is the time it takes for one-half of the original unstable radioactive parent element to decay into a new, more stable daughter element.
- The most accurate radiometric dates are obtained from long-lived radioactive isotope pairs in igneous rocks. The most reliable dates are those obtained by using at least two different radioactive decay series in the same rock.
- Carbon-14 dating can be used only on organic matter such as wood, bones, and shells and is effective back to approximately 70,000 years ago. Unlike the long-lived radioactive isotopic pairs, the carbon-14 dating technique determines age by the ratio of radioactive carbon 14 to stable carbon 12.
- The geologic time scale was developed primarily during the 19th century through the efforts of many people. It was originally a relative geologic time scale, but with the discovery of radioactivity and the development of radiometric dating methods, numerical ages were added at the beginning of the 20th century. Since then, refinement of the time-unit boundaries has continued.
- Stratigraphic terminology includes two fundamentally different kinds of units: those based on content such as lithostratigraphic and biostratigraphic units, and those related to geologic time, which include time-stratigraphic and time units.
- To reconstruct past climate changes and link them to possible causes, geologists must have a geologic calendar that is as precise and accurate as possible. Thus, they must be able to date geologic events and the onset and duration of climate changes as precisely as possible.

# Important Terms

angular unconformity   339

biostratigraphic unit   354

carbon-14 dating technique   352

correlation   344

disconformity   339

guide fossil   346

half-life   348

lithostratigraphic unit   354

nonconformity   340

numerical dating   332

principle of cross-cutting relationships   336

principle of fossil succession   338

principle of inclusions   338

principle of lateral continuity   336

principle of original horizontality   336

principle of superposition   335

radioactive decay   348

relative dating   332

time-stratigraphic unit   354

time unit   354

unconformity   339

# Review Questions

1. The basic lithostratigraphic unit is the
   a. group.
   b. formation.
   c. bed.
   d. member.
   e. supergroup.

2. If a rock is heated during metamorphism and the daughter atoms migrate out of a mineral that is subsequently radiometrically dated, an inaccurate date will be obtained. This date therefore
   a. will be the same as the actual date.
   b. will be younger than the actual date.
   c. will be older than the actual date.
   d. cannot be determined.
   e. none of these.

3. Placing geologic events in sequential order as determined by their position in the geologic record is
   a. numerical dating.
   b. correlation.
   c. historical dating.
   d. relative dating.
   e. uniformitarianism.

4. If a radioactive element has a half-life of 32 million years, what fraction of the original amount of parent material will remain after 96 million years?
   a. one-half
   b. one-eighth
   c. one-sixteenth
   d. one-fourth
   e. one-thirty-second

5. If a flake of biotite within a sedimentary rock (such as a sandstone) is radiometrically dated, the date obtained indicates when

   a. the biotite crystal formed.
   b. the sedimentary rock formed.
   c. the parent radioactive isotope formed.
   d. the daughter radioactive isotope(s) formed.
   e. none of these.

6. When geologists reconstruct the geologic history of an area, why is it important for them to differentiate between a sill and a lava flow? How could you tell the difference between a sill and a lava flow at an outcrop if both structures consisted of basalt? What features would you look for in an outcrop to positively identify the structure as either a sill or a lava flow?

7. Why can fossils be used to demonstrate the age equivalence of geographically separated and (often) lithologically dissimilar strata?

8. Given the current debate over global warming and the many short-term consequences for humans, can you visualize how the world might look in 10,000 years or even 1 million years from now? Use what you have learned about plate tectonics and the direction and rate of movement of plates, as well as how plate movement and global warming will affect ocean currents, weather patterns, weathering rates, and other factors, to help make your prediction. Do you think such short-term changes can be extrapolated to long-term trends in trying to predict what Earth will be like using a geologic time perspective?

9. A volcanic ash fall was radiometrically dated using the potassium 40 to argon 40 and rubidium 87 to strontium 87 isotope pairs. The isotope pairs yielded distinctly different ages. What possible explanation could be offered as to why these two isotope pairs yielded different ages? What would you do to rectify the discrepancy in ages?

# Creative Thinking Visual Question

This photograph (●**FIGURE 1**) was taken on Fish Creek Trail in the Sierra Nevada, California, and shows two black metavolcanic erratics on granodiorite bedrock that has been intruded by a white dike. Using your knowledge of igneous rocks and processes, glacial features, and relative dating principles, provide a geologic history of this area based on what is shown in this image.

● **FIGURE 1** Sierra Nevada, California

Reed Wicander

H. Mark Weidman Photography/Alamy Stock Photo

# 17

# EARTH HISTORY

The coal being mined in this aerial view of an open-pit coal mine northeast of Somerset, Pennsylvania, is Pennsylvanian in age. It formed as a result of tremendous accumulations of plant material deposited in a swampy environment as part of a cyclothem, or repetitive sequence of marine and nonmarine cycles of deposition. Notice the dragline (digging machine) on top of the coal deposit for scale.

# INTRODUCTION

Imagine a barren, lifeless, waterless, hot planet with a poisonous atmosphere. Cosmic radiation is intense, meteorites and comets crash to the ground, and volcanoes erupt almost continuously. Storms form in the turbulent atmosphere, lightning flashes much of the time, but no rain falls, because all water is in the form of vapor due to the high temperature, day and night. And because the atmosphere has no free oxygen, nothing burns, yet pools and streams of molten rock radiate a continuous red glow. This may seem like a description from a science fiction novel, but it is probably a reasonably accurate account of what Earth was like shortly after it formed (●**FIGURE 17.1A**).

Our sense of time is based on a human perspective, and as such, it is difficult to comprehend the magnitude of geologic time, or what some call *deep time*. Because we have no frame of reference for millions or billions of years, the fact that Earth is 4.6 billion years old can be hard to grasp. Consider this: Suppose that 1 second equals 1 year. If so, and you were to count to 4.6 billion, the task would take you and your descendants nearly 146 years! In fact, the time we designate as Precambrian alone constitutes about 88% of all geologic time.

We have before us the ambitious task of condensing Earth's 4.6 billion year geologic history into a single chapter. We are therefore limited to covering Earth's main geologic events from its formation to its current and quite different condition. For a more comprehensive coverage of the geologic history of Earth, the reader is referred to any current historical geology textbook.

In Chapter 1, we introduced the concept of a *system* as a combination of related parts that operate in an organized fashion, and we gave examples of interactions in discussing, for example, volcanism, plate tectonics, running water, and glaciation. After Earth formed some 4.6 billion years ago, its systems became operative, although not all at the same time or in their present form. For instance, Earth did not differentiate into a core and mantle until millions of years after it initially formed, and we know of no crust older than around 4.0 billion years. However, once Earth did differentiate into layers, internal heat drove plate movements and the crust began evolving—as it continues to do.

# PRECAMBRIAN EARTH HISTORY

**L01** Define the three eons of the Precambrian

**L02** Discuss the origin and evolution of the continents

**L03** Define a shield, platform, and craton

**L04** Discuss the significance of greenstone belts in terms of Archean Earth history

**L05** Briefly discuss the differences between Archean and Proterozoic Earth history

a)

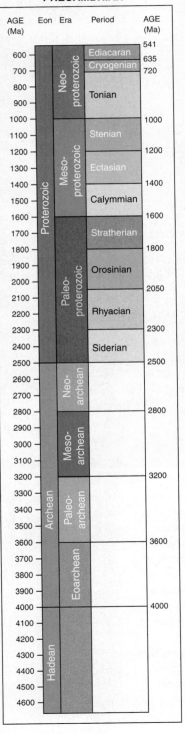

b)

●**FIGURE 17.1 The Precambrian Earth and Precambrian Timescale** (a) Earth as it may have appeared soon after it formed. No rocks are known from this time, but geologists can make reasonable inferences about the nature of the newly formed planet. (b) The Precambrian portion of the geologic timescale as published by the International Commission on Stratigraphy (ICS) in 2018. See Figures 1.17 and 16.1 for the complete timescale. Notice the use of the prefixes *eo* (early or dawn), *paleo* (old or ancient), *meso* (middle), and *neo* (new or recent). The age columns on the left and right sides of the timescale are in hundreds and thousands of millions of years (1,800 million years = 1.8 billion years, for example).

*Precambrian* is a widely used term that refers to both time and rocks. As a time term, it includes all geologic time from Earth's origin 4.6 billion years ago to the beginning of the Phanerozoic Eon 541 million years ago (●**FIGURE 17.1B**). The term also refers to all rocks lying below rocks of the Cambrian System. Unfortunately, no rocks are known for the first 600 million years of geologic time, so our geologic record begins about 4.0 billion years ago, with the oldest known rocks on Earth, the Acasta Gneiss of Canada (see Figure 7.1). The geologic record we do have for the Precambrian, particularly its older part, is difficult to decipher, because many of these ancient rocks (1) have been metamorphosed and complexly deformed; (2) in many areas, are deeply buried beneath younger rocks; and (3) contain few fossils useful for determining relative ages.

Geologists divide the Precambrian into three parts: the *Hadean* (an informal term) (4.6 to 4.0 billion years ago), the *Archean Eon* (4.0 to 2.5 billion years ago), and the *Proterozoic Eon* (2.5 billion to 541 million years ago). The eons are subdivided by using prefixes such as *paleo* (old or ancient), *neo* (new or recent), and so on (Figure 17.1b). Thus, the Precambrian lasted for more than 4.0 billion years, and Earth has existed for 4.6 billion years. Our assignment then is to cover this lengthy interval in only a few pages.

## The Origin and Evolution of Continents

Rocks 3.8 to 4.0 billion years old that are thought to represent continental crust are known from several areas, including Minnesota, Greenland, Canada, and South Africa.

Furthermore, these rocks are metamorphic, so they were altered from even older rocks.

According to one model for the origin of continents, the oldest crust was thin, unstable, and composed of ultramafic igneous rock. At first, this early crust was disrupted by rising basaltic magma at ridges and was consumed at subduction zones. A second stage in crustal evolution began when partial melting of earlier formed basaltic crust resulted in the formation of andesitic island arcs, and partial melting of lower crustal andesites yielded granitic magmas that were emplaced in the crust. As plutons were emplaced in these island arcs, they became more like continental crust. By approximately 4.0 billion years ago, plate movements accompanied by subduction and collisions of island arcs had formed several granitic continental nuclei (●**FIGURE 17.2**).

## Shields, Platforms, and Cratons

Each continent has a vast area of exposed Precambrian rocks known as a **shield**. Extending outward from shields are **platforms**, consisting of Precambrian rocks beneath more recent rocks. A shield and platform collectively form

**shield** A vast area of exposed ancient rocks on a continent; the exposed part of a craton.

**platform** The part of a craton that lies buried beneath flat-lying or mildly deformed sedimentary rocks.

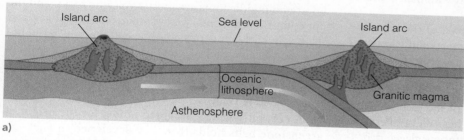
a)

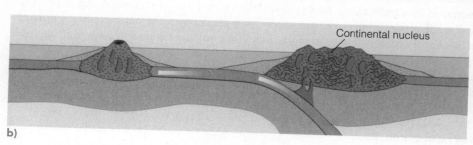
b)

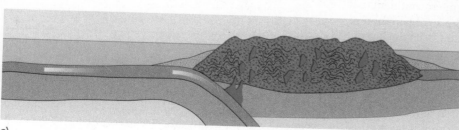
c)

● **FIGURE 17.2 Three Stages in the Origin of Granitic Continental Crust** Andesitic island arcs formed by the partial melting of basaltic oceanic crust are intruded by granitic magmas. As a result of plate movements, island arcs collide and form larger units, or cratons. (a) Two island arcs on separate plates move toward each other. (b) The island arcs shown in (a) collide, forming a small craton, and another island arc approaches this craton. (c) The island arc shown in (b) collides with the craton.

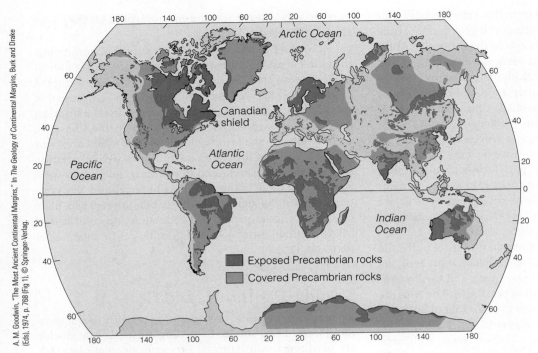

● **FIGURE 17.3 The Distribution of Precambrian Rocks** Areas of exposed Precambrian rocks constitute the shields, whereas the platforms consist of buried Precambrian rocks. A shield and its adjoining platform make up a craton.

a **craton**, which we can think of as the stable core or nucleus of a continent (●**FIGURE 17.3**). Many of the rocks within cratons have been strongly deformed, intruded by plutons, and altered by metamorphism, but they have experienced little or no deformation since the end of the Precambrian. Their stability since that time contrasts sharply with their Precambrian history of orogenic activity.

In North America, the exposed part of the craton is the **Canadian shield**, which occupies most of northeastern Canada, a large part of Greenland, the Adirondack Mountains of New York, and parts of the Lake Superior region in Minnesota, Wisconsin, and Michigan (Figure 17.3). Overall, the Canadian shield is a vast area of subdued topography, numerous lakes, and exposed Archean and Proterozoic rocks, thinly covered in places by Pleistocene glacial deposits. Beyond the Canadian shield, exposures of Precambrian rocks are limited to areas of deep erosion, such as the Grand Canyon, and areas of orogeny, such as the Appalachian and Rocky Mountains.

Geologists have delineated several smaller units within the Canadian shield, each of which is recognized by radiometric ages and trends of geologic structures. These smaller units, as well as others making up the platform, are the subunits that constitute the North American craton. Each smaller unit was likely an independent microcontinent that later assembled into a large craton. The amalgamation of these units took place along deformation belts during the Paleoproterozoic Eon.

## Archean Earth History

By far, the most common rocks of Archean age are complexes consisting of granite and gneiss, as well as subordinate, but reasonably common rock successions, known as **greenstone belts**. Although greenstone belts make up only about 10 percent of Archean rocks, they are important in unraveling some of the complexities of Archean tectonic events.

An ideal greenstone belt has three major rock units: The lower and middle units are mostly volcanic rocks, and the upper unit is sedimentary. Low-grade metamorphism and the origin of the minerals chlorite, epidote, and actinolite give the volcanic rocks a greenish color. Most greenstone belts have a synclinal structure and have been intruded by granitic plutons and complexly folded and cut by thrust faults (●**FIGURE 17.4A**).

A currently widely accepted model for the origin of greenstone belts holds that they developed in *back-arc basins* that first opened and then closed. An early stage of extension took place when the back-arc basin opened, during which time volcanism and sedimentation took place, and finally an episode of compression occurred as it closed (●**FIGURE 17.5**). During closure, the rocks were intruded by plutons and metamorphosed, and the greenstone belt took on a synclinal form as it was folded and faulted.

---

**craton** The stable nucleus of a continent consisting of a shield of Precambrian rocks and a platform of buried ancient rocks.

**Canadian shield** The exposed part of the North American craton.

**greenstone belt** A linear or podlike association of igneous and sedimentary rocks particularly common in Archean terrains.

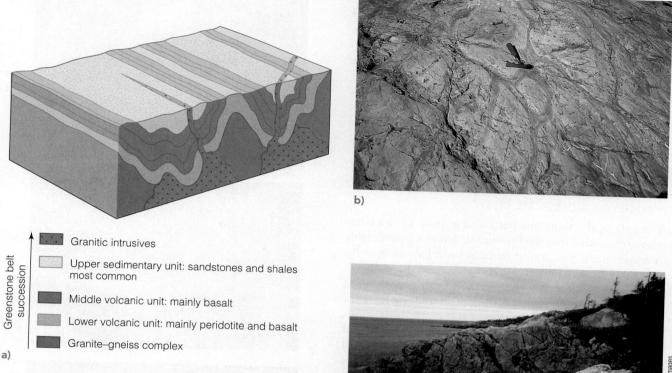

● **FIGURE 17.4 Greenstone Belts** (a) Two adjacent greenstone belts show their synclinal structure. The two lower units in greenstone belts are mostly volcanic rocks, whereas the upper unit is sedimentary. (b) Pillow lava of the Ishpeming greenstone belt in Michigan. (c) Granite dikes (pink) in a granite-gneiss complex near Cape Breton Highlands National Park, Nova Scotia, Canada.

**Critical Thinking Question** How did the pillow lava in (b) form?

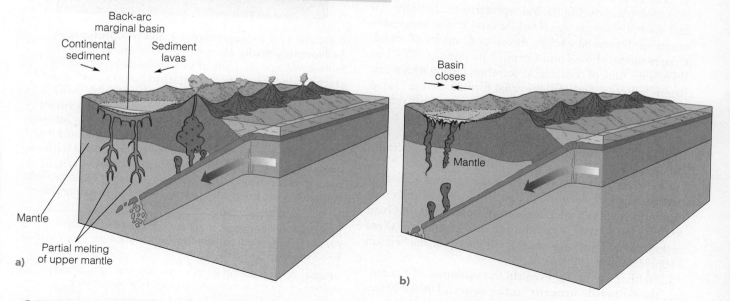

● **FIGURE 17.5 Origin of a Greenstone Belt in a Back-Arc Marginal Basin** (a) Basalt lavas and sediment derived from the continent and island arc fill the back-arc marginal basin. (b) Closure of the back-arc marginal basin causes compression and deformation. The evolving greenstone belt is deformed into a syncline-like structure into which granitic magmas intrude.

Undoubtedly, the present style of plate tectonics, involving the opening and closing of ocean basins, has been the primary agent of Earth evolution since at least the Paleoproterozoic Era. But many geologists are now convinced that some form of plate tectonics was operating during the Archean as well. However, Earth's radiogenic heat production has diminished through time, so during the Archean, when more heat was available, seafloor spreading and plate movements took place more rapidly. Nevertheless, by the Paleoproterozoic and since then, a plate tectonic style like that of the present has operated.

## Proterozoic Earth History

The origin of greenstone belts and granite-gneiss complexes continued into the Proterozoic but at a considerably reduced rate. Whereas most Archean rocks have been metamorphosed, many Proterozoic rocks have been little altered. Furthermore, the Proterozoic was a time of deposition of *banded iron formations*, consisting of alternating layers of iron minerals and chert (see Figure 6.26c); red beds, sandstones, siltstones, and shales with iron oxide cement; and sedimentary rocks indicating two episodes of widespread glaciation. Finally, widespread assemblages of sandstones, carbonates, and shales deposited on passive continental margins were common (●**FIGURE 17.6**).

In the preceding section, we noted that Archean crust assembled through a series of island-arc and microcontinent collisions, providing the nuclei around which Proterozoic continental crust accreted. One large landmass so formed, called **Laurentia**, consisted mostly of North America and Greenland, parts of northwestern Scotland, and perhaps parts of the Baltic shield of Scandinavia.

The first major episode of Laurentia's crustal evolution took place during the Paleoproterozoic, between 2.0 and 1.8 billion years ago (bya). Several major **orogens**—zones of deformed rocks—developed, many of which were metamorphosed and intruded by plutons. This was therefore a time of continental accretion during which collisions between Archean-age crust formed a large craton (●**FIGURE 17.7A**). Following this amalgamation of cratons, considerable accretion took place between 1.8 and 1.6 bya in what is now the southwestern and central United States, as successively younger belts were sutured to the craton (●**FIGURE 17.7B**).

Between 1.6 and 1.3 bya, extensive igneous activity—particularly the emplacement of granitic plutons and eruptions of rhyolite and ash flows—occurred that was unrelated to orogenic activity. According to one hypothesis, these rocks resulted from large-scale upwelling of magma beneath a supercontinent.

Another important event in the evolution of Laurentia, the **Grenville orogeny** in the eastern United States and Canada, took place between 1.3 and 1.0 bya (Mesoproterozoic Era) (●**FIGURE 17.7C**). Although geologists disagree on the cause of the Grenville orogeny, it was the

● **FIGURE 17.6 Paleoproterozoic and Neoproterozoic Sedimentary Rocks** Many of the sedimentary rocks of Proterozoic age are parts of sandstone-carbonate-shale assemblages that were deposited on passive continental margins. (a) Outcrop of the Paleoproterozoic Mesnard Quartzite in Michigan. The crests of the ripple marks point toward the observer. (b) The Paleoproterozoic Kona Dolomite of Michigan. The bulbous structures are stromatolites that were produced from the activities of cyanobacteria (blue-green algae). (c) This billion-year-old (Neoproterozoic) outcrop of sandstone and mudstone in Glacier National Park in Montana has been only slightly altered by metamorphism.

**Laurentia**  A Proterozoic continent composed of North America, Greenland, parts of Scotland, and perhaps part of the Baltic shield of Scandinavia.

**orogen**  A linear part of Earth's crust that was, or is, being deformed during an orogeny; part of an orogenic belt.

**Grenville orogeny**  An episode of deformation that took place in the eastern United States and Canada during the Neoproterozoic.

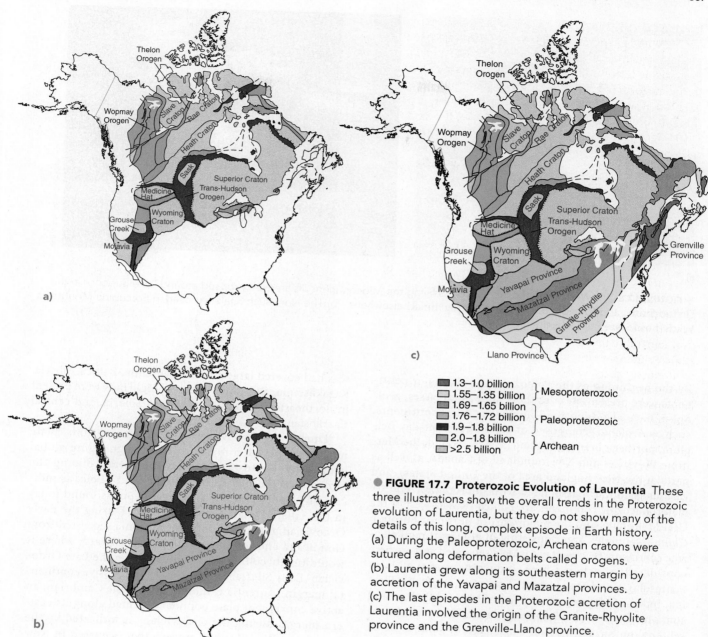

**● FIGURE 17.7 Proterozoic Evolution of Laurentia** These three illustrations show the overall trends in the Proterozoic evolution of Laurentia, but they do not show many of the details of this long, complex episode in Earth history. (a) During the Paleoproterozoic, Archean cratons were sutured along deformation belts called orogens. (b) Laurentia grew along its southeastern margin by accretion of the Yavapai and Mazatzal provinces. (c) The last episodes in the Proterozoic accretion of Laurentia involved the origin of the Granite-Rhyolite province and the Grenville-Llano province.

final episode of Proterozoic continental accretion of Laurentia. By then, about 75% of present-day North America existed. The remaining 25% accreted along its margins, particularly its eastern and western peripheries during the Phanerozoic Eon.

Beginning about 1.1 bya, tensional forces opened the **Midcontinent rift**, a long, narrow trough bounded by faults that outline two branches (**●FIGURE 17.8**). Although not all geologists agree, many think that the Midcontinent rift is a failed rift where Laurentia began splitting apart. Had rifting continued, Laurentia would have split into two separate landmasses, but the rifting ceased after about 20 million years. The central part of the rift is filled with hundreds of overlapping basaltic lava flows and sedimentary rocks forming a pile several kilometers thick.

# THE PALEOZOIC GEOGRAPHY OF EARTH

**LO6** Name the six major continents present at the beginning of the Paleozoic Era

**LO7** Discuss the movement of continents, that is, the paleogeography for each of the six periods during the Paleozoic Era

**LO8** Name the orogenies that occurred during the Paleozoic Era

---

**Midcontinent rift** A Late Proterozoic rift in Laurentia in which volcanic and sedimentary rocks accumulated.

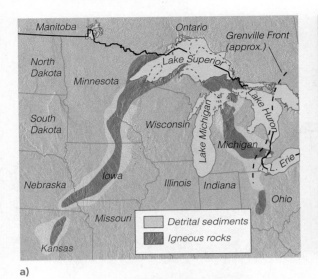

a)

b)

● **FIGURE 17.8 The Midcontinent Rift** (a) Rocks filling the Midcontinent rift are well exposed around Lake Superior in the United States and Canada, but they are deeply buried elsewhere. (b) The Nonesuch Shale exposed in Porcupine Mountains Wilderness State Park, Michigan.

By the beginning of the Paleozoic Era, six major continents were present. Besides these large landmasses, geologists have also identified numerous microcontinents, such as *Avalonia* (composed of parts of present-day Belgium, northern France, England, Wales, Ireland, the Maritime Provinces and Newfoundland of Canada, as well as parts of the New England area of the United States), and various island arcs associated with microplates. The six major Paleozoic continents are *Baltica* (Russia west of the Ural Mountains and the major part of northern Europe), *China* (a complex area consisting of at least three Paleozoic continents that were not widely separated and are here considered to include China, Indochina, and the Malay Peninsula), *Gondwana* (Africa, Antarctica, Australia, Florida, India, Madagascar, and parts of the Middle East and southern Europe), *Kazakhstania* (a triangular continent centered on Kazakhstan but considered by some to be an extension of the Paleozoic Siberian continent), *Laurentia* (most of present North America, Greenland, northwestern Ireland, and Scotland), and *Siberia* (Russia east of the Ural Mountains and Asia north of Kazakhstan and south of Mongolia). The paleogeographic reconstructions that follow are based on the methods used to determine and interpret location, geographic features, and environmental conditions on the paleocontinents.

In contrast to today's global geography, the Cambrian world consisted of these six continents dispersed around the globe at low tropical latitudes (●FIGURE 17.9A). Water circulated freely between ocean basins, and the polar regions were apparently ice free. By the Late Cambrian, shallow seas had covered large areas of Laurentia, Baltica, Siberia, Kazakhstania, and China, whereas highlands were present in northeastern Gondwana, eastern Siberia, and central Kazakhstania.

During the Ordovician and Silurian periods, plate movement played a major role in the changing global geography. Gondwana moved southward during the Ordovician and began to cross the South Pole, as indicated by Upper Ordovician glacial deposits found today in the Sahara Desert (●FIGURE 17.9B). During the Early Ordovician, the microcontinent Avalonia separated from Gondwana and began moving northeastward, where it would finally collide with Baltica during the Late Ordovician–Early Silurian. In contrast to the passive continental margin Laurentia exhibited during the Cambrian, an active convergent plate boundary formed along its eastern margin during the Ordovician, as indicated by the Late Ordovician *Taconic orogeny* that occurred in New England.

During the Silurian, Baltica, along with the newly attached Avalonia, moved northwestward relative to Laurentia and collided with it to form the larger continent of *Laurasia*. This collision, which closed the northern *Iapetus Ocean*, is marked by the *Caledonian orogeny* (*see* GEO-FOCUS). After this orogeny, the southern part of the Iapetus Ocean still remained open between Laurentia and Avalonia–Baltica. Siberia and Kazakhstania moved from a southern equatorial position during the Cambrian to north temperate latitudes by the end of the Silurian Period.

# GEO-FOCUS

## The Devonian Old Red Sandstone

The Devonian System was named by Roderick Murchison and Adam Sedgwick in 1839 for rock exposures in Devon County, England (•FIGURE 1). The Old Red Sandstone, which was deposited during the Devonian Period and crops out extensively throughout the British Isles, is predominantly of terrestrial origin.

The Old Red Sandstone consists of upward of 11,000 m of clastic sediments deposited in various structural basins throughout Scandinavia, the British Isles, Greenland, and northeastern Canada. The sediments are poorly sorted and variable, and consist of conglomerates, coarse-grained sandstones, cross-bedded sandstones, siltstones, shales, and thin limestones, deposited in several continental environments. The predominant red color comes from the presence of oxidized iron minerals, but colors also range from gray and green through orange and purple.

The collision of Baltica and Laurentia during the Caledonian orogeny (Late Silurian and Early Devonian) produced a mountain range along the western margin of Baltica that is referred to as the Caledonian Mountain chain, or Caledonian Highlands. The weathering and erosion of these highlands resulted in the formation of a large clastic wedge, whose sediments were named the Old Red Sandstone.

It was in 1788 at Siccar Point, Scotland, that James Hutton first recognized the significance of unconformities in interpreting Earth history. At this location, he observed the Devonian Old Red Sandstone overlying the steeply dipping Silurian poorly sorted sandstones (see Figure 16.10b).

Not only is the Old Red Sandstone important in British and global stratigraphy but it is also important for the many fossils of jawless armored fish (ostracoderms), jawed armored fish (placoderms), early bony fish, as well

● **FIGURE 1** Red cliffs composed of the Old Red Sandstone crop out along the shoreline in Devon County, England.

● **FIGURE 2** Remains of Goodrich Castle in Herefordshire, England, dated to the period between 1160 and 1270. Goodrich Castle is one of the many castles built from rocks quarried from the Old Red Sandstone, a Devonian-age formation. Identical red sandstone is found in the Catshill Mountains of New York. It too has been used as a building stone for many structures in the New York area.

as arthropods, and primitive seedless vascular plants found throughout its extent (see Chapter 18).

Lastly, the Old Red Sandstone has been widely used as a building stone. For example, many of the castles, cathedrals, and abbeys in Scotland, Wales, and England were built using the Old Red Sandstone. Some examples

include Muchalls Castle, Aberdeenshire, Scotland, St. Magnus Cathedral, Orkney, Scotland, and Goodrich Castle, Herefordshire, England (•FIGURE 2). Interestingly, the New York Life Insurance Building (also known as the Quebec Bank Building) in Montreal, Canada, was built in 1887–1889 using the Old Red Sandstone, imported from Dumfriesshire, Scotland.

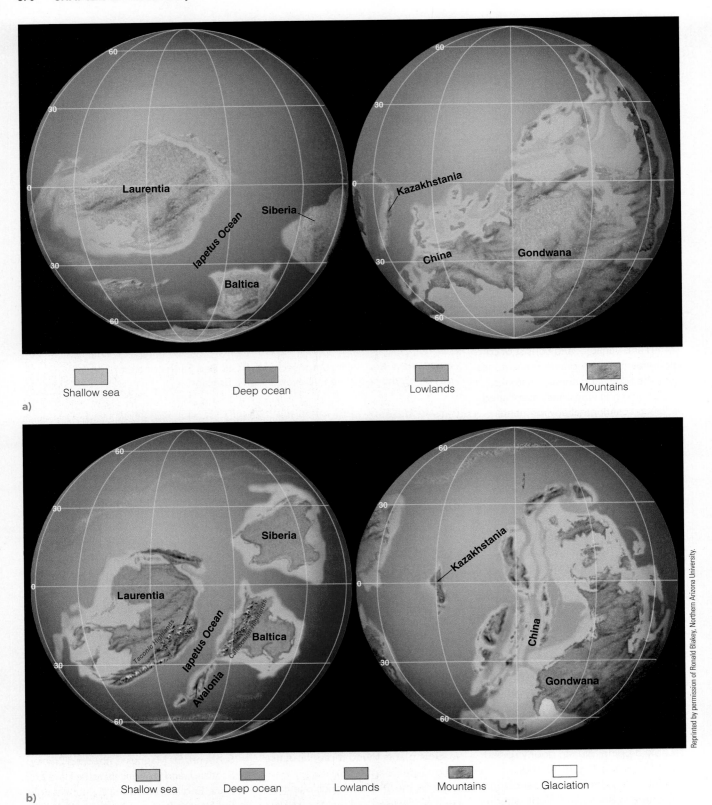

**Shallow sea**     **Deep ocean**     **Lowlands**     **Mountains**

a)

**Shallow sea**     **Deep ocean**     **Lowlands**     **Mountains**     **Glaciation**

b)

● **FIGURE 17.9 Paleogeography for the Late Cambrian and Ordovician Periods** (a) Late Cambrian Period.  (b) Late Ordovician Period.

During the Devonian, as the southern Iapetus Ocean narrowed between Laurasia and Gondwana, mountain building continued along the eastern margin of Laurasia with the *Acadian orogeny* (●**FIGURE 17.10A**). Erosion of the ensuing highlands spread vast amounts of reddish fluvial sediments over large areas of northern Europe and eastern North America.

Other Devonian tectonic events, probably related to the collision of Laurentia and Baltica, include the Cordilleran *Antler orogeny* and the change from a passive continental margin to an active convergent plate boundary in the Uralian mobile belt of eastern Baltica. The distribution of reefs, evaporites, and red beds, as well as the existence of similar floras throughout the world, suggest a rather uniform global climate throughout the Devonian Period.

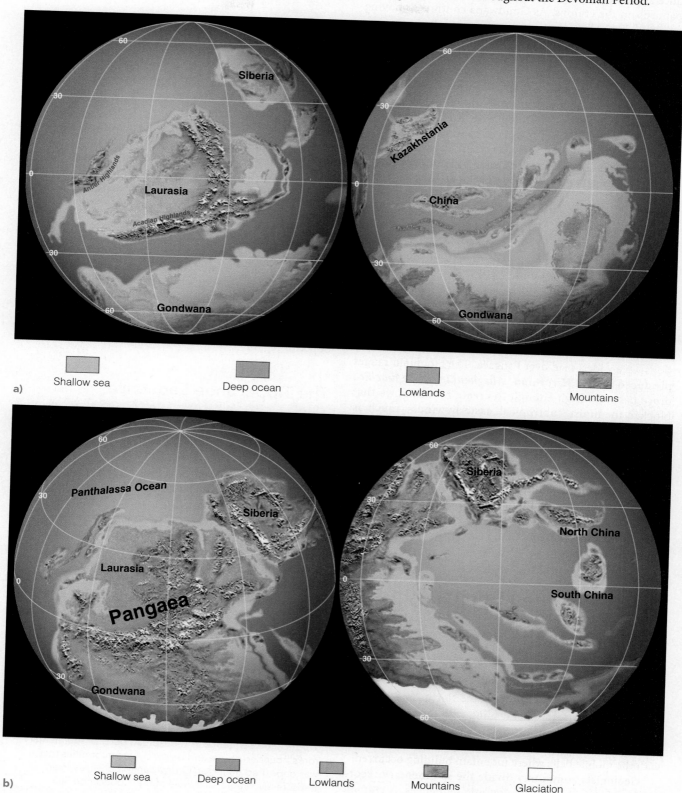

● **FIGURE 17.10 Paleogeography for the Late Devonian and Permian Periods** (a) Late Devonian Period. (b) Late Permian Period.

During the Carboniferous Period, southern Gondwana moved over the South Pole, resulting in extensive continental glaciation. The advance and retreat of these glaciers produced global changes in sea level that affected sedimentation patterns on the cratons. As Gondwana continued moving northward, it collided with Laurasia during the Early Carboniferous and continued suturing with it during the rest of the Carboniferous. The final phase of collision between Gondwana and Laurasia is indicated by the Ouachita Mountains of Oklahoma and Arkansas, formed by thrusting during the Late Carboniferous and Early Permian.

Elsewhere, Siberia collided with Kazakhstania and moved toward the Uralian margin of Laurasia (Baltica), colliding with it during the Early Permian. By the end of the Carboniferous, the various continental landmasses were fairly close together as the supercontinent Pangaea began taking shape.

The assemblage of Pangaea concluded during the Permian with the completion of many of the continental collisions that began during the Carboniferous (●**FIGURE 17.10B**). An enormous single ocean, *Panthalassa*, surrounded Pangaea and spanned Earth from pole to pole. Waters of this ocean probably circulated more freely than at present, resulting in more equable water temperatures.

The formation of a single landmass had climatic consequences for the terrestrial environment as well. Terrestrial Permian sediments indicate that arid and semiarid conditions were widespread over Pangaea. The mountain ranges produced by the *Hercynian*, *Alleghenian*, and *Ouachita orogenies* were high enough to create rainshadows that blocked the moist, subtropical, easterly winds—much as the southern Andes Mountains do in western South America today. This produced very dry conditions in North America and Europe, as indicated by the extensive Permian red beds and evaporites found in western North America, central Europe, and parts of Russia. Permian coals, indicative of abundant rainfall, were mostly limited to the northern temperate belts (latitude 40 to 60 degrees north), whereas the last remnants of the Carboniferous ice sheets continued their recession.

# THE PALEOZOIC EVOLUTION OF NORTH AMERICA

**LO9** Summarize the geologic history for each of the six North American sequences

We divide the Paleozoic history of the North American craton into two parts, the first dealing with the relatively stable continental interior over which shallow seas advanced (transgressed) and retreated (regressed), and the second with the mobile belts where mountain building occurred.

Geologists commonly divide the sedimentary record of North America into six cratonic sequences. A *cratonic sequence* is a major transgressive-regressive cycle bounded by craton-wide unconformities. The transgressive phase is usually well-preserved, whereas the regressive phase of each sequence is marked by an unconformity.

## The Sauk Sequence

Rocks of the **Sauk sequence** (Late Neoproterozoic–Early Ordovician) record the first major transgression onto the North American craton. During the Neoproterozoic and Early Cambrian, deposition of marine sediments was limited to the passive shelf areas of the Appalachian and Cordilleran borders of the craton. The craton itself was above sea level and experiencing weathering and erosion. Because North America was located in a tropical climate at this time (Figure 17.9a), and since there is no evidence of any terrestrial vegetation, we can conclude that weathering and erosion of the exposed Precambrian basement rocks must have proceeded rapidly.

During the Middle Cambrian, the transgressive phase of the Sauk began with shallow seas encroaching over the craton. By the Late Cambrian, the Sauk Sea had covered most of North America, leaving only a portion of the Canadian shield and a few large islands above sea level (●**FIGURE 17.11**). These islands, collectively named the *Transcontinental Arch*, extended from New Mexico to Minnesota and the Lake Superior region.

## The Tippecanoe Sequence

As the Sauk Sea regressed from the craton during the Early Ordovician, a landscape of low relief emerged. The exposed rocks were predominantly limestones and dolostones that were deeply eroded because North America was still located in a tropical environment (Figure 17.9b). The resulting craton-wide unconformity marks the boundary between the Sauk and Tippecanoe sequences.

Like the Sauk sequence, deposition of the **Tippecanoe sequence** (Middle Ordovician–Early Devonian) began with a major transgression onto the craton. This transgressing sea deposited clean, well-sorted quartz sands over most of the craton. The Tippecanoe basal sandstones were followed by widespread carbonate deposition. The limestones were formed by calcium carbonate–secreting marine organisms such as corals and brachiopods.

---

**Sauk sequence** A widespread association of sedimentary rocks bounded above and below by unconformities that was deposited during the latest Proterozoic to Early Ordovician transgressive–regressive cycle of the Sauk Sea.

**Tippecanoe sequence** A widespread body of sedimentary rocks bounded above and below by unconformities that was deposited during an Ordovician to Early Devonian transgressive–regressive cycle of the Tippecanoe Sea.

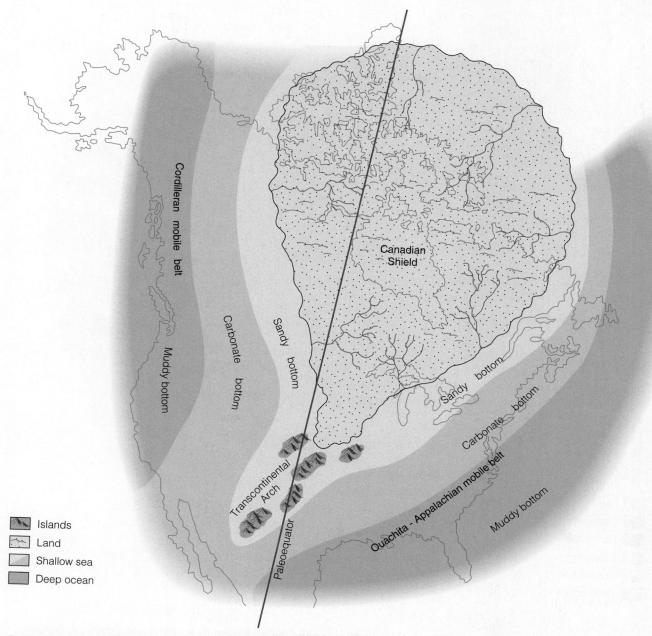

**FIGURE 17.11 Paleogeography of North America During the Cambrian Period** Note the position of the Cambrian paleoequator. During this time, North America straddled the equator, as indicated in Figure 17.9a.

As the Tippecanoe Sea gradually regressed from the craton during the Late Silurian, precipitation of evaporite minerals took place in the Appalachian, Ohio, and Michigan basins (●FIGURE 17.12). In the Michigan Basin alone, approximately 1,500 m of sediments were deposited, nearly half of which are halite and anhydrite.

By the Early Devonian, the regressing Tippecanoe Sea had retreated to the craton margin, exposing an extensive lowland topography. During this regression, marine deposition was initially restricted to a few interconnected cratonic basins and finally, by the end of the Tippecanoe, to only the margins surrounding the craton.

As the Tippecanoe Sea regressed during the Early Devonian, the craton experienced mild deformation, resulting in the formation of many domes, arches, and basins. These structures were mostly eroded during the time the craton was exposed, and deposits from the ensuing and encroaching Kaskaskia Sea eventually covered them.

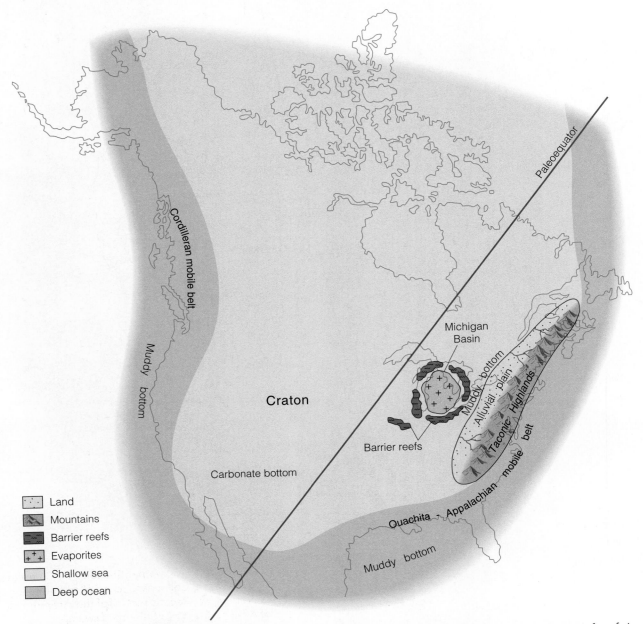

● **FIGURE 17.12 Paleogeography of North America During the Silurian Period** Note the development of reefs in the Michigan, Ohio, and Indiana-Illinois-Kentucky areas.

## The Kaskaskia Sequence

The boundary between the Tippecanoe sequence and the overlying **Kaskaskia sequence** (Middle Devonian–Late Mississippian) is marked by a major unconformity. As the Kaskaskia Sea transgressed over the low-relief landscape of the craton, most of the basal beds consisted of clean, well-sorted, quartz sandstones.

Except for widespread Upper Devonian and Lower Mississippian black shales, the majority of Kaskaskian rocks are carbonates, including reefs, and associated evaporite deposits. In many other parts of the world, such as southern England, Belgium, central Europe, Australia, and Russia, the

Middle and early Late Devonian epochs were times of major reef building.

During the Late Mississippian regression of the Kaskaskia Sea from the craton, carbonate deposition was replaced by vast quantities of detrital sediments. Before the end of the Mississippian, the Kaskaskia Sea had retreated to the craton margin, once again exposing the craton to widespread

---

**Kaskaskia sequence** A widespread sequence of Devonian and Mississippian sedimentary rocks bounded above and below by unconformities that was deposited during a transgressive–regressive cycle of the Kaskaskia Sea.

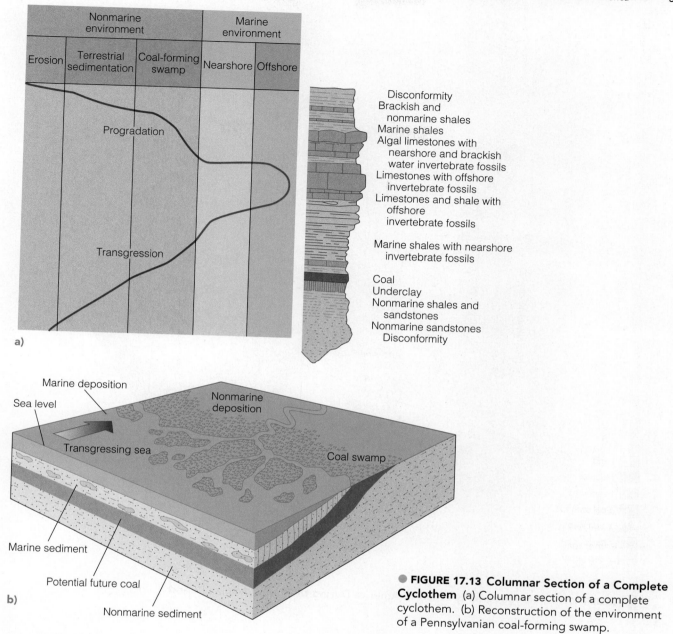

Disconformity
Brackish and
    nonmarine shales
Marine shales
Algal limestones with
    nearshore and brackish
    water invertebrate fossils
Limestones with offshore
    invertebrate fossils
Limestones and shale with
    offshore
    invertebrate fossils

Marine shales with nearshore
    invertebrate fossils

Coal
Underclay
Nonmarine shales and
    sandstones
Nonmarine sandstones
    Disconformity

a)

b)

● **FIGURE 17.13 Columnar Section of a Complete Cyclothem** (a) Columnar section of a complete cyclothem. (b) Reconstruction of the environment of a Pennsylvanian coal-forming swamp.

weathering and erosion, resulting in a craton-wide unconformity at the end of the Kaskaskia sequence.

## The Absaroka Sequence

The extensive unconformity separating the Kaskaskia and Absaroka sequences essentially divides the strata into the North American Mississippian and Pennsylvanian systems. These two systems are closely equivalent to the European Lower and Upper Carboniferous. The rocks of the **Absaroka sequence** (Late Mississippian–Early Jurassic) not only differ from those of the Kaskaskia sequence but also result from different tectonic regimes.

One characteristic feature of Pennsylvanian rocks is their repetitive pattern of alternating marine and nonmarine

strata known as **cyclothems** (●**FIGURE 17.13**). They result from repeated alternations of marine and nonmarine environments, usually in areas of low relief. Although seemingly simple, cyclothems reflect a delicate interplay between nonmarine deltaic and shallow marine interdeltaic and shelf environments.

**Absaroka sequence** Widespread Upper Mississippian to Lower Jurassic sedimentary rocks bounded above and below by unconformities; deposited during a transgressive–regressive cycle of the Absaroka Sea.

**cyclothem** A sequence of cyclically repeating sedimentary rocks resulting from alternating periods of marine and nonmarine deposition; commonly contains a coal bed.

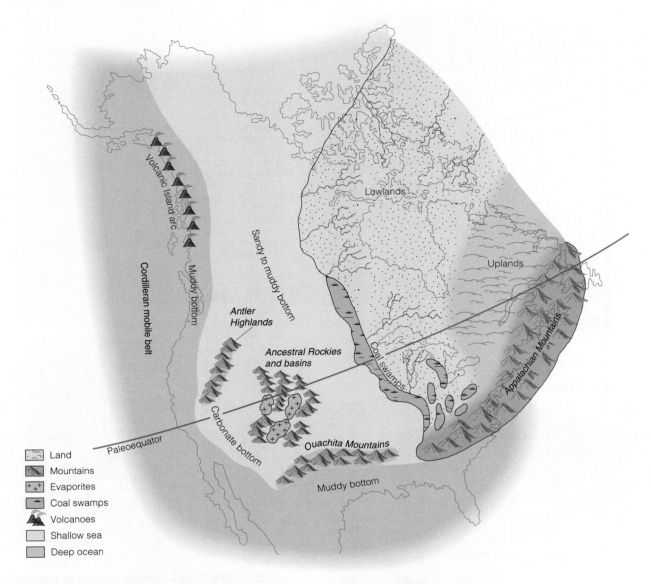

**● FIGURE 17.14 Paleogeography of North America During the Pennsylvanian Period**

Cyclothems represent transgressive and regressive sequences, with an erosional surface separating one cyclothem from another. Thus, an idealized cyclothem passes upward from fluvial–deltaic deposits, through coals, to detrital shallow-water marine sediments, and finally to limestones typical of an open marine environment (●FIGURE 17.13A).

Such repetitive sedimentation over a widespread area requires an explanation. The hypothesis currently favored by many geologists is a rise and fall of sea level related to advances and retreats of Gondwanan continental glaciers. When the Gondwanan ice sheets advanced, sea level dropped; when they melted, sea level rose. Late Paleozoic cyclothem activity on all of the cratons closely corresponds to Gondwanan glacial–interglacial cycles.

During the Pennsylvanian, the area of greatest deformation occurred in the southwestern part of the North

American craton, where a series of fault-bounded uplifted blocks formed the *Ancestral Rockies* (●FIGURE 17.14). These mountain ranges had diverse geologic histories and were not all elevated at the same time. Uplift of these mountains, some of which were elevated more than 2 km along near-vertical faults, resulted in the erosion of the overlying Paleozoic sediments and exposure of the Precambrian igneous and metamorphic basement rocks (●FIGURE 17.15). As the mountains eroded, tremendous quantities of coarse, red sediments were deposited in the surrounding basins, where they are preserved in many areas, such as the spectacular Garden of the Gods, Colorado.

While the various intracratonic basins were filling with sediment during the Late Pennsylvanian, the Absaroka Sea slowly began retreating from the craton. During the Middle and Late Permian, the Absaroka Sea was restricted

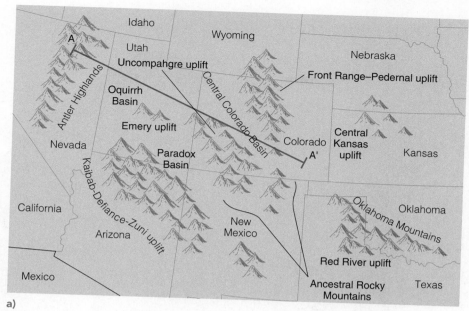

a)

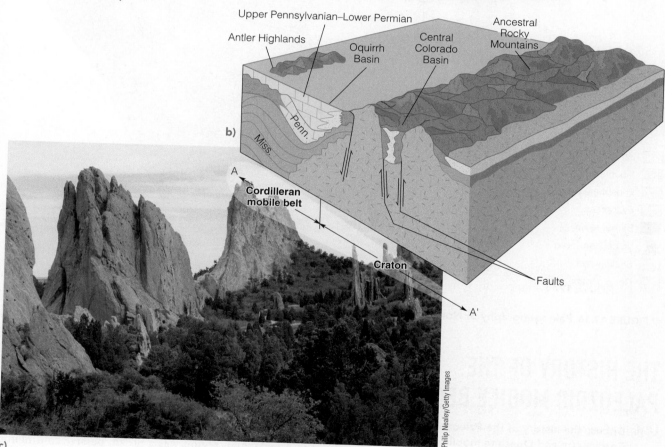

b)

c)

Philip Nealey/Getty Images

● **FIGURE 17.15 The Ancestral Rockies** (a) Location of the principal Pennsylvanian highland areas and basins of the southwestern part of the craton. (b) Block diagram of the Ancestral Rockies, elevated by faulting during the Pennsylvanian Period. Erosion of these mountains produced coarse, red sediments deposited in the basins adjacent to the Ancestral Rockies. (c) Garden of the Gods, Storm Sky View from Near Hidden Inn, Colorado Springs, Colorado.

to west Texas and southern New Mexico, forming an inter-related complex of lagoonal, reef, and open-shelf environments (●**FIGURE 17.16**). By the end of the Permian Period,

the Absaroka Sea had retreated from the craton, exposing continental red beds that had been deposited over most of the southwestern and eastern region.

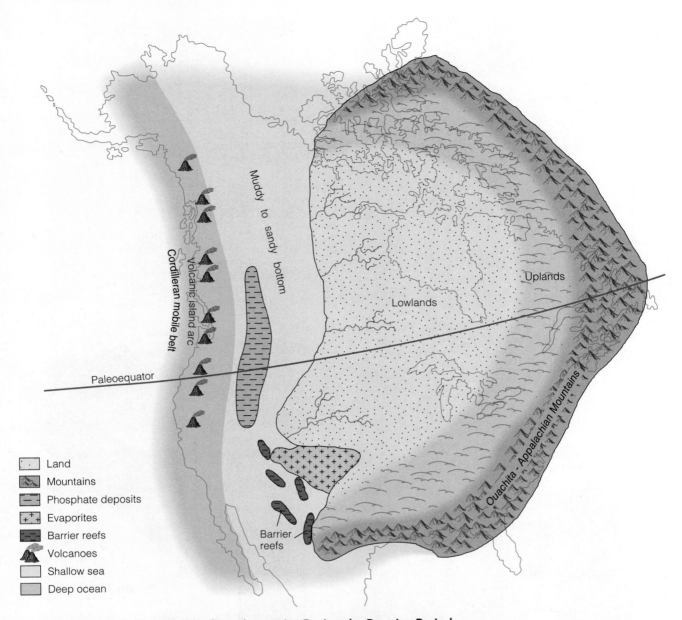

Land

Mountains

Phosphate deposits

Evaporites

Barrier reefs

Volcanoes

Shallow sea

Deep ocean

● **FIGURE 17.16** Paleogeography of North America During the Permian Period

# THE HISTORY OF THE PALEOZOIC MOBILE BELTS

**L010** Discuss the history of the Paleozoic Appalachian, Cordilleran, and Ouachita mobile belts in North America

Having examined the Paleozoic history of the craton, we now turn to the Paleozoic orogenic activity in the **mobile belts** (elongated areas of mountain-building activity along the margins of continents). The mountain building occurring during the Paleozoic Era had a profound influence on the climate and sedimentary history of the craton. In addition, it was part of the global tectonic regime that

sutured the continents together, forming Pangaea by the end of the Paleozoic.

## The Appalachian Mobile Belt

Throughout Sauk time (Neoproterozoic–Early Ordovician), the Appalachian region was a broad, passive continental margin. Sedimentation was closely balanced by subsidence as extensive carbonate deposits succeeded thick, shallow marine sands. During this time, movement along a divergent plate boundary was widening the Iapetus Ocean (●**FIGURE 17.17A**).

**mobile belt** An elongated area of deformation generally along the margins of a craton, such as the Appalachian mobile belt.

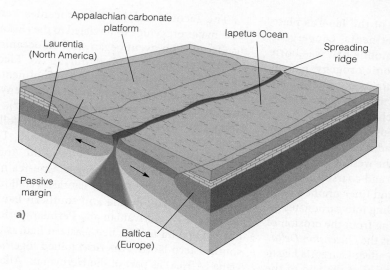

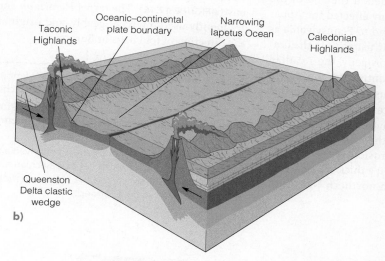

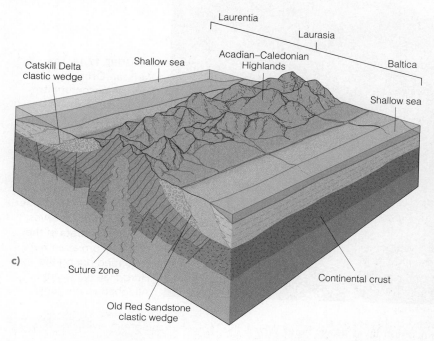

● **FIGURE 17.17 Evolution of the Appalachian Mobile Belt** (a) During the Neoproterozoic to the Early Ordovician, the Iapetus Ocean was opening along a divergent plate boundary. Both the east coast of Laurentia and the west coast of Baltica were passive continental margins where large carbonate platforms existed. (b) Beginning in the Middle Ordovician, the passive margins of Laurentia and Baltica became oceanic–continental convergent plate boundaries, resulting in orogenic activity. (c) By the Late Paleozoic, Laurentia and Baltica collided along a continental–continental convergent plate boundary, forming a larger continental landmass, Laurasia.

Beginning with the subduction of the Iapetus plate beneath Laurentia (an oceanic–continental convergent plate boundary), the Appalachian mobile belt was born (●FIGURE 17.17B). The resulting **Taconic orogeny**, named after the present-day Taconic Mountains of eastern New York, central Massachusetts, and Vermont, was the first of several orogenies to affect the Appalachian region.

A large *clastic wedge* (an extensive accumulation of mostly detrital sediments deposited adjacent to an uplifted area) formed in the shallow seas to the west of the Taconic orogeny. These deposits are thickest and coarsest near the highland area and become thinner and finer-grained away from the source area, eventually grading into carbonates on the craton. The clastic wedge resulting from the erosion of the Taconic Highlands is referred to as the *Queenston Delta*.

The second Paleozoic orogeny to affect Laurentia began during the Late Silurian and concluded at the end of the Devonian Period. The **Acadian orogeny** affected the Appalachian mobile belt from Newfoundland to Pennsylvania as sedimentary rocks were folded and thrust against the craton. As with the preceding Taconic orogeny, the Acadian orogeny occurred along an oceanic–continental convergent plate boundary. As the northern Iapetus Ocean continued to close during the Devonian, the plate carrying Baltica finally collided with Laurentia, forming a continental–continental convergent plate boundary along the zone of collision (●FIGURE 17.17C). Weathering and erosion of the Acadian Highlands produced the *Catskill Delta*, a thick clastic wedge named for the Catskill Mountains in northern New York, where it is well exposed.

The Taconic and Acadian orogenies were part of the same major orogenic event related to the closing of the Iapetus Ocean. This event began with an oceanic–continental convergent plate boundary during the Taconic orogeny and culminated with a continental–continental convergent plate boundary during the Acadian orogeny as Laurentia and Baltica became sutured (●FIGURE 17.17). After this, the **Hercynian–Alleghenian orogeny** began, followed by orogenic activity in the Ouachita mobile belt. The Hercynian mobile belt of southern Europe and the Appalachian and Ouachita mobile belts of North America mark the zone along which Europe (part of Laurasia) collided with Gondwana. While Gondwana and southern Laurasia collided during the Pennsylvanian and Permian periods in the area of the Ouchita mobile belt, eastern Laurasia (Europe and southeastern North America) joined together with Gondwana (Africa) as part of the Hercynian–Alleghenian orogeny (●FIGURE 17.18). The terms *Hercynian* and *Variscan* are frequently used interchangeably, and originally referred to the direction of fold belts in Europe.

---

**Taconic orogeny** An Ordovician episode of mountain building that resulted in the deformation of the Appalachian mobile belt.

**Acadian orogeny** An episode of Devonian deformation in the northern Appalachian mobile belt resulting from the collision of Baltica with Laurentia.

**Hercynian–Alleghenian orogeny** Pennsylvanian to Permian orogenic event during which the Appalachian mobile belt in eastern North America and the Hercynian mobile belt of southern Europe were deformed.

Courtesy of Reed Wicander

● **FIGURE 17.18 The Variscan Orogeny** The Variscan unconformity exposed at Telheiro Beach in southwestern Portugal shows the highly folded rocks of the Pennsylvanian Brejeira Formation, unconformably overlain by the red-colored terrestrial beds of the Upper Triassic "Grés de Silves." The folded strata of the Brejeira Formation were deformed during the Hercynian (Variscan)—Alleghenian orogeny.

**Critical Thinking Question** What type of unconformity is shown here? Why is this type of unconformity evidence of mountain building?

## The Cordilleran Mobile Belt

During the Neoproterozoic Era and Early Paleozoic, the Cordilleran area was a passive continental margin along which extensive continental shelf sediments were deposited. Beginning in the Middle Paleozoic, an island arc formed off the western margin of the craton. This eastward-moving island arc collided with the western border of the craton during the Late Devonian and Early Mississippian, resulting in a highland area termed the Antler Highlands (Figure 17.14). This orogenic event, the *Antler orogeny*, was the first in a series of orogenic events to affect the Cordilleran mobile belt.

## The Ouachita Mobile Belt

The Ouachita mobile belt extends for approximately 2,100 km from the subsurface of Mississippi to the Marathon region of Texas. During the Late Neoproterozoic to Early Mississippian, shallow-water detrital and carbonate sediments were deposited on a broad continental shelf, while in the deeper-water portion of the adjoining mobile belt, bedded cherts and shales were accumulating. Beginning in the Mississippian Period, the rate of sedimentation increased dramatically as the region changed from a passive continental margin to an active convergent plate boundary, marking the beginning of the **Ouachita orogeny**.

Approximately 80% of the former Ouachita mobile belt is buried beneath a Mesozoic and Cenozoic sedimentary cover. The two major exposed areas in this region are the Ouachita Mountains of Oklahoma and Arkansas, and the Marathon Mountains of Texas.

Thrusting of sediments continued throughout the Pennsylvanian and Early Permian, driven by the compressive forces generated along the zone of subduction as Gondwana collided with Laurasia. The collision of Gondwana and Laurasia is marked by the formation of a large mountain range, most of which eroded during the Mesozoic Era. Only the rejuvenated Ouachita and Marathon mountains remain of this once lofty mountain range.

## THE ROLE OF MICROPLATES

**L011**  Discuss the importance and significance of microplates in the formation of Pangaea

It is becoming increasingly clear that accretion along the continental margins is more complicated than the somewhat simple, large-scale plate interactions that we have described. Geologists now recognize that numerous microplates, such as Avalonia (Figure 17.9b), existed during the Paleozoic and were involved in the orogenic events that occurred during that time.

A careful examination of Paleozoic global paleogeographic maps shows numerous microplates, and their location and role during the formation of Pangaea must be taken into account. Thus, although the basic history of the formation of Pangaea during the Paleozoic remains the same, geologists now realize that microplates also played an important role in the formation of Pangaea. Furthermore, they help explain some previously anomalous geologic and paleontologic situations.

# THE BREAKUP OF PANGAEA

**L012**  Describe the four general stages in the breakup of Pangaea

Just as the formation of Pangaea influenced geologic and biologic events during the Paleozoic, the breakup of this supercontinent profoundly affected geologic and biologic events during the Mesozoic. The movement of continents affected global climatic and oceanic regimes as well as the climates of individual continents.

Because of the magnetic anomalies preserved in the oceanic crust (see Figure 2.15), geologists have a very good record of the history of Pangaea's breakup, and the direction of movement of the various continents during the Mesozoic and Cenozoic eras. Pangaea's breakup can be divided into four general stages. The first stage involved rifting between Laurasia and Gondwana during the Triassic (●FIGURE 17.19A). By the end of the Triassic, the newly formed and expanding Atlantic Ocean separated North America from Africa, and sometime during the Late Triassic and Early Jurassic, North America rifted from South America.

The second stage in Pangaea's breakup involved rifting and movement of the various Gondwana continents during the Late Triassic and Jurassic periods. As early as the Late Triassic, Antarctica and Australia, which remained sutured together, began separating from South America and Africa, while India began rifting from the Gondwana continent and moved northward.

The third stage of breakup started during the Late Jurassic, when South America and Africa began to separate (●FIGURE 17.19B). During this time, the eastern end of the Tethys Sea began closing as a result of the clockwise rotation of Laurasia and the northward movement of Africa. This narrow Late Jurassic and Cretaceous seaway between Africa and Europe was the forerunner of the present Mediterranean Sea.

By the end of the Cretaceous, Australia and Antarctica had detached from each other, and India had moved into low southern latitudes and was near to the equator. South America and Africa were now widely separated, and Greenland was essentially an independent landmass with only a shallow sea between it and North America and Europe (●FIGURE 17.19C).

A global rise in sea level during the Cretaceous resulted in worldwide transgressions onto the continents. Higher heat flow and rapid expansion of oceanic ridges were responsible for these transgressions. By the Middle Cretaceous, sea level was probably as high as at any time since the Ordovician, and about one-third of the present land area was inundated by widespread seas (Figure 17.19c).

---

**Ouachita orogeny** A period of mountain building that took place in the Ouachita mobile belt during the Pennsylvanian Period.

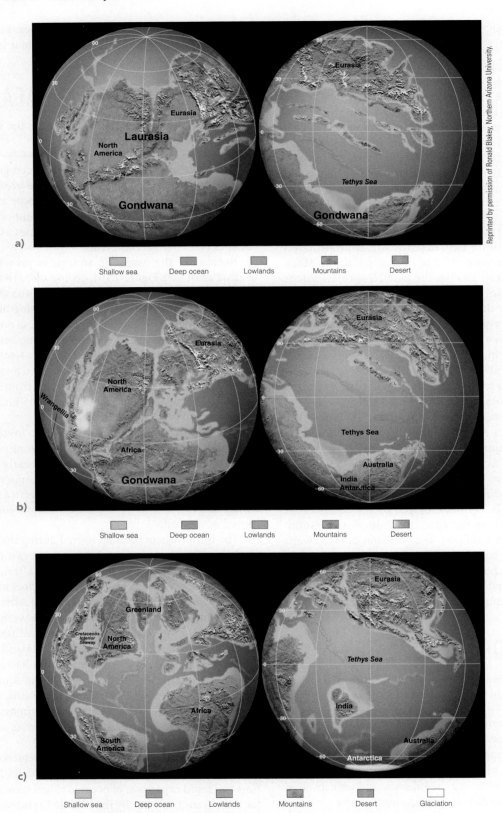

Reprinted by permission of Ronald Blakey, Northern Arizona University.

● **FIGURE 17.19 Paleogeography of the World During the Mesozoic Era**
(a) Triassic Period.  (b) Jurassic Period.  (c) Late Cretaceous Period.

**Critical Thinking Question**  Why would increased seafloor spreading, resulting in higher heat flow and an increase and rapid expansion in oceanic ridges during the Cretaceous, result in worldwide transgression onto the continents?

The final stage in Pangaea's breakup occurred during the Cenozoic Era. During this time, Australia continued moving northward, and Greenland was completely separated from Europe and North America and formed a separate landmass.

# THE MESOZOIC HISTORY OF NORTH AMERICA

**L013** Briefly discuss the Mesozoic geologic history for the Eastern Coastal and Gulf Coastal regions of North America

**L014** Define the Nevadan, Sevier, and Laramide orogenies of the Western region of North America

**L015** Discuss the significance of both the Sundance Sea and the Cretaceous Interior Seaway of North America

The beginning of the Mesozoic Era was essentially the same in terms of mountain building and sedimentation as the preceding Permian Period in North America (Figure 17.16). Terrestrial sedimentation continued over much of the craton, while block faulting and igneous activity began in the Appalachian region as North America and Africa began separating (●FIGURE 17.20). The newly forming Gulf of Mexico was the site of extensive evaporite deposition during the Late Triassic and Jurassic as North America separated from South America (Figures 17.19b and ●17.21).

A global rise in sea level during the Cretaceous resulted in worldwide transgressions onto the continents such that marine deposition was continuous over much of western North America (●FIGURE 17.22).

A volcanic island arc system that formed off the western edge of the craton during the Permian was sutured to North America sometime during the Permian or Triassic. During the Jurassic, the entire Cordilleran area was involved in a series of major mountain-building episodes, resulting in the

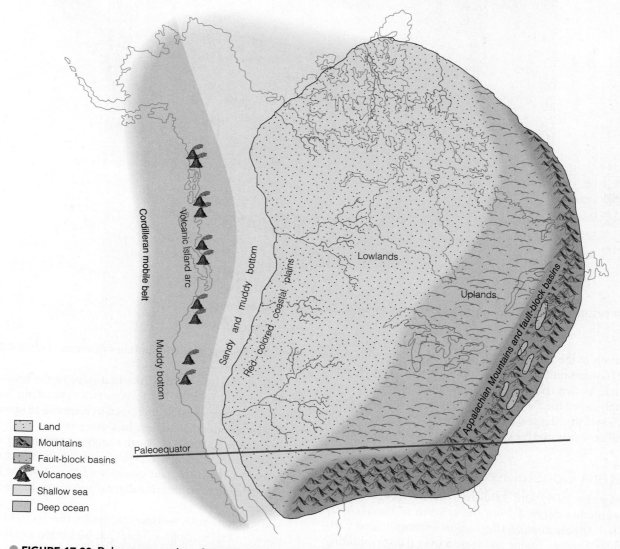

**● FIGURE 17.20 Paleogeography of North America During the Triassic Period**

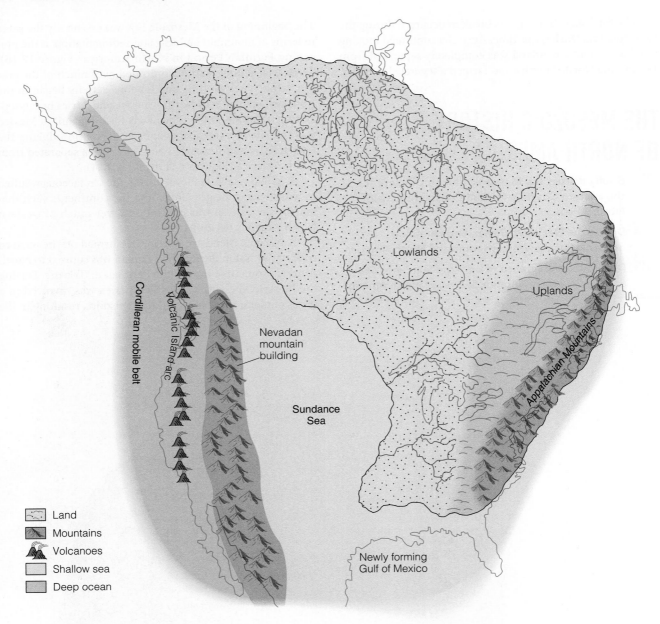

● **FIGURE 17.21 Paleogeography of North America During the Jurassic Period**

formation of the Sierra Nevada, the Rocky Mountains, and other lesser mountain ranges. Although each orogenic episode has its own name, the entire mountain-building event is simply called the *Cordilleran orogeny*. With this simplified overview of the Mesozoic history of North America in mind, we will now examine the specific regions of the continent.

## Eastern Coastal Region

During the Early and Middle Triassic, coarse detrital sediments derived from the erosion of the recently uplifted Appalachians (*Alleghenian orogeny*) filled various intermontane basins and spread over the surrounding areas. As weathering and erosion continued during the

Mesozoic, this once lofty mountain system was reduced to a low-lying plain.

During the Late Triassic, the first stage in the breakup of Pangaea began with North America separating from Africa. Fault-block basins developed in response to upwelling magma beneath Pangaea in a zone stretching from present-day Nova Scotia to North Carolina (●**FIGURE 17.23**). Erosion of the adjacent fault-block mountains filled the adjoining basins with great quantities (up to 6,000 m) of poorly sorted, red, nonmarine detrital sediments known as the *Newark Group*.

Concurrent with sedimentation in the fault-block basins were extensive lava flows that blanketed the basin floors, as well as intrusions of numerous dikes and sills.

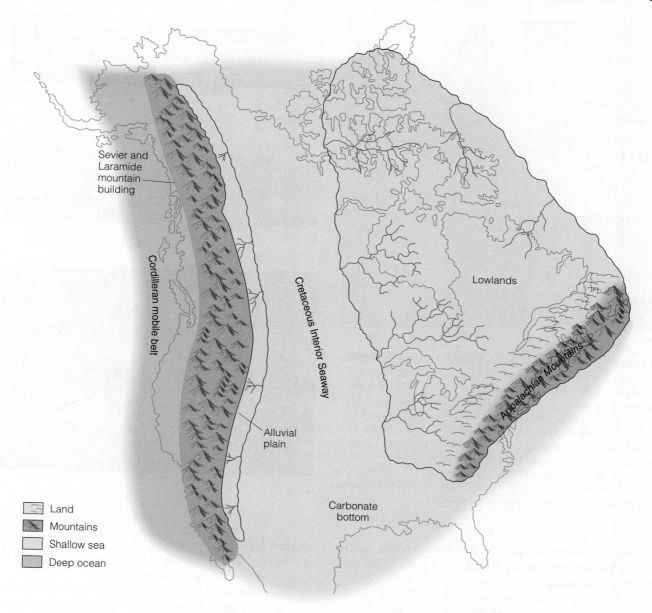

● **FIGURE 17.22 Paleogeography of North America During the Cretaceous Period**

The most famous intrusion is the prominent Palisades sill along the Hudson River in the New York–New Jersey area (●**FIGURE 17.23D**).

As the Atlantic Ocean grew, rifting ceased along the eastern margin of North America, and this once active convergent plate margin became a passive, trailing continental margin. The fault-block mountains produced by this rifting continued to erode during the Jurassic and Early Cretaceous until all that was left was an area of low-relief.

The sediments produced by this erosion contributed to the growing eastern continental shelf. During the Cretaceous Period, the Appalachian region was re-elevated and once again shed sediments onto the continental shelf,

forming a gently dipping, seaward-thickening wedge of rocks up to 3,000 m thick. These rocks are currently exposed in a belt extending from Long Island, New York, to Georgia.

## Gulf Coastal Region

The Gulf Coastal region was above sea level until the Late Triassic. As North America separated from South America during the Late Triassic and Early Jurassic, the Gulf of Mexico began to form (Figure 17.21). With oceanic waters flowing into this newly formed, shallow, restricted basin, conditions were ideal for evaporite formation. More than 1,000 m of evaporites were precipitated at this time, and

a)

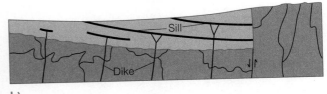

b)

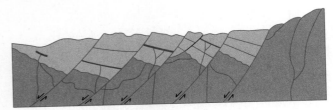

c)

d)

Courtesy of John Faivre

● **FIGURE 17.23 North America Triassic Fault-Block Basins**
(a) Areas where Triassic fault-block basin deposits crop out
in eastern North America. (b) After the Appalachians were
eroded to a low-lying plain by the Middle Triassic, fault-block
basins formed as a result of Late Triassic rifting between
North America and Africa. (c) These valleys accumulated tremendous thickness of sediments and were themselves broken
by a complex of normal faults during rifting. (d) Palisades of
the Hudson River. This sill was one of many intruded into the
Newark sediments during the Late Triassic rifting that marked
the separation of North America from Africa.

most geologists think that these Jurassic evaporites are the
source for the Cenozoic salt domes found today in the Gulf
of Mexico and southern Louisiana.

By the Late Jurassic, circulation in the Gulf of Mexico
was less restricted, and evaporite deposition ended. Normal marine conditions returned to the area with alternating
transgressing and regressing seas, resulting in sediments
that were in turn covered and buried by thousands of meters
of Cretaceous and Cenozoic sediments.

During the Cretaceous, the Gulf Coastal region,
like the rest of the continental margin, was flooded by
northward-transgressing seas forming a wide seaway, called
the *Cretaceous Interior Seaway* (Figure 17.22). This seaway
extended from the Arctic Ocean to the Gulf of Mexico.

## Western Region

During the Permian, an island arc and ocean basin formed
off the western North American craton (Figure 17.17), followed by subduction of an oceanic plate beneath the island
arc and the thrusting of oceanic and island arc rocks eastward against the craton margin. This event, similar to
the preceding Antler orogeny and known as the *Sonoma*

*orogeny*, occurred at or near the Permian–Triassic boundary
and resulted in the suturing of island-arc terranes along the
western edge of North America.

Following the Late Paleozoic–Early Mesozoic destruction of the volcanic island arc during the Sonoma orogeny,
the western margin of North America became an oceanic–
continental convergent plate boundary. During the Late
Triassic, a steeply dipping subduction zone developed along
the western margin of North America in response to the
westward movement of North America over the Pacific
plate. This newly created oceanic–continental plate boundary controlled Cordilleran tectonics for the rest of the Mesozoic Era and for most of the Cenozoic Era; this subduction
zone marks the beginning of the modern circum-Pacific
orogenic system.

The general term **Cordilleran orogeny** is applied to the
mountain-building activity that began during the Jurassic

---

**Cordilleran orogeny** An episode of deformation affecting
the western margin of North America from Jurassic to Early
Cenozoic time; divided into three separate phases called the
Nevadan, Sevier, and Laramide orogenies.

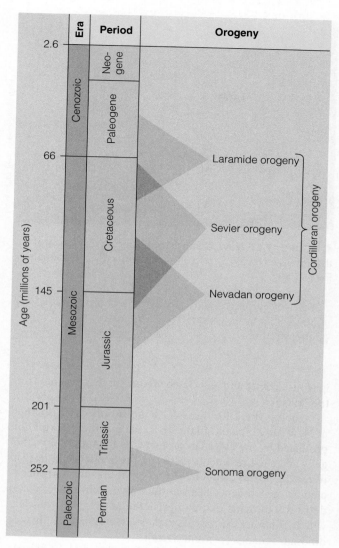

● **FIGURE 17.24 Mesozoic Cordilleran Orogenies** Mesozoic orogenies occurring in the Cordilleran mobile belt.

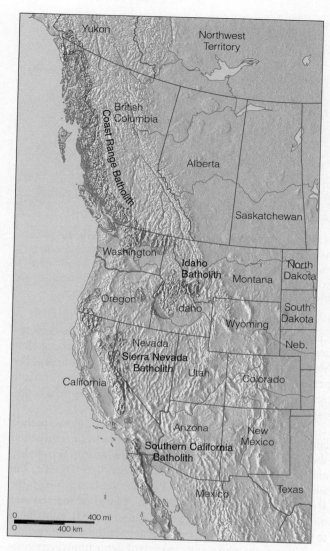

● **FIGURE 17.25 Cordilleran Batholiths** Location of the Jurassic and Cretaceous batholiths in western North America.

and continued into the Cenozoic (●**FIGURE 17.24**). The Cordilleran orogeny consisted of a series of individually named, but interrelated, mountain-building events that occurred in different regions at different times. Most of this Cordilleran orogenic activity is related to the continued westward movement of the North American plate as it overrode the Farallon plate (a small plate adjacent to the Pacific plate along the margin of western North America), and its history is highly complex.

The first pulse of the Cordilleran orogeny, the **Nevadan orogeny** (Figure 17.24), began during the Late Jurassic and continued into the Cretaceous as large volumes of granitic magma were generated at depth beneath the western edge of North America. These granitic masses were emplaced as huge batholiths that are now recognized as the Sierra Nevada, Southern California, Idaho, and Coast Range batholiths (●**FIGURE 17.25**).

The second pulse of the Cordilleran orogeny, the **Sevier orogeny**, affected western North America from Alaska to

Mexico and was mostly a Cretaceous event, even though it began in the Late Jurassic and is associated with the tectonic activity of the earlier Nevadan orogeny (Figure 17.24). Subduction of the Farallon plate beneath the North American plate continued during this time, resulting in numerous overlapping, low-angle thrust faults in which blocks of older rocks were thrust eastward on top of younger rocks. This deformation resulted in crustal shortening and produced generally north-south–trending mountain ranges stretching from Montana to western Canada.

During the Late Cretaceous to Early Cenozoic, the final pulse of the Cordilleran orogeny occurred (Figure 17.24).

**Nevadan orogeny** Late Jurassic to Cretaceous phase of the Cordilleran orogeny; most strongly affected the western part of the Cordilleran mobile belt.

**Sevier orogeny** Cretaceous phase of the Cordilleran orogeny that affected the continental shelf and slope areas of the Cordilleran mobile belt.

● **FIGURE 17.26 Petrified Forest National Park, Arizona**
All of the logs here are *Araucarloxylon*, the most abundant tree in the park. The petrified logs have been weathered from the Chinle Formation and are mostly in the position in which they were buried some 200 million years ago.

The **Laramide orogeny** developed east of the Sevier orogenic belt in the present-day Rocky Mountain areas of New Mexico, Colorado, and Wyoming. Most of the features of the present-day Rocky Mountains resulted from the Cenozoic phase of the Laramide orogeny, and for that reason, it is discussed later in this chapter.

Concurrent with the tectonism in the Cordilleran mobile belt, Early Triassic sedimentation on the western continental shelf consisted of shallow-water marine sandstones, shales, and limestones. During the Middle and Late Triassic, the western shallow seas regressed farther west, exposing large areas of former seafloor to erosion. Marginal marine and nonmarine Triassic rocks, particularly red beds, contribute to the spectacular and colorful scenery of the region.

These rocks represent a variety of depositional environments. The Upper Triassic *Chinle Formation*, for example, is widely exposed throughout the Colorado Plateau and is probably most famous for its petrified wood, spectacularly exposed in Petrified Forest National Park, Arizona (●**FIGURE 17.26**). Although best known for its petrified wood, the Chinle Formation has also yielded fossils of amphibians and various reptiles, including small dinosaurs.

Lower Jurassic deposits in a large part of the western region consist mostly of clean, cross-bedded sandstones indicative of wind-blown deposits. The thickest and most prominent of these is the *Navajo Sandstone*, a widespread cross-bedded sandstone that accumulated in a coastal dune environment along the southwestern margin of the craton. The sandstone's most distinctive feature is its large-scale cross-beds, some of them more than 25 m high (●**FIGURE 17.27**).

Marine conditions returned to the region during the Middle Jurassic when a seaway called the **Sundance Sea** twice flooded the interior of western North America (Figure 17.21). The resulting deposits were largely derived from erosion of the tectonic highlands to the west that paralleled the shoreline. These highlands resulted from intrusive

● **FIGURE 17.27 Navajo Sandstone** Large cross-beds of the Jurassic Navajo Sandstone exposed in Zion National Park, Utah.

**Critical Thinking Question** From the photo, how can you tell that these cross-beds are the result of wind deposition?

igneous activity and associated volcanism that began during the Triassic.

During the Late Jurassic, a mountain chain formed in Nevada, Utah, and Idaho as a result of the deformation produced by the Nevadan orogeny. As the mountain chain grew and shed sediments eastward, the Sundance Sea began retreating northward. A large part of the area formerly occupied by the Sundance Sea was then covered by multicolored detrital sediments that comprise the *Morrison Formation*, which contains the world's richest assemblage of Jurassic dinosaur remains (●**FIGURE 17.28**).

Shortly before the end of the Early Cretaceous, Arctic waters spread southward over the craton, forming a large inland sea in the Cordilleran area. By the beginning of the Late Cretaceous, this incursion joined the northward-transgressing waters from the Gulf area to create an enormous **Cretaceous Interior Seaway** that occupied the area east of the Sevier orogenic belt (Figure 17.22). Extending from the Gulf of Mexico to the Arctic Ocean and more than 1,500 km wide at its maximum extent, this seaway effectively divided North America into two large landmasses until just before the end of the Late Cretaceous.

As the Mesozoic Era ended, the Cretaceous Interior Seaway withdrew from the craton. During this regression,

**Laramide orogeny** Late Cretaceous to Early Cenozoic phase of the Cordilleran orogeny; responsible for many structural features of the present-day Rocky Mountains.

**Sundance Sea** A wide seaway that existed in western North America during the Middle Jurassic Period.

**Cretaceous Interior Seaway** A Late Cretaceous arm of the sea that effectively divided North America into two large landmasses.

Reed Wicander

● **FIGURE 17.28 Dinosaur Bones in Bas Relief** The north wall of the visitors' center at Dinosaur National Monument, Utah and Colorado showing dinosaur bones in bas relief, just as they were deposited 140 million years ago in the Morrison Formation.

marine waters retreated to the north and south, and marginal marine and continental deposition formed widespread coal-bearing deposits on the coastal plain.

# CENOZOIC EARTH HISTORY

**LO16**  Define the three periods of the Cenozoic Era

**LO17**  Define the two major Cenozoic orogenic belts

At 66 million years long, the Cenozoic Era is comparatively brief, constituting only 1.4% of all geologic time. Nevertheless, 66 million years is an extremely long time, certainly long enough for significant evolution of Earth and its biota to occur. Furthermore, Cenozoic rocks are at or near the surface and have been little altered, thereby making access to and interpretation of them easier than for rocks of previous eras.

Geologists divide the Cenozoic Era into three periods. The Paleogene Period (66 to 23 million years ago) includes the Paleocene, Eocene, and Oligocene epochs, the Neogene Period (23 to 2.6 million years ago) includes the Miocene and Pliocene epochs, and finally the Quaternary Period (2.6 million years ago to the present), which includes the Pleistocene and Recent (or Holocene) epochs (see Figure 16.1). Although you may see the term Tertiary Period for 66 to 2.58 million years ago, it is no longer recognized as a formal unit by the International Commission on Stratigraphy.

Many of Earth's features have long histories, but the present distribution of land and sea and the landforms of the continents developed recently in the context of geologic time. For instance, the Appalachian Mountain region began its evolution during the Precambrian, but its present expression is largely the product of Cenozoic uplift and erosion.

Likewise, distinctive landforms such as glacial valleys, badlands topography, and volcanoes of our national parks developed during the past few thousand to several millions of years.

## Cenozoic Plate Tectonics and Orogeny

The Late Triassic fragmentation of Pangaea (Figure 17.19a) began an episode of plate motions that continues even now. As a result, Cenozoic orogenic activity was concentrated in two major belts; the *Alpine-Himalayan belt*, which extends from southern Europe and North Africa eastward through the Middle East and Asia; and the *circum-Pacific belt*, which includes orogens along the west coasts of the Americas as well as the eastern margin of Asia and the islands north of Australia and New Zealand.

Within the Alpine-Himalayan orogenic belt, the *Alpine orogeny* began during the Mesozoic, but major deformation also occurred from the Eocene to Late Miocene as the African and Arabian plates moved northward against Eurasia. Deformation resulting from plate convergence formed mountains between Spain and France, the Alps of mainland Europe, and the mountains of Italy and North Africa. This orogenic belt remains geologically active.

Farther east in the Alpine-Himalayan orogenic belt, the *Himalayan orogeny* resulted from the collision of India with Asia (see Figure 9.18). Sometime during the Eocene, India's northward drift rate decreased abruptly, indicating the probable time of collision. In any event, an orogeny resulted during which two continental plates became sutured, which is why the present-day Himalayas are far inland rather than at a continental margin.

Plate subduction in the circum-Pacific orogenic belt took place throughout the Cenozoic, giving rise to orogenies in the Aleutians, the Philippines, and Japan, and along the west coasts of North, Central, and South America. The Andes Mountains in western South America, for example, formed as a result of convergence of the Nazca and South American plates (see Figure 9.17). Spreading at the East Pacific Rise and subduction of the Cocos and Nazca plates (see Figure 1.13) beneath Central and South America, respectively, account for continuing orogenic activity in these regions.

# THE NORTH AMERICAN CORDILLERA

**LO18**  Define the North American Cordillera

**LO19**  Discuss the history of the Laramide orogeny

**LO20**  Briefly discuss the Cenozoic geologic history of the continental interior, Gulf Coastal Plain, and Eastern North America

**LO21**  Discuss the Pleistocene history of North America

Part of the circum-Pacific orogenic belt is the **North American Cordillera**, a complex mountainous region in western North America extending from Alaska into central Mexico (●**FIGURE 17.29**). It has a long, complex geologic history involving accretion of island arcs along the continental margin, orogeny at an oceanic–continental boundary, vast outpourings of basaltic lavas, and block faulting. The most recent episode of large-scale deformation was the *Laramide orogeny*, which began during the Late Cretaceous, 85 to 90 million years ago. Like many other orogenies, it took place along an oceanic–continental boundary but was much farther inland than is typical (●**FIGURE 17.30**). Recall that the

Laramide orogeny was the final phase in a long episode of deformation referred to as the *Cordilleran orogeny*, which we previously discussed in this chapter.

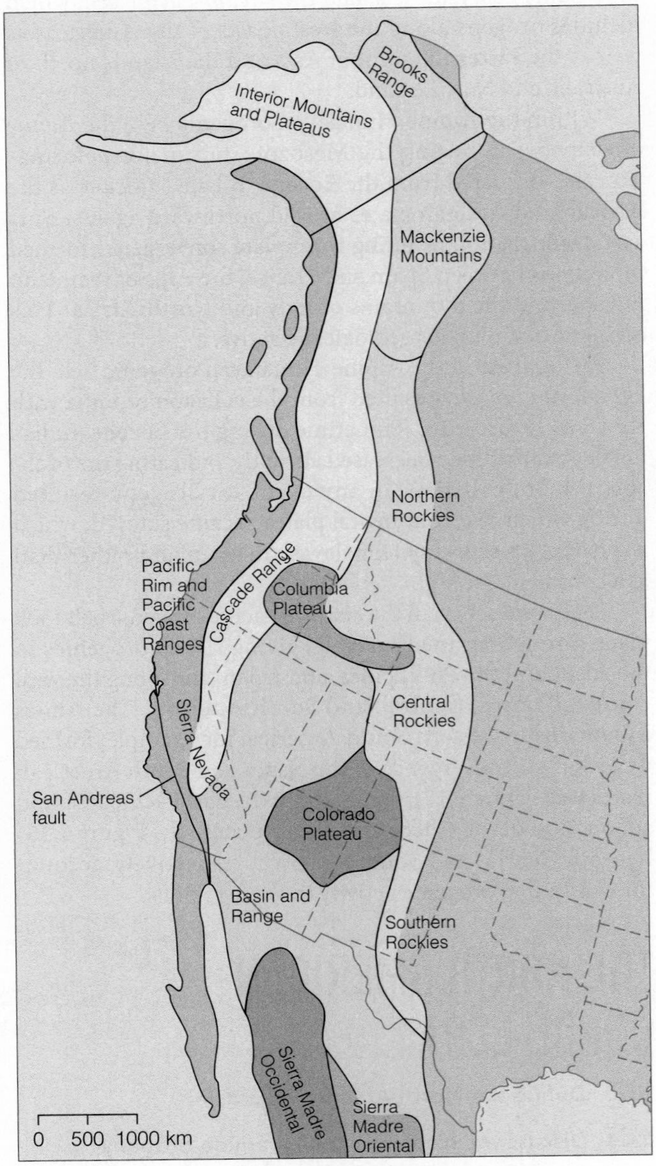

The Laramide orogeny ceased about 40 million years ago, but since that time the mountain ranges formed during the orogeny were eroded, and the valleys between the ranges filled with sediments. Many of the ranges were nearly buried in their own erosional debris, and their present-day elevations are the result of renewed uplift.

In other parts of the Cordillera, the Colorado Plateau was uplifted far above sea level, but the rocks were little deformed (●**FIGURE 17.31A**). In the Basin and Range Province, block faulting began during the Middle Cenozoic and continues to the present. At its western edge, the province is bounded by a large escarpment that forms the east face of the Sierra Nevada (●**FIGURE 17.31B**).

In the Pacific Northwest, an area of about 200,000 km², mostly in Washington, is covered by the Cenozoic Columbia River basalts (see Figure 5.15). Issuing from long fissures, these flows overlapped to produce an aggregate thickness of about 1,000 m. Widespread volcanism also took place in Oregon, Idaho, California, Arizona, and New Mexico (●**FIGURE 17.31C**).

The present-day elements of the Pacific Coast section of the Cordillera developed as a result of the westward drift of North America, the partial consumption of the oceanic Farallon plate, and the collision of North America with the Pacific–Farallon Ridge (●**FIGURE 17.32**). During the Early Cenozoic, the entire Pacific Coast was bounded by a subduction zone that stretched from Mexico to Alaska. Most of the Farallon plate was consumed at this subduction zone, and now only two small remnants exist: the Juan de Fuca and Cocos plates (Figure 17.32). The continuing subduction of these small plates accounts for seismicity and volcanism in the Cascade Range of the Pacific Northwest and Central America, respectively. Westward drift of the North American plate also resulted in its collision with the Pacific–Farallon Ridge and the origin of the Queen Charlotte and San Andreas transform faults (Figure 17.32).

## The Continental Interior and the Gulf Coastal Plain

The vast, shallow seas that had invaded the continents during the previous eras were largely absent during the Cenozoic. The notable exception was the Zuni Sea, which occupied a large area in the continental interior during part of the Cenozoic. Sediments derived from the Laramide highlands to the west and southwest were transported eastward and deposited in a variety of continental, transitional, and marine environments.

● **FIGURE 17.29 The North American Cordillera**
The North American Cordillera is a complex mountainous region extending from Alaska into central Mexico. It consists of the elements shown here.

**North American Cordillera**  A complex mountainous region in western North America extending from Alaska into central Mexico.

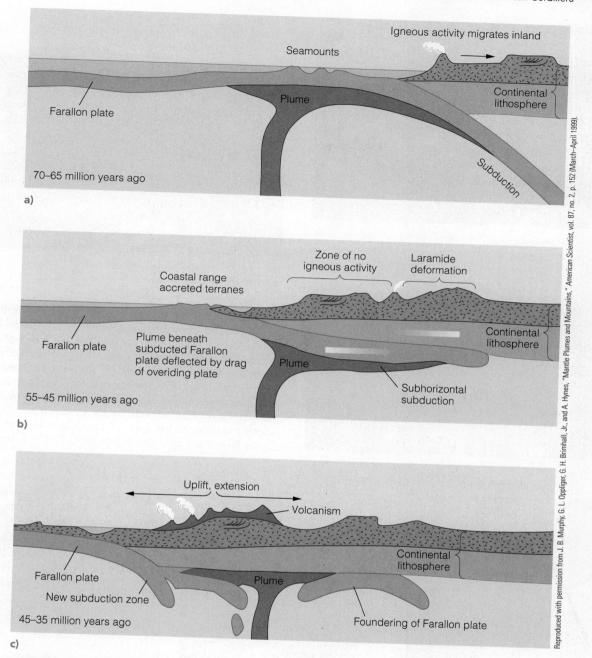

Reproduced with permission from J. B. Murphy, G. L. Oppliger, G. H. Brimhall, Jr., and A. Hynes, "Mantle Plumes and Mountains," *American Scientist*, vol. 87, no. 2, p. 152 (March–April 1999).

● **FIGURE 17.30 The Laramide Orogeny** The Laramide orogeny resulted when the Farallon plate was subducted beneath North America during the Late Cretaceous Period to Eocene Epoch. (a) As North America moved westward over the Farallon plate, beneath which was the deflected head of a mantle plume, the angle of subduction decreased and magmatism shifted inland. (b) With nearly horizontal subduction, magmatism ceased and the continental crust was deformed mostly by vertical forces. (c) Disruption of the oceanic plate by the mantle plume marked the onset of renewed volcanism.

The Gulf Coast sedimentation pattern was established during the Jurassic and persisted through the Cenozoic. Much of the sediment deposited on the Gulf Coastal Plain was detrital, but in the Florida section of the coastal plain and the Gulf Coast of Mexico, significant carbonate deposition occurred. A carbonate platform was established in Florida during the Cretaceous, and carbonate deposition continues at the present.

## Eastern North America

The eastern seaboard has been a passive continental margin since Late Triassic rifting separated North America from North Africa and Europe (Figure 17.19a). The present distinctive topography of the Appalachian Mountains is the product of Cenozoic uplift and erosion. By the end of the Mesozoic, the Appalachian Mountains had been eroded

Sue Monroe

a)

Sue Monroe

c)

Sue Monroe

b)

● **FIGURE 17.31 The Colorado Plateau, Basin and Range Province, and Cenozoic Lava Flows** (a) The Claron Formation in Bryce Canyon National Park in Utah is one of many exposed in the Colorado Plateau. This formation of gravel, sand, and mud was deposited mostly by streams during the Paleocene and Eocene epochs. (b) The Sierra Nevada at the western margin of the Basin and Range province has risen along normal faults, so that it is more than 3,000 m above the valley to the east. (c) Basalt lava flows of the Snake River Plain at Malad Gorge State Park in Idaho.

Reprinted with permission from W. R. Dickinson, "Cenozoic plate Techtonic setting of the cordilleran region in the western united states," *Cenozoic Paleogeography of the Western United States,* Pacific Coast Symposium 3, 1979, p. 2 (fig. 1).

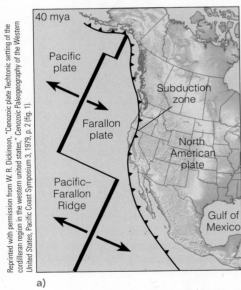

a)

b)

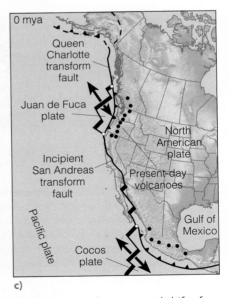

c)

● **FIGURE 17.32 Three Stages in Westward Drift of North America** Three stages (a), (b), and (c) in the westward drift of North America and its collision with the Pacific–Farallon Ridge. As the North American plate overrode the ridge, its margin became bounded by transform faults rather than a subduction zone.

to a plain. Cenozoic uplift rejuvenated the streams, which responded by renewed downcutting. As the streams eroded downward, they were superposed on resistant rocks and cut large canyons across these strata. The distinctive topography

of the Valley and Ridge Province is the product of Cenozoic erosion and preexisting geologic structures. It consists of northeast-southwest–trending ridges of resistant upturned strata and intervening valleys eroded into less-resistant strata.

## Pleistocene Glaciation

We know today that the Pleistocene (Ice Age) began 2.6 million years ago and ended about 11,700 years ago. During this time, several intervals of widespread continental glaciation took place, especially in the Northern Hemisphere, each separated by warmer interglacial periods (see Chapter 13). In addition, valley glaciers were more common at lower elevations and latitudes, and many extended much farther than they do today.

As you would expect, the climatic effects responsible for Pleistocene glaciers were worldwide. Nevertheless, Earth was not as frigid as it is portrayed in movies and cartoons, nor was the onset of the climatic conditions leading to glaciation rapid. Indeed, evidence from several types of investigations indicates that the climate cooled gradually from the Eocene through the Pleistocene. Furthermore, evidence from seafloor sediments shows that 20 major warm–cold cycles have occurred during the past 2 million years.

Geologists know that, at their greatest extent, Pleistocene glaciers covered about three times as much of Earth's surface

as they do now (●FIGURE 17.33A). Like the vast ice sheets now in Greenland and Antarctica, they were probably 3 km thick. Geologists have identified four major Pleistocene glacial episodes that took place in North America—the *Nebraskan, Kansan, Illinoian,* and *Wisconsinan* (●FIGURE 17.33B)—each named for the state in which the most southerly glacial deposits are well exposed. The three interglacial stages are named for localities of well-exposed soils and other deposits (Figure 17.33b).

Although the Pleistocene ended 11,700 years ago, we see the effects of glaciation in many parts of North America. Indeed, valley glaciers shaped and continue to modify mountains in Canada, Alaska, and several western states, and as a result have produced the majestic scenery seen in many areas (●FIGURE 17.34A). And even though we have emphasized glaciers in this section, the Pleistocene was also a time of continuing tectonism and volcanism (●FIGURE 17.34B). As we stated in the beginning of this book, Earth remains a dynamic and evolving planet.

●**FIGURE 17.33 Maximum Extent of Pleistocene Glaciers in North America** (a) Centers of ice accumulation and maximum extent of Pleistocene glaciers in North America. (b) Standard terminology for Pleistocene glacial and interglacial stages in North America.

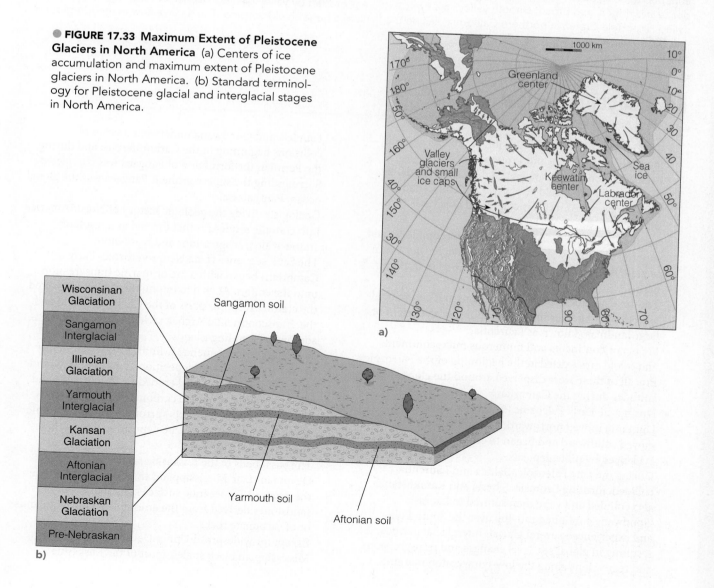

a)

b)

● **FIGURE 17.34 Ongoing Geologic Activity** Earth remains an active planet on which interactions among its systems continue to bring about change. (a) Valley glaciers continue to erode the Chugach Mountains in Alaska. The sharp, angular peaks and ridges, and broad, smooth valleys are typical of areas eroded by valley glaciers. (b) This 115-m-high cinder cone, called High Hole Crater, in northern California, lies on the flank of a huge shield volcano. The aa lava flow in the foreground is 1,100 years old.

# Key Concepts Review

- Geologists use a threefold division of the Precambrian—the Hadean, Archean, and Proterozoic.
- Each continent has an ancient, stable craton made up of a Precambrian shield and platform. The Canadian shield in North America is composed of several subunits.
- Archean rocks are mostly granite–gneiss complexes and subordinate greenstone belts. One model for the origin of greenstone belts holds that they formed in back-arc basins.
- The amalgamation of Archean cratons and continental accretion along their margins account for the origin of a large landmass known as Laurentia.
- Six major continents and numerous microcontinents and island arcs existed at the beginning of the Paleozoic Era; all of these were dispersed around the globe at low latitudes during the Cambrian.
- During the Early Paleozoic (Cambrian–Silurian), Laurentia moved northward, whereas Gondwana moved southward and began to cross the South Pole as evidenced by tillite deposits.
- During the Late Paleozoic, Baltica and Laurentia collided, forming Laurasia. Siberia and Kazakhstania also collided and were then sutured to Laurasia. Gondwana continued moving over the South Pole and experienced several glacial–interglacial periods, resulting in global sea-level changes and transgressions and regressions along the low-lying craton margins.

- Laurasia and Gondwana underwent a series of collisions beginning in the Carboniferous, and during the Permian the formation of Pangaea was completed. Surrounding the supercontinent Pangaea was the global ocean, Panthalassa.
- Geologists divide the geologic history of North America into cratonic sequences that formed as a result of craton-wide transgressions and regressions.
- The Sauk sequence (Late Neoproterozoic–Early Cambrian) began with a major marine transgression onto the craton. At its maximum, the Sauk Sea covered the craton except for parts of the Canadian shield and the Transcontinental Arch, a series of large, northeast-southwest–trending islands.
- The Tippecanoe sequence (Middle Ordovician–Early Devonian) began with deposition of an extensive sand unit over the exposed and eroded Sauk landscape and was followed by extensive carbonate deposition. In addition, large barrier reefs surrounded many cratonic basins, resulting in evaporite deposition within these basins.
- The basal beds of the Kaskaskia sequence (Middle Devonian–Late Mississippian) that were deposited on the exposed Tippecanoe surface consisted either of sandstones derived from the eroding Taconic Highlands or of carbonate rocks.
- Except for widespread Upper Devonian and Lower Mississippian black shales, most of the Kaskaskia

sequence is dominated by carbonates and associated evaporites. The Devonian Period was a time of major reef building in western Canada, southern England, Belgium, Australia, and Russia.

- During the Absaroka sequence (Late Mississippian–Early Jurassic) transgressions and regressions over the low-lying North American craton, probably caused by advancing and retreating Gondwanan ice sheets, resulted in cyclothems and the formation of coals during the Pennsylvanian Period.

- Cratonic mountain building, specifically the Ancestral Rockies, occurred during the Pennsylvanian Period and resulted in thick, nonmarine detrital sediments and evaporites being deposited in the intervening basins.

- By the Early Permian, the Absaroka Sea occupied a narrow zone of the south-central craton. Here, several large reefs and associated evaporites developed. By the end of the Permian Period, this shallow sea had retreated from the craton.

- The eastern edge of North America was a stable carbonate platform during Sauk time. During Tippecanoe time, an oceanic–continental convergent plate boundary formed, resulting in the Taconic orogeny, the first of three major orogenies to affect the Appalachian mobile belt.

- The newly formed Taconic Highlands shed sediments into the western epeiric sea, producing a clastic wedge that geologists call the Queenston Delta.

- The Acadian and Hercynian–Alleghenian orogenies were all part of the global tectonic activity that resulted in the assembly of Pangaea.

- The Cordilleran mobile belt was the site of the Antler orogeny, a minor Devonian orogeny, during which deep-water sediments were thrust eastward over shallow-water sediments.

- During the Pennsylvanian and Early Permian, mountain building occurred in the Ouachita mobile belt. In addition to the mountain range the Ouachita orogeny produced, this tectonic activity was partly responsible for the cratonic uplift in the southwest, which resulted in the Ancestral Rockies.

- During the Paleozoic Era, numerous microplates and terranes, such as Avalonia, existed and played an important role in the formation of Pangaea.

- We can summarize the breakup of Pangaea as follows:

1. During the Late Triassic, North America began separating from Africa. This was followed by the rifting of North America from South America.

2. During the Late Triassic and Jurassic periods, Antarctica and Australia—which remained sutured together—began separating from South America and Africa, and India began rifting from Gondwana.

3. South America and Africa began detaching during the Jurassic, and Europe and Africa started converging at this time.

4. The final stage in Pangaea's breakup occurred during the Cenozoic, when Greenland completely separated from Europe and North America.

- The Gulf Coastal region was the site of major evaporite accumulation during the Jurassic as North America rifted from South America. During the Cretaceous, it was inundated by a transgressing sea, which, at its maximum, connected with a sea transgressing from the north to create the Cretaceous Interior Seaway.

- Mesozoic rocks of the western region of North America were deposited in a variety of continental and marine environments. One of the major controls of sediment distribution patterns was tectonism.

- Western North America was affected by four inter-related orogenies: the Sonoma, Nevadan, Sevier, and Laramide. Each involved igneous intrusions, as well as eastward thrust faulting and folding.

- The cause of the Nevadan, Sevier, and Laramide orogenies was the changing angle of subduction of the oceanic Farallon plate beneath the continental North American plate. The timing, rate, and, to some degree, the direction of plate movement were related to seafloor spreading and the opening of the Atlantic Ocean.

- Cenozoic tectonism was concentrated in the Alpine-Himalayan and circum-Pacific belts.

- The Cenozoic evolution of the North American Cordillera included deformation during the Laramide orogeny, extensional tectonics that yielded basin-and-range structures, extensive intrusive and extrusive igneous activity, and uplift and erosion.

- One model for the Laramide orogeny involves near-horizontal subduction of the Farallon plate beneath North America, resulting in fault-bounded uplifts in the area of the present-day Rocky Mountains.

- As the North American plate drifted westward, it collided with the Pacific–Farallon Ridge, at which time subduction ceased and the continent became bounded by large transform faults, except in the Pacific Northwest where subduction continues.

- Sediments eroded from Laramide uplifts were deposited in intermontane basins and in the Great Plains, whereas a wedge of sediments pierced by salt domes is found on the Gulf Coastal Plain.

- Cenozoic uplift and erosion were responsible for the present topography of the Appalachian Mountains. As the Appalachians eroded, much of the sediment was deposited on the Atlantic Coastal Plain.

- Vast glaciers covered Earth's land surface during the Pleistocene. About 20 warm–cold Pleistocene climatic cycles are recognized from evidence found in deep-sea cores.

- Four major glacial and interglacial stages occurred in North America during the Pleistocene Epoch.

# Important Terms

Absaroka sequence   375

Acadian orogeny   380

Canadian shield   364

Cordilleran orogeny   386

Cretaceous Interior Seaway   388

craton   364

cyclothem   375

greenstone belt   364

Grenville orogeny   366

Hercynian–Alleghenian orogeny   380

Kaskaskia sequence   374

Laramide orogeny   388

Laurentia   366

Midcontinent rift   367

mobile belt   378

Nevadan orogeny   387

North American Cordillera   390

orogen   366

Ouachita orogeny   381

platform   363

Sauk sequence   372

Sevier orogeny   387

shield   363

Sundance Sea   388

Taconic orogeny   380

Tippecanoe sequence   372

# Review Questions

1. An elongated area marking the site of mountain building is a
   a. cyclothem.
   b. mobile belt.
   c. platform.
   d. shield.
   e. craton.

2. Which one of the following sequences of geologic time designations is in the correct order from oldest to youngest?
   a. Phanerozoic-Archean-Mesozoic
   b. Hadean-Archean-Proterozoic
   c. Archean-Hadean-Paleozoic
   d. Cenozoic-Proterozoic-Phanerozoic
   e. Proterozoic-Cenozoic-Precambrian

3. The time of greatest post-Paleozoic inundation of the craton occurred during which geologic period?
   a. Triassic
   b. Jurassic
   c. Cretaceous
   d. Paleogene
   e. Neogene

4. A complex part of the circum-Pacific orogenic belt in the United States is the
   a. Cordilleran orogeny.
   b. North American Cordillera.
   c. Rio Grande rift.

   d. Transcontinental Arch.
   e. Pacific–Farallon Ridge.

5. The economically valuable deposit in a cyclothem is
   a. gravel.
   b. metallic ore.
   c. coal.
   d. carbonate.
   e. evaporites.

6. From a plate tectonic perspective, how does the orogenic activity that occurred in the Cordilleran mobile belt during the Mesozoic Era differ from that which took place in the Appalachian mobile belt during the Paleozoic Era?

7. Paleogeographic maps of what the world looked like during the Paleozoic Era can be found in almost every Earth history book and in numerous scientific journals. What criteria are used to determine the location of ancient continents and ocean basins, and why are there minor differences in the location and size of these paleocontinents among the various books and articles?

8. Outline the events that led to the evolution of Laurentia during the Proterozoic.

9. How would you explain the Cenozoic evolution of the North American Cordillera in terms that a nongeologist could readily understand?

# Creative Thinking Visual Question

This view (●**FIGURE 1**) in Capitol Reef National Park, Utah, shows excellent exposures of various Triassic and Jurassic sedimentary rock formations.

a. What accounts for the reddish color of some of the rock layers?

b. The rocks visible on the skyline belong to the Navajo Formation, which is exposed over a large area in the southwest. It is composed of well-sorted, well-rounded quartz sandstone, it has tracks of land-dwelling animals, including dinosaurs, and the sandstone has cross-beds up to 30 m high. How do you think it was deposited?

Sue Monroe

●**FIGURE 1** This view of Capital Reef National Park, Utah, shows excellent exposures of several sedimentary rock formations that were deposited during the Triassic and Jurassic periods.

# 18
# LIFE HISTORY

The biota of the Middle Cambrian Burgess Shale, British Columbia, Canada, was composed of a number of strange-looking and now extinct animals that inhabited the seas of North America at that time.

# INTRODUCTION

Scientists have been seriously investigating the history of life on Earth for two centuries. Without science, our only knowledge of how the animals and plants around us have changed would be from written human records. Such records are interesting, but they simply do not go far enough back in time to help us understand anything more than the current diversity of organisms and a few very recent extinctions.

By looking at the geologic record, scientists have clearly established the outline of Earth history going back some 4 billion years. Although there is much we still do not know about life history, we have learned a great deal. For example, we now know that extinction is the rule, not the exception. But we have also learned that the variety and complexity of organisms have increased after episodes of mass extinction time and time again.

Just as in the previous chapter concerning Earth history, we have in this chapter the daunting task of condensing and summarizing what we know about the many and varied creatures that have inhabited our planet almost since Earth began. The evidence we rely on is the **fossils** that have endured through millions, even billions, of years. The quality of the *fossil record* varies considerably, depending on the types of organisms existing at a particular time and the environment in which they lived.

The preservation of any one organism as a fossil is rare, but fossils are nevertheless quite common. This apparent contradiction is easily explained when you consider that so many billions of organisms have existed during so many millions of years that if only a tiny fraction was preserved, the total number of fossils is phenomenal. In fact, fossils of many **vertebrate** animals (those with a segmented vertebral column), such as dinosaurs, are much more common than most people realize. In general, the fossil record does give a good overview of life history.

# PRECAMBRIAN LIFE HISTORY

**LO1**  Define stromatolites

**LO2**  Discuss their importance in the evolution of the early atmosphere

**LO3**  Explain the difference between prokaryotic and eukaryotic cells

**LO4**  Discuss the importance of the Ediacaran fauna

Prior to the mid-1950s, scientists had assumed that the fossils so abundant in Cambrian rocks must have had an earlier history, but they had little direct knowledge of Precambrian life. A few enigmatic Precambrian fossils had been reported, but they were mostly dismissed as inorganic structures rather than traces of organisms. In fact, the Precambrian was once called the *Azoic*, meaning "devoid of life."

Then, during the early 1900s, Charles Walcott proposed that layered, moundlike structures from the Paleoproterozoic-age Gunflint Iron Formation in Ontario, Canada, were ancient reefs constructed by algae. However, it was not until 1954 did scientists demonstrate that they were actually the products of organic activity. These structures, now called **stromatolites**, still form in a few areas such as Shark Bay, Australia, where they trap sediment on sticky mats of photosynthesizing cyanobacteria, commonly called *blue-green algae* (●**FIGURE 18.1**). Although widespread in Proterozoic rocks, they are now restricted to aquatic environments with especially high salinity where snails cannot live and graze on them. We now know that stromatolites are somewhat common in Proterozoic-age rocks. Until recently, the

**fossil**  Remains or traces of prehistoric organisms preserved in rocks.

**vertebrate**  Any animal that has a segmented vertebral column, as in fish, amphibians, reptiles, birds, and mammals.

**stromatolite**  A biogenic sedimentary structure, especially in limestone, produced by the entrapment of sediment on sticky mats of photosynthesizing bacteria.

a)

b)

James S. Monroe

Phil Playford, Geological Survey of Western Australia

●**FIGURE 18.1 Stromatolites**  (a) Stromatolites from the Neoproterozoic in Glacier National Park, Montana. Note the gently curved structures or layers in the rock.  (b) Present-day stromatolites displaying their pillow-shaped growth in Shark Bay, Australia, one of few places where they still live.

oldest stromatolites were from Western Australia and have been dated at roughly 3.5 billion years old (Paleoarchean). In 2016, it was announced that the oldest record of life was found in rocks of the Isua supracrustal belt in southwest Greenland. These rocks are dated at 3.7 billion years and appear similar to known ancient and recent stromatolites, although to some, the evidence is still inconclusive.

The importance of stromatolites in Earth history should not be overlooked. The atmosphere of the planet during the Paleoarchean had no free oxygen (no oxygen gas), but stromatolites, like plants, produce oxygen as a by-product of photosynthesis. Slowly but surely, as the Archean and especially the Proterozoic unfolded, increasing amounts of oxygen accumulated in Earth's atmosphere.

All known fossils from Archean- and Paleoproterozoic-age rocks are of single-celled bacteria lacking a cell nucleus and reproducing asexually, much as bacteria do today. Such cells are termed **prokaryotic cells** as opposed to **eukaryotic cells**, the latter having a cell nucleus and other internal structures not present in prokaryotic cells, and most reproduce sexually.

The origin of eukaryotic cells is one of the most important events in life history. No one doubts they were present by the Mesoproterozoic Era, and some evidence indicates that they first evolved even sooner. How then did eukaryotes originate? One theory, known as endosymbiosis and supported by evidence mostly from living organisms, holds that two or more prokaryotic cells entered into a beneficial symbiotic relationship, and the symbionts became increasingly interdependent until the unit could exist only as a whole.

The first eukaryotes (organisms composed of eukaryotic cells) were still single-celled, but by Paleoproterozoic or Mesoproterozoic time, the first multicelled organisms had made their appearance. Carbonaceous impressions of what appear to be multicelled algae are known from several areas, but the first multicelled animal fossils are found in Neoproterozoic-age rocks.

One of the most interesting and enigmatic fossil assemblages comes from 545- to 600-million-year-old rocks in South Australia. Named the *Ediacaran fauna*, and found worldwide, it consists of the impressions of organisms that look like present-day jellyfish, sea pens, segmented worms, and primitive arthropods (●**FIGURE 18.2**). Although they

---

**prokaryotic cell** A cell that lacks a nucleus and organelles such as mitochondria and plastids; the cells of bacteria and cyanobacteria.

**eukaryotic cell** A cell with internal structures such as mitochondria and an internal membrane-bounded nucleus containing chromosomes.

a)

b)

c)

d)

● **FIGURE 18.2 The Ediacaran Fauna of Australia (a and b) and Ediacaran-Like Fossils from Newfoundland (c)** (a) The affinities of *Tribrachidium heraldicum*, a member of the Ediacaran fauna, remain uncertain. It has been suggested that it could be a primitive cnidarian or echinoderm. (b) *Spriggina floundersi* was originally described as a segmented (possibly annelid) worm. Its relationship and affinities to living animals is still unclear, although it has been suggested that it could be closely related to the arthropods, or even an extinct phylum. (c) *Pteridinium carolinensis* is a member of the Ediacaran fauna from North Carolina. It was originally thought to be a primitive cnidarian related to modern-day sea pens. However, current research indicates that if it was a cnidarian, it is not closely related, and might instead be a phylum that became extinct. (d) Reconstruction of the Ediacaran fauna.

are superficially similar to modern invertebrate animals, a number of researchers think that they represent an early evolutionary development distinct from the ancestry of any present-day animals.

Although the Proterozoic fossil record is still sparse, there is increasing evidence that indicates animals with at least partial skeletons were present then. Minute scraps of shell-like material and spicules, presumably from sponges, suggest that hard skeletal elements had evolved during the Neoproterozoic. Nevertheless, animals with durable skeletons of chitin (a complex organic substance), silica ($SiO_2$), and calcium carbonate ($CaCO_3$) were not abundant until the beginning of the Paleozoic Era.

# PALEOZOIC LIFE HISTORY

**L05** Discuss the changes taking place in the invertebrate faunas for each of the periods of the Paleozoic Era

**L06** Explain what the Permian mass extinction event was

**L07** Discuss the possible causes for the Permian mass extinctions

**L08** Define a vertebrate

**L09** Discuss the evolutionary history of fishes

**L010** Discuss the transition between lobe-finned fish and the earliest amphibians

**L011** Explain what problems had to be surmounted to make the transition from a water to a land environment

**L012** Explain why the evolution of the amniote egg allowed reptiles to colonize all parts of the land

**L013** Discuss the diversification of the reptiles during the Paleozoic Era

**L014** Explain the problems that had to be solved for plants to make the transition to a land environment

**L015** Discuss the significance of the evolution of gymnosperms

At the beginning of the Paleozoic Era, animals with skeletons appeared rather abruptly in the fossil record. In fact, their appearance is described as an explosive development of new types of animals and is referred to as the "Cambrian explosion" by most scientists. This sudden appearance of new animals in the fossil record is rapid, although, only in the context of geologic time, having taken place over millions of years during the Early Cambrian Period.

The earliest of these animals with skeletons were *invertebrates*—that is, animals that lack a segmented vertebral column. The major invertebrate groups and their geologic ranges are given in ●TABLE 18.1. Rather than focusing on the history of each invertebrate group, we will survey the evolution of the Paleozoic marine invertebrate communities through time, concentrating on the major features and changes that took place.

## Marine Invertebrates

Although almost all of the major invertebrate phyla evolved during the Cambrian Period (Table 18.1), many were represented by only a few species. Whereas echinoderms were diverse, trilobites, brachiopods, and archaeocyathids (bottom-dwelling organisms that constructed reeflike structures and lived only during the Cambrian) comprised the

**TABLE 18.1** The Major Invertebrate Groups and Their Geologic Ranges

| Group | Range | Group | Range |
|---|---|---|---|
| **Phylum Protozoa** | Cambrian–Recent | **Phylum Mollusca** | Cambrian–Recent |
| Class Sarcodina | Cambrian–Recent | Class Monoplacophora | Cambrian–Recent |
| Order Foraminifera | Cambrian–Recent | Class Gastropoda | Cambrian–Recent |
| Order Radiolaria | Cambrian–Recent | Class Bivalvia | Cambrian–Recent |
| **Phylum Porifera** | Cambrian–Recent | Class Cephalopoda | Cambrian–Recent |
| Class Demospongea | Cambrian–Recent | **Phylum Annelida** | Precambrian–Recent |
| Order Stromatoporoida | Ordovician–Mississippian | **Phylum Arthropoda** | Cambrian–Recent |
| **Phylum Archaeocyatha** | Cambrian | Class Trilobita | Cambrian–Permian |
| **Phylum Cnidaria** | Cambrian–Recent | Class Crustacea | Cambrian–Recent |
| Class Anthozoa | Ordovician–Recent | Class Insecta | Silurian–Recent |
| Order Tabulata | Ordovician–Permian | **Phylum Echinodermata** | Cambrian–Recent |
| Order Rugosa | Ordovician–Permian | Class Blastoidea | Ordovician–Permian |
| Order Scleractinia | Triassic–Recent | Class Crinoidea | Cambrian–Recent |
| **Phylum Bryozoa** | Ordovician–Recent | Class Echinoidea | Ordovician–Recent |
| **Phylum Brachiopoda** | Cambrian–Recent | Class Asteroidea | Ordovician–Recent |
| Class Inarticulata | Cambrian–Recent | **Phylum Hemichordata** | Cambrian–Recent |
| Class Articulata | Cambrian–Recent | Class Graptolithina | Cambrian–Mississippian |

majority of Cambrian skeletonized life. It is important to remember, however, that the fossil record is biased toward organisms with durable skeletons and that we generally know little about the soft-bodied organisms of that time.

No discussion of Cambrian life would be complete without mentioning one of the best examples, a preserved soft-bodied fauna and flora: the *Burgess Shale biota*. As the Sauk Sea transgressed from the Cordilleran shelf onto the western edge of the craton (see Figure 17.11), Early Cambrian-age sands were covered by Middle Cambrian-age black muds that preserved members of this diverse soft-bodied assemblage as carbonaceous impressions (see the Chapter Opening photo).

In recent years, the reconstruction, classification, and interpretation of many of the Burgess Shale fossils have undergone a major change. Traditionally, the majority of Burgess Shale organisms were placed into existing phyla. Thus, the marine fauna of the Cambrian was viewed as being essentially the same in number as the phyla of the present-day world, but with fewer species in each phylum. According to this view, the history of life has been simply a gradual increase in the diversity of species within each phylum through time, and thus, the number of basic body plans has remained more or less constant since the initial radiation of multicelled organisms.

A different view holds that the initial explosion of varied life-forms in the Cambrian was promptly followed by a short period of experimentation and then extinction of many phyla. Accordingly, the richness and diversity of modern taxa are the result of repeated variations of the basic body plans that survived the Cambrian extinctions. In other words, life was much more diverse in terms of phyla during the Cambrian than it is today.

A major transgression that began during the Middle Ordovician (Tippecanoe Sequence) resulted in the most widespread inundation of the craton in Earth history. This vast, shallow sea, which was uniformly warm during this time, opened numerous new marine habitats that were soon filled by a variety of organisms such that the Ordovician is characterized by a dramatic increase in the diversity of the total shelly fauna. In fact, so spectacular was this increase in marine invertebrate diversity that it is rightly called the "Great Ordovician Biodiversity Event" (●FIGURE 18.3). The end of the Ordovician, however, was a time of mass extinctions in the marine realm. More than 100 families of marine invertebrates became extinct. Many geologists think these extinctions were the result of extensive glaciation at the end of the Ordovician as Gondwana moved over the South Pole region. This widespread glaciation resulted in decreased sea level, and a cooling of the surface waters, thus creating stressful conditions for the near-surface phytoplankton (primary producers) and the shallow-water marine faunas (consumers). The mass extinction at the end of the Ordovician Period was followed by rediversification and recovery of many of the decimated invertebrate groups. In fact, the Silurian and Devonian periods were times of major reef building in which organic reef builders diversified in new ways, building massive reefs larger than any produced during the Cambrian or Ordovician.

Near the end of the Devonian, another mass extinction occurred that resulted in a worldwide near-total collapse of the massive reef communities. On land, however, the *seedless vascular plants* (which will be discussed shortly) were seemingly unaffected. Thus, extinctions at this time were most extensive among marine life, particularly in the reef communities.

The Carboniferous invertebrate marine community responded to the Late Devonian extinctions in much the same way that the Silurian invertebrate marine community responded to the Late Ordovician extinctions—that is, by

● **FIGURE 18.3 Middle Ordovician Marine Community** Reconstruction of a Middle Ordovician seafloor fauna. Cephalopods, crinoids, colonial corals, bryozoans, trilobites, and brachiopods are shown.

**Critical Thinking Question**
What role did the transgressing seas during the Tippecanoe Sequence play in terms of opening up new ecologic niches for the marine invertebrate community, and what was the contribution of the invertebrates to the sedimentary rocks of the Tippecanoe Sequence?

Field Museum Library/Getty Images

● **FIGURE 18.4 Permian Patch-Reef Marine Community** Reconstruction of a Permian patch-reef community from the Glass Mountains of West Texas. Shown are algae, brachiopods, cephalopods, sponges, and corals.

renewed adaptive radiation and rediversification. However, large organic reefs like those existing earlier in the Paleozoic virtually disappeared and were replaced by small patch reefs that flourished during the Late Paleozoic.

The Permian invertebrate marine faunas (●**FIGURE 18.4**) resembled those of the Carboniferous. However, they were not as widely distributed because of the restricted size of the shallow seas on the cratons and the reduced shelf space along the continental margins (see Figure 17.10b).

## The Permian Mass Extinction

The greatest recorded mass extinction to affect Earth's biota occurred at the end of the Permian Period (●**FIGURE 18.5**). By the time the Permian ended, it is estimated that approximately 50% of all taxonomic families died out, 90–96% of all marine species as well as perhaps 70% of the land's insects, amphibians, and reptiles went extinct.

What caused such a crisis for both marine and land-dwelling organisms? Various hypotheses have been proposed, but no completely satisfactory answer has yet been found. Many scientists think an episode of deep-sea anoxia and increased oceanic $CO_2$ levels resulted in a highly stratified ocean during the Late Permian. In other words, there was very little, if any, circulation of oxygen-rich surface waters into the deep ocean. During this time, stagnant waters also covered the shelf regions, thus affecting the shallow marine fauna.

During the Late Permian, widespread volcanic and continental fissure eruptions, such as the Siberian Traps, where lava flows covered more than 2 million $km^2$, were also taking place. These eruptions released not only large amounts of additional carbon dioxide into the atmosphere but also high amounts of fluorine and chlorine, which could have damaged the ozone layer and helped contribute to increased climatic instability and ecologic collapse.

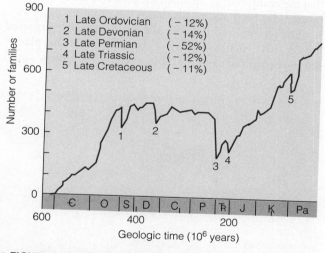

● **FIGURE 18.5 Phanerozoic Diversity of Marine Invertebrate and Vertebrate Families** Reproduced with permission from "Mass extinction in the marine fossil record," by D.M. Raup and J.J. Sepkoski, *Science*, vol. 215, p. 1502 (fig. 2). © 1982 American Association for the Advancement of Science.

**Critical Thinking Question** Note that the greatest extinction in Earth history occurred at the end of the Permian, yet the Late Cretaceous mass extinction is the one most people are familiar with. Why do you think most people have heard of the Late Cretaceous mass extinction but not the one at the end of the Permian?

By the end of the Permian, a near-collapse of both the marine and terrestrial ecosystem had occurred. Although the ultimate cause of such devastation is still being debated and investigated, it is safe to say that it was probably a combination of interconnected and related geologic and biologic events.

## Vertebrates

Besides the numerous invertebrate groups, vertebrates (animals with backbones) also evolved and diversified during the Paleozoic Era. Remains of some of the oldest and undisputed fish fossils are found in Upper Cambrian marine rocks in Wyoming. These fish, referred to as *ostracoderms*, were jawless, had poorly developed fins, an external covering of bony armor, and lived from the Early Cambrian to Late Devonian (●FIGURE 18.6A).

The evolution of jaws was a major evolutionary advance among primitive vertebrates. Although their jawless ancestors could only feed on detritus, jawed fish could chew food and become active predators, thus opening many new ecological niches.

The fossil remains of the first jawed fish are found in Lower Silurian rocks and belong to the *acanthodians*, a group of enigmatic fish characterized by large spines, scales covering much of the body, jaws, teeth, and reduced body armor (●FIGURE 18.6C). Although the relationship of acanthodians to other fish is not yet well established, many scientists think that the acanthodians included the probable ancestors of the present-day bony and cartilaginous fish groups.

The other jawed fish, the *placoderms* (heavily armored jawed fish), evolved during the Silurian. The placoderms exhibited considerable variety, including small bottom-dwellers (●FIGURE 18.6B), as well as some of the largest marine predators that ever existed.

The **bony fish** evolved during the Devonian and are the most varied and numerous of all the fishes (●FIGURE 18.6D). Because of this, and the fact that amphibians evolved from them, the evolutionary history of bony fish is especially important. There are two groups of bony fish: the common *ray-finned fish*, which include the familiar trout, bass, and perch, among others, and the lesser known *lobe-finned fish*.

The lobe-finned fish are particularly important, because this group included the probable ancestor of the amphibians, the first land-dwelling vertebrate animals. In fact, the similarity between the group of lobe-finned fish known as *crosspterygians* and the earliest amphibians, such as *Ichthyostega* (●FIGURE 18.7), is striking and one of the most widely cited examples of a transition from one major group to another. However, recent discoveries of older lobe-finned fish and newly published findings of fish ancestors of amphibians are filling in the gaps in the evolution from fish to amphibians.

Although amphibians were the first vertebrates to live on land, they were not the first land-living organisms. Land plants, which probably evolved from green algae, initially appeared during the Ordovician Period. Furthermore,

---

**bony fish** Fish with an internal skeleton of bone; the class Osteichthyes.

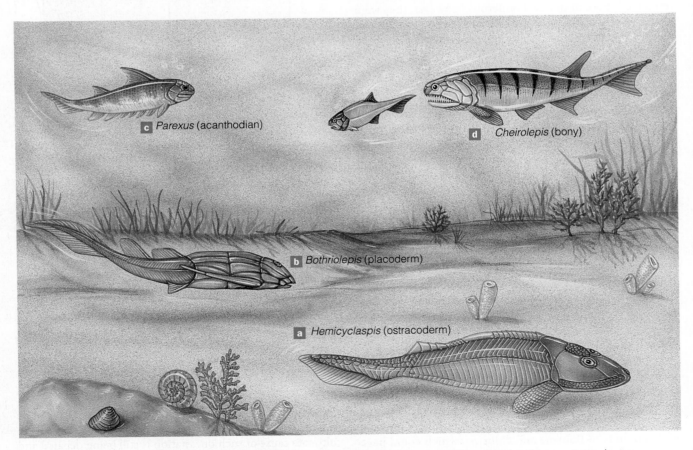

●**FIGURE 18.6 Recreation of a Devonian Seafloor** Recreation of a Devonian seafloor showing (a) an ostracoderm, (b) a placoderm, (c) an acanthodian, and (d) a bony fish.

● **FIGURE 18.7 Late Devonian Landscape** Shown is *Ichthyostega*, both in the water and on land. *Ichthyostega* was an amphibian that grew to a length of about 1 m. The flora of the time was diverse, consisting of a variety of small and large seedless vascular plants.

insects, millipedes, spiders, and even snails invaded the land before amphibians.

The transition from water to land required animals to surmount several barriers. The most critical were drying out, reproduction, the effects of gravity, and the extraction of oxygen from the atmosphere by lungs rather than from water by gills. These problems were partly solved in the transition from crossopterygians to intermediate forms that had limbs, albeit not strong enough for them to walk on land, to the earliest amphibians. The structural adaptations from crossopterygians to the earliest amphibians include a backbone, rib cage, pelvic and pectoral girdles, and limbs, capable of supporting the animal on land. Furthermore, the crossopterygians had lungs to extract oxygen from the atmosphere.

As just mentioned, the oldest known amphibians, such as *Ichthyostega* and related genera, had skeletal features that allowed them to spend their life on land. Because amphibians did not evolve until the Late Devonian, they were a minor element of the Devonian terrestrial ecosystem. Like other groups that moved into new and previously unoccupied niches, amphibians underwent rapid adaptive radiation and became abundant during the Carboniferous and Early Permian periods.

The Late Paleozoic amphibians did not at all resemble the familiar frogs, toads, newts, and salamanders that make up the modern amphibian fauna. In fact, one group were large, sluggish creatures, up to 2 m in length, that lived in swamps and streams eating fish, vegetation, insects, and other small amphibians (●**FIGURE 18.8**).

These large amphibians were abundant in the Carboniferous, when swampy conditions were widespread (see Figure 17.13), but they soon declined in abundance during the Permian, perhaps in response to changing climatic conditions. Only a few species survived into the Triassic.

Amphibians were limited in colonizing the land, however, because they had to return to water to lay their gelatinous eggs. The evolution of the **amniote egg** (●**FIGURE 18.9**)

● **FIGURE 18.8 Carboniferous Coal Swamp** The amphibian fauna of the Carboniferous was varied. Shown is the serpentlike *Dolichosoma* (foreground) and the large amphibian *Eryops*.

**Critical Thinking Question** What were the conditions that made the Late Carboniferous (Pennsylvanian) Period so favorable for the proliferation of amphibians and seedless vascular plants?

freed reptiles from this constraint. In such an egg, the developing embryo is surrounded by a liquid-filled sac and provided with both a food and waste sac. In this way, the emerging reptile, in essence, a miniature adult, bypasses the need for a larval stage in the water. The evolution of such an egg allowed vertebrates to colonize all parts of the land because they no longer had to return to the water as part of their reproductive cycle.

The oldest known reptiles evolved during the Mississippian Period and were small, agile animals that continued to diversify during the Pennsylvanian Period. One of the descendant groups of these early reptiles was the **pelycosaurs**, or finback reptiles, which became the

**amniote egg** Egg in which the embryo develops in a fluid-filled cavity called the amnion. The egg also contains a yolk sac and a waste sac.

**pelycosaur** Pennsylvanian to Permian reptiles that had some mammal characteristics; many species had large fins on their back.

dominant reptile group by the Permian (●FIGURE 18.10). An interesting feature of the pelycosaurs is their sail. It was formed by vertebral spines that, in life, were covered with skin. The sail has been variously explained as a type of sexual display, a means of protection, and a display to look more ferocious. The current consensus seems to be that the sail served as some type of thermoregulatory device, raising the reptile's temperature by catching the sun's rays or cooling it by facing the wind. Because pelycosaurs are considered to

be the group from which therapsids (mammal-like reptiles) evolved, it is interesting that they may have had some sort of body-temperature control.

The pelycosaurs became extinct during the Permian and were succeeded by the **therapsids**, mammal-like reptiles that evolved from the carnivorous pelycosaur lineage and rapidly diversified into herbivorous and carnivorous lineages. Therapsids were small- to medium-size animals that displayed the beginnings of many mammalian features: fewer bones in the skull because many of the small skull bones were fused; enlarged lower jawbone; differentiation of teeth for various function such as nipping, tearing, and chewing food; and more vertically placed legs for greater flexibility, as opposed to the way that the legs sprawled out to the side in primitive reptiles.

Furthermore, many paleontologists think that therapsids were *endothermic*, or warm-blooded, enabling them to maintain a constant internal body temperature. This characteristic would have allowed them to expand into a variety of habitats, and, indeed, the Permian rocks in which their fossil remains are found are distributed not only in low latitudes but in middle and high latitudes as well.

As the Paleozoic Era came to an end, the therapsids were the dominant large terrestrial animals and occupied a wide range of ecological niches. The mass extinction that decimated the marine fauna at the close of the Paleozoic had an equally great effect on the terrestrial population. By the end of the Permian, about 90–96% of all marine invertebrate species were extinct, compared with more than two-thirds of all amphibians and reptiles. Plants, in contrast, apparently did not experience as great a turnover as invertebrates and animals.

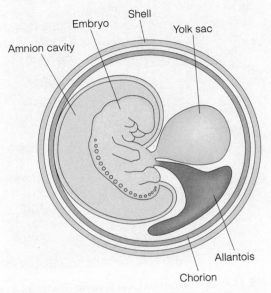

● **FIGURE 18.9 Amniote Egg** In an amniote egg, the embryo is surrounded by a liquid sac (amnion cavity) and provided with a food source (yolk sac) and waste sac (allantois). The evolution of the amniote egg freed reptiles from having to return to the water for reproduction and let them inhabit all parts of the land.

**therapsid** Permian to Triassic mammal-like reptiles; the ancestors of mammals are among one group of therapsids known as cynodonts.

● **FIGURE 18.10 Pelycosaurs** Most pelycosaurs, or finback reptiles, had a characteristic sail on their back. One hypothesis explains the sail as a type of thermoregulatory device. Other hypotheses are that it was a sexual display or a feature to make the reptile look more intimidating. Shown here are the carnivore *Dimetrodon* and the herbivore *Edaphosaurus*.

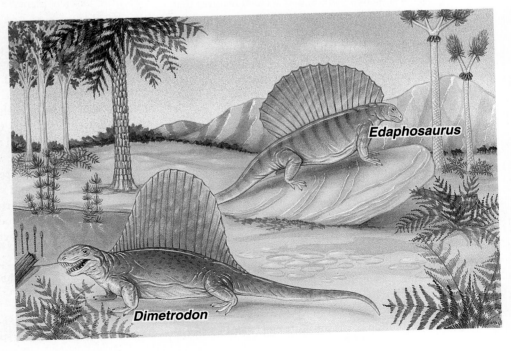

# Plants

Plants encountered many of the same problems animals faced when they made the transition to land: drying out, the effects of gravity, and reproduction. Plants adapted by evolving a variety of structural features that allowed them to invade the land during the Ordovician (●TABLE 18.2).

The most common and widespread land plants are **vascular plants**, which have a tissue system of specialized cells for the movement of water and nutrients. *Nonvascular plants*, such as mosses and fungi, lack these specialized cells and are typically small and usually live in low, moist areas. Nonvascular plants were probably the first to make the transition to land, but their fossil record is poor.

The earliest known vascular land plants are small, leafless, Y-shaped stems from the Middle Silurian of Wales and Ireland. They are known as **seedless vascular plants**, because they did not produce seeds nor did they have a true root system. Although these plants lived on land, they, like the amphibians, never completely solved the problem of drying out and were thus restricted to moist areas. Even their living descendants, such as ferns, are usually found in moist areas. During the Pennsylvanian Period, the seedless vascular plants became abundant and diverse because of the widespread coal-forming swamps (see Figure 17.13) that were ideally suited to their lifestyle (●FIGURE 18.11).

Along with the evolution of diverse seedless vascular plants, another significant floral event occurred during the Devonian. The evolution of the seed at this time liberated vascular plants from their dependence on moist conditions and allowed them to spread over all parts of the land. The first to do so were the flowerless seed plants, or **gymnosperms**, which include the living cycads, conifers, and ginkgoes. Whereas the seedless vascular plants dominated the flora of the Pennsylvanian coal-forming swamps, the gymnosperms made up an important element of the Late Paleozoic flora, particularly in the non-swampy areas.

---

**vascular plant**  A plant with specialized tissue for transporting nutrients and fluids.

**seedless vascular plant**  A plant with specialized tissues for transporting fluids and nutrients that reproduces by spores rather than seeds, as in ferns and horsetail rushes.

**gymnosperm**  A flowerless, seed-bearing plant.

---

**TABLE 18.2** Major Events in the Evolution of Land Plants

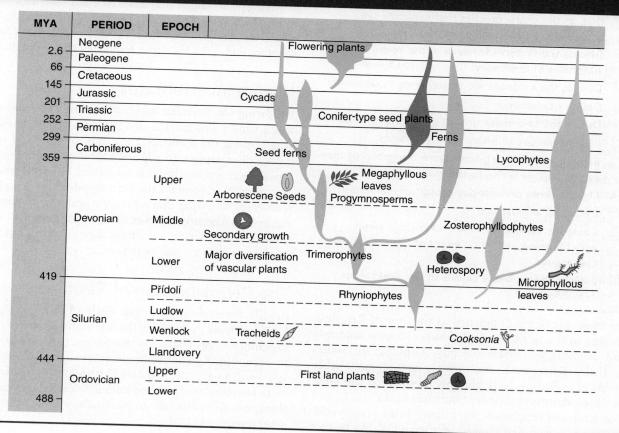

**● FIGURE 18.11 Pennsylvanian Coal Swamp** Reconstruction of a Pennsylvanian coal swamp with its characteristic vegetation of seedless vascular plants.

# MESOZOIC LIFE HISTORY

LO16 Define archosaurs

LO17 Discuss the characteristics that separate saurischian from ornithischian dinosaurs

LO18 Discuss the features for each of the seven suborders of dinosaurs

LO19 Discuss the evidence for endothermy in dinosaurs

LO20 Define pterosaurs

LO21 Define the two major groups of marine reptiles

LO22 Discuss the evolutionary history of birds

LO23 Explain why the cynodonts were the group of theraspids that gave rise to the mammals

LO24 List the three present-day groups of mammals

LO25 Define an angiosperm

LO26 Compare the Permian mass extinction event with the mass extinction event at the end of the Cretaceous

LO27 Discuss the probable causes of the Cretaceous extinction

The Mesozoic Era is designated as the "Age of Reptiles," alluding to the fact that reptiles were the most common land-dwelling vertebrate animals. Although dinosaurs and their relatives evoke considerable interest, many other groups of organisms were not only present but also quite common. Many invertebrates flourished in the seas as well as on land, and vertebrates other than reptiles proliferated. For instance, mammals evolved from mammal-like reptiles during the Triassic, and birds probably evolved from small carnivorous dinosaurs during the Jurassic.

Important changes also took place in plants when the first flowering plants (angiosperms) evolved and soon became the most common plants on land. The end of the Mesozoic witnessed another global extinction event that like its predecessor at the end of the Paleozoic, resulted in mass extinctions.

## Marine Invertebrates

Following the Permian mass extinctions, the Mesozoic was a time when marine invertebrates repopulated the seas. Among the mollusks, the clams, oysters, and snails became increasingly diverse and abundant, and the cephalopods were among the most important Mesozoic invertebrate groups. In contrast, the brachiopods never completely recovered from their near extinction and have remained a minor invertebrate group ever since. In areas of warm, clear, shallow marine waters, corals again proliferated, but these corals were of a new and more familiar type.

## The Diversification of Reptiles

Reptile diversification began during the Mississippian Period with the evolution of the first animals capable of laying amniote eggs which was one feature that allowed them to occupy many environments unavailable to amphibians. From a basic stock of Late Paleozoic *stem reptiles*, two main branches emerged, one leading to the pelycosaurs and therapsids and eventually to mammals, and the other to all other reptiles and birds.

## ARCHOSAURS AND THE ORIGIN OF DINOSAURS

Reptiles known as **archosaurs** (*archo*, "ruling," and *sauros*, "lizard") include crocodiles, pterosaurs (flying reptiles), dinosaurs, and birds (●FIGURE 18.12). Incorporating such diverse animals in a single group implies that they share a common ancestor, and indeed several characteristics unite them.

All **dinosaurs** possess several shared characteristics yet differ enough for us to recognize two distinct orders: **Saurischia** and **Ornithischia**. Each order has a distinctive pelvic structure. Saurischian dinosaurs have a lizardlike pelvis and are thus called *lizard-hipped dinosaurs*, whereas ornithischians have a birdlike pelvis and are called *bird-hipped dinosaurs* (●FIGURE 18.13). The probable ancestors of dinosaurs were archosaurs. The earliest dinosaurs (●FIGURE 18.14) had evolved by the Late Triassic and were small (about 1 m long), long-legged carnivores that walked and ran on their hind limbs, so they were *bipedal*, as opposed to *quadrupedal* animals that moved on all four limbs.

**DINOSAURS** Before discussing the dinosaurs, it is important to note that we can only give a basic outline, and by the time you read this some of the information might be dated. There is a tremendous amount of research being conducted by geologists, paleontologists, and biologists, such that new discoveries are constantly being made that provide greater insight into the history of these animals and their evolution.

The term *dinosaur* was proposed by Sir Richard Owen in 1842 to mean "fearfully great lizard," although now "fearfully" has come to mean "terrible"—thus the characterization of dinosaurs as "terrible lizards." Of course, we have no reason to think that they were any more terrible than animals

---

**archosaur** A term referring to the ruling reptiles—dinosaurs, flying reptiles (pterosaurs), crocodiles, and birds.

**dinosaur** Any of the Mesozoic reptiles belonging to the orders Saurischia and Ornithischia.

**Saurischia** An order of dinosaurs characterized by a lizardlike pelvis; includes theropods, prosauropods, and sauropods.

**Ornithischia** One of the two orders of dinosaurs, characterized by a birdlike pelvis; includes ornithopods, stegosaurs, ankylosaurs, pachycephalosaurs, and ceratopsians.

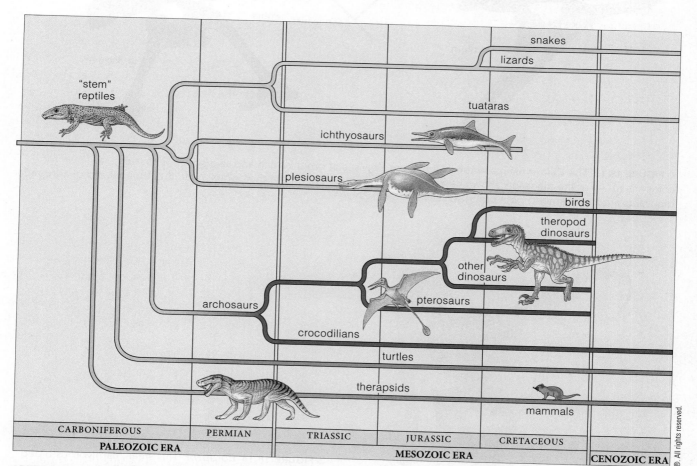

●**FIGURE 18.12 Evolution of the Amniotes** All of the organisms shown here are amniotes—that is, animals that lay eggs. The living amniotes are snakes, lizards, tuataras, birds, turtles, and mammals.

**Critical Thinking Question** Other than their pelvises, are there other anatomical feature you can cite to differentiate the two orders of dinosaurs?

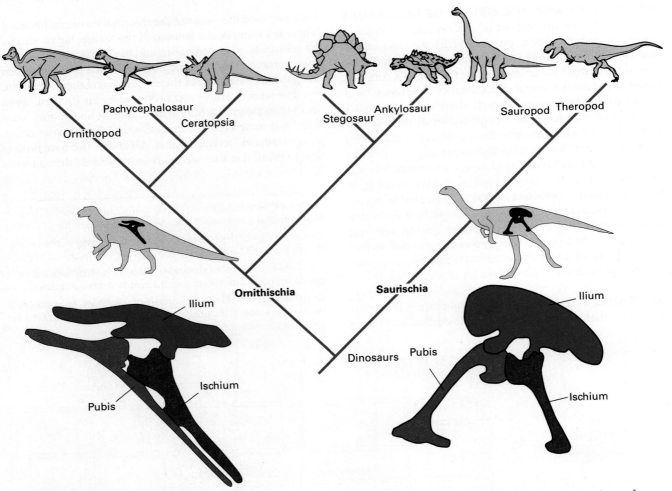

● **FIGURE 18.13 The Relationships among Dinosaurs** Pelvises of ornithischian and saurischian dinosaurs are shown for comparison. All of the dinosaurs shown here were herbivores, except for the theropods. Note that bipedal and quadrupedal dinosaurs are found among both ornithischians and saurischians.

De Agostini Picture Library/De Agostini/Getty Images

● **FIGURE 18.14 *Eoraptor*** This small, carnivorous biped was one of the earliest dinosaurs. These animals, or ones very much like them, are the probable ancestors of all later dinosaurs. *Eoraptor* lived during the Late Triassic (about 230 million years ago) in what is now Argentina. It was only about a meter long and likely weighed no more than 10 kg.

living today, nor were they lizards. Nevertheless, these ideas persist and the popularization of dinosaurs in movies and television shows has often been inaccurate and has contributed to misunderstandings. For instance, many people think that all dinosaurs were large. It is true that *many* were large, but, actually, dinosaurs varied from giants weighing several tens of metric tons to others no larger than a chicken.

Although various media portray dinosaurs as active animals, the misconception that they were lethargic, dimwitted beasts still persists. Evidence now available indicates that some were quite active and perhaps even warm-blooded. It also appears that various species cared for their young long after hatching, a behavioral characteristic most often found in birds and mammals. Many questions remain unanswered about dinosaurs, but their fossils and the rocks that contain them are revealing more and more about their evolutionary relationships and behavior.

Among the saurischian dinosaurs, scientists define two suborders: theropods and sauropods. All *theropods* were bipedal carnivores that ranged in size from tiny *Compsognathus* to comparative giants such as *Tyrannosaurus* and similar but even larger species (●**FIGURE 18.15A**). Some of the smaller theropods such as *Velociraptor* and its relative *Deinonychus*, with a large, sickle-shaped claw on each hind foot, likely used these claws in a slashing type of attack. Some remarkable discoveries beginning in 1996 and continuing today by Chinese paleontologists have yielded a number of species of small theropods with feathers.

Included among the sauropods are the giant, quadrupedal herbivores such as *Apatosaurus, Diplodocus,* and *Brachiosaurus,* the largest land animals of any kind (●**FIGURE 18.15B**). According to one estimate, *Brachiosaurus* weighed more than 75 metric tons, and partial remains indicate that even larger sauropods existed. Evidence from trackways shows that sauropods moved in herds.

The diversity of ornithischians is evident from the five distinct suborders recognized: ornithopods, pachycephalosaurs, ankylosaurs, stegosaurs, and ceratopsians (Figure 18.13). All ornithischians were herbivores, but some were bipeds, whereas others were quadrupeds.

a)

b)

● **FIGURE 18.15 Saurischian Dinosaurs** (a) *Gigantosaurus* was a huge theropod that lived during the Late Cretaceous in Argentina. It was 12 to 13 m long, weighed approximately 8.2 metric tons, and lived 30 million years before the more familiar *Tyrannosaurus rex*. (b) *Brachiosaurus* was a large herbivorous sauropod that lived in North America during the Late Jurassic. It is estimated to have been 26 to 30 m long, 12 to 16 m high, and had a calculated weight between 28 to 58 metric tons.

In fact, although ornithopods such as the well-known duck-billed dinosaurs were primarily bipeds, their well-developed forelimbs allowed them to walk on all fours as well (●**FIGURE 18.16A**). Also included among the ornithischians were the heavily armored ankylosaurs (●**FIGURE 18.16B**); the stegosaurs, with plates on their backs as well as spikes on their tails for defense (●**FIGURE 18.16C**); the horned ceratopsians (●**FIGURE 18.16D**); and the peculiar, domeheaded pachycephalosaurs (●**FIGURE 18.16E**).

a)

Douglas Henderson

d)

Renegade 9 /National Geographic Image Collection

b)

Warpaint/Shutterstock.com

e)

Sergey Krasovskiy/Stocktrek Images/Getty Images

c)

leonello calvetti/Alamy Stock Photo

● **FIGURE 18.16 Ornithischian Dinosaurs** (a) The ornithopod dinosaurs were abundant and varied. *Maisaura* nested in colonies and tended its young long after hatching. (b) *Ankylosaurus* is a genus of the heavily armored ankylosaurs that lived during the Late Cretaceous in North America. It was 6 to 8 m long, 1.7 m in height, and weighted between 4 and 7 metric tons. (c) *Stegosaurus* is noted for the rows of plates on its back and the bony spikes at the end of its tail. Stegosaurs died out by the end of the Jurassic Period. (d) *Styracosaurus* was one of several ceratopsian dinosaurs that were common in North America during the Late Cretaceous. This dinosaur was about 5.5 m long and weighed 2.7 metric tons. (e) The pachycephalosaurs, represented here by *Stegoceras*, had dome-shaped skulls from thickening of the bones. *Stegoceras* was small, measuring only 2 m long, but some of its relatives were up to 4.5 m long.

**WARM-BLOODED DINOSAURS?** Were dinosaurs *endotherms* (warm-blooded) like mammals and birds, or were they *ectotherms* (cold-blooded) like all of today's reptiles? Almost everyone now agrees that some compelling evidence exists for dinosaur endothermy.

Bones of endotherms typically have numerous passageways that, when the animals are alive, contain blood vessels, but considerably fewer passageways are present in bones of ectotherms. Proponents of dinosaur endothermy note that dinosaur bones are more similar to those of living endotherms. Crocodiles and turtles have this so-called endothermic bone, yet they are ectotherms, and some small mammals have bones more typical of ectotherms. It may be that bone structure is related more to body size and growth patterns than to endothermy, so this evidence is inconclusive.

Endotherms must eat more than comparable-size ectotherms, because their metabolic rates are so much higher. Consequently, endothermic predators require large prey populations and thus constitute a much smaller proportion of the total animal population than their prey, usually only a few percent. In contrast, the proportion of ectothermic predators to prey may be as high as 50%. Where data are sufficient to allow an estimate, dinosaur predators made up 3–5% of the total population. Nevertheless, uncertainties in the data make this argument less than convincing for many paleontologists.

Endothermy seems to be a prerequisite for having a large brain, because a complex nervous system needs a rather constant body temperature. Some dinosaurs did have large brains compared to body size, especially the small- and medium-sized carnivores, but many did not. So brain size for some dinosaurs may be a convincing argument, but even more compelling evidence for theropod endothermy comes from their probable relationship to birds, and the recent discoveries in China of theropods with feathers or a featherlike outer covering. Today, only endotherms have hair, fur, or feathers for insulation.

There are good arguments for endothermy for several types of dinosaurs, particularly theropods. Although the large sauropods were probably not endothermic, they nevertheless were capable of maintaining a rather constant body temperature. Large animals heat up and cool down more slowly than smaller ones, because they have a small surface area compared to their volume. With their comparatively smaller surface area for heat loss, sauropods most likely retained heat more effectively than their smaller relatives.

In general, a fairly good case can be made for endothermy in many theropods and in some ornithopods. However, disagreement exists, and for some dinosaurs the question is still open.

**FLYING REPTILES** Paleozoic insects were the first animals capable of flight, but the first among vertebrates were the **pterosaurs**, or flying reptiles. They were common in the skies from the Late Triassic until their extinction at the end of the Cretaceous (●**FIGURE 18.17**). Adaptations

---

**pterosaur** Any of the Mesozoic flying reptiles.

a)

b)

●**FIGURE 18.17 Pterosaurs (Flying Reptiles)**
(a) *Pterodactylus* from the Late Jurassic is a good example of the earlier pterosaurs. It has a long tail, numerous teeth, and a wingspan of about 1.5 m.
(b) Later pterosaurs such as *Pteranodon* of the Late Cretaceous were much larger, tailless, and had toothless beaks. *Pteranodon's* wingspan was about 6 m.

for flight include a wing membrane supported by an elongated fourth finger; light, hollow bones; and development of those parts of the brain associated with muscular coordination and sight.

Pterosaurs ranged from the size of a sparrow to *Quetzalcoatlus northropi*, one of the largest flying animals yet unearthed with a wingspan of 12 m, and nearly as tall as a giraffe. The comparatively large pterosaurs probably took advantage of thermal updrafts to stay airborne, mostly by soaring but occasionally by flapping their wings for maneuvering. Smaller pterosaurs probably stayed aloft by vigorously flapping their wings, just as present-day small birds do. The fact that new discoveries, especially in China, of specimens with the impressions of a coat of hair or hairlike feathers suggests that possibly all pterosaurs were endotherms.

A recent surge in fossil finds is changing our views of pterosaurs. For example, they display a wide variety of beak shapes, some of which have teeth, and others that don't, implying that such variation might be based on dietary needs. The purpose of the different head crests of pterosaurs, and even what their coloring and patterns were like, is still being debated. One suggestion is that they might have been useful for attracting mates. The discovery of a bone bed in Brazil that contained more than 45 pterosuars of a single species is thought to be convincing evidence that some pterosaurs lived in colonies.

**MARINE REPTILES** Probably the most familiar of the Mesozoic marine reptiles are the rather porpoiselike **ichthyosaurs** (●FIGURE 18.18A). These fully aquatic animals ranged in size from only 0.7 to 15 m. All ichthyosaurs had a streamlined body, a powerful tail for propulsion, and flipperlike forelimbs, which are simply modified forelimbs of their land-dwelling ancestors, for maneuvering.

Their numerous sharp teeth indicate they were fish eaters, although they undoubtedly preyed on other marine organisms as well. Ichthyosaurs were so completely aquatic that it is doubtful they could come onto land, so females probably retained eggs within their bodies and gave birth to live young. A few fossils with small ichthyosaurs within the appropriate part of the body cavity support this interpretation.

An interesting side note on ichthyosaurs is the story of Mary Anning (see GEO-FOCUS), who, when she was only 11 years old, discovered and excavated a nearly complete ichthyosaur in southern England.

Another well-known group, the **plesiosaurs**, belonged to one of two subgroups: short-necked and long-necked (●FIGURE 18.18B). Most were modest-sized 3.6 to 6 m long, but one species found in Antarctica measures 15 m. Short-necked plesiosaurs may have been bottom feeders, but their long-necked cousins probably used their necks in a snakelike fashion to capture fish with their numerous sharp teeth. These animals probably came ashore to lay their eggs.

## Birds

Since the discovery of feathered dinosaurs beginning in the mid-1990s, it has become increasingly difficult to make a distinction between dinosaurs and birds. Although living birds do not resemble any of today's reptiles, the best current evidence indicates that they evolved from a group of small theropods. Both reptiles and birds lay shelled amniote eggs, they also share several skeletal features, such as the way the jaw attaches to the skull, and several fossils point to a close relationship between birds and theropods.

Since 1860, about 10 fossils of a birdlike animal called *Archaeopteryx* (from the Greek *archaeo*, "ancient," and *pteryx*, "feather") have been recovered from the Jurassic Solnhofen Limestone in Germany (●FIGURE 18.19). *Archaeopteryx* had feather impressions and a wishbone and fused clavicles so typical of birds, but most of its skeletal features,

---

**ichthyosaur** Any of the porpoiselike, Mesozoic marine reptiles.

**plesiosaur** A type of Mesozoic marine reptile; short-necked and long-necked plesiosaurs existed.

a)

b)

● **FIGURE 18.18 Mesozoic Marine Reptiles** (a) Icthyosaur. (b) A long-necked plesiosaur.

# GEO-FOCUS

## Mary Anning's contributions to paleontology

Paleontologists use fossils to study life of the past, so part of their efforts are spent in finding and collecting fossils. Western European men dominated the early history of this field, but this situation no longer prevails. Men and women from many countries are now making significant contributions. Perhaps the most notable early exception is Mary Anning (1799–1847), who began a remarkable career as a fossil collector when she was only 11 years old.

Mary Anning was born in Lyme Regis on England's southern coast. When only 15 months old, she survived a lightning strike that, according to one report, killed three girls and, according to another report, killed a nurse tending

her. In 1810, Mary's father, a cabinet maker who also sold fossils part time, died, leaving the family nearly destitute. Mary Anning (●FIGURE 1) expanded the fossil business and became a professional fossil collector known to paleontologists of her time, some of whom visited her shop to buy fossils or gather information. She collected fossils from the Dorset coast near Lyme Regis and is reported to have been the inspiration for the tongue twister,

She sells seashells on the seashore
The shells she sells are seashells, I'm sure
So if she sells seashells on the seashore
Then I'm sure she sells seashore shells.

Soon after her father's death, Mary Anning made her first important discovery—a nearly complete skeleton of a Jurassic ichthyosaur, which was described in 1814 by Sir Everard Home. The sale of this fossil specimen provided considerable financial relief for her family.

In 1821, she made a second major discovery and excavated the remains of a plesiosaur. And in 1828, she found the first pterosaur in England, which was sent to the eminent geologist William Buckland at Oxford University.

By 1830, Mary Anning's fortunes began declining as

collectors and museums had fewer funds with which to buy fossils. In fact, she may once again have become destitute were it not for her geologist friend Henry Thomas de la Beche, also a resident of Lyme Regis. De la Beche drew a fanciful scene called *Duria antiquior*, meaning "An earlier Dorset," in which he brought to life the fossils Mary Anning had collected. The scene was made into a lithograph that was printed and sold widely, and its proceeds went directly to Mary Anning.

Mary Anning died of cancer in 1847, and although only 48 years old, she had a fossil-collecting career that spanned 37 years. Her contributions to paleontology are now widely recognized, but, unfortunately, soon after her death, she was mostly forgotten. Apparently, the people who purchased her fossils were credited with finding them.

It didn't occur to them to credit a woman from the lower classes with such astonishing work. So an uneducated little girl, with a quick mind and an accurate eye, played a key role in setting the course of the 19th-century geologic revolution. Then—we simply forgot about her.[*]

Following Mary Anning's death, her contributions to paleontology were mostly forgotten. Charles Dickens wrote about her life in 1865, but otherwise little notice was taken of her work.

However, since 2002, the Palaeontological Association has presented the Mary Anning Award to someone who is "not professionally employed within palaeontology but who made an outstanding contribution to the subject."[**] She was further recognized in 2010 when the Royal Society of London named her as one of the 10 British women who have most influenced the history of science.

Sue Monroe

● **FIGURE 1** Mary Anning, who lived in Lyme Regis on England's southern coast, began collecting and selling fossils when she was only 11 years old. She made several important discoveries during the early 1800s, but was largely forgotten after her death in 1847.

---

[*]   John H. Lienhard, Professor, College of Engineering, University of Houston.

[**]   The Palaeontological Association, http://www.palass.org/modules.php?name=palaeo &sec=Awards&page=122.

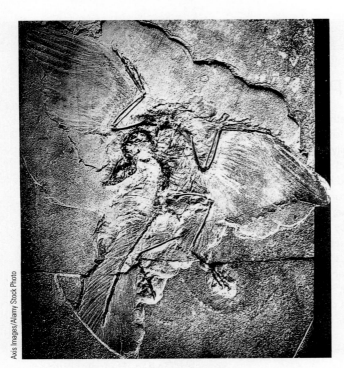

● **FIGURE 18.19 Fossils of *Archaeopteryx* from the Jurassic-Age Solnhofen Limestone of Germany** Notice the feather impressions on the wings and the long tail. This animal had feathers and a wishbone, making it a bird, but in most details of its anatomy, it resembled small theropod dinosaurs. For example, it had reptilian teeth, claws on its wings, and a long tail—none of which are found on birds today.

teeth, clawed wings, and a long tail are much more like those of small theropods.

Feathers and a wishbone may seem like definitive features, but even these are no longer sufficient to distinguish birds from theropods, because some theropods had wishbones consisting of fused clavicles, which is a feature that is typical of birds. Furthermore, recent discoveries of theropods in China show some kind of feathery covering, further reinforcing the connection between theropods and birds. Unfortunately, the fossil record of *Archaeopteryx* is not good enough to resolve whether it is the ancestor of today's birds or an animal that died out without leaving descendants. Of course, that in no way diminishes the fact that it shows both reptilian and avian characteristics.

## Mammals

In a previous section, we briefly mentioned therapsids, or advanced mammal-like reptiles. One particular group of therapsids known as **cynodonts** was the most mammal-like of all and gave rise to mammals during the Late Triassic. This transition is especially well documented by fossils and is so gradual that classification of some fossils as either reptile or mammal is difficult.

The first mammals retained several reptilian characteristics but had mammalian features as well. A good example is their lower jaw. In typical mammals, the lower jaw is a single bone, whereas in reptiles it consists of several bones. The first mammals retained more than one bone in the lower jaw but had teeth and a jaw–skull joint characteristic of mammals. In fact, the quadrate and articular bones that formed the jaw–skull joint in reptiles were modified and became the incus and amalleus, two small bones in the mammalian middle ear. Another typical mammalian feature is occlusion, which means that the chewing teeth meet surface to surface to allow grinding, a feature also found in some advanced cynodonts.

In short, some mammalian features evolved more rapidly than others, thereby accounting for animals that have characteristics of both reptiles and mammals. Even though mammals appeared at the same time as dinosaurs, their diversity remained low, and most of them were small animals during the rest of the Mesozoic Era.

Mammals today belong to one of three groups: The *monotremes*, or egg-laying mammals; the *marsupial mammals*, or pouched animals; and the *placental mammals*, which includes most mammals with which we are familiar. The history of the monotremes is uncertain, but the marsupials and placentals diverged from a common ancestor during the Late Cretaceous (●**FIGURE 18.20**).

## Plants

Just as during the Late Paleozoic, seedless vascular and gymnosperms dominated the Triassic and Jurassic land-plant communities, and in fact, representatives of both groups are still common today, but are less abundant than in the past. Among the gymnosperms, the large seed ferns became extinct at the end of the Triassic, cycads, which superficially resemble palms evolved, ginkgos remained abundant and still exist in isolated regions, and the conifers continued to diversify and are now widespread in some terrestrial habitats, particularly at high elevations.

The long dominance of seedless vascular plants and gymnosperms ended during the Early Cretaceous, when many were replaced by **angiosperms**, or flowering plants (Table 18.2). Studies of fossil and living gymnosperms show that they are closely related to angiosperms, but unfortunately, the early fossil record of angiosperms is sparse, so their precise ancestry remains obscure.

Since they first evolved, angiosperms have adapted to nearly every terrestrial habitat from mountains to deserts. Some have even adapted to shallow coastal waters. Their reproduction, involving flowers to attract animal pollinators and the evolution of enclosed seeds, largely accounts for their success. With 250,000 to 300,000 species, angiosperms comprise more than about 90% of all land plant species.

---

**cynodont** A type of carnivorous therapsid (advanced mammal-like reptile); ancestors of mammals are among cynodonts.

**angiosperm** Vascular plants that have flowers and seeds; the flowering plants.

a)

b)

● **FIGURE 18.20 Mesozoic Mammals** Both of these fossils come from Early Cretaceous-age rocks in China. (a) Restoration of the oldest known marsupial mammal, *Sinodelphym*, which measures 15 cm long. (b) Restoration of *Eomaia*, the oldest known placental mammal. It was only 12 or 13 cm long.

## Cretaceous Mass Extinctions

The mass extinctions at the close of the Mesozoic were second in magnitude only to those at the end of the Paleozoic. Casualties of the Mesozoic extinctions include dinosaurs, flying reptiles, marine reptiles, and several kinds of marine creatures such as ammonites.

A hypothesis for these extinctions is based on a discovery at the Cretaceous–Paleogene boundary in Italy—a clay layer 2.5 cm thick with an abnormally high concentration of the platinum-group element iridium. Since this discovery, high iridium concentrations have been identified at many other Cretaceous–Paleogene boundary sites (●**FIGURE 18.21**). The significance of this discovery lies in the fact that iridium is rare in crustal rocks but occurs in much higher concentrations in some meteorites. Several investigators proposed a meteorite impact to explain this iridium anomaly and further postulated that the impact of a large meteorite, perhaps 10 km in diameter, set in motion a chain of events that led to extinctions.

The meteorite-impact scenario goes something like this: On impact, about 60 times the mass of the meteorite was blasted from Earth's crust high into the atmosphere, and the heat generated at impact started raging fires that added more particulate matter to the atmosphere. Sunlight was blocked

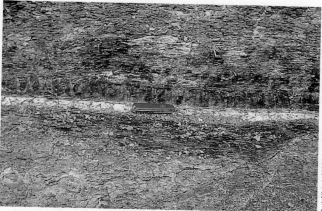

● **FIGURE 18.21 End of Mesozoic Extinctions** Close-up view of the iridium-rich Cretaceous–Paleogene boundary clay in the Raton Basin, New Mexico.

for several months, causing a temporary cessation of photosynthesis; food chains collapsed and extinctions followed. In addition, with sunlight greatly diminished, Earth's surface temperatures were drastically reduced and could have added to the biological stress.

Some now claim that a probable impact site has been found centered on the town of Chicxulub on the Yucatán Peninsula of Mexico. The structure is about 180 km in diameter and lies beneath layers of Cenozoic sedimentary rock.

Most geologists now concede that a large meteorite impact occurred, but we also know that vast outpourings of lava were taking place in what is now India. Perhaps these eruptions brought about detrimental atmospheric changes. Furthermore, the vast, shallow seas that covered large parts of the continents had mostly withdrawn by the end of the Cretaceous, and the mild, equable Mesozoic climates became harsher and more seasonal by the end of that era.

# CENOZOIC LIFE HISTORY

LO28    Discuss the importance of large birds during the Cenozoic Era

LO29    Define each of the three living mammal groups

LO30    Explain the relationship between therapsids and the earliest mammals

LO31    Discuss the evolutionary history of mammals during the Cenozoic Era

LO32    Describe the Pleistocene mammalian faunas

LO33    Define a primate

LO34    Explain the difference between the prosimians and anthropoids

LO35    Discuss the differences between a hominoid, hominid, and hominin

LO36    Discuss the evolutionary history of hominins

LO37    Explain the differences between Neanderthals and Cro-Magnons

Even though we emphasize the evolution of mammals in this section, you should be aware of other important life events. Flowering plants continued to dominate land plant communities, the present-day groups of birds evolved during the Early Paleogene, and some marine invertebrates continued to diversify, eventually giving rise to today's marine fauna.

Mammals coexisted with dinosaurs for more than 100 million years, yet their Mesozoic fossil record indicates that they were not abundant, diverse, or very large. The largest one known was 1 m long and weighed 12 to 14 kg. Extinctions at the end of the Mesozoic eliminated dinosaurs and some of their relatives, thereby creating the adaptive opportunities that mammals quickly exploited. The "Age of Mammals," as the Cenozoic Era is commonly called, had begun.

## Marine Invertebrates and Phytoplankton

Survivors of the Mesozoic extinctions populated the Cenozoic marine ecosystem. Especially abundant among the invertebrates were single-celled organisms such as foraminifera and radiolarians, corals, bryozoans, mollusks, and echinoids. Only a few species of phytoplankton survived into the Cenozoic, but those that did flourished and diversified.

Marine and freshwater plants called *diatoms*, with silica skeletons, were particularly abundant.

## Cenozoic Birds

Birds today vary considerably in diet, habitat, adaptations, and size. Nevertheless, their basic skeletal structure remained remarkably constant throughout the Cenozoic. Given that birds evolved from a creature very much like *Archaeopteryx* (Figure 18.19), this uniformity is not surprising because adaptations for flying limit variations in structure. Penguins adapted to an aquatic environment, and in some large, extinct and living flightless birds, the skeleton became robust and the wings were reduced to vestiges.

In fact, following the demise of the dinosaurs at the end of the Mesozoic Era, the dominant large, land-dwelling predators during the Paleocene and well into the Eocene were flightless birds. Among these predators were giants such as *Titanis*, which stood 2.5 m tall and had a huge head and beak, toes with large claws, and small, vestigial wings (●FIGURE 18.22). Its massive, short legs indicate that *Titanis* was not very fast, but neither were the mammals it preyed upon. This extraordinary bird and related genera were widespread in North America and Europe, and in South America, they were important predators until they died out 2 million years ago.

## Diversification of Mammals

Among living mammals, **monotremes**, such as the platypus, lay eggs, whereas marsupials and placentals give birth to live young. **Marsupial mammals** are born in an immature,

---

**monotreme** Any of the egg-laying mammals; includes only the platypus and spiny anteater of the Australian region.

**marsupial mammal** Pouched mammals, such as kangaroos and wombats, that give birth to their young in a very immature state.

●**FIGURE 18.22 Cenozoic Flightless Bird** Restoration of *Titanis walleri*, a predatory flightless bird that stood 2.5–3.0 m high and weighed about 150 kg. It lived during the Pliocene and Pleistocene, mostly in South America, but it migrated to North America, where its remains are found in Texas and Florida.

almost embryonic condition, and then develop further in their mother's pouch. **Placental mammals** have no shelled egg, but they have developed a placenta within the uterus across which nutrients and oxygen are carried from the mother to the developing embryo. As a result, placental mammals are much more fully developed than marsupials before birth.

A measure of the success of placental mammals is that more than 90% of all mammals, fossil and living, are placental. Judging from the fossil record, monotremes have never been very common; the only living ones are platypuses and spiny anteaters of the Australian region. Marsupials have been more successful in terms of number of species and geographic distribution, but even they have been largely restricted to South America and the Australian region.

As we have already noted, mammals first evolved during the Triassic, and by the Late Cretaceous the marsupials and placentals diverged from a common ancestor (●FIGURE 18.23). Then, following the Late Cretaceous extinctions, mammals began a major adaptive radiation that continued throughout the Cenozoic Era. Several groups of Paleocene mammals are *archaic*, meaning that they were holdovers from the Mesozoic Era or that they did not give rise to any of today's mammals (●FIGURE 18.24A). But also among these mammals were the first rodents, rabbits,

**placental mammal** All mammals with a placenta to nourish the developing embryo, as opposed to egg-laying mammals (monotremes) and pouched mammals (marsupials).

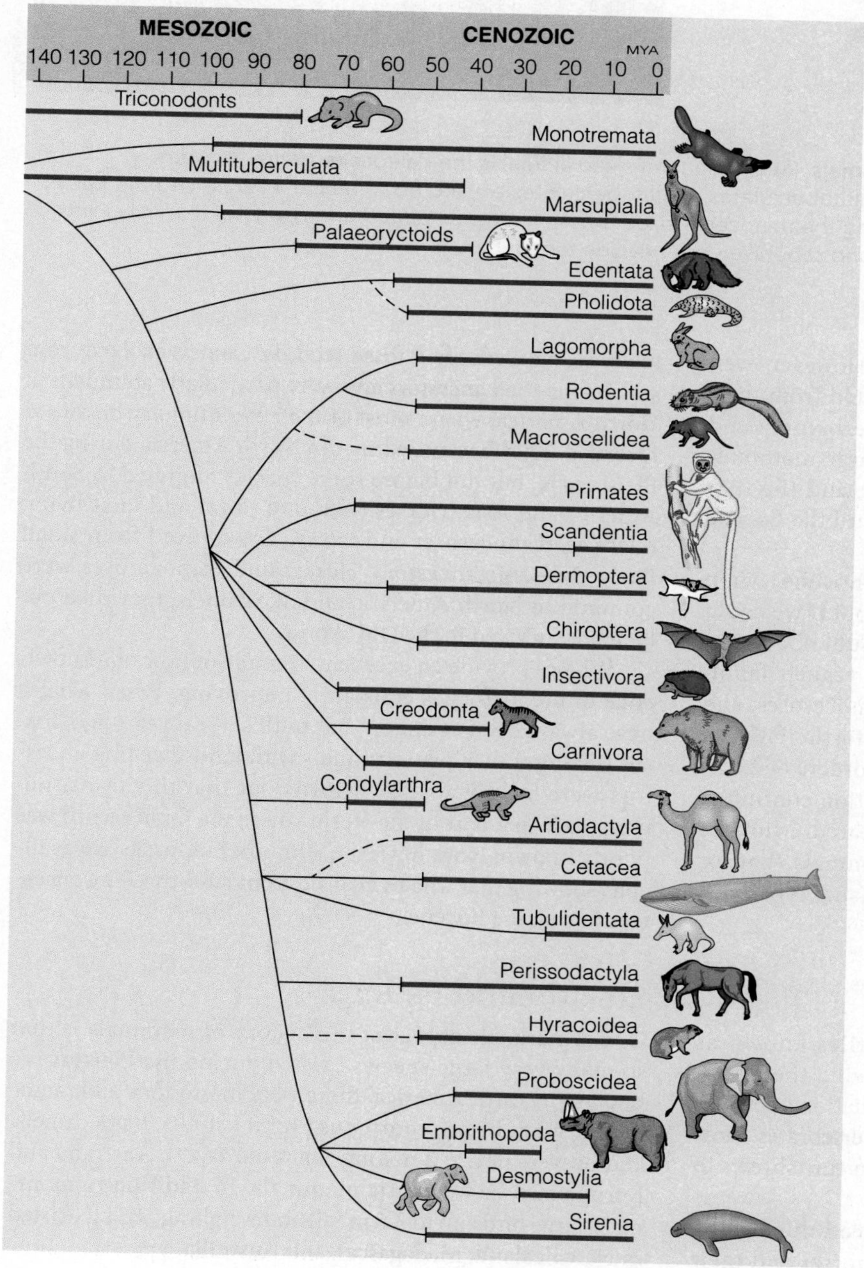

●**FIGURE 18.23 Diversification of Mammals** Mammals existed during the Mesozoic, but most placental mammals diversified during the Paleocene and Eocene epochs. Among the living orders of mammals, all are placentals except for the monotremes and marsupials. Several extinct orders are not shown. Bold lines indicate actual geologic ranges, whereas the thinner lines indicate the inferred branching of the groups.

**FIGURE 18.24 Paleocene and Eocene Mammals** (a) This squirrel-sized animal is the Paleocene genus *Ptilodus*, a member of a group of mammals known as multituberculates. Multituberculates existed from the Early Jurassic to the Early Oligocene. (b) Scene from the Eocene showing the rhinoceros-sized animal known as *Unitatherium*, one of the first giants among mammals. It had three pairs of horns and saberlike upper canine teeth.

primates, carnivores, and hoofed mammals. However, even these had not yet become clearly differentiated from their ancestors, and the differences between herbivores and carnivores were slight. Most were small; large mammals were not present until the Late Paleocene, and the first giant terrestrial mammals did not appear until the Eocene (**FIGURE 18.24B**).

Diversification continued during the Eocene, when several more types of mammals appeared, but if we could go back and visit that time, we probably would not recognize many of these animals. Some would be vaguely familiar, but the ancestors of horses, camels, rhinoceroses, and elephants would bear little resemblance to their living descendants. By Oligocene time, all of the orders of existing mammals were present, but diversification continued as more familiar families and genera appeared. Miocene and Pliocene mammals were mostly mammals that we could readily identify, although a few unusual types still existed (**FIGURE 18.25**).

## Cenozoic Mammals

Mammals arose from mammal-like reptiles known as *cynodonts* during the Late Triassic. Following the Mesozoic extinctions, they began an adaptive radiation and soon became the most abundant land-dwelling vertebrates. Now more than 4,000 species exist, ranging from tiny shrews to elephants and whales.

Numerous groups of mammals evolved during the Cenozoic, and some, such as camels and horses and their relatives, have excellent fossil records. Camels evolved from small, four-toed ancestors and were particularly abundant in North America, where most of their evolutionary history is recorded. They became extinct in North America during the Pleistocene, but not before some species migrated to South America and Asia. Horses (**FIGURE 18.26**) and their living relatives, rhinoceroses and tapirs, also evolved from small Early Cenozoic ancestors. Horses and rhinoceroses were common in North America, and like camels, they died out here but survived in the Old World.

Whales provide an excellent example of how our knowledge of life history is constantly improving. Fossil whales have always been common, but until several years ago, fossils linking fully aquatic animals with land-dwelling ancestors were largely absent. It turns out that this transition took place in a part of the world where the fossil record was poorly known. Now, however, a number of fossils are available, showing that whales evolved from land-dwelling ancestors during the Eocene.

## Pleistocene Faunas

One remarkable aspect of the history of mammals is that so many very large species existed during the Pleistocene Epoch. In North America, there were mastodons and mammoths, giant bison, enormous ground sloths, huge camels, and beavers nearly 2 m long (**FIGURE 18.27**). Australia and Europe also had gigantic mammals. In addition to mammals, large birds up to 3.5 m tall and weighing 585 kg existed in New Zealand, Madagascar, and Australia.

**FIGURE 18.25 Recreation of a Paleogene–Neogene North American landscape** Some of the animals shown include oreodonts (*Promerycochoerus*) in the foreground, camels (*Stenomylus*) on the right side, a group of mastodonts (*Mammut*), and a bird.

Spencer Sutton/Science History Images/Alamy Stock Photo

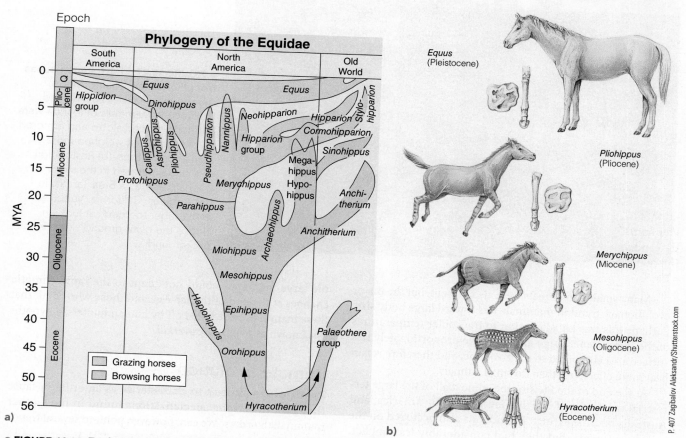

P. 407 Zagibalov Aleksandr/Shutterstock.com

**FIGURE 18.26 Evolution of Horses** (a) Summary chart showing the relationships between the genera of horses. During the Oligocene two lines emerged, one leading to three-toed browsers and the other to one-toed grazers, including the present-day horse, *Equus*. (b) Simplified diagram showing some trends in horse evolution. Trends include a size increase, a lengthening of the limbs, a reduction of the number of toes, and the development of high-crowned teeth with complex chewing surfaces.

a)

b)

● **FIGURE 18.27 Pleistocene Faunas from Florida and California** (a) Among the diverse Pliocene and Pleistocene mammals of Florida were 6-m-long ground sloths and armored glyptodonts that weighed more than 2 metric tons. (b) Restoration of a camel trapped in the sticky tar (asphalt) at the La Brea Tar Pits in Los Angeles, California. Vultures and a saber-toothed cat look on. Notice the giant ground sloths in the background.

Many smaller mammals were also present, but the major evolutionary trend in mammals was toward large body size. Perhaps this was an adaptation to the cooler temperatures of the Pleistocene. Large animals have proportionately less surface area compared to their volume, and therefore retain heat more effectively than do small animals.

At the end of the Pleistocene, almost all of the large terrestrial mammals of North America, South America, and Australia became extinct. Extinctions also occurred on the other continents, but they had considerably less impact. These Pleistocene extinctions were modest by comparison to earlier ones, but they were unusual in that they affected mostly large terrestrial mammals. The debate over the cause of this extinction continues between those who think that

the large mammals could not adapt to the rapid climatic changes at the end of the Ice Age and those who think that these mammals were killed off by human hunters, a hypothesis known as *prehistoric overkill*.

## Primate Evolution

**Primates** are difficult to characterize as an order because they lack the strong specializations found in most other mammalian orders. We can, however, point to several trends

---

**primate** Any of the mammals that belong to the order Primates; includes prosimians (lemurs and tarsiers), monkeys, apes, and humans.

in their evolution that help define primates and that are related to their *arboreal*, or tree-dwelling, ancestry. These trends include changes in the skeleton and mode of locomotion; an increase in brain size; a shift toward smaller, fewer, and less specialized teeth; stereoscopic vision; and the evolution of a grasping hand with opposable thumb. Not all of these trends took place in every primate group, nor did they evolve at the same rate in each group. In fact, some primates have retained certain primitive features, whereas others show all or most of these trends.

The Primate order is divided into two suborders. The *prosimians*, include the lemurs, lorises, tarsiers, and tree shrews. They are generally small, ranging from species that are the size of a mouse up to those as large as a house cat. They are arboreal, have five digits on each hand and foot with either claws or nails, and are typically omnivorous. The prosimians are the oldest primate lineage, with a fossil record extending back to the Paleocene.

Sometime during the Late Eocene, the *anthropoids*, primates that include monkeys, apes, and humans, evolved from a prosimian lineage, and by the Oligocene, the anthropoids were a well-established group with both Old World monkeys (Africa, Asia) and New World monkeys (Central and South America) having evolved during this epoch.

The *hominoids* consist of two families: one includes the gibbons and siamangs, and the other includes the orangutans, gorillas, chimpanzees, and humans and their extinct ancestors. The hominoid lineage diverged from Old World monkeys sometime before the Miocene, but exactly when is still being debated. It is, however, generally accepted that the hominoids evolved in Africa from an ancestral anthropoid group.

Beginning in the Late Eocene, the northward movement of the continents resulted in pronounced climatic shifts. In Africa, Europe, Asia, and elsewhere, a major cooling trend began and the tropical and subtropical rain forests slowly began to change to a variety of mixed forests separated by savannas and open grasslands as temperatures and rainfall decreased. As the climate changed, the primate population also changed. Prosimians and monkeys became rare, whereas hominoids diversified in the newly forming environments and became abundant. During the Miocene Epoch, Africa collided with Europe, producing additional changes in the climate as well as providing opportunities for migration of animals between the two landmasses. Two apelike groups evolved during the Miocene that ultimately gave rise to present-day hominoids.

## Hominids and Hominins

The term **hominids** is now generally used to comprise two subfamilies. The first is the Homininae, which includes gorillas, chimpanzees, humans and their extinct ancestors, and the second, the Ponginae, contains only the orangutans and their extinct ancestors. The Homininae are further subdivided into three tribes, only one of which, the Hominini (humans and their extinct ancestors), concerns

us here. The members of the Hominini are now referred to as **hominins**. Several features serve to distinguish hominins from the other two tribes (gorillas and chimpanzees). Hominins are bipedal; that is, they have an upright posture and walk on two legs rather than four. In addition, they show a trend toward a large and internally reorganized, complex brain. Other features include a reduced face and small canine teeth, omnivorous feeding, increased manual dexterity, and the use of sophisticated tools.

Presently, there is no clear consensus on the evolutionary history of the hominin lineage. This is due, in part, to the incomplete fossil record of hominins, as well as new discoveries, and also because some species are known only from partial specimens or fragments of bone. It is most likely that by the time you read this chapter, new discoveries may change some of the conclusions stated here.

A complete discussion of all of the proposed hominin species and the various competing schemes of their evolutionary history is beyond the scope of this section. We will, however, briefly discuss the generally accepted taxa (●FIGURE 18.28) and present some of the current theories of hominin evolution.

Remember that although the fossil record of hominin evolution is not complete, what exists is well documented. However, it is the interpretation of that fossil record that precipitates the often vigorous and sometimes acrimonious debates concerning our evolutionary history.

Discovered in northern Chad's Djurab Desert, the nearly 7-million-year-old skull and dental remains of *Sahelanthropus tchadensi* (●FIGURE 18.29) make it the oldest known hominin yet unearthed and one that existed at or very near the time when hominins and chimpanzees diverged from a common ancestor. Currently, most paleoanthropologists accept that the human–chimpanzee stock separated from gorillas about 8 million years ago and that hominins separated from chimpanzees about 7 million years ago.

The next oldest hominin is *Orrorin tugenensis*, whose fossils have been dated at 6 million years and consist of bits of jaw, isolated teeth, and finger, arm, and partial upper leg bones. There is currently still debate as to where *O. tugenensis* fits into the hominin lineage, but analysis of the leg bones indicates individuals of this species climbed trees, but also probably walked upright while on the ground.

Sometime between 5.8 and 5.2 million years ago, another hominin, *Ardipithecus kadabba*, was present in eastern Africa. The discovery of a second species of *Ardipithecus*, *A. ramidus*, has caused paleoanthropologists to once again rethink the hominin evolutionary record. This skeleton and skull of a female, standing 1.2 m tall and weighing approximately 50 kg, lived about 4.4 million years ago and shows an interesting mosaic of evolutionary characteristics.

---

**hominid** Abbreviated term for members of the subfamily Homininae, which includes the gorillas, chimpanzees, humans, and their extinct ancestors.

**hominin** Abbreviated term for members of the tribe Hominini, which includes humans and their extinct ancestors.

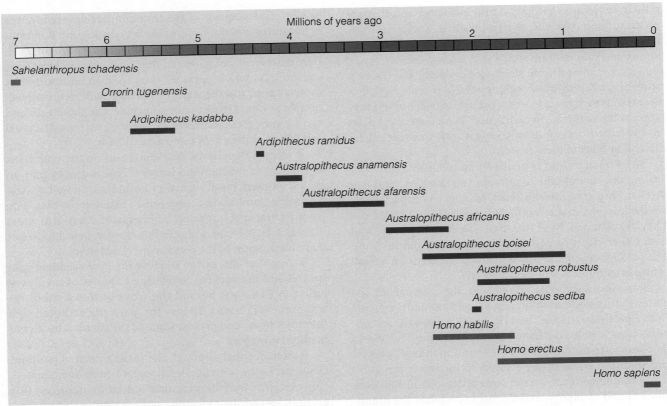

● **FIGURE 18.28 The Stratigraphic Record of Hominins** The geologic ranges for the commonly accepted species of hominins (the tribe of primates that includes present-day humans and their extinct ancestors).

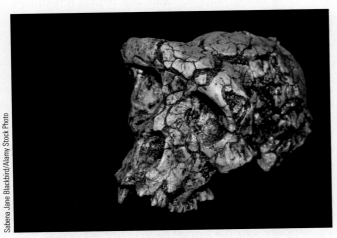

● **FIGURE 18.29** *Sahelanthropus tchadensis* Discovered in Chad in 2002 and dated at nearly 7 million years, this skull of *Sahelanthropus tchadensis* is presently the oldest known hominin.

These include a dexterous hand for grasping and a foot with an opposable big toe which apes use for climbing and maneuvering in trees. Based on these and other features, scientists have concluded that *A. ramidus* was able to walk upright on the ground while retaining some ability to climb and maneuver in trees.

**Australopithecine** is a collective term for all members of the genus *Australopithecus*. Currently, five species are

recognized. However, a recently discovered sixth species (*A. sediba*) may result in a new interpretation of the origin of our own genus, *Homo* (Figure 18.27). Notwithstanding this new discovery, many paleontologists accept the evolutionary scheme in which *Australopithecus anamensis*, the oldest known australopithecine, is ancestral to *A. afarensis* (●**FIGURE 18.30**), who in turn is ancestral to *A. africanus* and the genus *Homo*, as well as the side branch of australopithecines represented by *A. robustus* and *A. boisei*.

The earliest member of our own genus, **Homo**, is *Homo habilis*, who lived 2.5–1.6 million years ago (Figure 18.28). The evolutionary transition from *H. habilis* to *H. erectus* appears to have occurred in a short period of time, between 1.8 and 1.6 million years ago. However, evidence indicating that *H. habilis* and *H. erectus* may have coexisted for approximately 500,000 years has led some scientists to suggest that *H. habilis* and *H. erectus* evolved from a common ancestor and thus represent separate lineages of *Homo*, rather than the traditional linear view of *H. erectus* evolving from *H. habilis*.

**australopithecine** A collective term for all species of the extinct genus *Australopithicus* that existed in South Africa during the Pliocene and Pleistocene.

**Homo** The genus of hominins consisting of *Homo sapiens* and their ancestors *Homo erectus* and *Homo habilis*.

● **FIGURE 18.30 African Pliocene Landscape** Recreation of a Pliocene landscape showing members of *Australopithecus afarensis* gathering and eating various fruits and seeds.

In contrast to the australopithecines and *H. habilis*, which are unknown outside of Africa, *H. erectus* was a widely distributed species that migrated from Africa during the Pleistocene Epoch. The archaeological record indicates that *H. erectus* made tools, used fire, and lived in caves, an advantage for those living in more northerly climates (●**FIGURE 18.31**).

Debate still surrounds the transition from *H. erectus* to our own species, *Homo sapiens* (Figure 18.28). Paleoanthropologists are split into two camps. One camp supports the "out of Africa" hypothesis; that is, the idea that early modern humans originated in Africa and then migrated from Africa, perhaps as recently as 100,000 years ago, populating Europe and Asia and driving the earlier hominin populations there to extinction.

The alternative explanation, the "multiregional" view, maintains that early modern humans did not have an isolated origin in Africa, but rather that they established separate populations throughout Eurasia. Occasional contact and interbreeding between these populations enabled our species to maintain its overall cohesiveness while still preserving the regional differences in people that we see today. Regardless of which theory turns out to be correct, our species, *H. sapiens*, most certainly evolved from *H. erectus*.

Perhaps the most famous of all fossil humans are the *Neanderthals*, who inhabited Europe and the Near East from about 200,000 to 30,000 years ago, and according to the best estimates, never exceeded 15,000 individuals in western Europe. Some paleoanthropologists regard the Neanderthals as a variety or subspecies (*Homo sapiens*

PUBLIPHOTO/Science Source

● **FIGURE 18.31 Pleistocene Landscape with *Homo erectus*** Recreation of a Pleistocene setting in Europe showing members of *Homo erectus* using fire and stone tools at an encampment site.

*neanderthalensis*), whereas others consider them as a separate species (*Homo neanderthalensis*). In any case, their name comes from the first specimen found in 1856 in the Neander Valley near Düsseldorf, Germany.

The most notable difference between Neanderthals and present-day humans is in the skull. The Neanderthal skull was characterized by prominent heavy brow ridges, a projecting mouth, and a weak chin. In addition, the Neanderthal brain was slightly larger, on average, than that of present-day humans.

Based on specimens from more than 100 sites, as well as genetic analysis, we now know that Neanderthals are our closest extinct relative, albeit it is morphologically more robust. European Neanderthals were the first humans to move into truly cold climates, enduring miserably long winters and short summers as they pushed north into tundra country (●FIGURE 18.32). Their remains are found chiefly in caves and hutlike rock shelters, which also contain a variety of specialized stone tools and weapons. In addition, archeological evidence indicates that Neanderthals commonly took care of their injured and buried their dead, frequently with such grave items as tools, food, and perhaps even flowers.

The name *Cro-Magnons*, who lived in Europe and the Middle East from about 35,000 to 10,000 years ago is the term given to the successors of the Neanderthals. The name derives from the discovery in 1868 of parts of five skeletons found in a shallow cave at Cro-Magnon in the Dordogne region of southwestern France and has been part of the lexicon of hominin names ever since. Many paleoanthropologists today prefer to use the term *Anatomically Modern Humans* (AMH) or *Early Modern Humans* (EMH) to designate the hominin remains from this time interval that look very much like us but differ slightly in certain features. For example, the skeletons of AMH were more robust and had a slightly larger cranial volume (1,600 cc on average) than that of humans today. However, they also possessed features unique to present-day humans, such as a tall, rounded skull with a straight forehead, reduced brow ridges, and a prominent chin.

It was during the Late Paleolithic Period, which began approximately 40,000 years ago, that the development of art and technology far exceeded anything the world had previously seen. Using paints made from manganese and iron oxides, Cro-Magnon (AMH) people painted hundreds of scenes on the ceilings and walls of caves in France and Spain, where many of them are still preserved (●FIGURE 18.33).

Cro-Magnons (AMH) were also skilled nomadic hunters, following the herds in their seasonal migrations and

● **FIGURE 18.32 Pleistocene Cave Setting with Neanderthals** Archeological evidence indicates that Neanderthals lived in caves and participated in ritual burials as depicted in this painting of a burial ceremony such as occurred approximately 60,000 years ago at Shanidar Cave, Iraq.

Painting by Ronald Bowen/Robert Harding Picture Library

● **FIGURE 18.33 Cro-Magnon Cave Painting** Cro-Magnon (Anatomically Modern Humans) were very skilled cave painters. Shown is a painting of a horse from the cave of Niaux, France.

using a variety of specialized tools in their hunts, perhaps including the bow and arrow. They sought refuge in caves and rock-shelters and formed living groups of various sizes.

With the appearance of Anatomically Modern Humans, our evolution has become almost entirely cultural rather than biologic. Since the appearance of the Neanderthals approximately 200,000 years ago, humans have gone from a stone culture to today's culture with its emphasis on technology that has allowed us to visit other planets and land astronauts on the Moon. It remains to be seen how we will use this technology in the future and whether we will continue as a species, evolve into another species, or become extinct as many groups have before us.

# Key Concepts Review

- Stromatolites, which are pillow-shaped structures composed of sediment and blue-green algae, are among the earliest fossils on Earth. Because the blue-green algae produce oxygen as a by-product of photosynthesis, they undoubtedly contributed to the increasing amounts of oxygen in Earth's early atmosphere.
- Well-documented multicelled animals are found in several Neoproterozoic localities. Animals were widespread at this time, but because all lacked durable skeletons, their fossils are not common. These assemblages are referred to as the Ediacaran faunas, named after the location in Australia where they were first discovered.
- Invertebrates with hard parts "suddenly" appeared during the Early Cambrian in what is called the "Cambrian explosion." Skeletons provided such advantages as protection against predators and support for muscles, enabling organism to grow large and increase locomotor efficiency. Hard parts probably evolved as a result of various geologic and biologic factors rather than a single cause.
- The Middle Cambrian Burgess Shale contains one of the finest examples of a well-preserved, soft-bodied biota in the world and provides us with an important glimpse of not only rarely preserved organisms but also the soft-part anatomy of many extinct groups.
- The Ordovician marine invertebrate community marked the beginning of the dominance of the shelly fauna and the start of large-scale reef building. The end of the Ordovician Period was a time of major extinction of many invertebrate phyla.
- The Silurian and Devonian periods were periods of diverse faunas dominated by reef-building animals,

whereas the Carboniferous and Permian periods saw a great decline in invertebrate diversity.
- The greatest recorded mass extinction occurred at the end of the Permian Period, affecting the invertebrates as well as the vertebrates. Its cause is still the subject of debate.
- Fish are the earliest known vertebrates, with their first fossil occurrence in Cambrian rocks. They have had a long and varied history, including jawless and jawed armored forms (ostracoderms and placoderms), cartilaginous forms, and bony forms. It is from the lobe-finned bony fish group that amphibians evolved.
- The link between crossopterygian lobe-finned fish and the earliest amphibians is convincing and includes a close similarity of bone and tooth structures. New fossil discoveries, however, show that the transition between the two groups is more complicated than originally hypothesized and includes several intermediate forms.
- Amphibians evolved during the Late Devonian, but were a minor element until the Carboniferous, when swampy conditions were widespread, resulting in a diverse and abundant fauna. Amphibians declined in the Permian, perhaps in response to changing climatic conditions.
- The Late Mississippian marks the earliest fossil record of reptiles. The evolution of an amniote egg was the critical factor that allowed reptiles to completely colonize the land.
- Pelycosaurs were the dominant reptiles during the Early Permian, whereas therapsids dominated the landscape for the rest of the Permian Period.

- In making the transition from water to land, plants had to overcome the same basic problems as animals—namely, desiccation, reproduction, and gravity.
- The earliest fossil record of land plants is from Middle to Upper Ordovician rocks.
- The evolution of vascular tissue was an important event in plant evolution as it allowed nutrients and water to be transported throughout the plant and provided the plant with additional support.
- The earliest seedless vascular plants were small, leafless Y-shaped stems. The Late Devonian witnessed the evolution of gymnosperms (flowerless seed plants) whose reproductive style freed them from having to stay near water.
- The Carboniferous Period was a time of vast coal swamps, where conditions were ideal for the seedless vascular plants. With the onset of more arid conditions during the Permian, the gymnosperms became the dominant element of the world's forests.
- The marine invertebrate survivors of the Permian mass extinction diversified and gave rise to increasingly diverse Mesozoic marine invertebrate communities.
- Land plant communities of the Triassic and Jurassic consisted of seedless vascular plants and gymnosperms. The angiosperms, or flowering plants, evolved during the Early Cretaceous, diversified rapidly, and were soon the most abundant land plants.
- Dinosaurs evolved during the Late Triassic. The two distinct orders of dinosaurs, based on pelvic structure, are Saurischia (lizard-hipped) and Ornithischia (bird-hipped).
- Bone structure, predator–prey relationships, and other features have been cited as evidence of dinosaur endothermy. Most paleontologists think that at least some dinosaurs were endothermic.
- That some theropods had feathers indicates that they were warm-blooded and provides further evidence of their relationship to birds.
- Small pterosaurs were probably active, wing-flapping fliers, whereas large one may have depended on soaring to stay aloft. New discoveries indicate many, if not most, had hair or hairlike feathers suggesting that they were probably endothermic.
- The fish-eating, porpoiselike ichthyosaurs were thoroughly adapted to an aquatic life, whereas the plesiosaurs with their paddlelike limbs could in all likelihood come out of the water to lay their eggs on land.
- Birds probably evolved from small theropod dinosaurs. The oldest known bird, *Archaeopteryx*, appeared during the Jurassic.
- The earliest mammals evolved during the Late Triassic, but they are difficult to distinguish from advanced cynodonts.
- Several types of Mesozoic mammals existed, but most were small and their diversity was low.
- Mammals today belong to one of three groups: monotremes, or egg-laying mammals; marsupials, or pouched animals; and placentals, which include most mammals with which we are familiar.
- Among the victims of the Cretaceous mass extinctions were dinosaurs, pterosaurs, marine reptiles, and several groups of marine invertebrates. A meteorite impact may have caused these extinctions, but some paleontologists think that other factors also contributed.
- Marine invertebrate groups that survived the extinctions at the end of the Mesozoic continued to diversify, giving rise to the present-day marine fauna.
- The Paleocene mammalian fauna was composed of Mesozoic holdovers and several new orders. By Oligocene time, all of the orders of existing mammals were present.
- An important evolutionary trend in Pleistocene mammals and some birds was toward gigantism. Large species of mammals included mastodons and mammoths, huge ground sloths, giant camels, and beavers, and others. Many of these large species died out by the end of the Pleistocene.
- Primates, which evolved during the Paleocene, differ from other mammalian orders on the basis of overall skeletal structure and mode of locomotion, an increase in brain size, stereoscopic vision, and a grasping hand with opposable thumb.
- The primates are divided into two suborders: prosimians, which are the oldest primate lineage and include lemurs, lorises, tarsiers, and tree shrews and anthropoids, which include the Old World monkeys (Africa and Asia) and New World monkeys (Central and South America).
- The hominin lineage begins nearly 7 million years ago with *Sahelanthropus tchadensis*, followed by *Orrorin tugenensis* at 6 million years, and then two species of *Ardipithecus*, at 5.8 and 4.4 million years ago, respectively. These early hominins were succeeded by the australopithecines, a fully bipedal group that evolved in Africa 4.2 million years ago.
- The human lineage began about 2.5 million years ago in Africa with the evolution of *Homo habilis*, superseded by *H. erectus*, sometime between 1.8 and 1.6 million years ago, which was the first hominin to migrate out of Africa.
- Although the transition from *H. erectus* to *H. sapiens* is still unresolved, the most famous of all human fossils are the Neanderthals, who inhabited Europe and the Near East between 200,000 and 30,000 years ago and differed mainly from present-day humans in being more robust and having a long, low skull with heavy

brow ridges. Neanderthals also made specialized tools and weapons, apparently took care of their injured, and buried their dead.

- The Cro-Magnons, also referred to as "Anatomically Modern Humans" (AMH), were highly skilled hunters and cave painters who lived throughout Europe and the Middle East from about 35,000 to 10,000 years ago.
- Present-day humans succeeded the Cro-Magnons (AMH), having spread throughout the world as well as setting foot on the Moon.

# Important Terms

amniote egg  405
angiosperm  416
archosaur  409
australopithecine  424
bony fish  404
cynodont  416
dinosaur  409
eukaryotic cell  400
fossil  399
gymnosperm  407

hominid  423
hominin  423
*Homo*  424
ichthyosaur  414
marsupial mammal  418
monotreme  418
Ornithischia  409
placental mammal  419
pelycosaur  405
plesiosaur  414

primate  422
prokaryotic cell  400
pterosaur  413
Saurischia  409
seedless vascular plant  407
stromatolite  399
therapsid  406
vascular plant  407
vertebrate  399

# Review Questions

1. Which plant group first successfully invaded land?
   a. seedless vascular
   b. gymnosperms
   c. naked seed bearing
   d. angiosperms
   e. flowering

2. Which evolutionary innovation allowed reptiles to colonize all of the land?
   a. tear ducts
   b. additional bones in the jaw
   c. the middle-ear bone
   d. an egg that contained food and waste sacs and an embryo surrounded by a fluid-filled sac
   e. limbs and a backbone capable of supporting the animals on land

3. Which reptile group gave rise to the mammals?
   a. labyrinthodonts
   b. acanthodians
   c. pelycosaurs
   d. protothyrids
   e. therapsids

4. Which of the following evolutionary trends characterize primates?
   a. increase in brain size
   b. change in overall skeletal structure
   c. steroscopic vision
   d. grasping hand with opposable thumb
   e. all of these

5. The monotremes, one of the three basic groups of mammals, is the only one that
   a. gives birth to live young.
   b. is warm blooded.
   c. lays eggs.
   d. has a well-developed placenta.
   e. is extinct.

6. Based on what you know about Pennsylvanian geology (Chapter 17), why was this time period so advantageous to the evolution of both plants and amphibians?

7. What is the most popular hypothesis for the mass extinctions at the end of the Mesozoic, and what evidence supports this idea?

8. How do you explain the fact that much more is known about Cenozoic life history than about the previous intervals of geologic time?

9. Why has it become increasingly difficult to differentiate between small theropods and birds?

# Creative Thinking Visual Question

Until the recently discovered 395-million-year-old fossilized footprints in the Holy Cross Mountains in Poland, these fossilized footprints, which are more than 385 million years old and are part of a tetrapod trackway at Valentia Island, Ireland, represented evidence of some of the oldest land-dwelling tetrapods (four-legged animal). What evidence in this photo (●FIGURE 1) indicates that the sediments were deposited in a wet environment? How can you use the footprints to establish that the animal making them was a tetrapod? And lastly, how do you think scientists estimated the size of the tetrapod that made this trackway?

Courtesy of Ken Higgs, Department of Geology, University of Cork, Ireland

● **FIGURE 1  Devonian Fossilized Tetrapod Footprints** These fossilized footprints, which are more than 385 million years old, are part of a tetrapod trackway at Valentia Island, Ireland.

## English-Metric Conversion Chart

| | English Unit | Conversion Factor | Metric Unit | Conversion Factor | English Unit |
|---|---|---|---|---|---|
| Length | Inches (in) | 2.54 | Centimeters (cm) | 0.39 | Inches (in) |
| | Feet (ft) | 0.305 | Meters (m) | 3.28 | Feet (ft) |
| | Miles (mi) | 1.61 | Kilometers (km) | 0.62 | Miles (mi) |
| Area | Square inches (in$^2$) | 6.45 | Square centimeters (cm$^2$) | 0.16 | Square inches (in$^2$) |
| | Square feet (ft$^2$) | 0.093 | Square meters (m$^2$) | 10.8 | Square feet (ft$^2$) |
| | Square miles (mi$^2$) | 2.59 | Square kilometers (km$^2$) | 0.39 | Square miles (mi$^2$) |
| Volume | Cubic inches (in$^3$) | 16.4 | Cubic centimeters (cm$^3$) | 0.061 | Cubic inches (in$^3$) |
| | Cubic feet (ft$^3$) | 0.028 | Cubic meters (m$^3$) | 35.3 | Cubic feet (ft$^3$) |
| | Cubic miles (mi$^3$) | 4.17 | Cubic kilometers (km$^3$) | 0.24 | Cubic miles (mi$^3$) |
| Weight | Ounces (oz) | 28.3 | Grams (g) | 0.035 | Ounces (oz) |
| | Pounds (lb) | 0.45 | Kilograms (kg) | 2.20 | Pounds (lb) |
| | Short tons (st) | 0.91 | Metric tons (t) | 1.10 | Short tons (st) |
| Temperature | Degrees Fahrenheit (°F) | $-32° \times 0.56$ | Degrees centigrade (Celsius) (°C) | $\times 1.80 + 32°$ | Degrees Fahrenheit (°F) |

Examples:

10 inches = 25.4 centimeters; 10 centimeters = 3.9 inches

100 square feet = 9.3 square meters; 100 square meters = 1080 square feet

50°F = 10.1°C; 50°C = 122°F

To identify most minerals, geologists use physical properties such as color, luster, crystal form, hardness, cleavage, specific gravity, and several others (see Chapter 3, "Minerals"). This mineral identification table is divided into two parts: The first part comprises minerals with a metallic luster, and the second part covers minerals with a nonmetallic luster. After determining the luster of a mineral, ascertain its hardness and note that each part of the table is arranged with minerals of increasing hardness. Thus, if you have a nonmetallic mineral with a hardness of 6, it must be augite, hornblende, plagioclase, or one of the two potassium feldspars (orthoclase or microcline). If this hypothetical mineral is dark green or black, it must be augite or hornblende, and if it has two cleavage planes intersecting at nearly right angles, it is augite.

## Mineral Identification Tables

### Metallic Luster

| Mineral | Chemical Composition | Color | Hardness Specific Gravity | Other Features | Comments |
|---------|----------------------|-------|---------------------------|----------------|----------|
| Graphite | $C$ | Black | 1–2 2.09–2.33 | Greasy feel; writes on paper; 1 direction of cleavage | Used for pencil "leads"; mostly in metamorphic rocks |
| Galena | $PbS$ | Lead gray | 2.5 7.6 | Cubic crystals; 3 cleavages at right angles | The ore of lead; mostly in hydrothermal rocks |
| Chalcopyrite | $CuFeS_2$ | Brassy yellow | 3.5–4 4.1–4.3 | Usually massive; greenish black streak; iridescent tarnish | The most common copper mineral; mostly in hydrothermal rocks |
| Magnetite | $Fe_3O_4$ | Black | 5.5–6.5 5.2 | Strong magnetism | An ore of iron; an accessory mineral in many rocks |
| Hematite | $Fe_2O_3$ | Red brown | 6 4.8–5.3 | Usually granular or massive; reddish brown streak | Important iron ore; an accessory mineral in many rocks |
| Pyrite | $FeS_2$ | Brassy yellow | 6.5 5.0 | Cubic and octahedral crystals | Found in some igneous and hydrothermal rocks and in sedimentary rocks associated with coal |

### Nonmetallic Luster

| Mineral | Chemical Composition | Color | Hardness Specific Gravity | Other Features | Comments |
|---------|----------------------|-------|---------------------------|----------------|----------|
| Talc | $Mg_3Si_4O_{10}(OH)_2$ | White, green | 1 2.8 | 1 cleavage direction; usually in compact masses; soapy feel | Formed by the alteration of magnesium silicates; mostly in metamorphic rocks |
| Clay minerals | Varies | Gray, buff, white | 1–2 2.5–2.9 | Earthy masses; particles too small to observe properties | Found in soils, mudrocks, slate, phyllite |
| Chlorite | $(Mg,Fe)_3(Si,Al)_4 O_{10}(OH)_8$ | Green | 2 2.6–3.4 | 1 cleavage; occurs in scaly masses | Common in low-grade metamorphic rocks such as slate |
| Gypsum | $CaSO_4 \cdot 2H_2O$ | Colorless, white | 2 2.3 | Elongated crystals; fibrous and earthy masses | The most common sulfate mineral; found mostly in evaporite deposits |

## Nonmetallic Luster

| Mineral | Chemical Composition | Color | Hardness Specific Gravity | Other Features | Comments |
|---------|---------------------|-------|---------------------------|----------------|----------|
| Muscovite (Mica) | $KAl_2Si_3O_{10}(OH)_2$ | Colorless | 2–2.5 <br> 2.7–2.9 | 1 direction of cleavage; cleaves into thin sheets | Common in felsic igneous rocks, metamorphic rocks, and some sedimentary rocks |
| Biotite (Mica) | $K(Mg,Fe)_3AlSi_3O_{10}(OH)_2$ | Black, brown | 2.5 <br> 2.9–3.4 | 1 cleavage direction; cleaves into thin sheets | Occurs in both felsic and mafic igneous rocks, in metamorphic rocks, and in some sedimentary rocks |
| Calcite | $CaCO_3$ | Colorless, white | 3 <br> 2.7 | 3 cleavages at oblique angles; cleaves into rhombs; reacts with dilute HCl | The most common carbonate mineral; main component of limestone and marble |
| Anhydrite | $CaSO_4$ | White, gray | 3.5 <br> 2.9–3.0 | Crystals with 2 cleavages; usually in granular masses | Found in limestones, evaporite deposits, and the cap rock of salt domes |
| Halite | $NaCl$ | Colorless, white | 3–4 <br> 2.2 | 3 cleavages at right angles; cleaves into cubes; cubic crystals; salty taste | Occurs in evaporite deposits |
| Dolomite | $CaMg(CO_3)_2$ | White, yellow, gray, pink | 3.5–4 <br> 2.9 | Cleavage as in calcite; reacts with dilute hydrochloric acid when powdered | The main constituent of dolostone; also found associated with calcite in some limestones and marble |
| Fluorite | $CaF_2$ | Colorless, purple, green, brown | 4 <br> 3.2 | 4 cleavage directions; cubic and octahedral crystals | Occurs mostly in hydrothermal rocks and in some limestones and dolostones |
| Augite | $Ca(Mg,Fe,Al)(Al,Si)_2O_6$ | Black, dark green | 6 <br> 3.3–3.6 | Short 8-sided crystals; 2 cleavages; cleavages nearly at right angles | The most common pyroxene mineral; found mostly in mafic igneous rocks |
| Hornblende | $NaCa_2(Mg,Fe,Al)_5(Si,Al)_8O_{22}(OH)_2$ | Green, black | 6 <br> 3.0–3.4 | Elongated, 6-sided crystals; 2 cleavages intersecting at 56° and 124° | A common rock-forming amphibole mineral in igneous and metamorphic rocks |
| Plagioclase feldspars | Varies from $CaAl_2Si_2O_8$ to $NaAlSi_3O_8$ | White, gray, brown | 6 <br> 2.6 | 2 cleavages at right angles | Common in igneous rocks and a variety of metamorphic rocks; also in some arkoses |
| Potassium feldspars | | | | | |
| Microcline | $KAlSi_3O_8$ | White, pink, green | 6 <br> 2.6 | 2 cleavages at right angles | Common in felsic igneous rocks, some metamorphic rocks, and arkoses |
| Orthoclase | $KAlSi_3O_8$ | White, pink | 6 <br> 2.6 | 2 cleavages at right angles | |
| Olivine | $(Fe,Mg)_2SiO_4$ | Olive green | 6.5 <br> 3.3–3.6 | Small mineral grains in granular masses; conchoidal fracture | Common in mafic igneous rocks |
| Quartz | $SiO_2$ | Colorless, white, gray, pink, green, black | 7 <br> 2.7 | 6-sided crystals; no cleavage; conchoidal fracture | A common rock-forming mineral in all rock groups and hydrothermal rocks; also occurs in varieties known as chert, flint, agate, and chalcedony |
| Garnet | $Fe_3Al_2(SiO_4)_3$ | Dark red, green | 7–7.5 <br> 4.3 | 12-sided crystals common; uneven fracture | Found mostly in gneiss and schist |
| Zircon | $Zr_2SiO_4$ | Brown, gray | 7.5 <br> 3.9–4.7 | 4-sided, elongated crystals | Most common as an accessory in granitic rocks |
| Topaz | $Al_2SiO_4(OH,F)$ | Colorless, white, yellow, blue | 8 <br> 3.5–3.6 | High specific gravity; 1 cleavage direction | Found in pegmatites, granites, and hydrothermal rocks |
| Corundum | $Al_2O_3$ | Gray, blue, pink, brown | 9 <br> 4.0 | 6-sided crystals and great hardness are distinctive | An accessory mineral in some igneous and metamorphic rocks |

# ANSWERS

## Multiple-Choice Questions

**CHAPTER 1**
1. e; 2. c; 3. c; 4. e; 5. c.

**CHAPTER 2**
1. a; 2. e; 3. b; 4. c; 5. c.

**CHAPTER 3**
1. a; 2. d; 3. b; 4. e; 5. b.

**CHAPTER 4**
1. d; 2. a; 3. e; 4. d; 5. b.

**CHAPTER 5**
1. b; 2. c; 3. d; 4. c; 5. e.

**CHAPTER 6**
1. b; 2.d; 3. e; 4. a; 5. c.

**CHAPTER 7**
1. e; 2. d; 3. e; 4. d; 5. a.

**CHAPTER 8**
1. e; 2. c; 3. d; 4. c; 5. d.

**CHAPTER 9**
1. e; 2. a; 3. b; 4. b; 5. b.

**CHAPTER 10**
1. d; 2. b; 3. b; 4. b; 5. e.

**CHAPTER 11**
1. c; 2. a; 3. d; 4. c; 5. e.

**CHAPTER 12**
1. b; 2. c; 3. c; 4. d; 5. e.

**CHAPTER 13**
1. c; 2. b; 3. a; 4. d; 5. d.

**CHAPTER 14**
1. d; 2. b; 3. b; 4. a; 5. d.

**CHAPTER 15**
1. b; 2. c; 3. d; 4. a; 5. c.

**CHAPTER 16**
1. b; 2. b; 3. d; 4. b; 5. a.

**CHAPTER 17**
1. b; 2. b; 3. c; 4. b; 5. c.

**CHAPTER 18**
1. a; 2. d; 3. e; 4. e; 5. c.

# ANSWERS

## Selected Short Answer Questions

### CHAPTER 1

6. Plate tectonic theory provides a unifying explanation for many geologic features and events by relating various seemingly unrelated phenomena and providing a framework for interpreting Earth's composition, structure, and internal processes. It fits into a systems approach for the study of Earth because it shows how the continents and ocean basins are part of a lithosphere–atmosphere–hydrosphere system that evolved along with Earth's interior.

### CHAPTER 2

8. Many metallic mineral deposits are related to igneous and associated hydrothermal activity. The magma and associated hydrothermal fluids generated by subducting plates along convergent plate boundaries contain minute amounts of metallic elements. As this magma rises toward the surface, it cools and precipitates and concentrates metallic ores. In the same way, rising magma along divergent plate boundaries precipitates metallic elements adjacent to hydrothermal vents.

### CHAPTER 3

6. Some minerals have a range of chemical compositions because one element can substitute for another if they have the same charge and are of about the same size. In the plagioclase feldspars, the elements calcium (Ca) and sodium (Na) meet these criteria, and therefore the composition of plagioclase ranges from calcium-rich ($CaAl_2Si_2O_8$) to sodium-rich ($NaAlSi_3O_8$) varieties.

### CHAPTER 4

7. The most important controls on the viscosity of a lava flow are temperature and composition, although the presence of gases, mineral crystals, and the surface over which it flows have some influence. In general, the hotter the lava, the less viscous (more fluid) it is, but composition can also have a large effect as well. For example, silica-rich lava (>65% silica) has numerous chemical bonds between atoms that inhibit flow and thus the lava is viscous (not very fluid). In contrast, mafic lava with 45%–52% silica is much more fluid.

### CHAPTER 5

6. Geologists monitor several physical and chemical aspects of volcanoes to better anticipate eruptions. They monitor gas emissions, increases or decreases in hot spring activity, and temperature, as well as changes in the volcano such as bulging when magma is injected into it. Other important criteria include variations in the local magnetic and electrical fields. Lastly, geologists are keenly aware of volcanic tremor which indicates that magma is moving beneath the surface.

### CHAPTER 6

8. Mechanical weathering involves the breakdown of Earth materials with no change in composition, as when granite breaks down into small pieces of granite and individual minerals. Chemical weathering, in contrast, decomposes rocks and minerals by chemical changes in the parent material. Mechanical weathering contributes to chemical weathering because chemical alteration takes place faster on small particles.

### CHAPTER 7

9. Foliated texture results when rocks are subjected to heat and differential pressure causing the minerals to align in a parallel fashion. Nonfoliated texture shows a mosaic of roughly equidimensional minerals that result from contact metamorphism or regional metamorphism of rocks with no alignment of any platy or elongated minerals, if present.

### CHAPTER 8

8. Insurance companies use the Modified Mercalli Intensity Scale because it measures actual damage done to structures and areas. Therefore, insurance companies can determine the type of damage to expect from a given earthquake in a particular area based on the amount and type of damage that resulted from previous earthquakes in that area, and thus adjust their rates accordingly.

### CHAPTER 9

6. Convert 200 km to millimeters (20 million) and divide by 5 million years, giving an average rate of movement of 4 mm per year. In many segments of faults, such as the San Andreas Fault, blocks on opposite sides of the fault do creep past one another slowly, but on occasion they also show movements of many meters during a single earthquake. Accordingly, the average rate of movement does not adequately describe the history of movement along a fault.

### CHAPTER 10

8. The factors that influence mass wasting are slope angle, weathering and climate, water content, vegetation, and overloading. These factors are interconnected by the fact that the amount of rain is related to climate, and rain can increase the water content of a slope's material, leading to an increase in weight, which might cause

mass movement or decrease the friction between grains, leading to a loss of cohesion and sliding. Too little rain may cause plants to die, decreasing the vegetative root system which helps bind soil particles together.

## CHAPTER 11

9. Point bars are bodies of mostly sand that are deposited on the gently sloping inner bank of a meandering stream. They form in that location because that is where flow velocity is least, resulting in deposition of some of the stream's sediment load.

## CHAPTER 12

9. In arid regions, such as Nevada, the water table is very low, in some cases tens and hundreds of meters below the surface. Furthermore, rainfall is scant, and what little rain does fall and percolate into the ground usually evaporates before it gets very far. Thus, the likelihood of water reaching deeply buried canisters containing radioactive nuclear waste material where it could corrode the containers or become polluted by the radioactive material is virtually nonexistent for the foreseeable future.

However, if the climate should change and the region become humid, say, in the next several thousand years, water from rainfall will percolate freely through the zone of aeration into the zone of saturation, and eventually it will cause the water table to rise, possibly to the level of the buried radioactive nuclear waste material, where the groundwater system could then become polluted by the radioactive material.

The average rate of groundwater movement during the next 5,000 years would have to be 8 cm per year to reach radioactive waste canisters buried at a depth of 400 m (400 m = 40,000 cm [100 cm = 1 m; 400 m = 40,000 cm], thus 40,000 cm/5,000 years = 8 cm/year).

## CHAPTER 13

7. You can recognize a mountain area that was once glaciated by landforms that resulted from erosion or deposition. For instance, erosion scours out U-shaped valleys, abrades bedrock (striations), and leaves other features, such as cirques, arêtes, and horns. The easily recognized deposits of former valley glaciers include lateral and medial moraines, and perhaps silt and clay deposits that formed in lakes.

## CHAPTER 14

9. The main line of evidence that indicates wind was an important agent in the formation of the Martian surface includes desert pavement, ventifacts, and dunes.

## CHAPTER 15

7. Upwelling involves the vertical circulation of cold, nutrient-rich seawater from depth to the surface. It is important because it carries nutrients, especially phosphates and nitrates, into the photic zone that support large communities of organisms. Therefore, a sedimentary rock that is rich in phosphorous was deposited in an area of upwelling.

## CHAPTER 16

9. The potassium 40 to argon 40 and rubidium 87 to strontium 87 radiometric dating techniques could produce distinctly different ages for a volcanic ash fall due to the fact that if the volcanic ash was heated after deposition, some of the argon might have escaped, yielding an apparently younger age than that obtained from the rubidium 87 to strontium 87 isotopic pair.

To rectify the discrepancy, whole-rock analysis of the volcanic ash could be conducted, which might yield a more accurate potassium to argon ratio.

## CHAPTER 17

7. Reconstructing paleogeography requires the synthesis of all pertinent paleoclimatic, paleomagnetic, paleontologic, paleomagnetic, sedimentologic, stratigraphic, and tectonic data available for the time period of interest. Such information as the distribution of plants and animals yields insights into the possible latitudes of landmasses and ocean basins. Sedimentary structures can reveal something about depositional environments. Although most geology books and journals agree on the major features of paleogeography, some evidence preserved in the geologic record can be interpreted in different ways, leading to minor differences in paleogeographic reconstructions.

## CHAPTER 18

6. The Pennsylvanian Period was a time of repetitive transgressive–regressive cycles that resulted in widespread coal-forming swamps. These swamps provided the perfect environment for amphibians and seedless vascular plants, both of which required moist conditions as part of their reproductive cycle.

# GLOSSARY

**aa** Lava flow with a surface of rough, angular blocks and fragments.

**abrasion** The process whereby rock is worn smooth by the impact of sediment transported by running water, glaciers, waves, or wind.

**Absaroka sequence** Widespread Upper Mississippian to Lower Jurassic sedimentary rocks bounded above and below by unconformities; deposited during a transgressive–regressive cycle of the Absaroka Sea.

**abyssal plain** The flat surface covering vast areas of the seafloor. Abyssal plains are the flattest, most featureless areas on Earth, and their flatness is a result of sediment deposition covering the usually rugged topography of the seafloor.

**Acadian orogeny** An episode of Devonian deformation in the northern Appalachian mobile belt resulting from the collision of Baltica with Laurentia.

**active continental margin** A continental margin with volcanism and seismicity at the leading edge of a continental plate where oceanic lithosphere is subducted.

**alluvial fan** A cone-shaped accumulation of mostly sand and gravel deposited where a stream flows from a mountain valley onto an adjacent lowland.

**alluvium** A collective term for all detrital sediment transported and deposited by running water.

**amniote egg** Egg in which the embryo develops in a fluid-filled cavity called the amnion. The egg also contains a yolk sac and a waste sac.

**angiosperm** Vascular plants that have flowers and seeds; the flowering plants.

**angular unconformity** An unconformity below which older rocks dip at a different angle (usually steeper) than younger overlying strata.

**anticline** A convex upward fold in which the oldest exposed rocks coincide with the fold axis and all strata dip away from the axis.

**aphanitic texture** A texture in igneous rocks in which individual mineral grains are too small to be seen without magnification; results from rapid cooling of magma and generally indicates an extrusive origin.

**aphotic zone** The depth in the ocean below which sunlight does not penetrate.

**aquifer** A permeable layer in which groundwater flows. From the Latin *aqua*, "water."

**archosaur** A term referring to the ruling reptiles—dinosaurs, flying reptiles (pterosaurs), crocodiles, and birds.

**arête** A narrow, serrated ridge between two glacial valleys or adjacent cirques.

**artesian system** A confined groundwater system with high hydrostatic pressure that causes water to rise above the level of the aquifer.

**aseismic ridge** A ridge, or broad area, rising above the seafloor that lacks seismic activity.

**assimilation** A process whereby magma changes composition as it reacts with country rock.

**asthenosphere** The part of the mantle that lies below the lithosphere; it behaves plastically and flows slowly.

**atom** The smallest unit of matter that retains the characteristics of an element.

**atomic mass number** The number of protons plus neutrons in the nucleus of an atom.

**atomic number** The number of protons in the nucleus of an atom.

**aureole** A zone surrounding a pluton in which contact metamorphism took place.

**australopithecine** A term referring to several extinct species of the genus *Australopithecus* that existed during the Pliocene and Pleistocene epochs.

**barchan dune** A crescent-shaped sand dune with its tips pointing downwind.

**barrier island** A long, narrow island of sand parallel to a shoreline but separated from the mainland by a lagoon.

**basal slip** Movement involving a glacier sliding over its underlying surface.

**basalt plateau** A plateau built up by horizontal or nearly horizontal overlapping lava flows that erupted from fissures.

**base level** The level below which a stream or river cannot erode; sea level is ultimate base level.

**basin** An oval to circular fold in which all strata dip inward toward a central point and the youngest exposed strata are in the center.

**batholith** An irregularly shaped, discordant pluton with at least 100 km$^2$ of exposed surface area.

**baymouth bar** A spit that has grown until it closes off a bay from the open sea or lake.

**beach** Any deposit of sediment extending landward from low tide to a change in topography or where permanent vegetation begins.

**bed** An individual layer of rock, especially sediment or sedimentary rock.

**bed load** The part of a stream's sediment load, mostly sand and gravel, transported along its bed.

**Big Bang** A model for the evolution of the universe in which a dense, hot state was followed by expansion, cooling, and a less dense state.

**biochemical sedimentary rock** Any sedimentary rock produced by the chemical activities of organisms.

**biostratigraphic unit** An association of sedimentary rocks defined by its fossil content.

**bonding** The process whereby atoms join to other atoms.

**bony fish** Fish with an internal skeleton of bone; the class Osteichthyes.

**Bowen's reaction series** A series of minerals that form in a specific sequence in cooling magma or lava; originally proposed to explain the origin of intermediate and felsic magma from mafic magma.

**braided stream** A stream with multiple dividing and rejoining channels.

**breaker** A wave that steepens as it enters shallow water until its crest plunges forward.

**butte** An isolated, steep-sided, pillar-like hill formed when resistant cap rock is breached, allowing erosion of less resistant underlying rocks.

**caldera** A large, steep-sided, oval to circular depression usually formed when a volcano's summit collapses into an underlying partially drained magma chamber.

**Canadian shield** The exposed part of the North American craton.

**carbon-14 dating technique** A numerical dating technique relying on the ratio of carbon 12 to carbon 14 in an organic substance; useful back to about 70,000 years ago.

**carbonate mineral** A mineral with the carbonate radical $(CO_3)^{-2}$, as in calcite $(CaCO_3)$ and dolomite $[CaMg(CO_3)_2]$.

**carbonate rock** Any rock, such as limestone and dolostone, made up mostly of carbonate minerals.

**Cascade Range** A mountain range with several active volcanoes in northern California, Oregon, Washington, and southern British Columbia, Canada.

**cave** A natural subsurface opening generally connected to the surface and large enough for a person to enter.

**cementation** The process whereby minerals crystallize in the pore spaces of sediment and bind the loose particles together.

**chemical sedimentary rock** Sedimentary rock made up of minerals that were dissolved during chemical weathering and later precipitated from seawater, more rarely lake water, or extracted from solution by organisms.

**chemical weathering** The decomposition of rocks by chemical alteration of parent material.

**cinder cone** A small, steep-sided volcano made up of pyroclastic materials resembling cinders that accumulate around a vent.

**circum-Pacific belt** A zone of seismic and volcanic activity and mountain building that nearly encircles the Pacific Ocean basin.

**cirque** A steep-walled, bowl-shaped depression on a mountainside at the upper end of a glacial valley.

**cleavage** The breaking or splitting of mineral crystals along planes of internal weakness.

**columnar joints** Columns in igneous rocks bounded by fractures that formed when lava or magma cooled and contracted.

**compaction** Reduction in the volume of a sedimentary deposit that results from its own weight and the weight of any additional sediment deposited on top of it.

**complex movement** A combination of different types of mass movements in which no single type is dominant; usually involves sliding and flowing.

**composite volcano (stratovolcano)** A volcano composed of lava flows and pyroclastic layers, typically of intermediate composition, and mudflows.

**compound** Any substance resulting from the bonding of two or more different elements (e.g., water, $H_2O$, and quartz, $SiO_2$).

**compression** Stress resulting when rocks are squeezed by external forces directed toward one another.

**concordant** Igneous body whose boundaries parallel the layering in the country rock.

**cone of depression** A cone-shaped depression around a well where water is pumped from an aquifer faster than it can be replaced.

**contact (thermal) metamorphism** Metamorphism of country rock adjacent to a pluton and beneath a lava flow.

**continental accretion** The process in which continents grow by additions of Earth materials along their margins.

**continental–continental plate boundary** A convergent plate boundary along which two continental lithospheric plates collide.

**continental drift** The theory that the continents were joined into a single landmass that broke apart with the various fragments (continents) moving with respect to one another.

**continental glacier** A glacier that covers a vast area (at least 50,000 $km^2$) and is not confined by topography; also called an ice sheet.

**continental margin** The area separating the part of a continent above sea level from the deep seafloor.

**continental rise** The gently sloping part of the continental margin between the continental slope and the abyssal plain.

**continental shelf** The very gently sloping part of the continental margin between the shoreline and the continental slope.

**continental slope** The relatively steeply inclined part of the continental margin between the continental shelf and the continental rise or between the continental shelf and an oceanic trench.

**convergent plate boundary** The boundary between two plates that move toward each other.

**Cordilleran orogeny** An episode of deformation affecting the western margin of North America from Jurassic to Early Cenozoic time; divided into three separate phases called the Nevadan, Sevier, and Laramide orogenies.

**core** The interior part of Earth beginning at a depth of 2,900 km that probably consists mostly of iron and nickel.

**Coriolis effect** The apparent deflection of a moving object from its anticipated course because of Earth's rotation. Winds and oceanic currents are deflected clockwise in the Northern Hemisphere and counterclockwise in the Southern Hemisphere.

**correlation** Demonstration of the physical continuity of rock units or biostratigraphic units, or demonstration of time equivalence as in time-stratigraphic correlation.

**country rock** Any preexisting rock that has been intruded by a pluton or altered by metamorphism.

**covalent bond** A chemical bond formed by the sharing of electrons between atoms.

**crater** An oval to circular depression at the summit of a volcano resulting from the eruption of lava, pyroclastic materials, and gases.

**craton** The stable nucleus of a continent consisting of a shield of Precambrian rocks and a platform of buried ancient rocks.

**creep** A widespread type of mass wasting in which soil or rock moves slowly downslope.

**Cretaceous Interior Seaway** A Late Cretaceous arm of the sea that effectively divided North America into two large landmasses.

**cross-bedding** A type of bedding in which layers are deposited at an angle to the surface on which they accumulate, as in sand dunes.

**crust** Earth's outermost layer; the upper part of the lithosphere that is separated from the mantle by the Moho; divided into continental and oceanic crust.

**crystal** A naturally occurring solid of an element or compound with a specific internal structure that is manifested externally by planar faces, sharp corners, and straight edges.

**crystal settling** The physical separation and concentration of minerals in the lower part of a magma chamber or pluton by crystallization and gravitational settling.

**crystalline solid** A solid in which the constituent atoms are arranged in a regular, three-dimensional framework.

**Curie point** The temperature at which iron-bearing minerals in cooling magma or lava attain their magnetism.

**cyclothem** A sequence of cyclically repeating sedimentary rocks resulting from alternating periods of marine and nonmarine deposition; commonly contains a coal bed.

**cynodont** A type of carnivorous therapsid (advanced mammal-like reptile); ancestors of mammals are among cynodonts.

**debris flow** A type of mass wasting that involves a viscous mass of soil, rock fragments, and water that moves downslope; debris flows have larger particles than mudflows and contain less water.

**deflation** The removal of sediment and soil by wind.

**deformation** A general term for any change in shape or volume, or both, of rocks in response to stress; involves folding and fracturing.

**delta** An alluvial deposit formed where a stream or river flows into the sea or a lake.

**density** The mass of an object per unit volume; usually expressed in grams per cubic centimeter ($g/cm^3$).

**depositional environment** Any site such as a floodplain or beach where physical, biologic, and chemical processes yield a distinctive kind of sedimentary deposit.

**desert** Any area that receives less than 25 cm of rain per year and that has a high evaporation rate.

**desertification** The expansion of deserts into formerly productive lands.

**desert pavement** A surface mosaic of close-fitting pebbles, cobbles, and boulders found in many dry regions; results from wind erosion of sand and smaller particles.

**detrital sedimentary rock** Sedimentary rock made up of the solid particles (detritus) of preexisting rocks.

**differential pressure** Pressure that is not applied equally to all sides of a rock body.

**differential weathering** Weathering that occurs at different rates on rocks, thereby yielding an uneven surface.

**dike** A tabular or sheetlike discordant pluton.

**dinosaur** Any of the Mesozoic reptiles belonging to the orders of Saurischia and Ornithischia.

**dip** A measure of the maximum angular deviation of an inclined plane from horizontal.

**dip-slip fault** A fault on which all movement is parallel with the dip of the fault plane.

**discharge** The volume of water in a stream or river moving past a specific point in a given interval of time; expressed in cubic meters per second ($m^3/sec$) or cubic feet per second ($ft^3/sec$).

**disconformity** An unconformity above and below which the rock layers are parallel.

**discontinuity** A boundary across which seismic wave velocity or direction of travel changes abruptly, such as the mantle–core boundary.

**discordant** Igneous body with boundaries that cut across the layering in the country rock.

**dissolved load** The part of a stream's load consisting of ions in solution.

**divergent plate boundary** The boundary between two plates that are moving apart.

**divide** A topographically high area that separates adjacent drainage basins.

**dome** A rather circular geologic structure in which all rock layers dip away from a central point and the oldest exposed rocks are at the dome's center.

**downwelling** The slow transfer of ocean surface water to depth.

**drainage basin** The surface area drained by a stream or river and its tributaries.

**drainage pattern** The regional arrangement of channels in a drainage system.

**dropstones** Pieces of gravel, some of boulder size, found in otherwise very fine-grained deposits.

**drumlin** An elongated hill of till formed by the movement of a continental glacier or by floods.

**dune** A mound or ridge of wind-deposited sand.

**dynamic metamorphism** Metamorphism in fault zones where rocks are subjected to high differential pressure.

**earthflow** A mass-wasting process involving the downslope movement of water-saturated soil.

**earthquake** Vibrations caused by the sudden release of energy, usually as a result of displacement of rocks along faults.

**elastic rebound theory** An explanation for the sudden release of energy that causes earthquakes when deformed rocks fracture and rebound to their original undeformed condition.

**elastic strain** A type of deformation in which the material returns to its original shape when stress is relaxed.

**electron** A negatively charged particle of very little mass that encircles the nucleus of an atom.

**electron shell** Electrons orbit an atom's nucleus at specific distances in electron shells.

**element** A substance composed of atoms that all have the same properties; atoms of one element can change to atoms of another element by radioactive decay, but they cannot be changed by ordinary chemical means.

**emergent coast** A coast where the land has risen with respect to sea level.

**end moraine** A pile or ridge of rubble deposited at the terminus of a glacier.

**epicenter** The point on Earth's surface directly above the focus of an earthquake.

**erosion** The removal of weathered materials from their source area by running water, wind, glaciers, or waves.

**esker** A long, sinuous ridge of stratified drift deposited by running water in a tunnel beneath stagnant ice.

**eukaryotic cell** A cell with internal structures such as mitochondria and an internal membrane-bounded nucleus containing chromosomes.

**evaporite** Any sedimentary rock, such as rock salt, formed by inorganic chemical precipitation of minerals from evaporating water.

**Exclusive Economic Zone (EEZ)** An area extending 370 km seaward from the coast of the United States and its possessions in which the United States claims rights to all resources.

**exfoliation dome** A large, rounded dome of rock resulting when concentric layers of rock are stripped from the surface of a rock mass.

**fault** A fracture along which rocks on opposite sides of the fracture have moved parallel with the fracture surface.

**fault plane** A fault surface that is more or less planar.

**felsic magma** Magma with more than 65% silica and considerable sodium, potassium, and aluminum, but little calcium, iron, and magnesium.

**ferromagnesian silicate** Any silicate mineral that contains iron, magnesium, or both.

**fetch** The distance the wind blows over a continuous water surface.

**fiord** An arm of the sea extending into a glacial trough eroded below sea level.

**firn** Granular snow formed by partial melting and refreezing of snow; transitional material between snow and glacial ice.

**fissure eruption** A volcanic eruption in which lava or pyroclastic materials issue from a long, narrow fissure (crack) or group of fissures.

**floodplain** A low-lying, flat area adjacent to a channel that is partly or completely water-covered when a stream or river overflows its banks.

**fluid activity** An agent of metamorphism in which water and carbon dioxide promote metamorphism by increasing the rate of chemical reactions.

**focus** The site within Earth where an earthquake originates and energy is released.

**fold** A type of geologic structure in which planar features in rock layers such as bedding and foliation have been bent.

**foliated texture** A texture in metamorphic rocks in which platy and elongated minerals are aligned in a parallel fashion.

**footwall block** The block of rock that lies beneath a fault plane.

**fossils** Remains or traces of prehistoric organisms preserved in rocks.

**fracture** A break in rock resulting from intense applied pressure.

**frost action** The disaggregation of rocks by repeated freezing and thawing of water in cracks and crevasses.

**geologic structure** Any feature in rocks that results from deformation, such as folds, joints, and faults.

**geologic time scale** A chart that subdivides geologic time into a hierarchy of increasingly shorter time intervals, each of which has a specific name and duration.

**geology** The study of Earth, as well as the planets and moons in our solar system. It is generally divided into two broad areas—*physical geology*, which is the study of Earth's materials, such as minerals and rocks, as well as the processes operating within Earth and on its surface, and *historical geology*, which examines the origin and evolution of Earth, its continents, oceans, atmosphere, and biota.

**geothermal energy** Energy that comes from steam and hot water trapped within Earth's crust.

**geothermal gradient** Earth's temperature increase with depth; it averages 25°C/km near the surface but varies from area to area.

**geyser** A hot spring that periodically ejects hot water and steam.

**glacial budget** The balance between expansion and contraction of a glacier in response to accumulation versus wastage.

**glacial drift** A collective term for all sediment deposited directly by glacial ice (till) and by meltwater streams (outwash).

**glacial erratic** A rock fragment carried some distance from its source by a glacier and usually deposited on bedrock of a different composition.

**glacial ice** Water in the solid state within a glacier; forms as snow partially melts and refreezes and compacts so that it is transformed first to firn and then to glacial ice.

**glacial polish** A smooth, glistening rock surface formed by the movement of sediment-laden ice over bedrock.

**glacial striation** A straight scratch rarely more than a few millimeters deep on a rock caused by the movement of sediment-laden glacial ice.

**glacial surge** A time of greatly accelerated flow in a glacier. Commonly results in displacement of the glacier's terminus by several kilometers.

**glaciation** Refers to all aspects of glaciers, including their origin, expansion, and retreat, and their impact on Earth's surface.

**glacier** A mass of ice on land that moves by plastic flow and basal slip.

***Glossopteris* flora** A Late Paleozoic association of plants found only on the Southern Hemisphere continents and India; named for its best-known genus, *Glossopteris*.

**Gondwana** A major Paleozoic continent composed of South America, Africa, Australia, India, and Antarctica.

**graded bedding** A sedimentary layer that shows a decrease in grain size from bottom to top.

**graded stream** A stream that has an equilibrium profile in which a delicate balance exists among gradient, discharge, flow velocity, channel characteristics, and sediment load so that neither significant deposition nor erosion takes place within its channel.

**gradient** The slope over which a stream or river flows; expressed in meters per kilometer (m/km) or feet per mile (ft/mi).

**greenhouse effect** The retention of heat in the atmosphere that results when carbon dioxide and other gases, such as methane, nitrous oxide, chlorofluorocarbons, and water vapor, allow sunlight to pass through them but traps the heat that is reflected back from Earth's surface.

**greenstone belt** A linear or podlike association of igneous and sedimentary rocks particularly common in Archean terrains.

**Grenville orogeny** An episode of deformation that took place in the eastern United States and Canada during the Neoproterozoic.

**ground moraine** The layer of sediment released from melting ice as a glacier's terminus retreats.

**groundwater** Underground water stored in the pore spaces of soil, sediment, and rock.

**guide fossil** Any easily identified fossil with an extensive geographic distribution and short geologic range useful for determining the relative ages of rocks in different areas.

**guyot** A flat-topped seamount of volcanic origin rising more than 1 km above the seafloor.

**gymnosperm** A flowerless, seed-bearing plant.

**gyre** A system of ocean currents rotating clockwise in the Northern Hemisphere and counterclockwise in the Southern Hemisphere.

**half-life** The time necessary for half of the original number of radioactive atoms of an element to decay to a stable daughter product; for example, the half-life for potassium 40 is 1.3 billion years.

**hanging valley** A tributary glacial valley whose floor is at a higher level than that of the main glacial valley.

**hanging wall block** The block of rock that overlies a fault plane.

**hardness** A term used to express the resistance of a mineral to abrasion.

**headland** Part of a shoreline, commonly bounded by cliffs, that extends out into the sea or a lake.

**heat** An agent of metamorphism.

**Hercynian (Variscan)–Alleghenian orogeny** Pennsylvanian to Permian orogenic event during which the Appalachian mobile belt in eastern North America and the Hercynian mobile belt of southern Europe were deformed.

**hominid** Abbreviated term for members of the subfamily Homininae, which includes the gorillas, chimpanzees, humans, and their extinct ancestors.

**hominin** Abbreviated term for members of the tribe Hominini, which includes humans and their extinct ancestors.

***Homo*** The genus of hominids consisting of *Homo sapiens* and their ancestors *Homo erectus* and *Homo habilis*.

**horn** A steep-walled, pyramid-shaped peak formed by the headward erosion of at least three cirques.

**hot spot** A localized zone of melting below the lithosphere that probably overlies a mantle plume; detected by volcanism at the surface.

**hot spring** A spring in which the water temperature is warmer than the temperature of the human body (37°C).

**hydraulic action** The removal of loose particles by the power of moving water.

**hydrologic cycle** The continuous recycling of water from the oceans, through the atmosphere, to the continents, and back to the oceans, or from the oceans, through the atmosphere, and back to the oceans.

**hydrolysis** The chemical reaction between hydrogen ($H^+$) ions and hydroxyl ($OH^-$) ions of water and a mineral's ions.

**hydrothermal** A term referring to hot water as in hot springs and geysers.

**hypothesis** A provisional explanation for observations that is subject to continual testing. If well supported by evidence, a hypothesis may be called a theory.

**ice cap** A dome-shaped mass of glacial ice that covers less than 50,000 km².

**ice-scoured plain** A low relief bedrock surface with glacial striations and polish eroded by a glacier.

**ichthyosaur** Any of the porpoiselike, Mesozoic marine reptiles.

**igneous rock** Any rock formed by cooling and crystallization of magma or lava or the consolidation of pyroclastic materials.

**incised meander** A deep, meandering canyon cut into bedrock by a stream or river.

**index mineral** A mineral that forms within specific temperature and pressure ranges during metamorphism.

**infiltration capacity** The maximum rate at which soil or sediment absorbs water.

**intensity** The subjective measure of the kind of damage done by an earthquake as well as people's reaction to it.

**intermediate magma** Magma with a silica content between 53% and 65% and an overall composition intermediate between mafic and felsic magma.

**ion** An electrically charged atom produced by adding or removing electrons from the outermost electron shell.

**ionic bond** A chemical bond resulting from the attraction between positively and negatively charged ions.

**isostatic rebound** The phenomenon in which unloading of the crust causes it to rise until it attains equilibrium with the underlying upper mantle.

**joint** A fracture along which no movement has occurred or where movement is perpendicular to the fracture surface.

**Jovian planet** Any of the four planets (Jupiter, Saturn, Uranus, and Neptune) that resemble Jupiter. All are large and have low mean densities, indicating that they are composed mostly of lightweight gases, such as hydrogen and helium, and frozen compounds, such as ammonia and methane.

**kame** A conical hill of stratified drift originally deposited in a depression on a glacier's surface.

**karst topography** Landscape consisting of numerous caves, sinkholes, and solution valleys formed by groundwater solution of rocks such as limestone and dolostone.

**Kaskaskia sequence** A widespread sequence of Devonian and Mississippian sedimentary rocks bounded above and below by unconformities that was deposited during a transgressive–regressive cycle of the Kaskaskia Sea.

**laccolith** A concordant pluton with a mushroomlike geometry.

**lahar** A mudflow composed of pyroclastic materials such as ash.

**Laramide orogeny** Late Cretaceous to Early Cenozoic phase of the Cordilleran orogeny; responsible for many structural features of the present-day Rocky Mountains.

**lateral moraine** A ridge of sediment deposited along the margin of a valley glacier.

**laterite** A red soil, rich in iron or aluminum, or both, resulting from intense chemical weathering in the tropics.

**Laurasia** A Late Paleozoic Northern Hemisphere continent made up of North America, Greenland, Europe, and Asia.

**Laurentia** A Proterozoic continent composed of North America, Greenland, parts of Scotland, and perhaps part of the Baltic shield of Scandinavia.

**lava** Magma that reaches Earth's surface.

**lava dome** A bulbous, steep-sided volcano formed by viscous magma moving upward through a volcanic conduit.

**lava tube** A tunnel beneath the solidified surface of a lava flow through which lava moves; also, the hollow space left when the lava within a tube drains away.

**lithification** The process of converting sediment into sedimentary rock by compaction and cementation.

**lithosphere** Earth's outer, rigid part, consisting of the upper mantle, oceanic crust, and continental crust.

**lithostatic pressure** Pressure exerted on rocks by the weight of overlying rocks.

**lithostratigraphic unit** A body of rock, such as a formation, defined solely by its physical attributes.

**Little Ice Age** An interval from about 1500 to the mid- to late-1800s during which glaciers expanded to their greatest historic extent.

**loess** A wind-blown deposit of silt and clay.

**longitudinal dune** A long ridge of sand generally parallel to the direction of the prevailing wind.

**longshore current** A current resulting from wave refraction found between the breaker zone and a beach that flows parallel to the shoreline.

**Love wave (L-wave)** A surface wave in which the individual particles of material move only back and forth in a horizontal plane perpendicular to the direction of wave travel.

**luster** The appearance of a mineral in reflected light. Luster is metallic or nonmetallic, although the latter has several subcategories.

**mafic magma** Magma with between 45% and 52% silica and proportionately more calcium, iron, and magnesium than intermediate and felsic magma.

**magma** Molten rock material generated within Earth.

**magma chamber** A reservoir of magma within Earth's upper mantle or lower crust.

**magma mixing** The process whereby magmas of different composition mix together to yield a modified version of the parent magmas.

**magnetic anomaly** Any deviation, such as a change in average strength, in Earth's magnetic field.

**magnetic field** The area in which magnetic substances are affected by lines of magnetic force emanating from Earth.

**magnetic reversal** The phenomenon involving the complete reversal of the north and south magnetic poles.

**magnetism** A physical phenomenon resulting from moving electricity and the spin of electrons in some solids in which magnetic substances are attracted toward one another.

**magnitude** The total amount of energy released by an earthquake at its source.

**mantle** The thick layer between Earth's crust and core.

**mantle plume** A cylindrical mass of magma rising from the mantle toward the surface; recognized at the surface by a hot spot, an area such as the Hawaiian Islands where volcanism takes place.

**marine regression** The withdrawal of the sea from a continent or coastal area, resulting in the emergence of the land as sea level falls or the land rises with respect to sea level.

**marine terrace** A wave-cut platform now above sea level.

**marine transgression** The invasion of a coastal area or a continent by the sea, resulting from a rise in sea level or subsidence of the land.

**marsupial mammal** Pouched mammals, such as kangaroos and wombats, that give birth to their young in a very immature state.

**mass wasting** The downslope movement of Earth materials under the influence of gravity.

**matter** Anything that has mass and occupies space.

**meandering stream** A stream that has a single, sinuous channel with broadly looping curves.

**mechanical weathering** Disaggregation of rocks by physical processes that yields smaller pieces that retain the composition of the parent material.

**medial moraine** A moraine carried on the central surface of a glacier; formed where two lateral moraines merge.

**Mediterranean belt** A zone of seismic and volcanic activity extending through the Mediterranean region of southern Europe and eastward to Indonesia.

**mesa** A broad, flat-topped erosional remnant bounded on all sides by steep slopes.

**metamorphic facies** A group of metamorphic rocks characterized by particular minerals that formed under the same broad temperature and pressure conditions.

**metamorphic rock** Any rock that has been changed from its original condition by heat, pressure, and the chemical activity of fluids, as in marble and slate.

**metamorphic zone** The region between lines of equal metamorphic intensity known as isograds.

**metamorphism** The phenomenon of changing rocks subjected to heat, pressure, and fluids so that they are in equilibrium with a new set of environmental conditions. Metamorphism takes place in the solid state.

**Midcontinent rift** A Late Proterozoic rift in Laurentia in which volcanic and sedimentary rocks accumulated.

**Milankovitch theory** An explanation for the cyclic variations in climate and the onset of ice ages as a result of irregularities in Earth's rotation and orbit.

**mineral** A naturally occurring, inorganic, crystalline solid that has characteristic physical properties and a narrowly defined chemical composition.

**mobile belt** An elongated area of deformation generally along the margins of a craton, such as the Appalachian mobile belt.

**Modified Mercalli Intensity Scale** A scale with values from I to XII used to characterize earthquakes based on damage.

**Mohorovičić discontinuity (Moho)** The boundary between Earth's crust and mantle.

**monocline** A bend or flexure in otherwise horizontal or uniformly dipping rock layers.

**monotreme** The egg-laying mammals; includes only the platypus and spiny anteater of the Australian region.

**mud crack** A crack in clay-rich sediment that forms in response to drying and shrinkage.

**mudflow** A flow consisting mostly of clay- and silt-sized particles and up to 30% water that moves downslope under the influence of gravity.

**natural levee** A ridge of sandy alluvium deposited along the margins of a channel during floods.

**nearshore sediment budget** The balance between additions and losses of sediment in the nearshore zone.

**neutron** An electrically neutral particle found in the nucleus of an atom.

**Nevadan orogeny** Late Jurassic to Cretaceous phase of the Cordilleran orogeny; most strongly affected the western part of the Cordilleran mobile belt.

**nonconformity** An unconformity in which stratified sedimentary rocks overlie an erosion surface cut into igneous or metamorphic rocks.

**nonferromagnesian silicate** A silicate mineral that has no iron or magnesium.

**nonfoliated texture** A metamorphic texture in which there is no discernable preferred orientation of minerals.

**normal fault** A dip-slip fault on which the hanging wall block has moved downward relative to the footwall block.

**North American Cordillera** A complex mountainous region in western North America extending from Alaska into central Mexico.

**nucleus** The central part of an atom consisting of protons and neutrons.

**nuée ardente** A fast-moving, dense cloud of hot pyroclastic materials and gases ejected from a volcano.

**numerical dating** Uses various radioactive decay dating techniques to assign ages to rocks in years before the present.

**oblique-slip fault** A fault showing both dip-slip and strike-slip movement.

**oceanic–continental plate boundary** A convergent plate boundary along which oceanic lithosphere is subducted beneath continental lithosphere.

**oceanic–oceanic plate boundary** A convergent plate boundary along which two oceanic plates collide and one is subducted beneath the other.

**oceanic ridge** A mostly submarine mountain system composed of basalt and found in all ocean basins.

**oceanic trench** A long, narrow feature restricted to active continental margins and along which subduction occurs.

**ooze** Deep-sea sediment composed mostly of shells of marine animals and plants.

**organic evolution** The theory holding that all living things are related and that they descended with modification from organisms that lived during the past.

**Ornithischia** One of the two orders of dinosaurs, characterized by a birdlike pelvis; includes ornithopods, stegosaurs, ankylosaurs, pachycephalosaurs, and ceratopsians.

**orogen** A linear part of Earth's crust that was, or is, being deformed during an orogeny; part of an orogenic belt.

**orogeny** An episode of mountain building involving deformation, usually accompanied by igneous activity, metamorphism, and crustal thickening.

**Ouachita orogeny** A period of mountain building that took place in the Ouachita mobile belt during the Pennsylvanian Period.

**outwash plain** The sediment deposited by meltwater discharging from a continental glacier's terminus.

**oxbow lake** A cutoff meander filled with water.

**oxidation** The reaction of oxygen with other atoms to form oxides or, if water is present, hydroxides.

**pahoehoe** A type of lava flow with a smooth ropy surface.

**paleomagnetism** Residual magnetism in rocks, studied to determine the intensity and direction of Earth's past magnetic field.

**Pangaea** The name Alfred Wegener proposed for a supercontinent consisting of all of Earth's landmasses at the end of the Paleozoic Era.

**parabolic dune** A crescent-shaped dune with its tips pointing upwind.

**parent material** The material that is chemically and mechanically weathered to yield sediment and soil.

**passive continental margin** A continental margin within a tectonic plate as along the East Coast of North America where little seismic activity and no volcanism occur; characterized by a broad continental shelf and a continental slope and rise.

**pediment** An erosion surface of low relief gently sloping away from the base of a mountain range.

**pelagic clay** Brown or red deep-sea sediment composed of clay-sized particles.

**pelycosaur** Pennsylvanian to Permian reptiles that had some mammal characteristics; many species had large fins on their back.

**permafrost** Ground that remains permanently frozen.

**permeability** A material's capacity to transmit fluids.

**phaneritic texture** Igneous rock texture in which minerals are easily visible without magnification.

**photic zone** The sunlit layer in the oceans where plants photosynthesize.

**pillow lava** Bulbous masses of basalt, resembling pillows, formed when lava is rapidly chilled under water.

**placental mammal** All mammals with a placenta to nourish the developing embryo, as opposed to egg-laying mammals (monotremes) and pouched mammals (marsupials).

**plastic flow** The flow that takes place in response to pressure and causes deformation with no fracturing.

**plastic strain** Permanent deformation of a solid with no failure by fracturing.

**plate** An individual segment of the lithosphere that moves over the asthenosphere.

**plate tectonic theory** The theory holding that large segments of Earth's outer part (lithospheric plates) move relative to one another.

**platform** The part of a craton that lies buried beneath flat-lying or mildly deformed sedimentary rocks.

**playa** A dry lakebed found in deserts.

**plesiosaur** A type of Mesozoic marine reptile; short-necked and long-necked plesiosaurs existed.

**pluton** An intrusive igneous body that forms when magma cools and crystallizes within the crust, such as a batholith or sill.

**plutonic (intrusive igneous) rock** Igneous rock that formed from magma intruded into or formed in place within the crust.

**point bar** The sediment body deposited on the gently sloping side of a meander loop.

**porosity** The percentage of a material's total volume that is pore space.

**porphyritic texture** An igneous texture with minerals of markedly different sizes; results from slow cooling of magma and generally indicates an intrusive origin.

**pressure release** A mechanical weathering process in which rocks that formed under pressure expand on being exposed at the surface.

**primate** Any of the mammals that belong to the order Primates; includes prosimians (lemurs and tarsiers), monkeys, apes, and humans.

**principle of cross-cutting relationships** A principle holding that an igneous intrusion or fault is younger than the rocks it intrudes or cuts across.

**principle of fossil succession** A principle holding that fossils, and especially groups or assemblages of fossils, succeed one another through time in a regular and predictable order.

**principle of inclusions** A principle holding that inclusions or fragments in a rock unit are older than the rock unit itself; for example, granite inclusions in sandstone are older than the sandstone.

**principle of isostasy** The theoretical concept of Earth's crust "floating" on a dense underlying layer.

**principle of lateral continuity** A principle holding that rock layers extend outward in all directions until they terminate.

**principle of original horizontality** According to this principle, sediments are deposited in horizontal or nearly horizontal layers.

**principle of superposition** A principle holding that in a vertical sequence of undeformed sedimentary rocks, the relative ages of the rocks can be determined by their position in the sequence—oldest at the bottom followed by successively younger layers.

**principle of uniformitarianism** A principle holding that past events can be interpreted by understanding present-day processes; based on the idea that natural processes have always operated in the same way.

**prokaryotic cell** A cell that lacks a nucleus and organelles such as mitochondria and plastids; the cells of bacteria and cyanobacteria.

**proton** A positively charged particle found in the nucleus of an atom.

**pterosaur** Any of the Mesozoic flying reptiles.

**P-wave** A compressional, or push-pull, wave; the fastest seismic wave and one that can travel through solids, liquids, and gases; also called a primary wave.

**P-wave shadow zone** An area between 103 and 143 degrees from an earthquake's focus where little P-wave energy is recorded by seismographs.

**pyroclastic (fragmental) texture** A fragmental texture characteristic of igneous rocks composed of pyroclastic materials.

**pyroclastic material** Fragmental substances, such as ash, explosively ejected from a volcano.

**pyroclastic sheet deposit** Vast, sheetlike deposit of felsic pyroclastic materials erupted from fissures.

**quick clay** A clay deposit that spontaneously liquefies and flows when disturbed.

**radioactive decay** The spontaneous change of an atom to an atom of a different element by emission of a particle from its nucleus (alpha and beta decay) or by electron capture.

**rain-shadow desert** A desert found on the lee side of a mountain range because precipitation falls mostly on the windward side of the range.

**rapid mass movement** Any kind of mass wasting that involves a visible downslope displacement of material.

**Rayleigh wave (R-wave)** A surface wave in which individual particles of material move in an elliptical path within a vertical plane oriented in the direction of wave movement.

**recessional moraine** An end moraine that forms when a glacier's terminus retreats, then stabilizes, and a ridge or mound of till is deposited.

**reef** A moundlike, wave-resistant structure composed of the skeletons of organisms.

**regional metamorphism** Metamorphism that occurs over a large area, resulting from high temperatures, tremendous pressure, and the chemical activity of fluids within the crust.

**regolith** The layer of unconsolidated rock and mineral fragments and soil that covers most of the land surface.

**relative dating** The process of determining the age of an event as compared to other events; involves placing geologic events in their correct chronologic order, but does not involve consideration of when the events occurred in number of years ago.

**reserve** The part of the resource base that can be extracted economically.

**resource** A concentration of naturally occurring solid, liquid, or gaseous material, in or on Earth's crust in such form and amount that economic extraction of a commodity from the concentration is currently or potentially feasible.

**reverse fault** A dip-slip fault on which the hanging wall block has moved upward relative to the footwall block.

**Richter Magnitude Scale** An open-ended scale that measures the amount of energy released during an earthquake.

**rip current** A narrow surface current that flows out to sea through the breaker zone.

**ripple mark** A wavelike (undulating) structure produced in granular sediment, especially sand, by unidirectional wind and water currents or by oscillating wave currents.

**rock** A solid aggregate of one or more minerals, as in limestone and granite, or a consolidated aggregate of rock fragments, as in conglomerate, or masses of rocklike materials, such as coal and obsidian.

**rock cycle** A pictorial representation of events leading to the origin, destruction, or changes, and reformation of rocks as a consequence of Earth's internal and surface processes. It also shows how the three major rock groups are interrelated and how any rock type can be derived from any other rock type.

**rockfall** A type of extremely fast mass wasting in which rocks fall through the air.

**rock-forming mineral** Any mineral common in rocks that is important in their identification and classification.

**rock slide** Rapid mass wasting in which rocks move downslope along a more or less planar surface.

**runoff** The surface flow in streams and rivers.

**salinity** A measure of the dissolved solids in seawater, commonly expressed in parts per thousand.

**salt crystal growth** A mechanical weathering process in which salt crystals growing in cracks and pores disaggregate rocks.

**Sauk sequence** A widespread association of sedimentary rocks bounded above and below by unconformities that was deposited during the latest Proterozoic to Early Ordovician transgressive–regressive cycle of the Sauk Sea.

**Saurischia** An order of dinosaurs characterized by a lizardlike pelvis; includes theropods, prosauropods, and sauropods.

**scientific method** A logical, orderly approach that involves gathering data, formulating and testing hypotheses, and proposing theories.

**seafloor spreading** The theory that the seafloor moves away from spreading ridges and is eventually consumed at subduction zones.

**seamount** A submarine volcanic mountain rising at least 1 km above the seafloor.

**sediment** Loose aggregate of solids derived by weathering from preexisting rocks, or solids precipitated from solution by inorganic chemical processes or extracted from solution by organisms.

**sedimentary facies** Any aspect of a sedimentary rock unit that makes it recognizably different from adjacent sedimentary rocks of the same or approximately the same age.

**sedimentary rock** Any rock composed of sediment that forms at or near Earth's surface.

**sedimentary structure** Any feature in sedimentary rock that formed at or shortly after the time of deposition, such as cross-bedding, animal burrows, and mud cracks.

**seedless vascular plant** A plant with specialized tissues for transporting fluids and nutrients that reproduces by spores rather than seeds, as in ferns and horsetail rushes.

**seismograph** An instrument that detects, records, and measures the various waves produced by earthquakes.

**seismology** The study of earthquakes.

**Sevier orogeny** Cretaceous phase of the Cordilleran orogeny that affected the continental shelf and slope areas of the Cordilleran mobile belt.

**shear strength** The resisting forces that help maintain a slope's stability.

**shear stress** The result of forces acting parallel to one another but in opposite directions; results in deformation by displacement of adjacent layers along closely spaced planes.

**shield** A vast area of exposed ancient rocks on a continent; the exposed part of a craton.

**shield volcano** A dome-shaped volcano with a low, rounded profile built up mostly by overlapping basalt lava flows.

**shoreline** The area between mean low tide and the highest level on land affected by storm waves.

**silica** A compound of silicon and oxygen.

**silica tetrahedron** The basic building block of all silicate minerals; consists of one silicon atom and four oxygen atoms.

**silicate** A mineral that contains silica, such as quartz ($SiO_2$).

**sill** A tabular or sheetlike concordant pluton.

**sinkhole** A depression in the ground that forms by the solution of the underlying carbonate rocks or by the collapse of a cave roof.

**slide** Mass wasting involving movement of material along one or more surfaces of failure.

**slow mass movement** Mass movement that advances at an imperceptible rate and is usually detectable only by the effects of its movement.

**slump** Mass wasting that takes place along a curved surface of failure and results in the backward rotation of the slump mass.

**soil** Regolith consisting of weathered materials, water, air, and humus that can support vegetation.

**soil degradation** Any process leading to a loss of soil productivity; may involve erosion, chemical pollution, or compaction.

**soil horizon** A distinct soil layer that differs from other soil layers in texture, structure, composition, and color.

**solar nebula theory** A theory for the evolution of the solar system from a rotating cloud of gas.

**solifluction** Mass wasting involving the slow downslope movement of water-saturated surface materials.

**solution** A reaction in which the ions of a substance become dissociated in a liquid and the solid substance dissolves.

**specific gravity** The ratio of a substance's weight, especially a mineral, to an equal volume of water at 4°C.

**spheroidal weathering** A type of chemical weathering in which corners and sharp edges of rocks weather more rapidly than flat surfaces, thus yielding spherical shapes.

**spit** A fingerlike projection of a beach into a body of water such as a bay.

**spring** A place where groundwater flows or seeps out of the ground.

**stock** An irregularly shaped discordant pluton with a surface area smaller than 100 km².

**stoping** A process in which rising magma detaches and engulfs pieces of the country rock.

**storm surge** The surge of water onto a shoreline as a result of a bulge in the ocean's surface beneath the eye of a hurricane and wind-driven waves.

**strain** Deformation caused by stress.

**strata (singular, stratum)** Refers to layering in sedimentary rocks.

**stratified drift** Glacial deposits that show both stratification and sorting; deposited by streams that discharge from glaciers.

**stream terrace** An erosional remnant of a floodplain that formed when a stream was flowing at a higher level.

**stress** The force per unit area applied to a material such as rock.

**strike** The direction of a line formed by the intersection of an inclined plane and a horizontal plane.

**strike-slip fault** A fault involving horizontal movement of blocks of rock on opposite sides of a fault plane.

**stromatolite** A biogenic sedimentary structure, especially in limestone, produced by the entrapment of sediment on sticky mats of photosynthesizing bacteria.

**submarine hydrothermal vent** A crack or fissure in the seafloor through which superheated water issues.

**submergent coast** A coast along which sea level rises with respect to the land or the land subsides.

**Sundance Sea** A wide seaway that existed in western North America during the Middle Jurassic Period.

**superposed stream** A stream that once flowed on a higher surface and eroded downward into resistant rocks while maintaining its course.

**suspended load** Consists of the smallest particles of silt and clay, which are kept suspended above the channel's bed by fluid turbulence.

**S-wave** A shear wave that moves material perpendicular to the direction of travel, thereby producing shear stresses in the material it moves through; also known as a secondary wave; S-waves travel only through solids.

**S-wave shadow zone** Those areas more than 103 degrees from an earthquake's focus where no S-waves are recorded.

**syncline** A down-arched fold in which the youngest exposed rocks coincide with the fold axis and all strata dip toward the axis.

**system** A combination of related parts that interact in an organized fashion; Earth systems include the atmosphere, hydrosphere, biosphere, and solid Earth.

**Taconic orogeny** An Ordovician episode of mountain building that resulted in the deformation of the Appalachian mobile belt.

**talus** Accumulation of coarse, angular rock fragments at the base of a slope.

**tension** A type of stress in which forces act in opposite directions but along the same line, thus tending to stretch an object.

**terminal moraine** An end moraine consisting of a ridge or mound of rubble marking the farthest extent of a glacier.

**terrane** A small lithospheric block with characteristics quite different from those of surrounding rocks. Terranes probably consist of seamounts, oceanic rises, and other seafloor features accreted to continents during orogenies.

**terrestrial planet** Any of the four innermost planets (Mercury, Venus, Earth, and Mars). They are all small and have high mean densities, indicating that they are composed of rock and metallic elements.

**theory** An explanation for some natural phenomenon that has a large body of supporting evidence. To be scientific, a theory must be testable—for example, plate tectonic theory.

**therapsid** Permian to Triassic mammal-like reptiles; the ancestors of mammals are among one group of therapsids known as cynodonts.

**thermal convection cell** A type of circulation of material in the asthenosphere during which hot material rises, moves laterally, cools and sinks, and is reheated and continues the cycle.

**thermal expansion and contraction** A type of mechanical weathering in which the volume of rocks changes in response to heating and cooling.

**thrust fault** A type of reverse fault in which a fault plane dips less than 45 degrees.

**tide** The regular fluctuation of the sea's surface in response to the gravitational attraction of the Moon and Sun.

**till** All sediment deposited directly by glacial ice.

**time-stratigraphic unit** A body of strata that was deposited during a specific interval of geologic time; for example, the Devonian System (a time-stratigraphic unit) was deposited during the Devonian Period (a time unit).

**time unit** Any of the units such as eon, era, period, epoch, and age, used to refer to specific intervals of geologic time.

**Tippecanoe sequence** A widespread body of sedimentary rocks bounded above and below by unconformities that was deposited during an Ordovician to Early Devonian transgressive–regressive cycle of the Tippecanoe Sea.

**tombolo** A type of spit that extends out from the shoreline and connects the mainland with an island.

**transform fault** A fault along which one type of motion is transformed into another; commonly displaces oceanic ridges; on land, recognized as a strike-slip fault, such as the San Andreas Fault.

**transform plate boundary** Plate boundary along which plates slide past one another and crust is neither produced nor destroyed.

**transverse dune** A ridge of sand with its long axis perpendicular to the wind direction.

**tsunami** A large sea wave that is usually produced by an earthquake, but can also result from submarine landslides and volcanic eruptions.

**ultramafic magma** Magma with less than 45% silica.

**unconformity** A break in the geologic record represented by an erosional surface separating younger rocks from older rocks.

**upwelling** The slow circulation of ocean water from depth to the surface.

**U-shaped glacial trough** A valley with steep or vertical walls and a broad, rather flat floor formed by the movement of a glacier through a stream valley.

**valley** A linear depression bounded by higher areas such as ridges or mountains.

**valley glacier** A glacier confined to a mountain valley or an interconnected system of mountain valleys.

**valley train** A long, narrow deposit of stratified drift confined within a glacial valley.

**varve** Alternating finely laminated deposits of light and dark layers of sediment that represent an annual episode of deposition.

**vascular plant** A plant with specialized tissue for transporting nutrients and fluids.

**velocity** A measure of distance traveled per unit of time, as in the flow velocity in a stream or river.

**ventifact** A stone with a surface polished, pitted, grooved, or faceted by wind abrasion.

**vertebrate** Any animal that has a segmented vertebral column, as in fish, amphibians, reptiles, birds, and mammals.

**vesicle** A small hole or cavity formed by gas trapped in cooling lava.

**viscosity** A fluid's resistance to flow.

**volcanic ash** Pyroclastic materials that measure less than 2 mm.

**volcanic explosivity index (VEI)** A semiquantitative scale for the size of a volcanic eruption based on evaluation of criteria such as the volume of material explosively erupted and the height of the eruption cloud.

**volcanic neck** An erosional remnant of the material that solidified in a volcanic pipe.

**volcanic pipe** The conduit connecting the crater of a volcano with an underlying magma chamber.

**volcanic (extrusive igneous) rock** An igneous rock formed when magma is extruded onto Earth's surface where it cools and crystallizes, or when pyroclastic materials become consolidated.

**volcanic tremor** Ground motion lasting from minutes to hours, resulting from magma moving beneath the surface, as opposed to the sudden jolts produced by most earthquakes.

**volcanism** The processes whereby magma and its associated gases rise through the crust and are extruded onto the surface or into the atmosphere.

**volcano** A hill or mountain formed around a vent as a result of the eruption of lava and pyroclastic materials.

**water table** The surface that separates the zone of aeration from the underlying zone of saturation.

**water well** A well made by digging or drilling into the zone of saturation.

**wave** An undulation on the surface of a body of water, resulting in the water surface rising and falling.

**wave base** The depth corresponding to about one-half wavelength, below which water in unaffected by surface waves.

**wave-cut platform** A beveled surface that slopes gently seaward; formed by the erosion and retreat of a sea cliff.

**wave refraction** The bending of waves so that they move nearly parallel to the shoreline.

**weathering** The physical breakdown and chemical alteration of rocks and minerals at or near Earth's surface.

**zone of accumulation** The part of a glacier where additions exceed losses and the glacier's surface is perennially covered with snow. Also refers to horizon B in soil where soluble material leached from horizon A accumulates as irregular masses.

**zone of aeration** The zone above the water table that contains both air and water within the pore spaces of soil, sediment, or rock.

**zone of saturation** The area below the water table in which all pore spaces are filled with water.

**zone of wastage** The part of a glacier where losses from melting, sublimation, and calving of icebergs exceed the rate of accumulation.

# INDEX

**Note:** Page numbers in italics refer to illustrations; those followed by "t" refer to tables.

Aa (lava flow), 92, *93, 394*
Abrasion, 121, 226, *227, 290*
   by glaciers, 274, *274*
   shorelines and, 318, *318*
Absaroka Sea, 376
Absaroka sequence, 375–378
Absolute dating, 332
Abyssal plains, 28, *29, 31*
Acadian orogeny, 371, 380
Acanthodians, *404*
Acasta Gneiss, 138, *138, 363*
Accessory minerals, 64
Accretionary wedge, 193, *193, 195*
Acidity, 112
Acid rain, 112–113, *113*
Acids, 112, 118
Active continental margins, 31, *31*
Active volcanoes, 90
Adirondack Highlands, 142, *142*
Aeration zone. *See* Zone of aeration
Agriculture, 3, 119
Air circulation, *296*
Air pressure belts, 295, *296*
Airy, George, 196
Alaskan earthquakes, 154t, 165, *215*
Alkali soils, 118, *118*
Alleghenian orogeny, 372
Alluvial fans, *231*, 231–232, 299, *302*
Alluvium, 227, 230, 231, 232, 240
Alpha decay, 348, *349*
Alpine glaciers, 268. *See also* Valley
   glaciers
Alpine-Himalayan belt, 389
Alpine orogeny, 389
Alps, 176, *333*
Aluminum, 53, 57, *57*, 65, 66.
    *See also* Bauxite
Amethyst crystals, 51
Ammonites, *131*
Amniote egg, 405, *406*
Amorphous texture, 123t
Amphibians, 404–406
Amphibole, 59, *60*, 62, 64t, 145, *146*, 150
Amphibolite, 144t, 146, 148, *149*
Anak Krakatau volcano, *89*
Ancestral Rockies, 376, *376*
Anchorage quick clay slide, 212, *215*
Andesite, 45, 80, *80, 81*
Andes Mountains, 39, *40*, 194–195, *195*,
   372, 389
   andesite in, 80
   circum-Pacific belt and, 101

metallic ores in, 45
Andrews, Sarah, 4
Angiosperms, 416
Angle of repose, 202, *203*
Angular unconformity, *306*, 339, 341,
   *342, 345*
Anhydrite, 58t, 64t, *126*
Animals, *333*
   fossils and, 129
   land-dwelling, 129, 131
   soil and, 116, 118, 125
   weathering and, 111, 115
Antarctica, glaciers in, *267, 269, 273, 274,*
   *284, 285*
Anthracite, 144t, 147–148
Anthracite coal, 126, 132
Anthropocene Epoch, 334
Anthropoids, 423
Anticlines, *22*, 185, *185, 186, 187, 187*
Antler Highlands, *376*, 381
Antler orogeny, 371, *376*, 381
Apatite, 58t, 63t
Aphanitic texture, 77, *78, 79*, 80
Aphelion, 284
Aphotic zone, 309
Appalachian Basin, 256
Appalachian mobile belt, 378–380, *379*
Appalachian Mountains, 195, *333*, 372,
   *373*, 389
   Cenozoic and, 391
   continental drift and, 25
   Mesozoic and, 383, 391
   Precambrian and, 364
Aquicludes, 245, 247, *247*, 249, *249*
Aquifers, 245, 247–250, *249*, 254,
   *255, 258*
Arabian plate, *22*
Aragonite, 56, 61, 65
*Archaeopteryx*, 414, *416*
Archean Eon, *333, 362*, 364–366
   North America in, 180
   stromatolites in, 399–400
Arches, 109
Archosaurs, 409
Arctic Ocean, *307*
*Ardipithecus ramidus*, 423, *424*
Arêtes, 277, *277*
Argentite, 61
Argon, 54
Arid regions, 225, 231, *296*
Arkose, 124
Armenia earthquake, 154t

Artesian-pressure surface, 249, *249*
Artesian systems, *249*, 249–250
Asbestos, *150*, 150–151, *151*
Aseismic ridges, 29–30, *30*
Ash, 82. *See also* Volcanic ash
Ashfall, 353, *353*
Ash flow, 94
Assimilation, 76, *76*, 77, 78, 85
Asteroids, 11
Asthenosphere, 12, *13, 173*, 175
Atlantic Ocean, *307, 307*
*Atlantis, 307, 308*
Atmosphere, *1*, 295, *296*, 307
   as Earth subsystem, 2, *3*
   plate tectonics and, 12
   volcanism and, 90, 91
Atmospheric gases, 112
Atmospheric pollution, 112
Atolls, 309, *310*
Atomic mass number, 53–54, *54*
Atomic number, 53, *54*
Atoms, 52–54, *53*, 77
Augite, 62–63, *63*, 73
Aureole, 140, *141*
Australopithecine, 424, 425
*Australopithecus afarensis*, 424, *425*
*Australopithecus anamensis*, 424, *424*
Avalonia, 29, 368, *370*, 381
Axial planes, 185, *185, 186*, 187
Axial tilt, 283

Back-arc basin, 39, *193*, 194, 364, *365*
Bacon, Sir Francis, 23
Bacteria, 117–118, 126
Badlands National Park, *108*
Badlands topography, *108*
Bakken Shale, 256
Baltica, 368, 369, *370*, 380
Baltic Shield, 366
Banded iron formation, *133, 134*, 366
Barchan dunes, *293, 295*
Barnett Shale, 256
Barrier islands, 323, *323*, 325, 327, *327*
Barrier reef, 309, *310*
Barrow, George, 148
Basal slip, 271, *271, 273*
Basalt, 12, *15*, 27, 176, 196
   classification of, 79–80, *80*
   metamorphism and, 139, *141*, 146
   in oceanic ridges, 28
   weathering and, 115
Basaltic flows, 37, 390, *392*

Basalt plateaus, *100,* 100–101
Basalt porphyry, 77
Base level, 236–238, *237,* 240
Basin(s), 187–188, *188, 193,* 194
Basin and Range Province, 176, 189, 191, *192,* 390, *390, 391, 392*
Batholiths, 85, *86,* 194, 338, 387, *387*
Bauxite, 120
Baymouth bars, 321–323, *322*
Beaches, 319–321, *321,* 325, 327
Bed (of sedimentary rock), 128, *128*
Bedded chert, 126, *126*
Bed load, 226, *227,* 289–290
Bellini, Giovanni, 4
Belts, *157, 158, 158,* 159, *169*
Benching, *219,* 219–220
Benioff-Wadati zones, 158, *158*
Benioff zones, 37, 158, *158*
Bering Glacier, *267*
Berms, 319
Beta decay, 348, *349*
Big Bang, 8–9
Bighorn Mountains, 184, *184*
Bingham Copper Mine, *47,* 85
Biochemical sedimentary rocks, 123t, 124, 126, *126,* 245
Biosphere, *1,* 91
    as Earth subsystem, 2, *3*
Biostratigraphic unit, 354
Biotic provinces, 47
Biotite, *52,* 59, *60,* 62, 64t, 66
    metamorphism and, *141,* 142, *143,* 145, 148
Biozones, 354
Bipedal animals, 409, *410,* 411, 423
Bird-hipped dinosaurs, 409. *See also* Ornithischia
Birds, 414, 416, *416*
    Cenozoic, 418
Bishop Tuff, 101
Bituminous coal, 126, *126,* 132
Black Hills, *69, 81, 111,* 249, 261
Black smoker, 28, *29*
Block faulting, 191–192, *192,* 383, *386,* 390
Blocks (volcanic), 94, *95*
Block slide, 208, *209. See also* Rock slides
Blowouts, 290, *291. See also* Deflation hollows
Blue-green algae, 399
Blue Haze Famine, 91
Blueschist facies, 149, *149,* 193
Body fossils, 129
Body waves (seismic), *159,* 159–160
Bombs (volcanic), 94, *95*
Bonding, 54–55, *55, 56*
Bootlegger Cove Clay, 212, *215*
Bornite, 151
Bottomset beds, 230, *230*
Bowen's reaction series, 72–74, *73*
Braided streams, 227, *228,* 229
Breakers, 315, 319, 320

Brittle rocks, 183, 188
Brook, 225
Bryce Canyon National Park, *336,* 345, *347, 392*
Buffon, Georges Louis de, 333, 335
Bullard, Edward, 24
Bulldozing, 274
Buoyancy, 196
Burgess Shale, *398*
Burgess Shale biota, 402
Buttes, 303, *303*

Calcareous tufa, 262
Calcite, 56, *59,* 61, *61,* 62, 64t, 65, 114, 118, 121, 123t, 125
Calcium, *57,* 58
    in mafic magma, 71, 71t
Calcium carbonate, 56, 61, 63, 122, *125*
    hot springs/geysers and, 262, *263*
    in ooze, 309
Caldera, *89,* 91, 95, 96, *96, 97,* 98, 101
Caledonian orogeny, 368, 369
Caliche, 118
Calico Mountains, *185*
California
    earthquakes in, 154t, 158, *158,* 164
    mass wasting in, *204,* 205, *206, 207,* 208, 209, *209, 212,* 218
    San Andreas Fault in (*See* San Andreas Fault)
Calving, 270, *272*
"Cambrian explosion," 401
Cambrian Period, *333,* 367, *370, 373*
    history of life in, 401–402, 401t
    Sauk sequence in, 372
    vertebrates in, 404
Canadian shield, 278, 364, *364, 372, 373*
Canyons, 238, 240, 392
Capillary fringe, 246, *246*
Cap rock, 132
Carbon, 53–55, *54,* 56, *57,* 63, 329
Carbon-14 dating, 352, *353*
Carbonate minerals, 58t, 61, *61,* 378, 391
Carbonate radical, 58, *59,* 61, 124
Carbonate rocks, 124, 252, *252,* 366
Carbon cycle, 6
Carbon dioxide, 114, 403
Carbon dioxide cycle, 6
Carbonic acid, 114, 122
Carboniferous Period, 372, 375, 402–403, 407t
Carnotite, 133
Cascade Range, 102, *102,* 192, 236, *276,* 390
    andesite in, 80
    circum-Pacific belt and, 101
Catskill Delta, 380
Cave of Crystals, 252
Cavern, 252
Caves/cave deposits, 252–253, *253, 254*
Cementation, 122, *123*
Cenozoic birds, 418

Cenozoic Era, *333,* 345, 383, *383,* 386, *387,* 389
    Cordilleran orogeny in, 386–387
    eastern North America in, 391–392
    Gulf Coast in, 390–391
    history of life in, 418–427
    lava flows in, *392,* 418
    mammals in, 418–420
    natural resources and, 45
    Ouachita mobile belt in, 380
Cephalopoda, *131*
Cervantes, Miguel, 300
Chalcopyrite, 151
Chalk, 125
Challenger Deep, 29
Channel flow, 224–225
Chemical sediment, 121, 122
Chemical sedimentary rocks, *120,* 123t, 124–126, *125,* 245
Chemical weathering, 111, 112–115, *115, 120,* 124, 207, 209, 298. *See also* Weathering
    particle size and, *115*
    rate of, 114–115
Chert, 126, *126,* 366
Chile earthquake, 154t, 163
China (Paleozoic), 369, *370, 371*
Chinese earthquakes, 154t, 165, 167, *167,* 168, 170
Chinle Formation, 388
Chloride, 308
Chlorine, 54–55, *55, 57,* 328, 403
Chlorite, 140, 142, 143, 144t, 145, 147, 148
Chromium, 66
Chronostratigraphic units, 354
Chrysotile, 150, *151*
Chugach Mountains, *394*
Cinder cones, 95, 98, *98, 394*
Circum-Pacific belt, 101, *101, 157,* 158, *158,* 169
    in Cenozoic, 389
    in North American Cordillera, 389, 390, *390*
Cirques, 276–278, *277*
Claron Formation, *392*
Clastic texture, 123, 123t, 124
Clastic wedge, 380
Clasts, 123, *124*
Clay, 121, 205, 207t, 209, 211–213, *215,* 309
Clay minerals, 59
    chemical weathering and, 112, 114
    clay *vs.,* 121
    metamorphism and, 142
    soil and, 116, 118
    uses of, 65
Clay-rich sedimentary rocks, 146, 147, 148
Claystone, 124
Cleavage, 57, *62,* 62–63, *63,* 66
Climate, 2, 6, 114, 372. *See also* Global warming/cooling
    and breakup of Pangaea, 381
    data, sources of, *357*

deserts and, 289, 298
and distribution of life, 47
geologic time and, 355–356, *357*
glaciers and, 283, 284–285
mass wasting and, 204
soil and, *117*, 117–118
Coal, 123t, 126, *126*, 132, *361*
Coal swamps, *375*, 376, 407, *408*
Coast Mountains, *279*, *282*
Coast Range batholiths, 387, *387*
Coasts, 324–326, *325*. *See also* Beaches;
    Shorelines
    emergent *vs.* submergent, *325*,
        325–326
    Mesozoic, 384–386
    shorelines *vs.*, 310, *311*
Cobalt, 65, 329
Coke, 132
Colorado earthquakes, 171
Colorado Plateau, 345, *347*, 388, 390, *390*
Columbia River basalt, 100, *100*
Columnar joints, 93, *94*
Columns, 129, 244, 253, *254*
Comets, 9, 11
Compaction, 122, *123*
Complex movement, 207t, 217, *217*
Composite volcanoes, 98–99, *99*,
    102, 103
Compounds, 54–55, *55*, *56*
Compression (of rock layers), 182, *183*
    faults and, *183*, 190, 193
    folds and, *183*, 184, *185*, 187
    joints and, 189
Compressional waves. *See* P-waves
Concordant plutons, 84, *84*
Concurrent range zone, 354
Cone of ascension, 255, *258*
Cone of depression, 248, *248*, 255, *258*
Conglomerate, *15*, 123, *123*, *124*
Constancy of interfacial angles, 57
Contact metamorphism, 140–142, *141*,
    147, 149, 150, 151
Continent(s), 363
    breakup of, 37 (*See also* Pangaea)
    Cambrian, 369, *370*
    deformation and, 180
    and distribution of life, 47
    Earth's crust and, 196–198, *197*, *198*
    floating, 196
    igneous rocks and, 70
    sedimentary rocks and, 127
Continental accretion, 180, 196
Continental–continental boundaries,
    195, *195*
Continental crust, 12, 75, *75*, *76*,
    *173*, 176
    continental margins and, 27
    evolution of, 363, *363*
    mountains/continents and, 196–198,
        *197*, *198*
Continental deposition, 121, *122*
Continental divide, 234

Continental drift, 23–26, *26*, *27*, 32
    defined, 24
    early ideas about, 23
    evidence for, 24–26
    hypothesis, 24
Continental fit, *24*, 24–25
Continental glaciers, 269, *269*, *281*
    drumlins, 280
    erosional landforms and, 278
    glacial surge in, 274
    sediment in, 274–275
    stratified drift and, 280
Continental interior, 390–391
Continental margins, 27, *28*, 390
    active *vs.* passive, 31, *31*
    deformation and, 180, *193*, *194*,
        195, 196
    oceanic circulation and, 309
    passive, 31, *31*, 37, 366, *366*, 385, 391
Continental rifting, 37
Continental rise, 28, *28*, 31, 37
Continental shelf, 28, *28*, 31, 39, 323, 325
Continental slope, 28, *28*, 31, 37
Continuous branch, Bowen's reaction
    series, 73, *73*
Convection cells, *3*, 12, *13*, 43, *44*
Convergent boundaries, 37–41, *40*,
    180, 189
    andesite, 80
    batholiths and, 85
    continental–continental, 39, *40*, 41
    deformation and, 180, 191, 192
    granitic rocks at, 81
    igneous rocks and, 70
    magma, 72, 74
    metallic ores and, 45, *47*
    metamorphism and, 139, 142,
        149, *149*
    mountain building and, 180, 187,
        192, 196
    oceanic–continental, 39, *40*, 380
    oceanic–oceanic, 37, 39, *40*
    recognizing ancient, 41
    volcanoes and, 103, 180
Copper, 45, *47*, 52, 61, 65, 329
    batholiths and, 85
    metallic bonding in, 55
    metamorphism and, 151
    plutons and, 70
Coquina, 125, *125*
Coral reefs, 309, *310*
Cordilleran batholiths, 387, *387*
Cordilleran mobile belt, 381, *383*, *387*
Cordilleran orogeny, 371, 372, 384,
    386–388, *387*, 390
Core (of Earth), 11, *12*, 173–174
    density/composition of, *173*, 174
    as Earth subsystem, 2, *3*
Core-mantle boundary, 175, *175*
Coriolis effect, 295, *296*, 309
Correlation (of rock units), 344, *346*, 353
Corundum, 63t, 151

Cosmology, 8
Country rock, *76*, 76–77
    batholiths and, 85, *86*
    metamorphism and, 139, 140, 142
Covalent bonding, 55, *56*
Crater (of volcano), *92*, 95, 98
Crater Lake, 95, 96, *96*, 98
Cratonic sequences, 372–378
Cratons, *363*, 364, 366, *367*, 377
Creek, 225
Creep, 207t, 215–216, *216*
Cretaceous Interior Seaway, *385*,
    386, 388
Cretaceous-Paleogene boundary, 417
Cretaceous Period, *333*, 382, *383*,
    385, *387*
    Cordilleran orogeny (*See* Cordilleran
        orogeny)
    history of life in, 413–417
    mammals in, 416, *420*
    natural resource distribution
        and, 45
    plants in, 407t
Crevasses, 271, *271*
Crocidolite, 150
Cro-Magnons, 426
Cross-bedding, *128*, 129, *292*
Cross-cutting relationships, 336, *336*,
    341, 344
Crust (of Earth), 12, *12*, 176
    carbon dioxide cycle and, 6
    crystalline rock in, 109
    elements, *58*, *59*
    meteorites and, 417
    minerals in, 14, 58, *59*, 64, 70
    mountains/continents and, 196–198,
        *197*, *198*
    weathering and, 109 (*See also*
        Weathering)
Crutzen, Paul, 334
Crystalline rock, 109, 123t
Crystalline solids, 52
Crystals, 56–57, 62, *62*, *243*, 244
    hot springs, 261, 261–262, *262*
    radioactive dating and, 350, *351*
    in weathering, 110
Crystal settling, 75–78, *76*
Cultured granite, 83
Curie, Pierre and Marie, 346
Curie point, 33, *34*
Current ripple marks, 129, *130*
Currents, 308, 315–318, *317*, 320
Cut-and-fill method, 219, *219*
Cut bank, 229, *229*
Cyanobacteria, *261*
Cyclothems, *375*, 375–376
Cynodonts, 416, *420*
*Cynognathus*, *27*

Dakota Sandstone, 249
Dams, 223, 234, *235*, 324
Darwin, Charles, 16, 17

Dating techniques. *See* Numerical dating; Relative dating
Daughter element, 347–350, *349, 351*
da Vinci, Leonardo, 4
Dead Horse State Park, *240*
Death Valley, *288, 291, 292, 299, 299, 302, 305*
    alluvial fans in, *231, 302*
    desert vegetation in, *298*
    playa lakes in, *299, 299*
Debris flow, 207t, 211–212, *214,* 217, *217*
Deep-focus earthquakes, 157, *157*
"Deep time,", 362. *See also* Geologic time
*Deepwater Horizon, 328, 328*
Deflation hollows, 290–291, *291*
Deformation, 133, 180–196, 270
    convergent boundaries and, 37
    geologic structures and, 180, 181, 183, 184–190, 192
    metamorphism and, 140, 142
    mountain building and, 191–196
    strike and dip, *183,* 183–184
Deltas, 230–231, *231, 232,* 232–233, *233,* 325
Denali National Park, 228
Dendritic drainage, 234, *236*
Density (of mineral), 63
Deposition, 120, 121, 127–128, *128,* 131. *See also* Sediment transport
    along coasts, 325
    along shorelines, 318–324, 325
    Earth's crust and, *198*
    fossils and, 131
    by glaciers, 269, 272 (*See also* Glacial deposits)
    by running water, 227–232
    by turbidity currents, 129
    types of, *113,* 121, *122*
Depositional environment, 121, *122,* 131, 227
Deranged drainage, 236, *236,* 278
Descartes, René, 10
Desertification, 289
Desert pavement, 291, *291*
Deserts, 289, *290*
    characteristics of, 297–299
    defined, 297
    distribution of, 297
    landforms, 299–303, *299–303*
    loess and, 295
Detrital sediment, 121, 325, 391
    Mesozoic, 384, 388
    running water, 227
Detrital sedimentary rocks, *123,* 123–124, *124,* 244
Devils Postpile National Monument, *274*
Devonian landscape, *405*
Devonian Period, *333,* 355, 368–369, *371,* 402
    Appalachian mobile belt and, 380
    Cordilleran mobile belt in, 381
    Kaskaskia sequence in, 374–375

plants in, 407t
seafloor in, *404*
Tippecanoe sequence in, 372–373
vertebrates in, 404, *405*
Devonian System, 355
Diamond, 55, 56, *56,* 61–63, 65
Diatoms, 418
Differential erosion, 109
Differential pressure, 139, 140, *140,* 142, *142,* 143, *144*
Differential weathering, 109
Dikes, 84, *84,* 344
    plate tectonics and, 37
    volcanoes and, 103
Dimension stone, 82, *83*
*Dimetrodon, 406*
Dinosaur National Monument, *389*
Dinosaurs, *333,* 409–413, *410–412*
    Jurassic, 388, *389*
    mammals and, 418
    Mesozoic, 413
    Ornithischian, *412*
    Triassic, 388
    warm-blooded, 413
Diorite, 80, *80,* 81, 82, 85
Dip, *183,* 183–184, 185, *186*
Dipolar magnetic field, 31
Dip-slip faults, 189, *190, 191*
Disappearing streams, *251,* 252
Discharge (running water), 225–226, 238, 278, 280
Disconformity, 339, 341, *341, 345*
Discontinuity, 172, 174, 175, *175,* 340
Discontinuous branch, Bowen's reaction series, 73, *73*
Discordant plutons, 84, *84*
Dissolved load, 226, *227*
Distributary channels, 230, 231, 232, *232*
Divergent boundaries, 36–37, *38, 39,* 378. *See also* Spreading ridges
    earthquakes and, 157
    igneous rocks and, 70
    metamorphism and, 149
    volcanoes and, 103
Divide, 234, *236,* 239
Djurab Desert, 423
Dolomite, 61, 64, 64t, 123t, 125, 143, 144t, 147
Dolostone, 15, 61, 64, 123t, 125, 144t, 147, 244, 245, 245t
    marble from, 138
    quartzite from, 149
Domes, 187–188, *188*
*Don Quixote de la Mancha* (Cervantes), 300
Doppler effect, 8
Dormant volcanoes, 90
Double refraction, 63
Downcutting, 239, *240,* 392
Downwelling, 309
Drainage basin, 234, *236,* 238, 240
Drainage pattern, 234, 236, *236,* 281

Drainage systems, 234–236, *236*
Drainpipes, 218, *218*
Drip curtains, 253, *254*
Dripstone, 252
Dropstones, 281, *282*
Drowned coasts, 325
Drumlin field, 280
Drumlins, 278, 280, *281*
"Dry fog," 91
Ductile rocks, 183, 184
Dunes, *288, 292–294, 292–295. See also* Sand dunes
    cross-bedding on, *128*
    formation of, 292
    migration of, 292
    types of, 292–295
Durand, Asher Brown, 4
Dust Bowl, 119, *119*
du Toit, Alexander, 24
Dwarf planet, 9, *9*
Dynamic equilibrium, 202
Dynamic metamorphism, 140, 142, *142*

Earth
    age of, 333–335
    asthenosphere of, 12, *13,* 173, 175
    core of (*See* Core (of Earth))
    crust of (*See* Crust (of Earth))
    differentiated, *11*
    as dynamic, evolving system, 2, *3,* 11–14, 154, 167, 180, 393
    evolution of, 366
    interior of, 172–173, *173,* 346, 348
    magnetic field of, 31–32, *32*
    mantle of (*See* Mantle)
    orbit of, 283, *283,* 314
    at Paleocene Epoch, *357*
    solar system and, *9,* 9–11, *10*
    subsystems of, *1, 3,* 289, 307
Earthflows, 207t, 211–212, *214,* 217, *217*
Earth history, 333–335, 362–394. *See also* Geologic record; Geologic time
    Archean (*See* Archean Eon)
    Cenozoic (*See* Cenozoic Era)
    deformation and, 180–181
    fossil record and, 17 (*See also* Fossil record)
    Hadean, *333, 362*
    plate tectonics and, 12, *13,* 35–36, 363–364, 366, 389
    Precambrian (*See* Precambrian)
    Proterozoic (*See* Proterozoic Eon)
    rock cycle and, 14–16
    sedimentary rocks and, 109
    volcanism and, 90
Earthquake-resistant structures, *166,* 166–167
Earthquakes, 101, 153–176. *See also specific ones*
    aftershocks of, 154, 170, *171*
    continental margins and, 31
    control of, 171, *172*

convergent boundaries and, 37
destructive effects of, 164–168
elastic rebound theory of, 154–155, *155*
focus/epicenter of, *156, 157,* 157–158, 161, *162, 171*
intensity of, 162–163, 163t
list of significant, 154t
locating, 160–162, *161*
mass movements and, 206
measuring strength of, 162–164, *163,* 163t
occurrence of, 158–159
plate boundaries and, *157,* 157–158
plate tectonics and, 12, 23, 35–36, 158–159
precursors, *171*
predicting, *169,* 169–170, *171*
tsunami following, 166–168, *168*
East African Rift Valley, 37, *38, 39,* 103
East Pacific Rise, 103
Ebb tide, 310
*Edaphosaurus, 406*
Ediacaran fauna, *400, 400*
EEZ. *See* Exclusive Economic Zone (EEZ)
Einstein, Albert, 8
Elasticity, 160
Elastic rebound theory, 154–155, *155*
Elastic strain, 182
Electricity generation, 313
Electromagnetic force, and Big Bang, 8
Electrons, 53, *53,* 348, *349*
Electron shells, 53, *53*
Elements, 53–54
    common, in Earth's crust, *58, 59*
    in magma, 70
Eluviation, *116,* 117
Emergent coasts, *325,* 325–326
Emperor Seamount-Hawaiian Island chain, *30, 42, 75*
End moraines, 278–279, *279, 280*
Energy
    from oceans, 312–313
    resources, 65, 226, 263, 285
Environmental problems, 5–8, 289
"Environmental refugees," 289
Eocene Epoch, *306, 333, 389, 392, 393*
    mammals in, 418, *419*
    primates in, 423
Eons, *17*
Epidote, 140, 144t, 147, *149*
Epochs, *17*
Equinoxes, 283
Eras, *17*
Erosion, 109, *290,* 318
    along coasts, 325
    along shorelines, *318–320,* 318–324, 328
    badlands topography, *108*
    deserts and, 299
    differential, 109
    Earth's crust and, 197, *198*
    by glaciers, 274–278
    by groundwater, 250–253

mass wasting and, *208,* 217
mountain building and, 192, *198*
rill, 119, *119*
by running water, 224, 234, *237,* 238, 239, *239*
sheet, 119, *119,* 225
soil degradation and, 119, *119*
stream, *203,* 208, 298, 299
by wind, *290,* 290–291, *291*
Erta Ale volcano, 105
Eskers, 280, *281*
Estuaries, 313, 325
Etna, Mount, 95, 102, 105
Eukaryotes, 400
Eurasian plate, *22*
Evans, David M., *171*
Evaporites, 123t, 125, 126, *126,* 132, 308, 372, 385
Evapotranspiration, 224
Evolution. *See* Organic evolution, theory of
Excavations, 203, *204*
Exclusive Economic Zone (EEZ), 328–329
Exfoliation domes, 110, *111*
Extinction, 399, 402
    Mesozoic, 408, 417, 418
    Paleozoic, 417
    Permian, 403
    Pleistocene, 420
Extinct volcanoes, 90
Extrusive igneous rock, 15, *15,* 70, 77
Eyjafjallajökull, 90, *90,* 105

Facies
    metamorphic, 148–149, *149*
    sedimentary, 127–128
Falls, 206–207, *207,* 207t, *208,* 211
Farallon plate, 387, 390, *391*
Fault breccia, 188
Fault plane, 188, 189, *189, 190*
Faults, 188–190, *189, 191,* 191–195, *192,* 244. *See also* Block faulting
    earthquakes and, 154–155, *155,* 157, 159, 164, 169–171, *170, 171*
    metamorphism and, 142
    plate boundaries and, 37, 190
    San Andreas (*See* San Andreas Fault)
    types of, 188–190, *189, 191, 192*
Fault scarp, *169,* 188, *189*
Faunal and floral succession (principle of), 338
Feldspars, 59, *60,* 62, 64, 65
    metamorphism and, 144t, 145, 146
    weathering and, *109,* 110, 112, 114, 116, 120, 122
Felsic magma, 71, 71t, 73, 76, *76,* 81
    granite and, 80, 85
    lava domes and, 99
    volcanoes and, 91, 99, 102
Ferromagnesian silicate, 59, *60,* 61–63, 64t, 73, *73, 75, 76, 79, 80, 173, 175*
    in Bowen's reaction series, 73, *73*

metamorphism and, 146
    oxidation and, 114
Fiords, 275–276
Fire (following earthquake), 165, *168*
Firn, 270
    limit, 272, *273*
Fish, 404, *404*
Fissure eruptions, *100,* 100–101
Flint, 126
Flood(s), *201, 202, 222, 223, 223*
    coastal, 307, 326
    control and predictions of, 233–234
    rivers/streams and, 227, 230, 234
Flood basalts, 75, *75*
Floodplains, *222,* 229–230, *230, 239,* 239–240, 295
Flood tide, 310
Floodwalls, 234, *235*
Floodways, 234, *235*
Flows, 207t, 211–216, *214, 215, 216*
Flow velocity (running water), 225, 226, 227, 229, *229,* 230, 238, 273, *273*
Fluid activity, metamorphism and, 138, 139–140, 141–142
Fluorine, 57, 61, 403
Fluorite, 61–62, *62,* 63
Folds, 184–188, *185, 186*
    mountain building and, 192
    plunging *vs.* nonplunging, 187, *187*
Foliated metamorphic rock, 15, *15,* 143–146, *144,* 144t
    amphibolite as, 146
    gneiss as, *15, 138,* 143, 145–146, *146*
Foliated texture, 143, *144,* 146
Footwall block, 189, *190, 191*
Foraminifera, 418
Forceful injection, 85
Fossil(s), 17, 128, 129, *131,* 399–401
    of birds, 414, 416, *416*
    body *vs.* trace, 129, *131*
    Ediacaran-like, *400*
    guide, 346, *348*
    of mammals, 416, 420
    Precambrian, 399–401
    skeleton of dinosaur *Allosaurus, 131*
    superposition and, 338, *339,* 346
Fossil record, 17, 26, *333,* 399, 401–402, 419–420, 423
    birds and, 416
    continental drift and, 26, *27*
    hominids and, 423
    mammals and, 418
Fossil succession, 338, *339,* 346, 353
Fracking, 256–257
Fractures, 62–63, 110, 114, *115, 132,* 244, 245, *245,* 262, *263. See also* Columnar joints
    deformation and, 180, 182, 183, 184
    as faults, 188–189, *189 (See also* Faults)
    joints and, 188, *188*
Fracturing, hydraulic, 256–257
Framework silicates, 112, 114

Frank (Canada) rock slide, 209, 211, *213*
Fringing reef, 309, *310*
Frost action, 110
Frost wedging, 110, *110*, 207, 298
Fumaroles, 91, *92*
Fumeroles, 261, *261*

Gabbro, 12, 27, 173t, 176, 196
  classification of, 78, *79*, 79–80, *80*
  in oceanic ridges, 28
  plate boundaries and, 41
  volcanoes and, 103
Galena, 56, *61*, 61–63, 151
Gamma rays, 176
Ganges-Brahmaputra delta, *231*
Garden of the Gods, 376, *377*
Garnet, 62, 138, *140*, 142, 144t, *145*,
    148, 151
Gas, 53, 112, 132–133, *133*. See also Natural
    gas; Volcanic gases
  noble, 54, 55
Gemstones, 52
Geodes, *51*
Geodimeters, 105
Geological Survey of India, 173
Geologic issues. See Environmental
    problems
Geologic principles, 335–339. See also
    specific ones
Geologic record, 17, 332, 335, *347*
  discontinuities in, 339
  dunes and, 292
  glaciation and, 283
Geologic structures, 180, 181, 183,
    184–190, 192
Geologic time, 17–18, 331–357. See also
    Uniformitarianism
  numerical dating in, 332
  Anthropocene Epoch and, 334
  climate and, 355–356, *357*
  and correlating rock units,
    344–346, *346*
  early concepts of, 332–335
  numerical dating in, 332, 353
  numerical dating methods, 346,
    348–353
  relative dating in, 332, 335–344
  stratigraphy and stratigraphic
    terminology, 354–355, 354t
Geologic time scale, *17*, 17–18, 332, *333*,
    353–354, 354t
  development of, 333, 353–354
  relative, 335, 353
Geology, 180–181
  beneficial for lives, 18
  definition of, 2
  economic and political power, 5
  everyday life and, 5
  father of modern, 335
  formulation theories, 3–4
  role in human history/ culture, 5–8
  slope stability and, *205*, 205–206

Geometry (of rocks), 128
Geothermal energy, 263, 264
Geothermal gradient, 74, 142, 176, 261
Geothermal power plants, 264, *264*
*The German Legends of the Brothers
    Grimm,* 4
Gettysburg National Military Park, 113
Geyserite, 263, *263*
Geysers, 262–264, *262–264*. See also
    Hot springs
Gibson Desert, *293*
Glacial budget, *272*, 272–274
Glacial deposits, 131, 278–282
  glacial lakes, 280–282
  loess and, 295
  sand *vs.,* 128
Glacial drift, 278, *279*
Glacial erratics, 278, *279*
Glacial ice, 270, *270*
Glacial lakes, 280–282
Glacial polish, 274, *274*
Glacial striations, 274, *274*
Glacial surge, 274
Glaciation, 267–285, 393, *394*, 402
  Pleistocene, 393, *394*
Glacier(s), 267–285. See also Ice Age(s)
  continental (See Continental glaciers)
  continental drift and, 25–26, *26*
  distribution of, 271–272
  Earth's crust and, 197
  global warming and, 7
  Gondwanan, 376
  Milankovitch theory and, 283–284
  mountain building and, 192
  movement of, 270–272
  Paleozoic, 376
  Pleistocene, 268, 275, 282–284, 285
  sand dunes *vs.,* 128
  sediment/sedimentary rocks, 109, 120
  shorelines and, 326
  stagnant, 273
  U-shaped glacial troughs, 275, *276*
  valley (See Valley glaciers)
Glacier National Park, *130*, 284, *285*,
    *366*, 399
Global Seismic Hazard Assessment Map,
    169, *169*
Global warming/cooling, 6–7, *7*
  deserts and, 289
  geologic time and, 355–356, *357*
  glaciers and, 268, 284–285, 393
  greenhouse effect and, 6–7, *7*
  mass extinctions and, 404
  predicted, 393
*Glossopteris* flora, *23*, 23–24, 26, *27*
Gneiss, *15*, 143, 144t, 145–146, *146*
  Acasta, 138, *138*, 363
  Archean, 364, *365*
  Proterozoic, 366
Gold, 45, 54, 55, 57, 61, 63, 65, 85, 120, 132
Gondwana, *23*, 23–24, 369, *370*, *371*, *382*
  Appalachian mobile belt and, 380

and breakup of Pangaea, *382*, 383
  fossils and, 26–27, *27*
  glaciers and, 25–26, *26*, 376
Gorges, 238
Grabens, 192, *192*
Graded bedding, 129, *129*
Graded stream, *237*, 238
Gradient (running water), *225*, 225–226
Grand Canyon, 4, 131, *132*, 224, *331*
  geologic time in regard to, 332, 345
  nonconformities in, *343*
  Precambrian and, 364
  unconformities in, *339*
Granite, *15*, 52, 64, 80–81, *81*, 110
  Archean, 364
  dimension stone, 82, *83*
  Harney Peak, *69*
  inclusions of, 338, *338*
  pegmatite and, 81, *81*
  Proterozoic, 366
  uses of, 70, 82
  weathering of, *109*, 110
Granite porphyry, 80
Granitic rocks, *76*, 81, 82, 85, 176
Granodiorite, 81, 82, 173
Granulite facies, 148, *149*
Graphite, 55, *56*, 57, 61, 63, 143,
    144t, 151
Gravel, 121, *121*, 123, *123*, 132, 245t
  glacial drift and, 278
  running water and, 226, *227*, *228*, 230
Gravity, 196, *197*
  Big Bang and, 8
  groundwater and, 246
  and history of life, 405
  mass wasting and, 202, *203*
  mountain building and, 196, *197*
  tides and, 310
Great Flood of 1993, 233, 234, *235*
Great Ordovician Biodiversity Event,
    402, *402*
Great Sand Dunes National
    Monument, *294*
Greenhouse effect, 6–7, *7*
Greenschist facies, 148, *149*
Greenstone, 144t, 147
Greenstone belts, 364, *365*
Grenville orogeny, 366–367
Grimes Point Archaeological Site, *111*
Grofé, Ferde, 4, *4*
Groins, 320
Ground failure, 168, *169*
Groundmass, 77
Ground moraines, 279
Ground shaking, 164–165, *165*, *166*
Groundwater, 243–264
  coastal flooding and, 327
  contamination of, 258–259, *260*
  deposition by, 250–253
  deserts and, 289, 298
  erosion by, 250–253
  glaciers and, 270, 278

hydrologic cycle and, 244
modification of groundwater system, 253–255, 258–259
movement of, 246–247, *247*
sediment/sedimentary rocks, 114, 121, 125, 133
springs and, 247, *248*
Guide fossil, 346, *348*
Gulf Coast, *384*, 385–386, 390–391
Gulf of California, 28, 37, 42, *42*
Gulf Stream, 309
Gullies, 119, *119*, 238, *238*
Guyots, 29–30, *30*
Gymnosperms, 407, 416
Gypsum, 61, *61*, 63, 64t, 65, *243*, 252
Gyres, 309

Hadean Eon, *333*, 362
Haiti earthquake, *153*, 154, 154t, 164
Half-life, 348–350, 352t
Halides, 58t, 61, 64t
Halite, 54–55, *55*, *56*, 57, 61, *61*, 64t
    cleavage, 62, *62*
    as crystal, 62, *62*
    solution of, *113*
Hanging valleys, 276, *276*
Hanging wall block, 188–189, *190*, *191*
Hansen, Em, 4
Hardness (of mineral), 63, 63t
Harmonic tremors, 105
Harney Peak Granite, *69*
Hawaiian tsunami, 168
Hawaiian-type eruptions, 95
Hawaiian volcanoes, 71, *71*, 95, 103, 192. *See also* Emperor Seamount-Hawaiian Island chain; *specific volcanoes*
Headland, 319, *320*
Headward erosion, 239, *239*
Heat. *See also* Temperature
    Earth's internal, 176
    metamorphism and, 138–139
*Hebrides Overture* (Mendelssohn), 4
Helium, 53, *53*
Hematite, 61, 63, 114, 122, 151
Hercynian-Alleghenian orogeny, 380
Hercyninan orogeny, 372
Hess, Harry, 33–35
Hiatus, 339, *340*
High Hole Crater, *394*
High Plains aquifer, 254, *255*
Himalayan orogeny, 389
Himalayas, *40*, 41, 159, 176, *179*, 195, *195*, 197, *333*
Historical geology, 2
Holocene Epoch, *333*, 389. *See also* Recent Epoch
Holocene Maximum, 268
Hominids, 423–427, *424*
Hominins, 423–427, *424*
Hominoids, 423
*Homo*, 424

Homogenous accretion theory, *11*
Hoodoos, 109
Horizontal-motion seismograph, *156*
Hornblende, *52*, *60*, *62*, *63*, 64t, *79*, 80, *80*, 144t, 145, *145*, 146, 147
Hornfels, 141, 144t, 147, 149
Horns, 276, *277*, 278
Horses, evolution of, *421*
Horsts, 192, *192*
Hot-spots, 30, *30*, 36, 192
    mantle plumes, 42
    and origin of magma, 75, *75*
    plate tectonics and, 42
Hot springs, *261*, 261–262, *262*
Hraunfossar waterfalls, *248*
Hubble, Edwin, 8
Humans
    desertification and, 289
    in geologic time scale, *333*
    history/culture of, 5–8
    importance of minerals to, 56
    as part of Earth system, 2
    plate tectonics and, 23
    prehistoric overkill by, 422
    primate evolution and, 423
    soil degradation and, 119
Humus, 116, *116*
Hurricane Harvey, 326, *327*
Hurricanes, *327*. *See also specific hurricanes*
    shorelines and, 307, 326
Hutton, James, 335, *335*, 340, 348
Hydraulic action, 226, 318, *318*
Hydraulic fracturing, 133, 256–257
Hydrocarbons, 132–133
Hydrologic cycle, 223–224, *224*, 244, *267*, *268*, 270
Hydrolysis, 114
Hydrosphere, *1*
    as Earth subsystem, 2, *3*, 307
    plate tectonics and, 12
    volcanism and, 90, 91
Hydrostatic pressure, 139
Hydrothermal activity, 261–264
Hydrothermal alteration, 142
Hydrothermal vents, 28, *29*, 65, 308
Hydroxide, 58t, 61, 114
Hydroxyl, 58, *59*
Hypocenter, 157
Hypothesis, definition of, 4

Iapetus Ocean, 369, *370*, *379*, 380
Ice Age(s), 237, 268, *333*, 422. *See also* Pleistocene Epoch
    causes of, 282–285
    Little, 268, 284
Ice caps, 269, *269*
    global warming and, 7
Icefall, 271, *275*
Ice-scoured plains, 278
Ice sheets, 269, *269*, 393. *See also* Continental glaciers

Ice shelves, 269, *269*
Ichthyosaurs, 414, *414*
*Ichthyostega*, 405, *405*
Igneous rocks, 69–86, 245, 261, 262, *263*
    classification of, 78–82, *79*
    composition of, 78
    crystalline rock and, 109
    dating of, 350
    extrusive, 15, *15*, 70
    forming, 64
    geologic time and, 340
    intrusive (*See* Intrusive igneous activity; Intrusive igneous rocks)
    metamorphic mixed with, 138
    metamorphic *vs.*, 146, 147
    peridotite as, 12, 175
    pillow lava and, 92, *94*, 103
    plate tectonics and, 15, *15*, 16
    Precambrian, 376
    rock cycle and, 14–15, *15*, 16
    textures, 77–78, *78*, 83
Illinoian glaciation, 393, *393*
Inactive volcanoes, 90
Incised meanders, 240, *240*
Inclined fold, 185, *186*
Inclusions, 76, *76*, 338, *338*
    batholiths and, 85, *86*, 338
    nonconformity and, 340, *341*
Index fossils, 346
Index minerals, 142–143, *143*, 148
India, earthquakes in, 154t
Indian Ocean, currents in, *308*
Indian Ridge, 103
Indonesia earthquake, 154t, 159, 164, 166, 167
Infiltration capacity, 224
Interglacial stage, 283
Intermediate-focus earthquakes, 157
Intermediate magma, 71, 71t, 73–76, *76*, *79*, 99
    andesite/diorite and, 80, *80*
    lava domes and, 99
Intraplate earthquakes, 159
Intraplate volcanism, 103
Intrusive igneous activity, 69–86, 337, 340, 341, 388
    geologic time scale and, 353
    metamorphism and, 139
    volcanism and, 70, 388
Intrusive igneous rocks, 15, *15*, 70, 77–82, *79*, 80, *81*, 83. *See also* Pluton(s); Plutonic rock
    granite as, *15*
    rifting and, 37
Invertebrates, 401–403, 401t, 406
    Cenozoic, 418
    geologic ranges of, 401t
    marine, 401–403, *402*, 408, 418
    Mesozoic, 408
Ionic bonding, 54–55, *55*
Ions, 54, 55, *57*, 59, 114
Iran earthquake, 154t

Iridium, 417
Iron, 52, 53, 57, 57, 59, 65, 114, 147, 151, 366
    in core of Earth, 174, 176
    oxidation and, 114
Iron oxide, 114, 118, 122, 126, 329, 366
Island arcs, 363, 363–364, 365, 366, 383, 390
Isograd, 148, 148
Isostasy, 196–197, 197, 326
Isostatic rebound, 197–198, 198
Isotopes, 54, 54, 176, 351–352, 352t
    radioactive, 54

Japanese earthquakes, 154, 154t, 158, 164, 165, 165, 166
Japanese tsunami, 154, 166, 168, 168
Jasper, 126
Jeffreys, Harold, 174
Johnstown flood, 223, 223
Joints, 93, 94, 188, 188. See also Columnar joints
Joshua Tree National Park
Journey to the Center of the Earth (Verne), 4
Jovian planets, 10
Juan de Fuca plate, 102
Jupiter, 91
Jurassic Period, 381, 382, 383, 387, 389
    Absaroka sequence in, 375
    Cordilleran orogeny in, 384, 387, 388
    Gulf Coast in, 391
    history of life in, 408, 414, 416, 416
    plants in, 407t

Kames, 280, 281
Kansan glaciation, 393, 393
Kaolinite, 114, 120
Karakoram Range, 179, 195
Karst topography, 250, 250–252, 251
Kaskaskia sequence, 372–373
Katmai National Park, 72
Katrina (hurricane), 326
Kazakhstania, 369, 370, 371
Kenai Fjords National Park, 271, 279
Kettles, 280, 281, 282
Key beds, 346
Kilauea volcano, 92, 92, 95, 103
Kindred Spirits (Durand), 4
Kinetic energy, 226
Kluane National Park, 215, 273, 274
Komatiites, 79
Kona Dolomite, 366
Krakatau volcano, 89, 168
Kuiper Belt, 9, 9
Kyanite, 143, 148, 151

Laccolith, 84, 84–85
Lahar, 95, 98, 99, 103, 104
Laki fissure, 91, 101
Land bridges, 24
Landforms, 389
    coasts and, 325
    desert, 299–303, 299–303

glaciers and, 275, 278–280 (See also Drumlins; Moraines)
    volcanoes as, 91, 100–101
    Brazil, 200, 201, 202
Landslides, 103, 201, 202, 204, 217
    Brazil, 201, 201
    mountain building and, 192
    retaining walls and, 220
    Southern California, 210, 210–211, 211
    triggers of, 206, 206
Lapilli, 94, 95
Laramide orogeny, 387, 388, 390, 391
Lassen Peak, 99, 102, 102
Lassen Volcanic National Park, 99
Lateral continuity, 336, 336
Lateral erosion, 239, 239, 240
Lateral moraines, 279–280, 280
Laterite, 118, 118, 120
Laurasia, 24, 369, 371, 380, 382
Laurentia, 366–367, 367, 369, 370, 379, 380
Lava, 64, 70
    viscosity of, 71, 72
Lava domes, 71, 72, 99, 99–100
Lava flows, 70, 90, 394. See also Basaltic flows
    Cenozoic, 392, 418
    geologic time scale and, 353
    hazards of, 103, 104
    mafic, 71, 72, 79
    metamorphism and, 140–141, 141
    mineral grains and, 77
    paleomagnetism and, 33
    pillow, 37
    plate boundaries and, 37, 38, 103
    plutons and, 70
    Siberian Traps and, 403
    soil and, 118
    ultramafic, 79
Lava tube, 92, 93
Lead, 61, 61, 63, 151, 351, 352t
Leaning Tower of Pisa, 255, 259
Left-lateral strike-slip fault, 190, 190
Lehmann, Inge, 174
Levees, 234, 235
Libyan Desert, Egypt, 290
Lichens, weathering and, 111, 112
Life, distribution of, 47–48, 48
Life, history of
    Cenozoic, 418–427
    Mesozoic, 408–414
    Pleistocene, 420–422, 422, 426
    Precambrian, 399–401
Lignite, 126, 132
Limb (of folded rock), 185, 185, 186
Limestone, 15, 61, 123t, 144t, 149, 245t
    calcite in, 114, 125
    foraminifera and, 418
    fossiliferous, 125
    groundwater and, 245, 250
    hot springs/geysers and, 262
    marble from, 138, 147, 147
    oolitic, 125, 125

Solnhofen, 414, 416
    uses of, 132
    weathering and, 114, 125
    world distribution of, 244
Limonite, 114, 122
Limpet (Land Installed Marine Powered Energy Transformer), 312
Liquefaction, 165, 165, 167
Liquids, 53
Lithification, 121, 122, 123
Lithosphere, 1, 12, 13, 157, 158, 173, 175, 176
    as Earth subsystem, 2, 3
Lithostatic pressure, 139, 139
Lithostratigraphic units, 354, 355, 356
Little Ice Age, 268, 284
Lizard-hipped dinosaurs, 409. See also Saurischia
Lobe-finned fish, 404
Local base level, 237, 237, 238
Loess, 168, 292, 295, 295
Longitudinal dunes, 293, 293
Longshore current, 316, 319, 320
Longshore drift, 320
Long Valley caldera, 101, 106
Love, A. E. H., 160
Love (L-) waves, 160, 160
Lower mantle, 12
Low-velocity zone, 175
Lung cancer, 150
Luster, 61–62, 66
Lyell, Charles, 335, 348
Lystrosaurus, 27

Mackinac Breccia, 180, 180
Mackinac Bridge, 180, 180
Mafic lava flows, 71, 72, 79, 93, 94, 95, 103
Mafic magma, 70–71, 71t
    origin of, 74, 74
    volcanoes and, 91, 102, 103
Magma, 12, 64, 70–77, 72, 74, 76
    batholiths and, 85, 86
    Bowen's reaction series, 72–74, 73
    common types of, 71t
    composition of, 70–71, 75–77, 76
    felsic, 71, 71t, 91, 99
    hot springs and, 261
    igneous rocks, 15, 15, 70
    intermediate, 71, 71t, 99
    mafic (See Mafic magma)
    metamorphism and, 139, 140, 141, 142, 146, 149, 150
    plate boundaries and, 37
    radioactive dating and, 350, 351
    at spreading ridges, 74, 74
    subduction zones and, 74, 74–75
    ultramafic, 70, 71t
    viscosity of, 71, 72
    volcanoes and, 23, 85, 86, 91, 94–95, 99, 101–103
Magma chambers, 72, 73, 76, 85
Magma mixing, 76, 77
Magnesium, 57, 57, 59, 308, 328

Magnetic reversals, 33–35, *35*
Magnetism, 31–35, *32*
    plate movement and, 43, *43, 44*
    remnant, 32, 33, *34*
Magnetite, 61, 63, 151
Malad Gorge State Park, *392*
Mammals
    "Age of,", 418
    Cenozoic, 420
    Cretaceous, 416, *420*
    Eocene, 420, *420*
    Mesozoic, 416, *417, 419*
    Miocene, 420, *421*
    Paleocene, *419*, 420
    plate tectonics and, 48
    Pleistocene, 420–422, *422*
    types of, 416
Mammoth Cave National Park, 244
Mammoth Cave, Western Australia, 243
Mammoth Mountain volcano, *104,* 106
Mand River, *22*
Manganese, 65
Manganese nodules, 329
Mantle, 11–12, *12,* 175, *175*
    discontinuities in, *173,* 174, 175, *175*
    Earth's crust and, 197, *198*
    as Earth subsystem, 2, *3*
    geothermal gradient in, 176
    magma in, 70, 72
Mantle plume, *30,* 42, 75, *75*
Marathon Mountains, 381
Marble, 15, 114, *137,* 138, 144t, 147, *147,* 149, 151
Marcellus Shale, 256
Mariana Trench, 29
Marine deposition, 121, *122,* 373
Marine invertebrates, 401–403, *402, 403,* 408, 418
Marine regression, *127,* 128, 373
Marine reptiles, 414
Marine terraces, 319, *325*
Marine transgression, *127,* 127–128, *132,* 402
Marine vertebrates, *403*
Marsupial mammals, 416, 418
Mass extinction, 399, 402, 403, 417–418
Mass spectrometer, 350
Mass wasting/movement, 201–220. *See also* Landslides
    classification of, 206
    deserts and, 298
    glaciers and, 275, 279
    minimizing effects of, 217–220
    mountain building and, 192
    rapid *vs.* slow, 206
    running water and, 226
    shorelines and, 318
    slope stability and, 202, *203,* 204, *205,* 205–206, 217 (*See also entries beginning* Slope)
    types of, 206–217
    valleys and, 238, 239, *239*
Matter, 53–55

Mauna Loa volcano, *72,* 95, *97,* 99, 103
Mayacamas Mountains, 264, *264*
Mayon volcano, 91, 98, 103
Meandering streams, 227, *228,* 229
Meander scars, *229,* 230
Mechanical weathering, 110–111, *111,* 298
Medial moraines, 279–280, *280*
Medicine Lake volcano, *107*
Medieval Warm Period, 268
Mediterranean-Asiatic belt, *157,* 159, *169*
Mediterranean belt, *101,* 102, 103
Mélange, 41
Mendelssohn, Felix, 4
Mesa, 303
*Mesosaurus,* 27
Mesothelioma, *150*
Mesozoic Era, *333,* 345, *382,* 383–389, *387*
    continental fit and, 24
    dinosaurs in, *412*
    eastern North America in, 391
    history of life in, 408–414, 418, 426
    mammals in, 416, *419*
    natural resources and, 45
    Ouachita mobile belt in, 381
    Pangaea in, 381, *382*
    reptiles in, 408, 414, *414*
Mesquite Flat Sand Dunes, *288*
Metallic bonding, 55
Metallic ores, 45, 46
Metallic resources, 65
Metamorphic aureole, 140, *141*
Metamorphic facies, 148–149, *149*
Metamorphic grade, 142–143, *143,* 144t
Metamorphic minerals, 138, 139, 140, 142, 148, 150, 151
Metamorphic zones, 148, 148–149
Metamorphism/metamorphic rock, 137–151, 244, 245, *245*
    classification of, 143–148, 144t
    contact, 140–142, *141,* 147, 149, 150, 151
    convergent boundaries and, 37
    crystalline rock and, 109
    dating of, 350, *351*
    deformation and, 180, 194
    dynamic, 140, 142, *142*
    Earth history and, 363
    foliated (*See* Foliated metamorphic rock)
    and forming of minerals, 65
    geologic time and, 340, *343*
    glacial ice as, 270
    igneous rock mixed with, 138
    igneous rock *vs.,* 146, 147
    index minerals and, 142–143, *143,* 148
    minerals and, 138, 140, 142, 150, 151
    mountain building and, 192, 193, *193,* 198
    natural resources and, 150–151
    nonfoliated, 15, *15,* 146–148
    plate tectonics and, 15, *15, 16, 149,* 149–140

Precambrian, 376, *377*
    rock cycle and, 14–15, *15, 16*
    types of, 140–143
Meterorites, 417
Methane, 257
Methane hydrate, 329
Mexico, earthquake in, 154t, 170
Micas, 59, *60*
    uses of, 66, *66*
Microcline, 59, 64
Microcontinents, 29, 367
Microfossils, 129
Microplates, 381
Mid-Atlantic Ridge, 28, *31,* 37, *38, 43,* 103
Midcontinent Rift, 367
Middle Ordovician seafloor fauna, *402*
Migmatites, *143,* 144t, 146, *146,* 149
Milankovitch theory, 283–284
Milky Way galaxy, 9–10
Mineral(s), 14, 51–66
    in Bowen's reaction series, 72–74, *73*
    carbonate, 58t, 61, *61*
    chemical composition of, 57
    convergent boundaries and, 37
    crystals, 56–57, 62, *62*
    dating of, 350, 352t
    definition of, 2, 52
    formation of, 64–65
    geode and, *51*
    hot springs/geysers and, 262, 263
    human usage of, 6
    importance of, 52
    index, 142–143, *143,* 148
    magma and (*See* Lava; Lava flows; Magma)
    mantle and, 175
    metallic *vs.* nonmetallic, 65
    metamorphic, 138, 140, 142, 150, 151
    as naturally occurring inorganic substances, 56, *56*
    natural resources and reserves, 65–66
    other groups, 61
    physical properties of, 57, 61–64
    plate tectonics and deposition of, 45, *47*
    radicals and, 58, *59*
    rocks and, 52, *52,* 64
    silicate (*See* Silicates)
    weathering and, *109,* 109–110, 112, 114–115, *115*
Mineral grains, 77
Minoan eruption, 96, *97*
Miocene Epoch, *333,* 389, 420, *421,* 423
Mississippian Period, *333,* 341
    Absaroka sequence in, 375
    Cordilleran mobile belt in, 381
    history of life in, 405–406, 408
    Kaskaskia sequence in, 374
    Ouachita mobile belt in, 381
    vertebrates in, 405–406
Mississippi River, 223, 224, 229, 231, 232–233, *232–233,* 234, *235*
"Mixed rocks," 146

Mobile belts, 378–381
Modified Mercalli Intensity Scale, 162, 163t
Mohorovičić, Andrija, 175
Mohorovičić discontinuity (Moho), 175, *175*
Mohs, Friedrich, 63
Mohs hardness scale, 63t
Moine Thrust Zone, 142
Mojave Desert, *299*
Molecules, 55
Molybdenum, 46, *46*
Monocline, *184*, 184–185
Monotremes, 416, 418
Montana earthquake, 168, *169*
Monument Valley, *84, 303*
Moon, 310
Moraines, 278–280, *279, 280*
Morrison Formation, 388, *389*
Mountain building. *See* Orogeny/mountain building
Mountain glaciers, 268. *See also* Valley glaciers
Mountain ranges, 16, *179*, 180, 191, 192, *192, 197*
    continental drift and, 25, *25*
    convergent boundaries and, *40*, 189
    in desert regions, 297
    glaciers and, 269
Mountain system, 191, 195, 196
Mount Rushmore, 69, 82
Mud, 121, 122
Mud cracks, 129, *130*
Mudflows, 207t, 211–212, *214*
Mudrock, *123*, 124, 126, 133
Mudslides, 202, 205, 209
Mudstone, *123*, 124
Muscovite, *52*, 59, *60*, 64t, 66
Mylonites, 142, *142*

Native elements, 57, 58t, 61
Natural gas, 52, 65, 132–133, *133*
    from anticlines, 187
    geologic structures and, 181, 187
Natural glass, 70, 77, *83*
Natural levees, 230, *230*
Natural resources, 5, 52, 65–66, 120
    energy, 65, 263, 329
    groundwater as, 244, 253
    metamorphism and, 150–151
    from oceans, 328–329
    plate tectonics and, 23, 45, *47*
    in sedimentary rocks, 132–133
    soil and, 117
    weathering and, 120
Natural selection, 17
Navajo Sandstone, 131, *132, 388*
Nazca plate, 39
Neanderthals, 425, 426, *426*
Nearshore currents, 315–318, 320
Nearshore sediment budget, 323–324, *324*
Nebraskan glaciation, 393, *393*

Neogene Period, *333, 387,* 389, 407t
Neon, 53, 54
Neoproterozoic Eon, *366*
Neptune, 9, 91
Neptunism, 82
Neutron, 53, 54
Nevada, *111, 118*
Nevadan orogeny, 387, *387,* 388
Nevado del Ruiz, 98, 103
New Orleans hurricane, 326
Newton, Isaac, 196
Nickel, *173,* 174, 329
Nile River, 222, 223, *231*
Nitrates, 309
Noble gases, 54, 55
Nonconformity, 340, 341, *343*
Nonferromagnesian silicate, 59, 80
    in Bowen's reaction series, 74
    and classifying igneous rocks, *80, 81*
Nonfoliated metamorphic rock, 15, *15,* 146–148
Nonfoliated texture, 143, 144t, 146, *147*
Nonmetallic resources, 65
Nonplunging folds, 187, *187*
Nonrenewable resources, 6, 65, 119
Normal fault, 189, *190, 191, 192*
North America, evolution of, 372–389, *384,* 391–392
North American Cordillera, 389–394
North American craton, 364
Northridge earthquake, 181
No-till planting, 119
Novarupta lava dome, *72*
Novarupta volcano, 105
Nuclear forces, 8
Nucleus (of atom), 53, *53, 54*
Nuée ardente, 99–100, *100,* 101, 103
Numerical ages, 18
Numerical dating, 346, 348–353, *353*

Oblique-slip fault, 190, *190*
Obsidian, 70, 77, *78,* 82, *83*
Ocean(s), 307, *307,* 308–309, 328–329
    energy from, 312–313
Ocean basins, 28, 29, 35, 37, *38,* 47, 309
Ocean currents, 47, 307, *307,* 309
Oceanic circulation, *307,* 309
Oceanic–continental boundaries, 39, *40,* 380, 384–386
    metamorphic facies and, 149, *149*
    orogenies at, *194,* 194–195, 390
Oceanic crust, 12, 29, 34–35, *35,* 41, *41,* 75, *75,* 80, *173,* 175, 176
    continental margins and, 27
    igneous rocks and, 70
    magnetism and, 34–35
    mountains/continents and, 196
    plate boundaries and, 37, 39
    volcanoes and, 90
Oceanic–oceanic boundaries, 37, 39, *40,* *193,* 193–194
Oceanic ridges, 28, *28, 35,* 37, 196

Oceanic trenches, *28,* 29, 37
Ocean thermal energy conversion (OTEC), 312
Offshore wind farms, 300–301
Ogallala aquifer. *See* High Plains aquifer
Oil, 133, *133,* 264. *See also* Petroleum
Oil spills, *328*
Old Faithful geyser, *262*
Oldham, R. D., 173
Oligocene Epoch, *333,* 389, 420, *420,* 423
Olivine, 57, 59, *60,* 64t, 79, *79*
    Bowen's reaction series and, *73,* 73–74
    crystal settling and, 75, *76*
    plate boundaries and, 41
Olympic National Park, *320*
*On the Origin of Species by Means of Natural Selection* (Darwin), 16
Ooids, 125, *125,* 131
Oolitic limestone, 125, *125*
Ooze, 309
Ophiolites, 41, *41*
Orbital eccentricity, 283
Ordovician Period, *333,* 346, 356, 369, *370,* 378
    invertebrates in, 401t
    plants in, 407t
    Sauk sequence in, 372
    Tippecanoe sequence in, 372–373, 402
Organic evolution, theory of, 16–17
    fossils and, 131 (*See also* Fossil(s); Fossil record)
    of horses, 420, *421*
    of humans, 422, 423
    "out of Africa" *vs.* "multiregional,", 425
    plate tectonics and, 47–48, *48*
    of primates, 422–423
Original horizontality, 183, 336, *336,* 340
*Origin of Continents and Oceans, The,* 24
Ornithischia, 409, *410*
Ornithischian dinosaurs, *412*
Orogeny/mountain building
    batholiths and, 85
    at continental–continental boundaries, 195, *195*
    at convergent boundaries, 37, *40*
    deformation and, 191–196
    Earth's crust and, 196–198, *197, 198*
    Mesozoic, 383–389
    metamorphism and, 140, *148*
    North American Cordillera and, 389–394, *390*
    at oceanic–continental boundaries, *194,* 194–195
    at oceanic–oceanic boundaries, *193,* 193–194
    Paleozoic, 369, 378–381
    plate tectonics and, 36, 192–195
    Precambrian, 364
    Proterozoic, 366
Orthoclase, 57, 59, 63t, 64t, 114
Ostracoderms, *404*
Ouachita mobile belt, 381

Ouachita orogeny, 261, 372, *373*
*Our Wandering Continents,* 24
Outgassing, 308
Outlet glaciers, 273
Outwash plains, 280, *281*
Overloading, 205
Overpopulation, 6, *6*
Overturned fold, 185, *186,* 187
Owen, Richard, 409
Oxbow lakes, 229, *229,* 230
Oxidation, 114
Oxide, 114
Oxygen, 53, *53,* 54, 57, *57,* 59, *60,* 61,
    *173,* 174
    and history of life, 405
    weathering by, 112, 114, 126
Ozone, 112, 403

Pacific-Farallon Ridge, 390, *392*
Pacific Ocean, currents in, *307*
Pahoehoe, 92, *93*
Pakistan, earthquakes in, 154t
Paleocene–Eocene Thermal
    Maximum, 356
Paleocene Epoch, 389, *392, 419, 420,* 423
    earth at, *357*
Paleogene Period, *333,* 387, 389, 407t
Paleomagnetism, 32–33, *33*
Paleoproterozoic, 364, 366, *366, 367*
Paleoseismology, 170
Paleozoic, 401–408
Paleozoic Era, *333,* 343, 345, 367–381, *387*
    continental drift and, 25–26, *27*
    Cordilleran mobile belt in, 381
    glaciers and, 25–26, *27*
    history of life in, 401–408, 413, 426
    insects in, 413
    invertebrates in, 401–403, *406*
    microcontinents in, 29
    mobile belts, 378–381
    North America in, 372–381
    Pangaea in, 381
    reefs in, 402–403, *403*
    Sonoma orogeny in, 386
    vertebrates in, 404–406
Panama, Isthmus of, 47–48, *48*
Pangaea, 14, 24–25, *27,* 43, *333,* 371, 372
    breakup of, 381–383, *382*
    microplates and, 381
    orogeny and, 194, 378, 389
Panthalassa Ocean, *371,* 372
Parabolic dunes, 293, *294*
Parent element, 348–350, *349, 351*
Parent material, 109, 115, *116,* 117,
    118, 120
Paricutín volcano, *98*
Passive continental margins, 31, *31,* 37,
    366, *366,* 385, 391
Peat, 126
Pediments, 299, *302*
Pegmatite, 81, *81,* 120
Pelagic clay, 309

Pelée, Mount, 99, *100,* 101
Pelycosaurs, 405, *406*
Pennsylvanian Period, 132, *333, 376,* 381,
    405, 407
    Absaroka sequence in, 375–378, *377*
    continental drift and, 25, *25,* 26, *27*
    Ouachita mobile belt in, 381
Penzias, Arno, 8
Peridotite, 12, 41, 79, *79, 173,* 175
Perihelion, 284
Periods, *17*
Permafrost, 213, 215, *215, 216,* 329
Permeability, 133, 244–245, 248
Permian Period, *333, 371,* 381, 383
    Absaroka sequence in, 376, 378, *378*
    continental drift and, 25, *25,* 26, *27*
    mass extinction in, 403
    Ouachita mobile belt in, 381
    plants in, 407t
    reefs in, 402–403, *403,* 405
Peru, earthquake in, 168
Peru-Chile Trench, 39
Petrified Forest National Park, 388, *388*
Petroleum, 65, 121
    from anticlines, 187
    geologic structures and, 181
    oil spills of, 328, *328*
    plate tectonics and, 45
    sediment/sedimentary rocks, 132–133
Phaneritic texture, 77, 78, *78, 79,* 80
Phanerozoic Eon, *333*
Phenocrysts, 77, *78,* 80
Phosphates, 58t, 61, 309
Phosphorite, 329
Photic zone, 309
pH value, 112, *112*
Phyllite, 143, *143,* 144t, 145
Physical geology, 2
Phytoplankton, 418
Pillow lava, 37, 92, *94,* 103, *365*
Pinatubo, Mount, 90, 91, 95, 103, *104,* 106
Pitted outwash plains, 280
Placental mammals, 416, 419
Placer deposits, 132
Placoderms, 404
Plagioclase feldspars, *52,* 59, *60,* 64t, *73,* 74,
    80, 146
Planetesimals, 10
Planets
    dwarf, 9, *9*
    Jovian *vs.* terrestrial, 10
    volcanism and, 91
Plants, *333,* 407, 407t, *408,* 416
    Cenozoic, 418
    seedless vascular, 402, *405, 408,* 416
    soil and, 116–119
    uranium and, 133
    weathering and, 111, 112
Plastic flow, 271, *271*
Plastic strain, 182, 183, 184
Plate(s), 12, *13,* 35
Plate boundaries, 36–42, 157, *157,* 158, 167

continental–continental, 195, *195*
convergent (See Convergent
    boundaries)
divergent (See Divergent boundaries)
earthquakes and, *157,* 157–158
igneous rocks and, 70
mineral deposits and, 45, *47*
transform, *41,* 41–42, *42,* 157
Plate tectonics (theory of), 4, 12, *13,* 14,
    22–48, 363, *363,* 366
    Cenozoic, 389
    coasts and, 325
    continental crust and, 363, *363*
    continental drift and
        (See Continental drift)
    driving mechanism of, 43–45, *44*
    faults and, 190
    glaciers and, 283, 284
    life, distribution of, and, 47–48, *48*
    magnetism and, 31–35, *32*
    Mesozoic and, 387, 389
    metamorphism and, *149,* 149–140
    mountain building and, 36, 192–195
    natural resources and, 45, *47*
    organic evolution and, 16–17
    Paleozoic and, 366, 375
    plate movement and, 43, *43, 44*
    Pleistocene Epoch and, 393
    rock cycle and, 15–16, *16*
    seafloor and, 27–31
    as unifying theory, 35–36
    volcanoes and, *102,* 102–103
Platforms, 363–366
Platinum, 57, 61, 65
Playa lakes, 299, *299*
Pleistocene Epoch, 268, *333,* 356, 389, 393.
    *See also* Ice Age(s)
    Canadian shield and, 364
    coasts and, 325, 326
    cooling during, 393
    glaciers during, 282–284, 326, 393, *394*
    history of life in, 420–422, *422,* 426
    loess and, 295
    mammals in, 420, *422*
    sea level during, 275, 285, 325
    water table and, 254
Pleistocene glaciers, sea level during, 237
Plesiosaurs, 414, *414*
Pliocene Epoch, *333,* 389
Plucking, 274
Plug domes, 99. *See also* Lava domes
Plumb line, *197*
Plunging breaker, 315, *316*
Plunging folds, 187, *187*
Pluto, *9*
Pluton(s), 39, 70
    deformation and, 180, 188, 194, *194*
    mountain building and, 192, 195,
        196, 198
    pegmatite and, 81
    plate tectonics and, 102–103
    as sills, 75

Plutonic rock, 70, 77. *See also* Intrusive
    igneous rocks
Plutonism, *194*
Plymouth Rock, 82–83, *83*
Pocket beaches, *321*
Point bar, 229, *229*
Polar wandering, 32–33, *33*
Pollution, 112, 119, 244, 256, 257
Popping, 110
Pore spaces, 122, 244–246, *246*
Porosity, 244–245, *245*, 245t
Porphyritic texture, 77, *78*
Porphyry, 77, *80*
    granite, 80, 85
Portugal, earthquake, 154t
Postojna Cave, Slovenia, 254
Postolonnec Formation, *336*
Potassium, 53, *57*, 174, 176, 352t
Potassium feldspar, *52*, 59, *60*, 64t
Potential energy, 226
Potholes (in streambeds), 226, *227*
Powell, John Wesley, 332
Pratt, J. H., 196
Precambrian, *333, 343, 345, 362,* 362–367,
    363–364, *364*
    Ancestral Rockies in, 376, *377*
    Canadian shield and, 364
    history of life in, 399–401
    rocks of, 363–364, *364*
    Sauk sequence in, 372
Precipitation/rain, 245, *249,* 253, 259,
    262, *263*
    deserts and, 298, *298*
    glaciers and, *272,* 283
    hydrologic cycle and, 223, *224*
    mass wasting and, *201, 202, 204,* 206,
        209–212, *211*
Prehistoric overkill, 422
Pressure, metamorphism and, 139, *139,*
    *140,* 142, *142, 144,* 148, *149*
Pressure release, 110
Primary waves. *See* P-waves
Primates, 422–423
Principles. *See* specific ones
*Principles of Geology,* 335
Progradation, 230, *230,* 232
Prokaryotic cell, 400
Prosimians, 423
Proterozoic, 366, *366*
Proterozoic Eon, 36, *148, 333, 362,* 364,
    366–367, *367*
    Appalachian region in, 378
    Canadian shield and, 364
    Cordilleran mobile belt in, 381
    history of life in, 426
    Ouachita mobile belt in, 381
    plate tectonics and, 36
    Sauk sequence in, 372
    stromatolites in, 399, *399*
Proton, 53, *53,* 54
P–S time interval, 161, *161, 162*
Pterosaurs, *413,* 413–414

Pumice, 82, *83*
Push-pull waves. *See* P-waves
P-waves, *159,* 159–161, *161, 163,* 172, *173,*
    *174,* 174–175
    core-mantle boundary and, 174
    mantle and, 175
P-wave shadow zones, 174, *174*
Pyrite, 56, 62, *62,* 151
Pyritohedron crystals, *56*
Pyroclastic flows, 71
Pyroclastic materials, 70, 72, 81, 82, *83,* 90,
    91, 94, *95, 98,* 98–99, *99,* 101–103, *104*
Pyroclastic sheet deposits, 101
Pyroclastic (fragmental) texture, 77, *78*
Pyroxene, 59, *60,* 62, 64t, 75, 79, *79*
Pyroxenite, 79

Quadrupedal animals, 409, *410,* 411
Quarrying, 274
Quartz, 52, *52,* 59, *59*
    chemical composition, 57
    color of, 62
    crystal shapes, *56*
    hardness of, 63
    metamorphism and, *143,* 144t, 145, 147,
        *147,* 149
    quartzite and, 147, *147*
    uses of, 65
Quartzite, *15,* 115, 118, 144t, 146, 147,
    *147,* 149
Quartz sandstone, 123–124
Quaternary Period, *333,* 389
Queen Charlotte transform fault, *392*
Queenstone Delta, 380
Quick clay, 207t, 212, 213, *215*
Quick clay slides, *215*

Radial drainage, 234, 236, *236*
Radicals, 58, *59*
Radioactive decay, 43, *44,* 53, 176,
    348–350, *349*
Radioactive isotopes, 54, 351–352, 352t
Radioactivity, discovery of, 332, 353
Radiometric dating, 18, 332, 352t, 353
Rance tidal power plant, France, 313
Rapid mass movement, 206
Rayleigh (R-) waves, 160, *160*
Recent Epoch, *333,* 389
Recessional moraines, 279, *281*
Rectangular drainage, 234, *236*
Recumbent folds, *186, 187,* 192
Recurved spits, *322,* 323
Red beds, 366, 372
Redoubt volcano, 94
Red Sea, 37, *38, 39*
Redwall Limestone, 354
Reefs, 309, *310,* 402–403, *403*
Reflection
    of seismic waves, 172, *173*
Refraction, 316, *317,* 318
    of seismic waves, 172, *173*
    of water waves, 316, *317,* 319, *322*

Regional metamorphism, 140, 142, 143,
    145, 146, 147, 148
Regolith, 116, 211, 212, *214*
Reid, H. F., 154
Relative dating, 332–344, *345*
Relative geologic time scale, 335, 353
Relative movement, 189
Relativity, theory of, 8
Relief (of terrain), 118
Reptiles, 408–414, *414*
Reserves (*vs.* resources), 65
Reservoir, 223, *235,* 237
Reservoir rock, 132–133
Residual bonds, 55
Residual concentrations, 120
Residual soil, 116
Resources, 65–66. *See also* Natural
    resources
Retaining walls, 215, *220*
Reverse fault, 189, *190, 191*
Rhombohedrons, 62
Rhyolite tuff, 82
Richter, Charles F., 163
Richter Magnitude Scale, *163,* 163–164
Rifting, 29, 37, 176, 385, 391
Rift valleys, 37, *38, 39*
Rift Valleys of East Africa, 176
Right-lateral strike-slip fault, 190, *190*
Rill erosion, 119, *119*
Rills, 119, *119,* 225
Ring of Fire, 101, *101*
Rip currents, 316–318, *317,* 319
Ripple marks, 129, *130*
Rivers/streams, 222–241. *See also* Flood(s);
    Water, running
    base level of, 236–238, *237,* 240
    braided, 227, *228,* 229
    deltas and, 230–231, *231, 232,*
        232–233, *233*
    erosion by, *203, 208,* 298, 299
    glaciers and, 268
    longitudinal profile of, 238
    meandering, 227, *228,* 229
    potential *vs.* kinetic energy of, 226
    superposed, *240,* 240–241
    valleys and, 238–241
Rock(s)
    Archean, 364
    Cambrian, 363
    Cenozoic, 389
    definition of, 2, 52, 64
    deformation of (*See* Deformation)
    density of, 63
    ductile *vs.* brittle, 183
    geologic time scale and, 353
    geometry of, 128
    granitic, *76,* 81, 82, 85, 176
    igneous (*See* Igneous rocks)
    metamorphic (*See* Metamorphism/
        metamorphic rock)
    minerals and, 52, *52*
    porosity of, 244, *245*

Proterozoic, 364
radioactive isotopes and, 54
sedimentary (*See* Sedimentary rocks)
silicates in, 64, 64t
soil and, 118
ultramafic, 79, *79*
Rock bursts, 110
Rock cycle, 14–16
Rockfalls, 206–207, *207*
Rock flour, 274, *274*
Rock-forming minerals, *60*, 61, 64, 64t
Rock gypsum, 126, *126*, 308
Rock salt, 64, 126, *126*, 308
Rock sequences, continental fit and, 25, *25*
Rock slides, 207t, 209, *210–213*, 211
Rock unit correlation, 344–346, *346*, 353
Rocky Mountain National Park, *146*, 214
Rocky Mountains, 159, 176, 191, 272, 364, 384, 388
Ross Ice Shelf, 269
Rotational slides. *See* Slumps
Rounding (of sediment), 121, *121*, 128, 131
Rubidium, 352t
Runoff, 224, *224*, 270, *272*
glaciers and, 270, *272*
seawater and, 308
valleys and, 238–239

Sahara, 289
*Sahelanthropus tschadensi*, 423, *424*
*Saint Francis in Ecstasy* (Bellini), 4
Salinity, 308
Salinization, 119
Salt, 328. *See also* Halite
Saltation, 226, *227*, 289, 292
Salt crystal growth, 110, 298
Saltwater incursion, 255, *258*, 327
San Andreas Fault, 41–42, *42*, 142, 155, 181, 390
seismic gaps along, 170, *171*
as strike-slip fault, 190
Sand, 121, 123, 245t. *See also* Deserts; Dunes
beaches and, 320–321
glacial drift and, 278
running water and, 226, *227*, 229, 230
Sand dunes, 129, 131, *292*. *See also* Dunes
beaches and, 319, 323
in Death Valley, California, *288*, *291*, *292*
glacial deposits *vs*., 128
Sand seas, 293
Sandstone, 123, *124*, 127, *128*, 245t, 303, *306*, 366
inclusions of, 338, *338*
Navajo, 131, *132*, 388
San Francisco earthquake, 154, 154t, *155*, 165, *171*
Sangamon interglacial, *393*
Santorini volcano, 96–97, *97*, 102
Satellite-laser ranging techniques, 43
Saturation zone. *See* Zone of saturation

Saturn, 91
Sauk sequence, 372
Saurischia, 409, *410*, 411
Schist, *140*, 143, *143*, 144t, 145, *145*
Schistose foliation, 145
Schistosity, 145
Scientific method, 4
Scoria, 82, *83*
Sea(s), 37, 307, 315
Sea arches, 319, *320*, 326
Sea caves, 319, *320*
Sea cliffs, 318, 325
Seafloor, 27–31, 109, 121, 132
Devonian, *404*
fractures in, 41
oceans and, 307
Ordovician, *402*
sediments on, *129*, 309
Seafloor spreading, 33–35, *35*
magnetic anomalies, *35*
Sea level
coasts and, 324, 325, 326
in Cretaceous, 383
glaciers at, 272, 276, 285
in Paleozoic, 376
in Pleistocene, 237, 275, 285
shorelines and, 307
Seamounts, 29–30, *30*
Sea of Japan, 39
Seashore. *See* Shorelines
Sea stacks, *320*, 325, *326*
Seawater, 124–125, *130*, 308–309, 328–329
Secondary body waves, *159*, 160
Secondary waves. *See* S-waves
Sediment, 15, 120, 244, 245t, 255, 258
base level and, 237
in continental *vs*. valley glaciers, 274–275
deep-sea, 41, *41*
deserts and, 298, 299
detrital *vs*. chemical, 121, *122*
Earth's crust and, 197
erosion and, 120
geologic time and, 336
origin/transport of, *120* (*See also* Sediment transport)
plate tectonics and, 15, 45
on seafloor, *129*, 309
sedimentary rock from, 120–122
waves and, 316, *317*, 319
weathering and, 120–122, *122*, 124
Sedimentary breccia, 123, *123*, *124*
Sedimentary facies, 127–128
Sedimentary iron deposits, 120
Sedimentary rocks, 120, 244, 245, *245*, 250, 255
Archean, *365*
Cenozoic, 389, 418
chemical, *120*, 123t, 124–126, *125*
dating of, 353–354
deformation and, 183, *183*

deposition of, 120, 121, 127–128, *128*, 131
deposits in, 121, *122*
deserts and, 303
detrital, *123*, 123–124, *124*, 244
geologic time and, 335, *336*, 340, *343*
geologic time scale and, *353*, 353–354
geometry of, 128
metamorphism and, 138, 140, 141, 147, *147*, *148*
minerals in, 61, 64
mountain building and, 192, 195
oceanic circulation and, 309
ocean resources from, 329
plate tectonics and, 15, *15*, *16*
Proterozoic, 366, *366*
rock cycle and, 15, *15*, *16*
rounding/sorting and, 121, *121*, 128, 131
from sediment, 120–122
textures of, 123t, 128, 131 (*See also* Texture)
transport of, *120*
types of, 122–126, 123t
weathering of, 120–122, *122*, 124 (*See also* Weathering)
Sedimentary structures, *128*, 128–129
Sediment transport, *120*. *See also* Deposition
by glaciers, 274–278, *275*
shorelines and, 319, 324
by water, *227*
by wind, 289–290
Seedless vascular plants, *402*, *405*, *408*, 416
Seif dunes, 293
Seismic activity, deformation and, 180
Seismic discontinuity, 175, *175*
Seismic gaps (locked regions), 170, 171, *171*
Seismic-moment magnitude scale, 164
Seismic risk maps, 169
Seismic sea wave, 168
Seismic wave amplitude, *165*
Seismic waves, 156, *159*, 159–160, *160*, *165*, 172–173, *173*, 175, *175*
Seismic wave velocity, 172, *173*, 175
Seismogram, 156, *161*, 163, 164
Seismographs, 156, 161, *161*, *162*, 164, 168, 174
Seismology, 156–158
Semiarid regions, 225, 231, 296, 297, 303
Serpentine asbestos, 150
Sevier orogeny, 387, *387*
Shadow zones, 174, *174*
Shale, *123*, 124, *124*, 127, 142, 143, *143*, 245, 245t, 247, 256–257, 366
Shallow-focus earthquakes, 157, *157*, 170
Shasta, Mount, 99
Shear strength, 202, *203*, 204, 205, 209, 217
Shear stress, 182, *183*
Shear waves. *See* S-waves
Sheep Mountain, *187*

Sheet erosion, 119, 225
Sheet flow, 224–225
Sheet joints, 110, *111*
Sheet silicates, 112, 114, 121
Shelf-slope break, 28
Shields, 363–364, 366. *See also* Canadian
    shield
Shield volcanoes, 95, *97*, 394
Shore Acres State Park, *306*
Shorelines, 307, 310, *311*, 314–318,
    *318, 320*
    beaches and, 319–321
    coasts *vs.*, 310, *311* (*See also* Coasts)
Siberia (Paleozoic), 369, *370, 371*, 372
Siberian Traps, 403
Siccar Point, 339, *342*
Sierra Nevada, *76*, 82, 85, 189, 272, *333*,
    384, 387, *387*, 390, *390, 391, 392*
Silica, 59, *59*, 70, 114, 121, 126, 263,
    309, 418
Silicates, 58t, 59, *60*, 61–64, 64t, 138, 148,
    *148*, 150. *See also* Ferromagnesian
    silicate; Nonferromagnesian silicate
Silica tetrahedron, 59, *60*
Siliceous sinter, 263
Silicon, *53*, 54, 57, *57*, 59, 173t, 174
Silicon dioxide, 122
Sillimanite, 140, 142, *143*, 148, *148*, 151
Sills, 37, 75, 84–85
Silt, 121, 245t
Siltstone, *123*, 124, 366
Silurian Period, *333*, 369, *374*, 380,
    402, 404
    plants in, 407, 407t
    Tippecanoe sequence in, 373
Silver, 55, 57, 61
Sinkholes, 250–252, *251*
Sinks, 250. *See also* Sinkholes
Skylight, 92, *93*
Slate, 138, 143, *143*, 144t, *145*
Slaty cleavage, 143, *145*
Slide-flow, 217
Slides, 207–211, 207t, *210–213*
Slope angle, 118, 202–203, *203*, 219, *219*
Slope direction, 118
Slope(s), 118, 202–206, *203, 205*, 217,
    219, *219*
Slope stability, 202, *203*, 204, *205*,
    205–206, 217
Slope-stability maps, 217, *218*
Slow mass movement, 206
Slumps, 207t, 208, *208, 209*, 211
Snake River Plain, 100, *100*
Sodium, 54–55, *55, 57*, 308, 328
Soil, 109, 116–119
    characteristics of, *116*, 116–117
    as natural resource, 117
    weathering and, 109–115, 298
Soil degradation, 119, *119*
Soil horizons, 116, *116*
Solar energy, 285
Solar nebula, 10

Solar nebula theory, 10, *10*
Solar system, *9*, 9–11, *10*
Solids, 53
Solifluction, 207t, 213, *215*
Solnhofen Limestone, 414, *416*
Solution, 112, *113*
Solution valleys, 251, *251*
Sonoma orogeny, 386, *387*
Sorting (of sediment), 121, *121*, 128, 131
Source rock, 132, 133
Space-time continuum, 8
Specific gravity, 63
Spectral lines, 8
Sphalerite, 62, 151
Spheroidal weathering, 115, *115*
Spilling breaker, 315, *316*
Spires, 109
Spits, 321–323, *322*
Spreading ridges, 36. *See also* Divergent
    boundaries
    earthquakes at, 157, 159
    heat in, 176
    origin of magma at, 74, *74*
    volcanism and, 102, 103
*Spriggina, 400*
Springs, 247, *248*. *See also* Hot springs
Stalactites, 253, *254*
Stalagmites, 253, *254*
Staurolite, *143*, 144t, 148, *148*
Stem reptiles, 408
Steno, Nicholas, 57, 335, 336
St. Helens, Mount, 95, 98, *99*, 102, 103,
    105, 106
Stocks, 85
Stone Mountain, *111*
Stoping, 85, *86*
Stoppani, Antonio, 334
Storms, 326
Storm surge, 326
Strain, 181–183, *182*, 184, 270
Strata, 128, *128*
Stratification, *128*, 131. *See also* Bed (of
    sedimentary rock)
Stratified drift, 278, 280
Stratigraphic record of hominins, *424*
Stratigraphic traps, 133, *133*
Stratigraphic units, 354t
Stratovolcanoes, 98–99, *99*
Stratum, 128
Stream(s). *See* Rivers/streams
Stream-dominated deltas, 230–231, *231*
Stream piracy, 239, *239*, 241
Stream terraces, *239*, 239–240
Stress, 139, *140*, 270, 271
    deformation and, 181–182, *182, 183*
    joints and, 188
Striations, 25–26, *26*
Strike, *183*, 183–184, 185, *186*
Strike-slip faults, 189–190, *190*
Stromatolites, 399, *399*
Stromboli volcano, 105
Strontium, 352t

Structural traps, 133, *133*
Subduction, 29, 31, *36*
    magma and, *74*, 74–75
    metamorphism and, 139, 142, 149
    plate boundaries and, 36, 193, *193*, 195
    zones of (*See* Convergent boundaries)
Sublimation, 270, *272*
Submarine canyons, 323, 324
Submarine hydrothermal vents, 28, *29*
Submarine lava flows, 103
Submergent coasts, *325*, 325–326
Subsidence, 255, 258, *259, 260*
Subsoil, *116*, 117
Suess, Edward, 23–24
Sulfates, 58t, 61, 64t
Sulfide, 58t, 61, *61*, 151
Sulfur, *57*, 65, 172, 174
Sulfur dioxide, 92, 112–113, *113*
Sumatra earthquake, 164, 166
Sun
    glaciers and, 283, 284, 285
    origin of, 9, *9*
    tides and, 310
Sundance Sea, 388
Supercontinents, 366
Superposed streams, *240*, 240–241
Superposition, 336, *337*, 340, 344
    correlation and, 346
    geologic time scale and, 353
Surface area, 114, *115*
Surface waves (seismic), 159, 160,
    *160, 161*
Suspended load, 226, *227*, 290
Suspended water, 245
Sutter Buttes volcano, 90
S-waves, *159*, 159–161, *161, 163*, 172, *173*,
    *174*, 174–175
    mantle and, 175, 197
S-wave shadow zones, 174, *174*
Swells, 315, 321
Synclines, *22*, 185, *185, 186*, 187, *187*
System, definition of, 2

Taconic orogeny, 369, *379*, 380
Taj Mahal, India, *137*
Talc, 63, 63t, 138, 140, 144t, 151
Talus, 110, *110*, 206, *207*
Tambora volcano, 91, 105
Tectonism. *See* Plate tectonics (theory of)
Temperature
    deserts and, 298, *298*
    geothermal gradient and, 176
    metamorphism and, 142, 143, 146,
        148–149, *149*
Temporary base level, 237
Tension, 182, *183*, 188
Tephra, *104*
Terminal moraines, 279, *281*
Terranes, 196
Terrestrial planets, 10
Tertiary Period, 389
Tethys Sea, 381

Teton Range, 191
Texas, earthquakes in, 171
Texture, 77–78, *78, 83,* 123t, 128, 131
   clastic, 123, 123t, 124
   foliated, 143, *144,* 146
   nonfoliated, 143, 144t, 146, *147*
Theories. *See also specific ones*
   definition of, 3
   Homogenous accretion theory, 11
   Milankovitch theory, 283–284
   relativity, 8
Therapsids, 406
Thermal convection cell, 34
Thermal expansion and contraction, 110
Thermal metamorphism, 140–142, *141*
Thermal spring, 261. *See also* Hot springs
Thorium, 176, 352t
Thrust faults, 189, 192, 193, 194, 195,
   364, *365*
Tidal bulges, *314*
Tidal power generation, 313
"Tidal wave," 167, 168
Tides, 307, 310, *314,* 314–315, 319
   beaches and, 320
   deltas and, 231, *231*
Tidewater glaciers, 268, 270
Till, 25, *26*
   glacial drift, 278
   landforms composed of, 278–280
Tiltmeters, 105
Time, geologic. *See* Geologic time
Time–distance graphs, 161, *161*
Time-stratigraphic units, 354
Time units, 354
Tippecanoe sequence, 372–373, 402
Toba caldera, 95
Tombolos, *322,* 323
Tonga Trench, 157, *158*
Topaz, 63t
Topset beds, 230, *230*
Topsoil, 116, *117*
Tourmaline, *52*
Trace fossils, 129, 399
Traction, 226
Transcontinental Arch, 372, *373*
Transform boundaries, *41,* 41–42, *42,* 157
Transform faults, 31, 41–42, *42,* 190
Transitional deposition, 121, *122*
Transported soil, 116
Transverse dunes, 293, *294*
Travertine, 262, *263*
Travertine terraces, 253, *253*
Trellis drainage, 234, *236*
Trench rocks, *193,* 194
Triassic fault basins, 37, *386*
Triassic fault-block basins, *386*
Triassic Period, *333,* 381, *382,* 383, *383,*
   386, *387,* 389, 405
   dinosaurs in, 409
   eastern North America and, 391
   history of life in, 408, 413, 416
   plants in, 407t

*Tribrachidium, 400*
Triton, 91
Tsunami, 23, 154, 166–168, *168*
Tuff, 82, *83*
Tungsten, 151
Turbidity currents, 129, *129*
Turkey earthquake, 154t, 159, 165, *165*
Turtle Mountain rock slide, 209, 211, *213*

Ultimate base level, 236–237, *237*
Ultramafic magma, 70, 71t, 74, 78
Ultramafic rocks, 79, *79*
Unconformities, 339–340, *340, 342*
   cratonic sequence and, 372, *373,* 375
Uniformitarianism, 17–18, 131, 335
United States Energy Administration,
   256, *257*
United States Geological Survey, *71,* 106,
   170, 254
Universe, 8–11, *9*
Upper mantle, 12
Upright fold, 185
Upwelling, 309
Ural Mountains, 195
Uraninite, 133
Uranium, 53, 65, 133, 176
   plate tectonics and, 43
   in radioactive dating, 348, *349,* 352t
U.S. Geological Survey, 65–66
U-shaped glacial troughs, 275, *276*

Valley(s), 238–241, 276, *276*
Valley and Ridge Province, 392
Valley glaciers, 268–269, *269,* 393, *394*
   erosion by, 269, 275–278
   flow velocity, 273, *273*
   glacial surge in, 274
   landforms and, 275, *275*
   moraines and, 279, *279*
   sediment in, 275, 280
Valley trains, 280, *282*
Van der Waals bonds, 55, *56*
Varves, 281, *282*
Vegetation
   in deserts, 289, 298, *298*
   slope stability and, 205
VEI. *See* Volcanic explosivity index (VEI)
Velocity (running water), 225, 226, *227,*
   229, *229,* 230, 238, 273, *273*
Ventifacts, 290, *291*
Venus, 91
Verne, Jules, 4
Vertebrates, 399, *403,* 404–406, –418
Vertical-motion seismograph, *156*
Vesicles, 77, *78,* 244
Vesicular basalt, 77, *78*
Vesuvius, Mount, 90, *91,* 102
*Virgin of the Rocks* (da Vinci), 4
Viscosity, 71, *72*
Vog, 92, *92*
Volcanic ash, 90, *90,* 94, *141,* 143, 144t,
   211. *See also* Ash

Volcanic breccia, 82
Volcanic domes, 99. *See also* Lava domes
Volcanic explosivity index (VEI), 103, *104*
Volcanic gases, 91–92, *92,* 103, *104*
Volcanic glass, *78,* 82
Volcanic hazards, 103, *104*
Volcanic island arc, 39, *40,* 193, *193,* 194,
   383, 386. *See also* Island arcs
Volcanic islands, 16, 309, *310*
Volcanic mudflows, 91, 98
Volcanic necks, *84,* 85
Volcanic pipes, 85
Volcanic rock, 70, 75, 77, 82, 85, 353, *365*
Volcanic tremors, 105
Volcanic vents, 90, 91. *See also* Fumeroles
Volcano(es)/volcanism, 89–106, 393, 403.
    *See also specific volcanoes*
   andesite and, 80
   convergent boundaries and, 37
   deformation and, 180, 194
   distribution of, 101–102
   dormant, 90
   forecasting, *105,* 105–106
   glaciers and, 274, 285
   guyots and, 29, *30*
   hot-spots, 75, *75* (See also Hot-spots)
   intraplate, 103
   intrusive igneous activity and, 70, 388
   lava and (See entries beginning Lava)
   mantle plumes and, 42
   monitoring, *105,* 105–106
   mountain building and, 192, 194,
    195, 196
   North American Cordillera and,
    389–390, *391*
   plate tectonics and, 12, 23, *102,*
    102–103
   porosity, 244, *245*
   seawater and, 308
   springs, *248*
   types of, 95–100
   VEI and, 103, *104*

Walcott, Charles, 399
Warm spring, 261. *See also* Hot springs
Water, 109–110, *110,* 223–224. *See also*
   Groundwater; Precipitation/rain
   frozen, 329
   glaciers and, 268
   mass wasting and, 204–205
   sediment/sedimentary rocks, 120
   subsurface, 217, 218, *218*
   weathering by, *113,* 114
Water, running, 222–241. *See also* Erosion
   deltas and, 230–231, *231, 232,*
    232–233, *233*
   deposition by, 227–232
   drainage systems and, 234–236, *236*
   mountain building and, 192
   rivers/streams and, 224–226
    (See also Rivers/streams)
   valleys and, 238–241

Waterfalls, 276
Water gaps, *240*, 240–241
Water table, 245–246, 254–255
    artesian system and, 249–250
    deserts and, 298
    perched, 247, *247*
Water wells, 248–249. *See also* Wells
Wave(s), 315, *316, 317, 318*
    beaches and, 320–321
    deltas and, 231, *231*
    energy, 312
    mass wasting and, 202–203, *204*, 208,
        *212, 318*
Wave base, 315, *316*
Wave-cut platforms, 318–319,
    *319*, 325
Wave-formed ripple marks, 129, *130*
Wave rays, 172, *173*
Wave refraction, 316, *317*, 319, *322*
Weathering, 109–115
    badlands topography, *108*
    chemical (*See* Chemical
        weathering)
    deserts and, 302

differential, 109
    mass wasting and, 204
    mechanical, 110–111, *111*, 302
    organisms and, *111*
    sediment/sedimentary rocks, 120–122,
        *122*, 124
    volcanism and, 91
Wegener, Alfred, 24
Welded tuff, 82, 101
Well cuttings, 131
Wells, *248*, 248–249, *260*, 264
Werner, Abraham Gottlob, 82
Wilson, J. Tuzo, 75
Wilson, Robert, 8
Wind, 289–296
    deposition by, 291–295, 323
    dunes and, 292–295 (*See also* Dunes;
        Sand dunes)
    erosion by, 119, *119*, 290, 290–291,
        *291*, 299
    global patterns of, 295, *296*
    loess and, 292, 295, *295*
    plate tectonics and, 47
    power, 300–301

sediment/sedimentary rocks, 120
    waves and, 323
Wind farms, 300–301
Wind gap, 241
Windmills, 300–301
Wisconsinan glaciation, 393, *393*

Yarmouth interglacial, *393*
Yellowstone National Park, *261*, 262, *262,
    263, 263*
Yosemite National Park, *207*, 276, *276*

Zagros Mountains, *22*, 195
Zinc, 62, 151
Zion National Park, *132*, 345, *347*
    cross-bedding in, 292, *388*
Zone of accumulation, *116*, 117, 272,
    273, 279
Zone of aeration, 246, *246, 247*, 252, 253,
    258, *260*
Zone of saturation, 246, *246*, 247, *247*, 248,
    252, 253, 258, *260*
Zone of wastage, 272, *272*
Zuni Sea, 390